iOS 8
案例
开发大全

朱元波 陈小玉 胡汉平 张晨洁 编著

人民邮电出版社
北京

图书在版编目（CIP）数据

iOS 8案例开发大全 / 朱元波等编著. -- 北京 : 人民邮电出版社, 2015.2
ISBN 978-7-115-37374-8

Ⅰ. ①i… Ⅱ. ①朱… Ⅲ. ①移动终端－应用程序－程序设计 Ⅳ. ①TN929.53

中国版本图书馆CIP数据核字(2014)第270561号

内 容 提 要

本书实例全面、典型，几乎囊括了所有和 iOS 应用开发相关的知识。全书分为 14 章，内容包括搭建开发环境，界面布局，iOS 控件应用，文本和表格处理，屏幕显示，图形、图像和动画，多媒体应用，互联网应用，地图定位应用，传感器，游戏开发，移动 Web 应用，Swift 实战。全书讲解细致，内容通俗易懂。

本书适合 iOS 开发初学者、程序员作为参考工具书，也可以作为相关培训学校和大专院校相关专业的教学用书。

◆ 编　著　朱元波　陈小玉　胡汉平　张晨洁
　责任编辑　张　涛
　责任印制　张佳莹　彭志环

◆ 人民邮电出版社出版发行　北京市丰台区成寿寺路 11 号
　邮编　100164　电子邮件　315@ptpress.com.cn
　网址　http://www.ptpress.com.cn
　北京铭成印刷有限公司印刷

◆ 开本：787×1092　1/16
　印张：37.75
　字数：1 126 千字　　　　　　2015 年 2 月第 1 版
　印数：1－3 500 册　　　　　　2015 年 2 月北京第 1 次印刷

定价：79.00 元

读者服务热线：(010)81055410　印装质量热线：(010)81055316
反盗版热线：(010)81055315
广告经营许可证：京崇工商广字第 0021 号

前　言

iOS 最早于 2007 年 1 月 9 日的苹果 Macworld 展览会上公布，随后苹果公司于同年 6 月发布了第一版 iOS 操作系统。当前的市场显示，搭载 iOS 系统的 iPhone 手机仍然是最受欢迎的一款智能手机，搭载 iOS 系统的 iPad 仍然是当前最受欢迎的一款平板电脑。

本书特色

本书内容丰富，实例覆盖全面。我们的目标是通过一本图书提供多本图书的价值，读者可以根据自己的需要有选择地阅读。在内容的编写上，本书具有以下特色。

1. Objective-C 和 Swift 双剑合璧

在本书涵盖的实例中，不但演示了用传统 Objective-C 语言开发 iOS 应用程序的方法，而且演示了用苹果最新语言——Swift 开发 iOS 应用程序的方法，实现了 Objective-C 和 Swift 的开发应用对比，能够给读者以启迪。

2. 实例全面

实例涉及 UI、控件、游戏、网络、多媒体、地图定位等，几乎涵盖了所有的 iOS 应用领域，每个实例讲解细致，让读者真正明白具体原理和实现的方法。

3. 结构合理

从开发者的实际需要出发，科学安排知识结构，内容由浅入深，叙述清楚。

4. 易学易懂

本书条理清晰、语言简洁，可帮助读者快速掌握每个知识点。读者既可以按照本书编排的章节顺序进行学习，也可以根据自己的需求对某一章节进行有针对性地学习。

5. 实用性强

本书彻底摒弃枯燥的理论和简单的操作，注重实用性，详细讲解了各个实例的具体实现，使用户在掌握相关的开发技能的同时，还能学习到相应的技巧。

读者对象

- 初学 iOS 编程的自学者
- Objective-C 开发人员
- Swift 开发人员
- 大中专院校的老师和学生
- 进行毕业设计的学生
- iOS 编程爱好者
- 相关培训机构的老师和学员
- 从事 iOS 开发的程序员

本书在编写过程中，得到了人民邮电出版社工作人员的大力支持，正是各位编辑的耐心和效率，才使得本书能在这么短的时间内出版。南阳理工学院的陈小玉编写了第 3 章、第 6 章、第 9 章。另外也十分感谢我们的家人，在我们写作的时候给予了巨大的支持。水平毕竟有限，纰漏和不足之处在所难免，恳请读者提出意见或建议，以便修订并使之更臻完善，编辑联系邮箱：zhangtao@ptpress.com.cn。

源程序下载地址：www.toppr.net。

读者交流答疑 QQ 群：283166615。

编 者

目　　录

第1章　搭建开发环境实战 1
- 实例 001　下载并安装 Xcode 1
- 实例 002　改变 Xcode 工程的公司名称 4
- 实例 003　通过搜索框缩小文件范围并格式化代码 5
- 实例 004　代码缩进和提示处理 6
- 实例 005　设置项目快照以及恢复到快照 7
- 实例 006　实现复杂的查找和替代工作 8
- 实例 007　使用书签 10
- 实例 008　实现断点调试 11
- 实例 009　启动模拟器 12
- 实例 010　使用第三方工具 iPhone Simulator 15

第2章　界面布局实战 16
- 实例 011　使用 Interface Builder 的故事板 16
- 实例 012　设置 UIView 的位置和尺寸 22
- 实例 013　隐藏指定的 UIView 区域 24
- 实例 014　改变背景颜色 25
- 实例 015　实现背景透明 27
- 实例 016　定位屏幕中的图片 29
- 实例 017　旋转和缩放视图 31
- 实例 018　伸缩屏幕中的视图 34
- 实例 019　实现视图的大小自适应 36
- 实例 020　实现视图嵌套 38
- 实例 021　插入或删除视图中的子元素 41
- 实例 022　设置视图位置互换显示（1）...... 42
- 实例 023　设置视图位置互换显示（2）...... 44
- 实例 024　获得屏幕内视图的坐标 46
- 实例 025　实现视图外观的自动调整 48
- 实例 026　自动调整视图中的子元素 51
- 实例 027　实现不同界面之间的跳转处理 54
- 实例 028　通过列表实现不同界面之间的跳转 56
- 实例 029　通过 UITabBarController 选项卡实现不同界面之间的跳转 58
- 实例 030　在布局中实现一个模态对话框 60
- 实例 031　实现仿 iPhone 的底部选项卡 61
- 实例 032　实现导航条效果 64
- 实例 033　在导航条中添加一个滑动条 65
- 实例 034　在屏幕中显示一个工具条 66
- 实例 035　在工具条中添加系统按钮 68
- 实例 036　在工具条中自定义按钮（1）...... 70
- 实例 037　在工具条中自定义按钮（2）...... 71
- 实例 038　改变状态栏的颜色 72

第3章　iOS 控件应用实战 74
- 实例 039　使用文本、键盘和按钮（1）...... 74
- 实例 040　使用文本、键盘和按钮（2）...... 77
- 实例 041　在屏幕中显示一个指定的文本 79
- 实例 042　设置屏幕中文本的对齐方式 79
- 实例 043　设置屏幕中标签的颜色和文本的颜色 80
- 实例 044　设置屏幕中显示不同字体的文本 81
- 实例 045　自动调整屏幕中的文本大小 83
- 实例 046　在一个 UILabel 控件中显示多行文本 84
- 实例 047　设置文本的换行和省略模式 85

实例 048	实现文本的阴影效果	86
实例 049	高亮显示屏幕中的文本	87
实例 050	定制一个文本绘制方法	88
实例 051	按下按钮后触发一个事件	89
实例 052	在屏幕中显示不同的按钮	90
实例 053	点击按钮后改变按钮的文字	92
实例 054	点击按钮后实现阴影反转	93
实例 055	点击按钮时实现闪烁效果	94
实例 056	在按钮中添加图像	95
实例 057	调整屏幕中按钮的边间距	97
实例 058	设置按钮中文本的换行和省略格式	99
实例 059	在屏幕中显示一个文本输入框	99
实例 060	设置文本输入框的边框线样式	100
实例 061	设置文本输入框的字体和颜色	101
实例 062	在文本输入框中设置一个清空按钮	102
实例 063	为文本输入框设置背景图片	103
实例 064	在文本输入框中添加 UIView 元素	104
实例 065	监视文本输入框的状态	105
实例 066	实现一个开关效果	106
实例 067	改变 UISWitch 文本和颜色	108
实例 068	显示具有开关状态的开关	110
实例 069	在屏幕中显示一个分段选项	112
实例 070	选择一个分段卡后可以改变屏幕的背景颜色	114
实例 071	设置分段卡的显示样式	115
实例 072	设置不显示分段卡的选择状态	116
实例 073	改变分段卡的显示颜色	117
实例 074	选择某个选项时在此分段卡中显示一幅图片	118
实例 075	设置指定图片作为分段卡的选项	119
实例 076	修改分段卡标题的位置	119
实例 077	设置某个选项不可用	120
实例 078	插入\删除分段卡中的选项（1）	121
实例 079	插入\删除分段卡中的选项（2）	122
实例 080	滑动滑块时显示对应的值	128
实例 081	滑动滑块控制文字的大小	129
实例 082	自定义一个滑块	131
实例 083	实现一个日期选择器	131
实例 084	获取当前的时间	138
实例 085	设置日期选择器中的时间间隔	139
实例 086	设置日期选择器框的显示样式（1）	140
实例 087	设置日期选择器框的显示样式（2）	141
实例 088	实现自动倒计时功能	142
实例 089	使用选择器视图	143
实例 090	自定义一个选择器	146
实例 091	实现一个数字选择器	153
实例 092	突出显示选择器中的某一行	154
实例 093	向选择器中添加 UIView 子类	155
实例 094	设置选择器框行和列尺寸	157
实例 095	实现一个播放器的活动指示器	158
实例 096	实现一个蓝色进度条效果	160
实例 097	在进度条中显示进度百分比	162
实例 098	在屏幕中实现一个检索框效果	163
实例 099	实现一个实时显示检索框效果	165
实例 100	设置检索框的背景颜色	166
实例 101	在检索框中添加一个书签按钮	167
实例 102	在检索框中添加一个范围条	169
实例 103	添加或删除屏幕中的翻页数目	170

实例 104　使用滚动的方式查看屏幕
　　　　　中的内容……………………172
实例 105　使用滚动的方式查看图片…175
实例 106　设置滚动条的颜色…………177
实例 107　将滚动条设置为分页的
　　　　　形式………………………178

第 4 章　文本和表格处理实战…………180

实例 108　在屏幕中换行显示文本……180
实例 109　在屏幕中显示可编辑的
　　　　　文本………………………181
实例 110　将屏幕中的文本实现编辑
　　　　　状态和非编辑状态之间的
　　　　　切换………………………182
实例 111　设置屏幕中文本的对齐方式，
　　　　　确定文本的选择范围………184
实例 112　自动处理屏幕中文本的 URL
　　　　　地址和电话号码……………187
实例 113　在屏幕文本中显示密码
　　　　　黑点"."……………………187
实例 114　自定义 UITableViewCell……188
实例 115　拆分表视图…………………192
实例 116　列表显示 18 条数据…………194
实例 117　分段显示列表中的数据……195
实例 118　删除单元格…………………197
实例 119　添加新的单元格……………198
实例 120　移动单元格的位置…………200
实例 121　实现单元格的编辑模式和
　　　　　非编辑模式的切换…………201
实例 122　编辑分组单元格（1）………202
实例 123　编辑分组单元格（2）………204
实例 124　设置单元格的尺寸和颜色…205
实例 125　在单元格中添加图片………206
实例 126　为单元格中的图片
　　　　　添加注释……………………207
实例 127　在单元格中添加附件………208
实例 128　在单元格中添加
　　　　　自定义附件…………………209
实例 129　设置只在编辑模式下
　　　　　显示附件……………………210
实例 130　向单元格中添加其他控件…211
实例 131　自定义单元格的背景………213

实例 132　设置被选中单元格的
　　　　　背景颜色……………………214
实例 133　自动滚动到被选中
　　　　　单元格………………………215
实例 134　在单元格中自动排列指定的
　　　　　数据…………………………216
实例 135　为每行单元格设置展开
　　　　　子项…………………………218
实例 136　实现气泡样式的聊天对话
　　　　　框效果………………………220
实例 137　在搜索框中实现下
　　　　　拉列表效果…………………222
实例 138　实现一个高度自动适应
　　　　　性的输入框…………………223

第 5 章　屏幕显示实战…………………226

实例 139　在屏幕中显示一段文本……226
实例 140　绘制字符串…………………227
实例 141　设置屏幕中文本的横向
　　　　　对齐方式……………………228
实例 142　缩小文本并设置纵向对齐
　　　　　方式…………………………229
实例 143　设置屏幕中的字符串
　　　　　自动缩小……………………231
实例 144　获取绘制文本所需要的
　　　　　空间范围……………………232
实例 145　显示系统中的字体…………233
实例 146　列表显示系统中所有的
　　　　　字体…………………………234
实例 147　在屏幕中显示不同的
　　　　　颜色…………………………236
实例 148　使用系统颜色………………238
实例 149　在屏幕中自定义颜色………239
实例 150　使用背景图片创建
　　　　　特殊背景……………………240
实例 151　在屏幕中绘制指定颜色的
　　　　　文字…………………………241
实例 152　在屏幕中显示图像…………242
实例 153　在屏幕中绘制一幅图像……243
实例 154　在屏幕中绘图时设置
　　　　　透明度………………………244
实例 155　限制图像的缩放区域………246

实例 156	使用 UIImageView 实现动画效果	246
实例 157	在屏幕中实现日历效果	248
实例 158	在屏幕中自定义一个导航条	254
实例 159	在屏幕中实现仿 iPhone 锁定界面效果	255

第 6 章 图形、图像和动画实战 ……257

实例 160	在屏幕中实现一个简单的动画效果	257
实例 161	设置在屏幕中的动画延迟	258
实例 162	设置在屏幕中动画的透明度	258
实例 163	设置屏幕中的动画实现放大/缩小/旋转效果	260
实例 164	检测屏幕中动画的状态	261
实例 165	在屏幕中实现过渡动画效果	262
实例 166	联合使用滑块和步进控件实现动画效果	264
实例 167	实现全屏显示效果	268
实例 168	实现渐变样式的全屏效果切换	270
实例 169	设置屏幕中的元素随着设备旋转而自动适应	271
实例 170	设置界面旋转时自动调整图像尺寸	272
实例 171	定制屏幕中的旋转图像	273
实例 172	同时实现屏幕自适应功能和全屏功能	275
实例 173	创建可旋转和调整大小的界面	276
实例 174	屏幕旋转时调整控件的框架	279
实例 175	屏幕旋转时切换视图	284
实例 176	实现一个图片浏览工具	287
实例 177	实现"烟花烟花满天飞"效果	289
实例 178	实现"漫天飞雪"效果	291
实例 179	在屏幕中绘制一个三角形	293

实例 180	在屏幕中实现颜色选择器/调色板功能	295
实例 181	在屏幕中实现滑动颜色选择器/调色板功能	297
实例 182	在屏幕中实现网格化视图效果	300

第 7 章 多媒体应用实战 ……304

实例 183	使用 MediaPlayer Framework 框架播放视频	304
实例 184	使用 Core Image 框架处理照片	306
实例 185	创建一个多功能播放器	308
实例 186	使用系统内的相册	320
实例 187	实现录制视频功能	322
实例 188	设置屏幕中视频的画面	324
实例 189	剪辑系统内的视频	326
实例 190	开发一个音频播放器	328
实例 191	在屏幕中实现一个电子琴效果	329
实例 192	在屏幕中实现一个 DJ 混音器	331
实例 193	在屏幕中实现一个音乐选择器	333
实例 194	在屏幕中听声音	336
实例 195	播放本地的视频文件	337
实例 196	在播放界面中叠加视频	339

第 8 章 互联网应用实战 ……342

实例 197	使用 Web 视图获取网络信息	342
实例 198	在屏幕中显示指定的网页	350
实例 199	控制屏幕中的网页	351
实例 200	在网页中加载显示 PDF、Word 和 JPEG 图片	353
实例 201	在网页中加载 HTML 代码	355
实例 202	在网页中实现触摸处理	356
实例 203	在屏幕中显示 CSDN 主页	359
实例 204	一个简单的网页浏览器	361

实例 205	下载并显示远程 URL 地址的 JPEG 图片	364
实例 206	解析指定的 XML 文件	365
实例 207	实时检测 Wi-Fi 状况	368
实例 208	断点续传下载后实现播放	372

第 9 章 地图定位应用实战 379

实例 209	获得当前所在位置和苹果公司总部的距离	379
实例 210	使用磁性指南针	383
实例 211	在屏幕中实现一个定位系统	388
实例 212	在屏幕中使用谷歌地图	391
实例 213	在收集地图中实现定位和位置标示	396
实例 214	在地图中实现标注	399
实例 215	在地图中灵活标注	404
实例 216	实现复杂的地图标注	407

第 10 章 传感器、触摸和交互 416

实例 217	实现一个可触摸识别程序	416
实例 218	触摸按钮	423
实例 219	同时滑动两个滑块	424
实例 220	触摸屏幕检测	425
实例 221	触摸屏幕中的文字标签	426
实例 222	演示一次触摸和两次触摸	427
实例 223	演示 3 次触摸	428
实例 224	拖曳方式移动屏幕中的图片	429
实例 225	可以检测上、下、左、右 4 个方向的触摸	431
实例 226	检测触摸滑动的方向	433
实例 227	实现屏幕的多点触摸	434
实例 228	检测双指滑动	435
实例 229	通过触摸方式放大或缩小屏幕中的图片	436
实例 230	通过触摸方式放大或缩小屏幕中的图片	437
实例 231	使用加速传感器	438
实例 232	触摸屏幕后插入一幅图片	441
实例 233	触摸后实现开花效果	442
实例 234	使用轻扫手势触摸	443
实例 235	双指触摸放大或缩小屏幕中图片	444
实例 236	自定义触摸手势删除屏幕中的图片	445
实例 237	触摸屏幕时发出声音	447

第 11 章 和设备之间的操作实战 449

实例 238	在屏幕中添加标记	449
实例 239	调用外部程序	450
实例 240	使用接近传感器	451
实例 241	获取电池的状态	453
实例 242	获取系统信息	454
实例 243	获取设备的终端识别符	455
实例 244	设置一个复制菜单	456
实例 245	复制/剪切/粘贴屏幕中的图片	458
实例 246	在粘贴板中保存自定义类	459
实例 247	获取电池的详细信息	460
实例 248	获取 iPhone 的硬件版本以及系统信息	464

第 12 章 游戏应用实战 467

实例 249	实现一个连连看游戏	467
实例 250	实现一个移动老虎机游戏	485
实例 251	实现一个移动打砖块游戏	488

第 13 章 移动 Web 实战 495

实例 252	实现页眉定位	495
实例 253	在页眉中使用按钮	497
实例 254	在页眉中使用分段控件	498
实例 255	在 iOS 网页中使用页脚	500
实例 256	在 iOS 系统中使用工具栏	501
实例 257	使用带有标准图标的标签栏	503
实例 258	在 iOS 中使用链接按钮	504
实例 259	在 iOS 中使用分组按钮	505

实例 260	在 iOS 中创建并使用动态按钮	507
实例 261	在 iOS 中使用表单	510
实例 262	在 iOS 中使用选择菜单	511
实例 263	在 iOS 中使用单选按钮	512
实例 264	在 iOS 中水平放置复选框	513
实例 265	在 iOS 中使用列表	515
实例 266	在 iOS 中使用两列表格	516
实例 267	在 iOS 中实现可折叠内容块效果	517
实例 268	搭建 PhoneGap 开发环境	519
实例 269	在 iOS 平台创建基于 PhoneGap 的程序	521
实例 270	使用通知 API	522
实例 271	使用确认 API	524

第 14 章 Swift 实战 ... 526

实例 272	使用 Xcode 创建 Swift 程序	526
实例 273	使用 Swift 实现 UITextField 控件	528
实例 274	基于 Swift 使用 UITextView 控件	534
实例 275	使用 Swift 实现 UISlider 控件效果	535
实例 276	使用 Swift 实现 Imageview 控件效果	538
实例 277	基于 Swift 控制是否显示密码明文	540
实例 278	基于 Swift 使用 UISegmentedControl 控件	543
实例 279	基于 Swift 使用 UIScrollView 控件	545
实例 280	基于 Swift 使用 UIPageControl 控件	547
实例 281	基于 Swift 使用 UIAlertView 控件	553
实例 282	基于 Swift 在表视图中使用其他控件	557
实例 283	使用 Swift 实现自定义进度条效果	560
实例 284	基于 Swift 使用 UIViewController 控件	565
实例 285	基于 Swift 综合使用界面视图	568
实例 286	基于 Swift 语言实现 ImagePicker 功能	575
实例 287	基于 Swift 实现一个音乐播放器	586

第 1 章 搭建开发环境实战

都说"工欲善其事，必先利其器！"，在进行 iOS 开发之前，开发者也同样需要先为自己准备一个好的开发工具，并预先搭建一个合适的开发环境。本章将以具体实例来详细介绍搭建 iOS 开发环境中的知识，让读者从实例中体会搭建 iOS 开发环境的方法和技巧，为步入本书后面知识的学习打下基础。

实例 001　下载并安装 Xcode

实例 001	下载并安装 Xcode
源码路径	无

实例说明

要开发 iOS 的应用程序，需要一台安装有 Xcode 工具的 Mac OS X 电脑。Xcode 是苹果提供的开发工具集，提供了项目管理、代码编辑、创建执行程序、代码级调试、代码库管理和性能调节等功能。这个工具集的核心就是 Xcode 程序，它提供了基本的源代码开发环境。

Xcode 是一款强大的专业开发工具，可以简单、快速而且以开发者熟悉的方式执行绝大多数常见的软件开发任务。相对于创建单一类型的应用程序所需要的能力而言，Xcode 要强大得多，它的设计目的是使开发者可以创建任何想要得到的软件产品类型，从 Cocoa 及 Carbon 应用程序，到内核扩展及 Spotlight 导入器等各种开发任务，Xcode 都能完成。Xcode 独具特色的用户界面可以帮助开发者以各种不同的方式来漫游工具中的代码，并且可以访问工具箱下面的大量功能，包括 GCC、javac、jikes 和 GDB，这些功能都是制作软件产品需要的。它是一个由专业人员设计的又由专业人员使用的工具。

由于能力出众，Xcode 已经被 Mac 开发者社区广为采纳。而且随着苹果电脑向基于 Intel 的 Macintosh 迁移，转向 Xcode 变得比以往任何时候更加重要。这是因为使用 Xcode 可以创建通用的二进制代码，这里所说的通用二进制代码是一种可以把 PowerPC 和 Intel 架构下的本地代码同时放到一个程序包的执行文件格式。事实上，对于还没有采用 Xcode 的开发人员，转向 Xcode 是将应用程序连编为通用二进制代码的第一个必要的步骤。

具体实现

其实对于初学者来说，只需安装 Xcode 即可。通过使用 Xcode，既能开发 iPhone 程序，也能够开发 iPad 程序。并且 Xcode 还是完全免费的，通过它提供的模拟器就可以在电脑上测试 iOS 程序。如果要发布 iOS 程序或在真实机器上测试 iOS 程序，就需要 99 美元。

（1）下载的前提是先注册成为一名开发人员，然后进入苹果开发页面主页 https://developer.apple.com/，如图 1-1 所示。

（2）登录 Xcode 的下载页面 http://developer.apple.com/devcenter/ios/index.action，如图 1-2 所示。

（3）单击图 1-2 左上方所示的"iOS 8 beta 2"链接，在新界面中显示下载 Xcode 6 Beta 链接界面。在此需要注意，开发者必须拥有付费账号，否则不能下载 Xcode 6 Beta，如图 1-3 所示。

（4）如果是付费账户，可以直接在苹果官方网站中下载获得。如果不是付费会员用户，可以从网络中搜索热心网友们的共享信息，以此来达到下载 Xcode 6 Beta 的目的。

（5）下载完成后单击打开下载的".dmg"格式文件，然后双击 Xcode 文件开始安装，如图 1-4 所示。

图 1-1　苹果开发页面主页

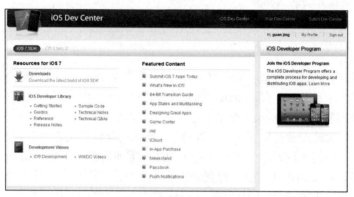

图 1-2　Xcode 的下载页面

图 1-3　提示必须拥有付费账号

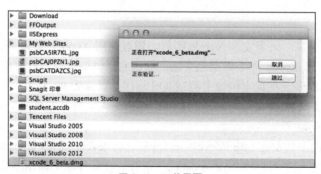

图 1-4　下载界面

（6）在弹出的对话框中单击"Continue"按钮，如图 1-5 所示。
（7）在弹出的欢迎界面中单击"Agree"按钮，如图 1-6 所示。
（8）在弹出的对话框中单击"Install"按钮，如图 1-7 所示。
（9）在弹出的对话框中输入用户名和密码，然后单击按钮"好"，如图 1-8 所示。

实例 001　下载并安装 Xcode

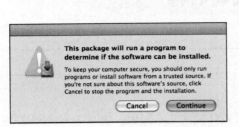

图 1-5　单击"Continue"按钮

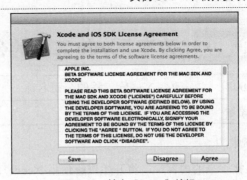

图 1-6　单击"Agree"按钮

图 1-7　单击"Install"按钮

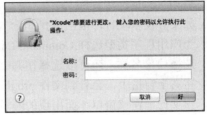

图 1-8　单击"好"按钮

（10）在弹出的新对话框中显示安装进度，进度完成后的界面如图 1-9 所示。

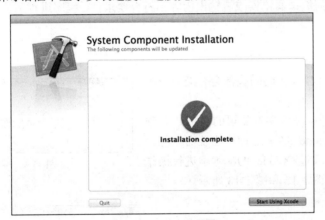

图 1-9　完成安装界面

（11）单击"Start Using Xcode"后可以启动 Xcode，如图 1-10 所示。

图 1-10　启动 Xcode 后的界面

当然如果使用的是苹果系统，我们完全可以使用 App Store 来获取 Xcode，这种方式的优点是完全自动，操作方便，比较适合于初学者。

实例 002　改变 Xcode 工程的公司名称

实例 002	改变 Xcode 工程的公司名称
源码路径	无

实例说明

在接下来的内容中，将开始讲解使用 Xcode 开发环境的基本知识。使用 Xcode 创建程序的基本步骤如下。

（1）启动 Xcode 应用程序。
（2）如果开发新项目，依次选择 File>New Project 命令。
（3）为应用程序类型选择 Command Line Utility、Foundation Tool，然后点击"Choose"命令。
（4）选择项目名称，还可以选择在哪个目录中存储项目文件，然后点击"Save"按钮。
（5）在右上窗格中，会看到文件 progl.m（或者是开发人员为项目起的其他名称，后面是.m）。突出显示该文件，在该窗口下面出现的编辑窗口中输入你的程序。
（6）依次选择 File>Save，保存已完成的更改。
（7）选择 Build、Build and Run，或者点击 Build and Go 按钮构建并运行程序。
（8）如果出现任何编译器错误或者输出内容不符合要求，对程序进行所需的更改并重复执行步骤（6）和步骤（7）。

具体实现

通过 Xcode 编写代码，代码的头部会有类似于图 1-11 所示的内容。

图 1-11　头部内容

在此需要将这部分内容改为公司的名称或者项目的名称，注意在 Xcode 3.2.x 之前，需要命令行设置变量。之后就可以通过 Xcode 的配置项进行操作了，操作步骤分别如图 1-12 和图 1-13 所示。

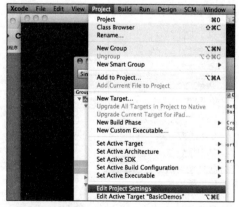

图 1-12　选择"Edit Project Settings"选项

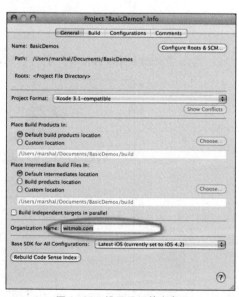

图 1-13　设置显示的内容

实例 003　通过搜索框缩小文件范围并格式化代码

这样如果再创建文件，就会产生图 1-14 所示的类似的效果了。

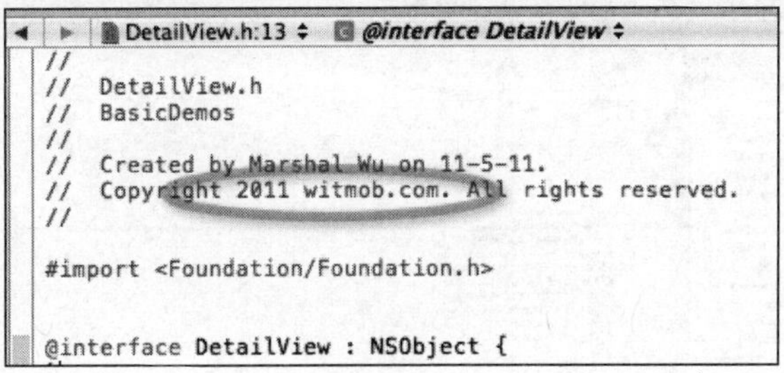

图 1-14　新创建文件时自动生成的内容

实例 003　通过搜索框缩小文件范围并格式化代码

实例 003	通过搜索框缩小文件范围并格式化代码
源码路径	无

实例说明

当项目开发到一段时间后，源代码文件会越来越多，如果再从 Groups & Files 的界面去点选，效率则比较差。此时开发人员可以借助 Xcode 的浏览器窗口缩小文件范围，此窗口的界面效果如图 1-15 所示。

图 1-15　Xcode 的浏览器窗口

具体实现

如果不想显示这个窗口，则可以通过快捷键"Shift+Command+E"来进行切换。在图 1-15 所示的搜索框中可以输入关键字，这样浏览器窗口里只显示带关键字的文件了，比如只想看 Book 相关的类，如图 1-16 所示。

例如在下面图 1-17 所示的界面中，有很多行都顶格了，此时需要进行格式化处理。

选中需要格式化的代码，然后在上下文菜单中进行查找，这是比较规矩的办法，如图 1-18 所示。

Xcode 没有提供快捷键，当然开发人员自己可以设置。用快捷键的做法是："Ctrl+A"（全选文字）、"Ctrl+X"（剪切文字）、"Ctrl+V"（粘贴文字）。Xcode 会对粘贴的文字格式化。

第1章 搭建开发环境实战

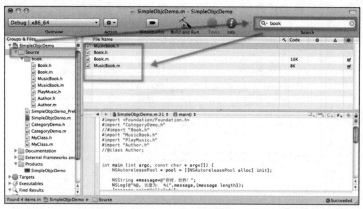

图 1-16 输入关键字

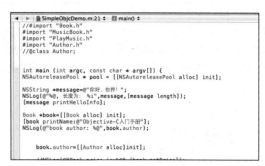

图 1-17 多行都顶格

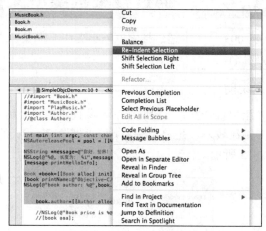

图 1-18 在上下文菜单中进行查找

实例 004 代码缩进和提示处理

实例 004	代码缩进和提示处理
源码路径	无

实例说明

有的时候代码需要缩进，有的时候又要做相反的操作。另外，使用 IDE 工具的一大好处是，工具能够帮助开发人员自动完成比如冗长的类型名称。

具体实现

单行缩进和其他编辑器类似，在 Xcode 中只需使用"Tab"键即可实现缩进。如果选中多行则需要使用快捷键，其中"Command+]"表示缩进；"Command+["表示反向缩进。

至于自动提示功能，假如有下面所示的输出日志：

```
NSLog(@"book author: %@",book.author);
```

如果开发人员都自己输入会很麻烦，可以先输入 NS，然后使用快捷键"Ctrl+."，会自动出现如下代码：

```
NSLog(NSString * format)
```

然后填写参数即可。快捷键"Ctrl+."的功能是自动给出第一个匹配 NS 关键字的函数或类型，而 NSLog 是第一个。如果继续使用"Ctrl+.",则会出现比如 NSString 的形式。依次类推,会显示所有 NS 开头的类型或函数,并循环往复。或者,也可以用"Ctrl+,"快捷键,比如还是 NS,那么会显示全部 NS 开头的类型、函数、常量等的列表,可以在这里选择。其实,Xcode 也可以在输入代码的过程中自动给出建议,比如要输入 NSString。当在编辑界面输入 NSStr 时:

```
NSString
```

此时在后面会自动出现提示,然后只需直接按"Tab"键确认即可。如果开发人员输入的是 NSStream,那么可以继续按。另外也可按"Esc"键,这时就会出现结果列表供选择了,如图 1-19 所示。
如果是正在输入方法,那么会自动完成图 1-20 所示的结果。

图 1-19 出现结果列表　　　　图 1-20 自动完成的结果

开发人员可以使用"Tab"键确认方法中的内容,或者通过快捷键"Ctrl+/"将方法中的参数来回切换。

实例 005　设置项目快照以及恢复到快照

实例 005	设置项目快照以及恢复到快照
源码路径	无

实例说明

在不用 Xcode 之前,开发人员使用 Eclipse 作为开发工具,那时习惯把代码提交到 SVN 上,并借助 SVN 的 Copy 功能实现服务器端的快照。在 Xcode 上不方便使用版本控制,因此本地快照功能还是很值得使用的。

快照(Snapshot)的主要作用是好比给开发人员的项目拍照,然后就可以随便修改代码了,从而不必担心改乱了无法回退到之前的版本。如果确实改乱了,只需恢复到快照就可以了,恢复后好像什么也没发生过。

具体实现

可以使用"Make Snapshot"命令创建快照,如图 1-21 所示。另外也可以使用快捷键"Ctrl+Command+S"来完成。

恢复的时候使用"Snapshots"命令实现,如图 1-22 所示。

然后选中做快照的版本,如图 1-23 所示。

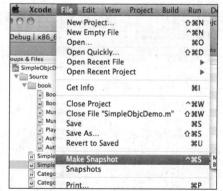

图 1-21 使用"Make Snapshot"命令创建快照

按下"Make"按钮可以拍照当前项目,并生成新的快照。可以在 Comments 框中写下该快照的备注信息,便于以后恢复时辨别,按下"Delete"按钮可以删除不必要的快照,按下"Restore"按钮可以用选中的快照覆盖当前项目,按下"Show Files"按钮可以列出选中快照和当前项目文件的差异。

图 1-22　使用"Snapshots"命令恢复　　　　图 1-23　选中做快照的版本

例如在图 1-24 所示的界面中列出了两个不同的文件，再选中文件可以看到不同的地方给出了标注，如图 1-25 所示。

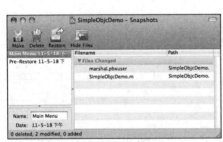
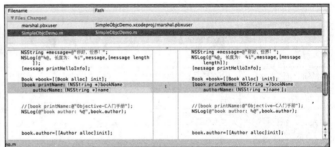

图 1-24　两个不同的文件　　　　　　　　图 1-25　不同的地方给出了标注

实例 006　实现复杂的查找和替代工作

实例 006	实现复杂的查找和替代工作
源码路径	无

实例说明

在编辑代码的过程中经常会做查找和替代的操作。如果只是查找，则直接按"Command+f"组合键即可，在代码的右上角会出现图 1-26 所示的对话框，只需在里面输入关键字，不论大小写，代码中所有命中的文字都高亮显示。

具体实现

其实在 Xcode 中也可以实现复杂的查找和替换工作，比如是否大小写敏感；是否使用正则表达式等。此功能的具体设置界面如图 1-27 所示。

通过图 1-28 中所示的"Find & Replace"可以切换到替代界面。

实例 006　实现复杂的查找和替代工作

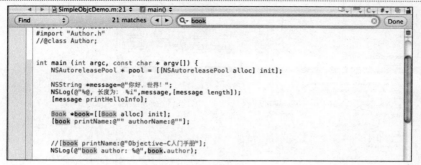
图 1-26　查找界面

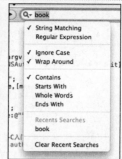
图 1-27　复杂查找设置

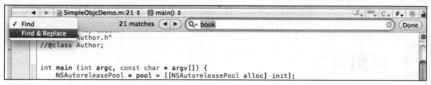
图 1-28　"Find & Replace" 替换

例如，图 1-29 所示的界面将查找设置为大小写敏感，然后替代为 myBook。

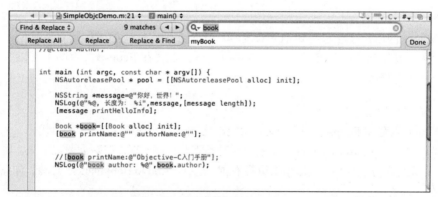
图 1-29　替代为 myBook

另外，也可以点击按钮是否全部替代，还是查找一个替代一个等。如果需要在整个项目内查找和替代，则依次单击 "Edit" > "Find" > "Find in Project..." 命令，如图 1-30 所示。

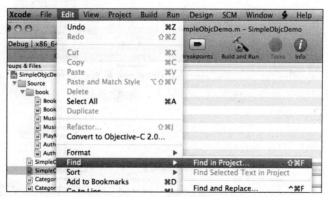
图 1-30　"Find in Project..." 命令

以查找关键字 book 为例，实现界面如图 1-31 所示。
替代操作的过程也与之类似，在此不再进行详细讲解。

第1章 搭建开发环境实战

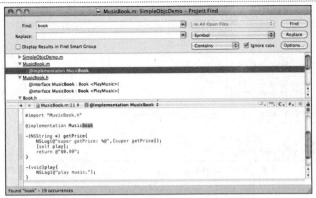

图1-31 在整个项目内查找"book"关键字

实例 007　使用书签

实例 007	使用书签
源码路径	无

实例说明

使用 Eclipse 的用户会经常用到 TODO 标签，比如在编写代码的时候需要做其他事情，或者提醒用户以后再实现的功能时，可以写一个 TODO 注释，这样在 Eclipse 的视图中可以找到，方便以后找到这个代码并修改。

具体实现

其实 Xcode 也有和 Eclipse 书签类似的功能，比如存在一段图 1-32 所示的代码。

这段代码的方法 printInfomation 是空的，暂时不需要具体实现。但是需要记下来，便于以后能找到并补充。让光标在方法内部，然后单击鼠标右键，选择 "Add to Bookmarks" 命令，如图 1-33 所示。

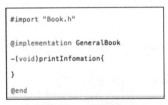

图 1-32　一段代码

图 1-33　选择 "Add to Bookmarks" 命令

此时会弹出对话框，可以在里面填写标签的内容，如图 1-34 所示。

这样就可以在项目的书签节点找到这个条目了，如图 1-35 所示。此时点击该条目，可以回到刚才添加书签时光标的位置。

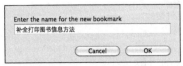

图 1-34　填写标签的内容

图 1-35　在项目的书签节点找到这个条目

实例 008　实现断点调试

实例 008	实现断点调试
源码路径	无

实例说明

在调试 iOS 项目程序时，最简单的调试方法是通过 NSLog 打印出程序运行中的结果，然后根据这些结果判断程序运行的流程和结果值是否符合预期。对于简单的项目，通常使用这种方式就足够了。但是，如果开发的是商业项目，它往往非常复杂，需要借助 Xcode 提供的专门调试工具。所有的编程工具的调试思路都是一样的。首先，开发人员要在代码中设置断点。想象一下，程序的执行是顺序的，可能怀疑某个地方的代码出了问题（引发 Bug），那么就在这段代码开始的地方，比如是方法的第一行，或者循环的开始部分，设置一个断点。那么程序在调试时会在运行到断点时中止，接下来可以一行一行地执行代码，判断执行顺序是否是自己预期的，或者变量的值是否和自己想的一样。

具体实现

在 Xcode 工程中设置断点的方法很简单，比如想对红框表示的行设置断点，就单击该行左侧的红圈位置，如图 1-36 所示。

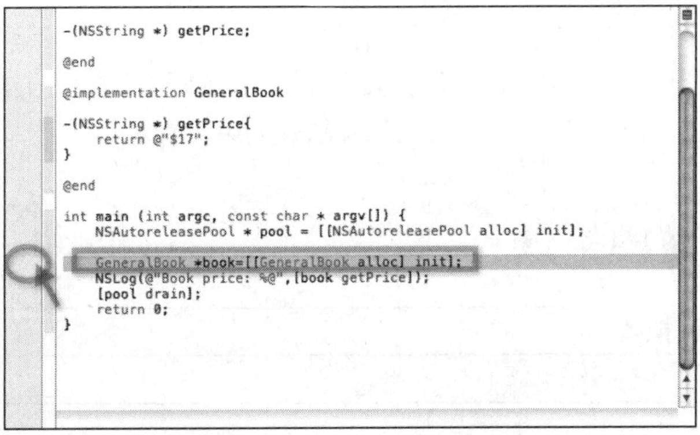

图 1-36　点击该行左侧红圈位置

单击后会出现断点标志，如图 1-37 所示。

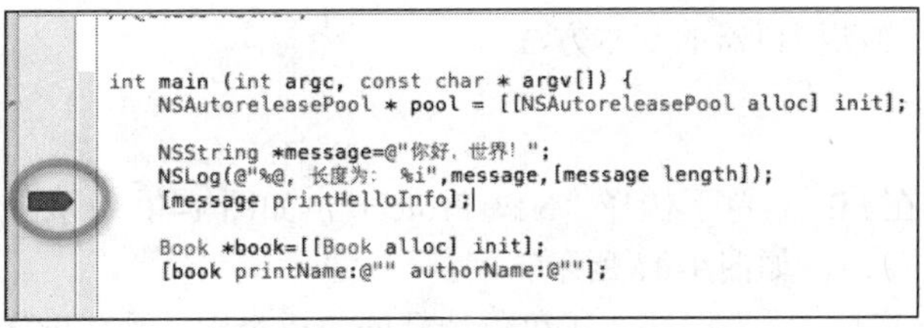

图 1-37　出现断点标志

然后运行代码，比如使用"Command+Enter"命令，这时将运行代码，并且停止在断点处，如图 1-38 所示。

可以通过"Shift+Command+Y"命令调出调试对话框，如图 1-39 所示。

这和其他语言 IDE 工具的界面大同小异，因为都具有类似的功能。下面是主要命令的具体说明。

（1）Continue：继续执行程序。

（2）Step over/ Step into/ Step out：用于单步调试，三者的具体说明如下所示。

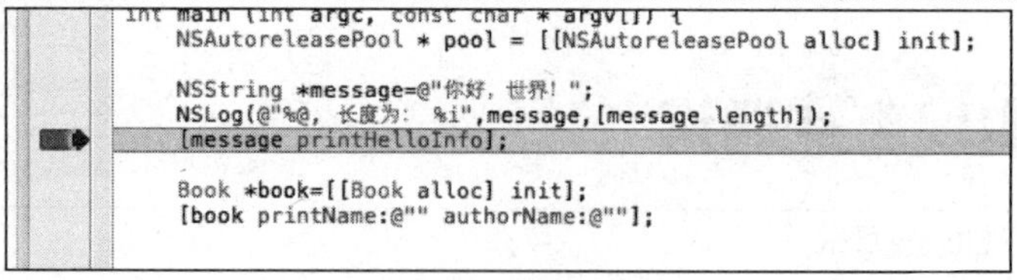

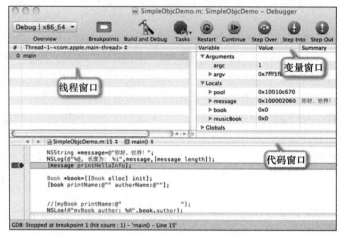

图 1-38　停止在断点处

图 1-39　调试对话框

- Step over：将执行当前方法内的下一个语句。
- Step into：如果当前语句是方法调用，将单步执行当前语句调用方法内部第一行。
- Step out：将跳出当前语句所在方法，到方法外的第一行。

实例 009　启动模拟器

实例 009	启动模拟器
源码路径	无

实例说明

Xcode 是一款功能全面的应用程序，通过此工具可以轻松输入、编译、调试并执行 Objective-C 程序。如果想在 Mac 上快速开发 iOS 应用程序，则必须学会使用这个强大工具的方法。在本实例中，将演示使用 Xcode 启动模拟器的基本方法。

具体实现

（1）启动 Xcode，在 File 菜单下选择"New Project"，如图 1-40 所示。

（2）此时出现一个窗口，如图 1-41 所示。

（3）在 New Project 窗口的左侧，显示了可供选择的模板类别，因为开发人员的重点是类别 iOS Application，所以在此需要确保选择了它。而在右侧显示了当前类别中的模板以及当前选定模板的描述。就此而言，请单击模板"Empty Application（空应用程序）"，再单击 Next（下一步）按钮。窗口界面效果如图 1-42 所示。

（4）选择模板并点击"Next"按钮后，在新界面中 Xcode 将要求开发人员指定产品名称和公司标识符。产品名称就是应用程序的名称，而公司标识符创建应用程序的组织或个人的域名，但按相反的顺序排列。这两者组成了束标识符，它将开发人员的应用程序和其他 iOS 应用程序区分开来，如图 1-43 所示。

实例009 启动模拟器

图1-40 启动一个新项目

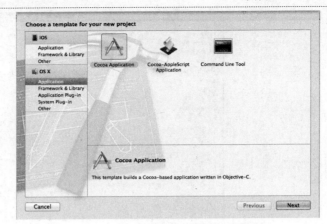
图1-41 启动一个新项目：选择应用程序类型

图1-42 单击模板"Empty Application（空应用程序）"

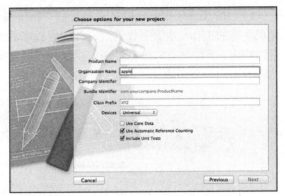
图1-43 Xcode文件列表窗口

例如，创建一个名为"Hello"的应用程序，这是产品名。设置域名是teach.com，因此将公司标识符设置为com.teach。如果开发人员没有域名，则开始开发时可使用默认标识符。

（5）将产品名设置为Hello，再提供选择的公司标识符。保留文本框Class Prefix为空。从下拉列表Device Family中选择开发人员使用的设备（iPhone或iPad），并确保选中了复选框Use Automatic Reference Counting（使用自动引用计数）。不要选中复选框Include Unit Tests（包含单元测试），界面效果如图1-44所示。

图1-44 指定产品名和公司标识符

（6）对设置满意后，单击"Next"按钮。Xcode要求指定项目的存储位置。切换到硬盘中合适的文件夹，确保没有选中复选框Source Control，再单击"Create（创建）"按钮。Xcode将创建一个名称与项目名相同的文件夹，并将所有相关联的模板文件都放到该文件夹中，如图1-45所示。

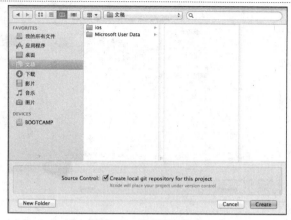

图 1-45　选择保存位置

（7）在 Xcode 中创建或打开项目后，将出现一个类似于 iTunes 的窗口，开发人员将使用它来完成所有的工作，从编写代码到设计应用程序界面。如果开发人员第一次接触 Xcode，令人眼花缭乱的按钮、下拉列表和图标将令人感到恐惧。为让开发人员对这些东西有大致的认识，下面首先介绍该界面的主要功能区域，如图 1-46 所示。

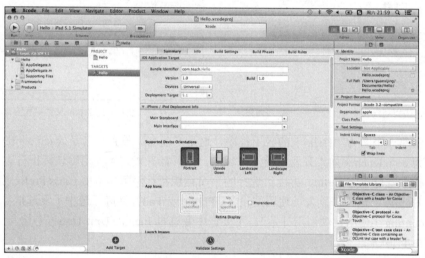

图 1-46　Xcode 界面

（8）运行 iOS 模拟器的方法十分简单，只需单击左上角的 按钮即可。iPad 模拟器的运行效果如图 1-47 所示。

iPhone 模拟器的运行效果如图 1-48 所示。

图 1-47　iPad 模拟器的运行效果

图 1-48　iPone 模拟器的运行效果

实例 010　使用第三方工具 iPhone Simulator

实例 010	使用第三方工具 iPhone Simulator
源码路径	无

实例说明

在 iOS 开发应用中，为了提高开发效率，需要借助第三方开发工具。例如测试程序需要模拟器 iPhone Simulator，设计界面需要 Interface Builder。iPhone Simulator 是 iPhone SDK 中的最常用工具之一，无需使用实际的 iPhone/iPod Touch 就可以测试应用程序。iPhone Simulator 位于如下文件夹中。

/Developer/iPhone OS <version> /Platforms/iPhoneSimulator.platform/Developer/Applications/

具体实现

在日常应用中，我们通常不需要直接启动 iPhone Simulator，它在 Xcode 运行（或是调试）应用程序时会自动启动。Xcode 会自动将应用程序安装到 iPhone Simulator 上。iPhone Simulator 是一个模拟器，并不是仿真器。模拟器会模仿实际设备的行为。iPhone Simulator 会模仿实际的 iPhone 设备的真实行为。但模拟器本身使用 Mac 上的 QuickTime 等库进行渲染，以便效果与实际的 iPhone 保持一致。此外，在模拟器上测试的应用程序会编译为 X86 代码，这是模拟器所能理解的字节码。与之相反，仿真器会模仿真实设备的工作方式。在仿真器上测试的应用程序会编译为真实设备所用的实际的字节码。仿真器会把字节码转换为运行仿真器的宿主计算机所能执行的代码形式。

iPhone Simulator 可以模拟不同版本的 iPhone OS。如果需要支持旧版本的平台以及测试并调试特定版本的 OS 上的应用程序所报告的错误，该功能就很有用。

启动 Xcode 后选择左边的 iPhone OS 下面的 Application，再依次选择"View"→"based Application"，然后为项目命名，如图 1-49 所示。

在新建的项目中不作任何操作，直接单击"Build and Run"按钮后即可在模拟器中运行程序，如图 1-50 所示。

图 1-49　Xcode 界面

图 1-50　模拟器界面

第 2 章 界面布局实战

对于网站开发人员来说，网站结构和界面设计是影响浏览用户第一视觉的关键。而对于 iOS 应用开发来说，除了功能强大的应用程序外，屏幕界面效果也是影响程序质量的重要元素。因为用户永远喜欢的是界面既美观又功能强大的软件产品。在设计优美界面之前，一定要先对屏幕进行布局。在本章的内容中，将以具体实例来介绍布局 iOS 屏幕的知识，为读者步入本书后面知识的学习打下基础。

实例 011　使用 Interface Builder 的故事板

实例 011	使用 Interface Builder 的故事板
源码路径	\daima\011\

实例说明

通过使用 Interface Builder（IB），确实能帮助开发人员快速地创建一个应用程序界面，但是它不仅是一个 GUI 绘画工具，还可以帮助开发人员在不编写任何代码的情况下添加应用程序功能。IB 向 Objective-C 开发者提供了包含一系列用户界面对象的工具箱，这些对象包括文本框、数据表格、滚动条、弹出式菜单等控件。IB 的工具箱是可扩展的，也就是说，所有开发者都可以开发新的对象，并将其加入 IB 的工具箱中。开发者只需要从工具箱中简单地向窗口或菜单中拖曳控件即可完成界面的设计。然后，用连线将控件可以提供的"动作"（Action）、控件对象分别和应用程序代码中对象"方法"（Method）、对象"接口"（Outlet）连接起来，就完成了整个创建工作。与其他图形用户界面设计器，例如 Microsoft Visual Studio 相比，这样的过程减小了 MVC 模式中控制器和视图两层的耦合，提高了代码质量。

当把 Interface Builder 集成到 Xcode 中后，和原来的版本相比主要有如下几点不同。

（1）在导航区选择 xib 文件后，在编辑区会显示 xib 文件的详细信息，这说明 Interface Builder 和 Xcode 确实是整合在一起了，如图 2-1 所示，选择 xib 文件时双击会弹出新窗口，这和以前的步骤基本一样。

（2）在工具栏选择 View 控制按钮，点击图 2-2 所示中最右边的按钮可以调出工具区，如图 2-3 所示。

在图 2-3 的工具区中，最上面的按钮分别是 4 个 Inspector：Identity、Attributes、Size、Connections。工具区下面就是可以往 View 上拖的控件。

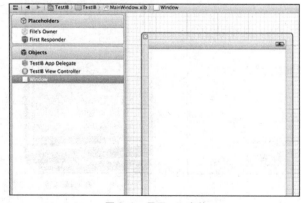

图 2-1　显示 xib 文件

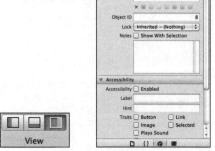

图 2-2　View 控制按钮　　图 2-3　工具区

实例 011　使用 Interface Builder 的故事板

（3）隐藏导航区。

若专心设计 UI，导航区就显得多余了，除非屏幕特别大。在刚才提到的"View 控制按钮"中点击第一个，将导航区隐藏，如图 2-4 所示。

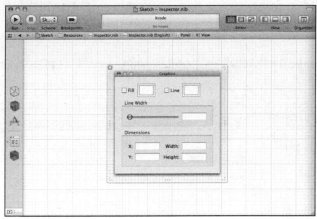

图 2-4　隐藏导航区

（4）关联方法和变量。

这是一个所见即所得功能，刚才已经打开隐藏很多区了，现在要涉及一个新 View：Assistant View，它是编辑区的一部分。此时只需将按钮（或者其他控件）拖到代码指定地方即可。在"拖"时需要按住"Ctrl"键。怎么让 Assistant View 显示要对应的".h"文件呢？使用这个 View 上面的选择栏进行选择。通过使用 Xcode 和 Cocoa 工具集，可手工编写生成 iOS 界面的代码：实例化界面对象，指定它们出现在屏幕的什么位置，设置对象的属性以及使其可见。例如通过下面的代码，可以在 iOS 设备屏幕设备的一角中显示文本"Hello Xcode"。

```
- (BOOL)application:(UIApplication *)application
     didFinishLaunchingWithOptions:(NSDictionary *)launchOptions
{
   self.window = [[UIWindow alloc]
                 initWithFrame:[[UIScreen mainScreen] bounds]];
   // Override point for customization after application launch.
   UILabel *myMessage;
   UILabel *myUnusedMessage;
   myMessage=[[UILabel alloc]
           initWithFrame:CGRectMake(30.0,50.0,300.0,50.0)];
   myMessage.font=[UIFont systemFontOfSize:48];
   myMessage.text=@"Hello Xcode";
   myMessage.textColor = [UIColor colorWithPatternImage:
                        [UIImage imageNamed:@"Background.png"]];
   [self.window addSubview:myMessage];
   self.window.backgroundColor = [UIColor whiteColor];
   [self.window makeKeyAndVisible];
   return YES;
}
```

如果要创建一个包含文本、按钮、图像以及数十个其他控件的界面，会耗费很多的事件。甚至如果做一些细微的修改，将需要阅读所有的代码。在过去的一段时间内，诞生了采用多种方法的图形界面生成器，其中最常见的实现方法之一是让开发人员"绘制"界面，并在幕后创建生成该界面的代码。在这种情况下，要做任何调整都必须手工编辑代码，这令人难以接受。

另一种常见的策略是维护界面定义，但将实现功能的代码直接关联到界面元素。不幸的是，这意味着如果要修改界面或将功能从一个 UI 元素切换到另一个，则必须移动代码。

而 Interface Builder 的工作原理与此不同，它不是自动生成界面代码，也不是将源代码直接关联到界面元素，而是生成实时的对象，并通过称为连接（Connection）的简单关联将其连接到应用程序代码。当开发人员需要修改应用程序功能的触发方式时，只需修改连接即可。要改变应用程序使用创建的对象的方式，只需连接或重新连接即可。

本实例是通过 Interface Builder 的故事板实现的，Storyboarding（故事板）是从 iOS 5 开始新加入的 Interface Builder（IB）的功能。主要的功能是在一个窗口中显示整个 app 用到的所有或者部分的页面，并且可以定义各页面之间的跳转关系，大大增加了 IB 的便利性。

具体实现

首先打开项目"lianjie"，并双击文件"ianjie.xcworkspace"，这将在 Xcode 中打开该项目，如图 2-5 所示。

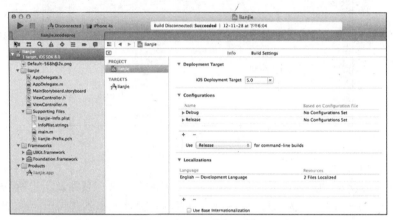

图 2-5　在 Xcode 中打开项目

加载该项目后，展开项目代码编组（Disconnected），并单击文件 MainStoryboard.storyboard，这个故事板文件包含该应用程序将把它显示为界面的场景和视图。Xcode 将刷新，并在 Interface Builder 编辑器中显示场景，如图 2-6 所示。

由图 2-6 所示的效果可知，该界面包含了如下 4 个交互式元素。
- 一个按钮栏（分段控件）。
- 一个按钮。
- 一个输出标签。
- 一个 Web 视图（一个集成的 Web 浏览器组件）。

这些控件将与应用程序代码交互，让用户选择花朵颜色并单击"给我花朵"按钮时，文本标签将显示选择的颜色，并从网站 http://www.floraphotographs.com 随机取回一朵这种颜色的花朵。预期的执行结果如图 2-7 所示。

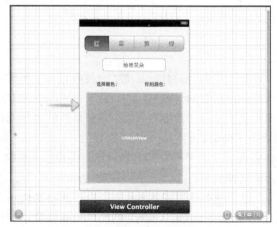

图 2-6　显示应用程序的场景和相应的视图

图 2-7　执行效果

但是到目前为止，这个演示程序没有任何功能，这是因为现在还没有将界面连接到应用程序代码，因此它不过是一张漂亮的图片。为让应用程序能够正常运行，将创建到应用程序代码中定义的输出口和操作的连接。

（1）输出口和操作。

输出口（Outlet）是一个通过它可引用对象的变量，假如 Interface Builder 中创建了一个用于收集用户姓名的文本框，在代码中为它创建一个名为 userName 的输出口。这样便可以使用该输出口和相应的属性获取或修改该文本框的内容。

另一方面，操作（Action）是代码中的一个方法，在相应的事件发生时调用它。有些对象（如按钮和开关）可在用户与之交互（如触摸屏幕）时通过事件触发操作。通过在代码中定义操作，Interface Builder 可使其能够被屏幕对象触发。

将 Interface Builder 中的界面元素与输出口或操作相连，这样就可以创建一个连接。为了让应用程序 Disconnected 能够成功运行，需要创建到如下所示的输出口和操作的连接。

- ColorChoice：一个对应于按钮栏的输出口，用于访问用户选择的颜色。
- GetFlower：这是一个操作，它从网上获取一幅花朵图像并显示它，然后将标签更新为选择的颜色。
- ChosedColor：对应于标签的输出口，将被 getFlower 更新以显示选定颜色的名称。
- FlowerView：对应于 Web 视图的输出口，将被 getFlower 更新以显示获取的花朵图像。

（2）创建到输出口的连接。

要想建立从界面元素到输出口的连接，可以先按住"Control"键，并同时从场景的 View Controller 图标（它出现在文档大纲区域和视图下方的图标栏中）拖曳到视图中对象的可视化表示或文档大纲区域中的相应图标。读者可以尝试对按钮栏（分段控件）进行这样的操作。按住"Control"键的同时，再单击文档大纲区域中的 View Controller 图标，并将其拖曳到屏幕上的按钮栏。拖曳时将出现一条线，这样能够轻松地指向要连接的对象。当松开鼠标时，将出现一个下拉列表，其中列出可供选择的输出口，如图 2-8 所示，再次选择 colorChoice。

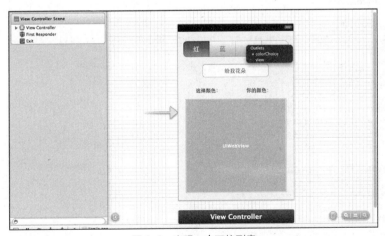

图 2-8　出现一个下拉列表

Interface Builder 知道什么类型的对象可连接到给定的输出口，因此它只显示适合当前要创建的连接的输出口。对文本为 Your Color 的标签和 Web 视图重复上述过程，将它们分别连接到输出口 chosenColor 和 flowerView。

在这个演示工程中，其核心功能是通过文件 ViewController.m 实现的，其主要代码如下所示。

```
#import "ViewController.h"
@implementation ViewController
@synthesize colorChoice;
```

```objc
@synthesize chosenColor;
@synthesize flowerView;
-(IBAction)getFlower:(id)sender {
    NSString *outputHTML;
    NSString *color;
    NSString *colorVal;
    int colorNum;
    colorNum=colorChoice.selectedSegmentIndex;
    switch (colorNum) {
        case 0:
            color=@"Red";
            colorVal=@"red";
            break;
        case 1:
            color=@"Blue";
            colorVal=@"blue";
            break;
        case 2:
            color=@"Yellow";
            colorVal=@"yellow";
            break;
        case 3:
            color=@"Green";
            colorVal=@"green";
            break;
    }
    chosenColor.text=[[NSString alloc] initWithFormat:@"%@",color];
    outputHTML=[[NSString alloc] initWithFormat:@"<body style='margin: 0px; padding: 0px'><img height='1200' src='http://www.floraphotographs.com/showrandom.php?color=%@'></body>",colorVal];
    [flowerView loadHTMLString:outputHTML baseURL:nil];
}

- (void)didReceiveMemoryWarning
{
    [super didReceiveMemoryWarning];
    // Release any cached data, images, etc that aren't in use.
}
#pragma mark - View lifecycle

- (void)viewDidLoad
{
    [super viewDidLoad];
    // Do any additional setup after loading the view, typically from a nib.
}
- (void)viewDidUnload
{
    [self setFlowerView:nil];
    [self setChosenColor:nil];
    [self setColorChoice:nil];
    [super viewDidUnload];
    // Release any retained subviews of the main view.
    // e.g. self.myOutlet = nil;
}
- (void)viewWillAppear:(BOOL)animated
{
    [super viewWillAppear:animated];
}
- (void)viewDidAppear:(BOOL)animated
{
    [super viewDidAppear:animated];
}
- (void)viewWillDisappear:(BOOL)animated
{
    [super viewWillDisappear:animated];
}
- (void)viewDidDisappear:(BOOL)animated
{
    [super viewDidDisappear:animated];
}
- (BOOL)shouldAutorotateToInterfaceOrientation:(UIInterfaceOrientation)interfaceOrientation
```

```
{
    // Return YES for supported orientations
    return (interfaceOrientation != UIInterfaceOrientationPortraitUpsideDown);
}
@end
```

(3) 创建到操作的连接。

连接到操作的方式稍有不同。对象的事件触发代码中的操作（方法），因此连接的方向正好相反：从触发事件的对象连接到场景的 View Controller。虽然可以像连接输出口那样按住"Control"键并拖曳来创建连接，但是再次建议不要这样做，因为用户无法指定哪个事件将触发它，究竟轻按按钮触发还是用户的手指离开按钮触发呢？

其实这个操作可以被很多不同的事件触发，因此需要确保选择了正确的事件，而不是让 Interface Builder 去决定。所以选择将调用操作的对象，并单击 Utility 区域顶部的箭头图标以打开 Connections Inspector（连接检查器）。也可以选择菜单 View→Utilities→Show Connections Inspector（Option+Command+6）。

Connections Inspector 显示了当前对象（这里是按钮）支持的事件列表，如图 2-9 所示。

每个事件旁边都有一个空心圆圈，要将事件连接到代码中的操作，可单击相应的圆圈并将其拖曳到文档大纲区域中的 View Controller 图标。假如要将按钮 Get Flower 连接到方法 getFlower，可选择该按钮并打开 Connections Inspector（Option+Command+6）。然后将 Touch Up Inside 事件旁边的圆圈拖曳到场景的 View Controller 图标，再松开鼠标。当系统询问时选择操作 getFlower，如图 2-10 所示。

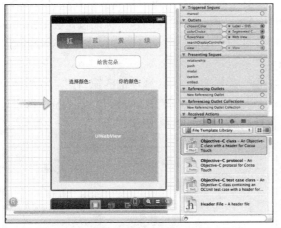

图 2-9 使用 Connections Inspector 操作连接

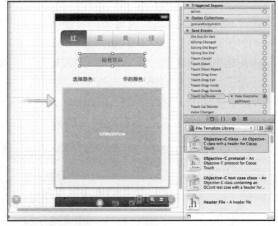

图 2-10 选择希望界面元素触发的操作

在建立连接后检查器会自动更新，以显示事件及其调用的操作，如图 2-11 所示。如果单击了其他对象，Connections Inspector 将显示该对象到输出口和操作的连接。

到现在为止，已经将界面连接到了支持它的代码。单击 Xcode 工具栏中的 Run 按钮，在 iOS 模拟器或 iOS 设备中便可以生成并运行该应用程序。执行效果如图 2-12 所示。

其实开发人员无需编写代码便可建立连接。虽然在 Interface Builder 中建立的大部分连接都位于对象和在代码中定义的输出口或操作之间，但有些对象实现了一些内置操作，不需要编写任何代码。例如，Web 视图实现了包括 goForward 和 goBack 在内的操作。通过使用这些操作，可以给该视图添加基本的导航功能。为此，只需将按钮的 Touch Up Inside 事件拖曳到 Web 视图对象（而不是 View Controller 图标）。正如前面指出的，系统将要求开发人员指定要连接到哪个操作，但这次连接的操作并非开发人员自己编写的。

(4) 使用快速检查器编辑连接。

在连接到界面时，常犯的一种错误是创建的连接并非我们需要的。经过一些拖曳操作后，界面

第 2 章　界面布局实战

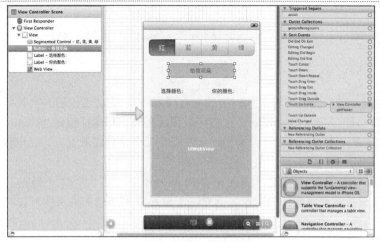

图 2-11　更新后的界面效果　　　　　图 2-12　执行效果

突然间不正确，不能正常运行了。要检查已建立的连接，可选择一个对象，并打开前面讨论过的 Connections Inspector，也可右击 Interface Builder 编辑器或文档大纲区域中的任何对象，以打开快速检查器（Quick Inspector）。这将出现一个浮动窗口，其中列出了与该对象相关联的所有输出口和操作，如图 2-13 所示。

（5）对象身份。

当将对象拖放到界面中时，实际上是在创建现有类（按钮、标签等）的实例。在这种情况下，需要给 Interface Builder 提供帮助，指出它应使用的子类。例如，假设创建了标准按钮类 UIButton 的一个子类，并将其命名为 ourFancyButtonClass。然后将一个按钮拖放到场景中以表示自定义按钮，但加载故事板文件时，创建的却是 UIButton 对象。

为了修复这种问题，可选择加入到视图中的按钮，单击 Utility 区域顶部的窗口图标或选择菜单 View→Utilities→Show Identity Inspector（Option+ Command+3）打开 Identity Inspector，再通过下拉列表/文本框指定在运行阶段加载界面时要实例化的类，如图 2-14 所示。

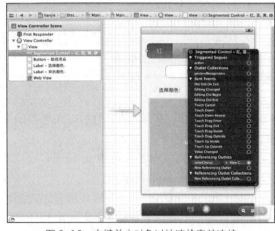
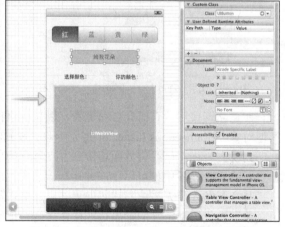

图 2-13　右键单击对象以快速检查其连接　　　　　图 2-14　在 InterfaceBuilder 中设置对象的身份

实例 012　设置 UIView 的位置和尺寸

实例 012	设置 UIView 的位置和尺寸
源码路径	\daima\012\

实例 012 设置 UIView 的位置和尺寸

实例说明

曾经有人这么说过，在 iOS 里看到的和摸到的都是 UIView，所以 UIView 在 iOS 开发里具有非常重要的作用。视图和窗口展示了应用的用户界面，同时负责界面的交互。UIKit 和其他系统框架提供了很多视图，开发人员可以就地使用而几乎不需要修改。当展示的内容与标准视图允许有很大的差别时，开发人员也可以定义自己的视图。无论是使用系统的视图还是创建自己的视图，需要理解类 UIView 和类 UIWindow 所提供的基本结构。这些类提供了复杂的方法来管理视图的布局和展示。理解这些方法的工作非常重要，使开发人员在应用发生改变时可以确认视图有合适的行为。

在 iOS 应用中，绝大部分可视化操作都是由视图对象（即 UIView 类的实例）进行的。一个视图对象定义了一个屏幕上的一个矩形区域，同时处理该区域的绘制和触屏事件。一个视图也可以作为其他视图的父视图，同时决定着这些子视图的位置和大小。UIView 类做了大量的工作去管理这些内部视图的关系，但是需要的时候也可以定制默认的行为。

在本实例中，使用方法 initWithFrame 可以依照 Frame 建立新的 View，建立出来的 View 要通过 addSubview 加入到父 View 中。本实例的最终目的是，分别在屏幕中间和屏幕右上角设置两个区域。

具体实现

实例文件 UIkitPrjFrame.h 的实现代码如下所示。

```objc
#import "SampleBaseController.h"
@interface UIKitPrjFrame : SampleBaseController
{
 @private
}
@end
```

实例文件 UIkitPrjFrame.m 的实现代码如下所示。

```objc
#import "UIKitPrjFrame.h"
@implementation UIKitPrjFrame
#pragma mark ----- Override Methods -----
- (void)viewDidLoad {
  [super viewDidLoad];
  self.view.backgroundColor = [UIColor blackColor];
  UILabel* label1 = [[[UILabel alloc] initWithFrame:CGRectZero] autorelease];
  label1.text = @"右上方";
  // 将 label1 的 frame 修改成任意的区域
  CGRect newFrame = CGRectMake( 220, 20, 100, 50 );
  label1.frame = newFrame;

  UILabel* label2 = [[[UILabel alloc] initWithFrame:[label1 frame]] autorelease];
  label2.textAlignment = UITextAlignmentCenter;
  label2.text = @"中心位置";
  // 将 label2 的 center 调整到画面中心
  CGPoint newPoint = self.view.center;
  // 空出状态条高度大小
  newPoint.y -= 20;
  label2.center = newPoint;
  // 向画面中追加 label1 与 label2
  [self.view addSubview:label1];
  [self.view addSubview:label2];
 UILabel* label = [[[UILabel alloc] initWithFrame:CGRectZero] autorelease];
  // frame 的设置
  label.frame = CGRectMake( 0, 0, 200, 50 );
  // center 设置
  label.center = CGPointMake( 160, 240 );
  // frame 的参照
  NSLog( @"x = %f", label.frame.origin.x );
  NSLog( @"y = %f", label.frame.origin.y );
  NSLog( @"width = %f", label.frame.size.width );
  NSLog( @"height = %f", label.frame.size.height );
  // center 的参照
  NSLog( @"x = %f", label.center.x );
  NSLog( @"y = %f", label.center.y );
```

```
}
- (void)touchesEnded:(NSSet*)touches withEvent:(UIEvent*)event {
  [self.navigationController setNavigationBarHidden:NO animated:YES];
}
@end
```

执行效果如图 2-15 所示。

图 2-15　执行效果

实例 013　隐藏指定的 UIView 区域

实例 013	隐藏指定的 UIView 区域
源码路径	\daima\012\

实例说明

本实例的功能是使用 UIView 的属性 hidden 来隐藏指定的区域。当属性 hidden 的值为 YES 时隐藏 UIView，当属性 hidden 的值为 NO 时显示 UIView。当单击本实例中的"点击"按钮时，会实现文本隐藏和文本显示的转换。

具体实现

实例文件 UIkitPrjHide.h 的实现代码如下所示。

```
#import "SampleBaseController.h"
@interface UIKitPrjHide : SampleBaseController
{
 @private
   UILabel* label_;
}
@end
```

实例文件 UIkitPrjHide.m 的实现代码如下所示。

```
#import "UIKitPrjHide.h"
@implementation UIKitPrjHide
#pragma mark ----- Override Methods -----
// finalize
- (void)dealloc {
  [label_ release];
  [super dealloc];
}
- (void)viewDidLoad {
  [super viewDidLoad];
  self.view.backgroundColor = [UIColor blackColor];
  label_ = [[UILabel alloc] initWithFrame:CGRectMake( 0, 0, 320, 200 )];
  label_.textAlignment = UITextAlignmentCenter;
  label_.backgroundColor = [UIColor blackColor];
  label_.textColor = [UIColor whiteColor];
  label_.text = @"I'm here!";
  [self.view addSubview:label_];
```

实例 014 改变背景颜色

```
  UIButton* button = [UIButton buttonWithType:UIButtonTypeRoundedRect];
  button.frame = CGRectMake( 0, 0, 100, 40 );
  CGPoint newPoint = self.view.center;
  newPoint.y = self.view.frame.size.height - 70;
  button.center = newPoint;
  [button setTitle:@"点击" forState:UIControlStateNormal];
  [button addTarget:self action:@selector(buttonDidPush) forControlEvents:UIControlEventTouchUpInside];
  [self.view addSubview:button];
}

- (void)buttonDidPush {
  label_.hidden = !label_.hidden;
}
- (void)touchesEnded:(NSSet*)touches withEvent:(UIEvent*)event {
  [self.navigationController setNavigationBarHidden:NO animated:YES];
}
@end
```

执行后的效果如图 2-16 所示，按下图 2-16 中的"点击"按钮后会隐藏文本，如图 2-17 所示。

图 2-16　执行效果　　　　图 2-17　隐藏了文本

实例 014　改变背景颜色

实例 014	改变背景颜色
源码路径	\daima\012\

实例说明

本实例的功能是使用 UIView 的属性 backgroundColor 来改变背景颜色。首先在屏幕上方设置 label 区域，在下方设置 3 个按钮，当单击不同的按钮后，会改变上方 label 区域的背景颜色。

具体实现

实例文件 UIkitPrjBackground.h 的实现代码如下所示。

```
#import "SampleBaseController.h"
@interface UIKitPrjBackground : SampleBaseController
{
 @private
  UILabel* label_;
  CGFloat redColor_;
  CGFloat greenColor_;
  CGFloat blueColor_;
}
@end
```

实例文件 UIkitPrjBackground.m 的实现代码如下所示。

```
#import "UIKitPrjBackground.h"
#pragma mark ----- Private Methods Definition -----
@interface UIKitPrjBackground ()
```

```objc
- (void)redDidPush;
- (void)greenDidPush;
- (void)blueDidPush;
- (void)changeLabelColor:(CGFloat*)pColor;
@end
#pragma mark ----- Start Implementation For Methods -----
@implementation UIKitPrjBackground
// finalize
- (void)dealloc {
  [label_ release];
  [super dealloc];
}
- (void)viewDidLoad {
  [super viewDidLoad];
  self.view.backgroundColor = [UIColor blackColor];
// 画面上方追加标签
  label_ = [[UILabel alloc] initWithFrame:CGRectMake( 0, 0, 320, 200 )];
  label_.textAlignment = UITextAlignmentCenter;
  redColor_ = 0.0;
  greenColor_ = 0.0;
  blueColor_ = 0.0;
  label_.backgroundColor = [[[UIColor alloc] initWithRed:redColor_
                                                  green:greenColor_
                                                   blue:blueColor_
                                                  alpha:1.0] autorelease];
  label_.textColor = [UIColor whiteColor];
  label_.text = @"染上新的颜色···";
  [self.view addSubview:label_];
  // 追加红色按钮
  UIButton* redButton = [UIButton buttonWithType:UIButtonTypeRoundedRect];
  redButton.frame = CGRectMake( 0, 0, 50, 40 );
  CGPoint newPoint = self.view.center;
  newPoint.x -= 50;
  newPoint.y = self.view.frame.size.height - 70;
  redButton.center = newPoint;
  [redButton setTitle:@"红" forState:UIControlStateNormal];
  [redButton setTitleColor:[UIColor redColor] forState:UIControlStateNormal];
  [redButton addTarget:self
              action:@selector(redDidPush)
      forControlEvents:UIControlEventTouchUpInside];
  // 追加绿色按钮
  UIButton* greenButton = [UIButton buttonWithType:UIButtonTypeRoundedRect];
  greenButton.frame = redButton.frame;
  newPoint.x += 50;
  greenButton.center = newPoint;
  [greenButton setTitle:@"绿" forState:UIControlStateNormal];
  [greenButton setTitleColor:[UIColor greenColor] forState:UIControlStateNormal];
  [greenButton addTarget:self
              action:@selector(greenDidPush)
      forControlEvents:UIControlEventTouchUpInside];
  // 追加蓝色按钮
  UIButton* blueButton = [UIButton buttonWithType:UIButtonTypeRoundedRect];
  blueButton.frame = redButton.frame;
  newPoint.x += 50;
  blueButton.center = newPoint;
  [blueButton setTitle:@"蓝" forState:UIControlStateNormal];
  [blueButton setTitleColor:[UIColor blueColor] forState:UIControlStateNormal];
  [blueButton addTarget:self
              action:@selector(blueDidPush)
      forControlEvents:UIControlEventTouchUpInside];
  [self.view addSubview:redButton];
  [self.view addSubview:greenButton];
  [self.view addSubview:blueButton];
}
#pragma mark ----- Private Methods -----
- (void)redDidPush {
  [self changeLabelColor:&redColor_];
}
- (void)greenDidPush {
```

```
    [self changeLabelColor:&greenColor_];
}
- (void)blueDidPush {
    [self changeLabelColor:&blueColor_];
}
- (void)changeLabelColor:(CGFloat*)pColor {
    if ( !pColor ) return;
// 对指定色以 0.1 为单位递增
    // 1.0 时回复为 0.0
    if ( *pColor > 0.99 ) {
        *pColor = 0.0;
    } else {
        *pColor += 0.1;
    }
    // 更新标签的颜色
    label_.backgroundColor = [[[UIColor alloc] initWithRed:redColor_
                                                    green:greenColor_
                                                     blue:blueColor_
                                                    alpha:1.0] autorelease];
}
- (void)touchesEnded:(NSSet*)touches withEvent:(UIEvent*)event {
    [self.navigationController setNavigationBarHidden:NO animated:YES];
}
@end
```

执行后的效果如图 2-18 所示，按下不同的按钮会显示对应的背景颜色。

图 2-18　执行效果

实例 015　实现背景透明

实例 015	实现背景透明
源码路径	\daima\012\

实例说明

本实例的功能是，使用 UIView 的属性 alpha 来改变指定视图的透明度。首先在屏幕上方设置了 label 区域，并显示了指定的文本"将白色文字背景设为红"。在下方设置了 1 个"透明化"按钮。每当单击一次"透明化"按钮，会设置上方 label 标签的 alpha 值以 0.1 为单位递减，从而逐渐实现透明效果。

具体实现

实例文件 UIkitPrjAlpha.h 的实现代码如下所示。

```
#import "SampleBaseController.h"
@interface UIKitPrjAlpha : SampleBaseController
{
    @private
```

```
    UILabel* label_;
}
@end
```

实例文件 UIkitPrjAlpha.m 的实现代码如下所示。

```
#import "UIKitPrjAlpha.h"
@implementation UIKitPrjAlpha
- (void)dealloc {
  [label_ release];
  [super dealloc];
}
- (void)viewDidLoad {
  [super viewDidLoad];
  // 背景设为白
  self.view.backgroundColor = [UIColor whiteColor];
  // 追加画面上方的标签
  label_ = [[UILabel alloc] initWithFrame:CGRectMake( 0, 0, 320, 200 )];
  label_.textAlignment = UITextAlignmentCenter;
  label_.backgroundColor = [UIColor redColor];
  label_.textColor = [UIColor whiteColor];
  label_.adjustsFontSizeToFitWidth = YES;
  label_.text = @"将白色文字背景设为红";
  [self.view addSubview:label_];
  // 追加染色按钮
  UIButton* alphaButton = [UIButton buttonWithType:UIButtonTypeRoundedRect];
  alphaButton.frame = CGRectMake( 0, 0, 100, 40 );
  CGPoint newPoint = self.view.center;
  newPoint.y = self.view.frame.size.height - 70;
  alphaButton.center = newPoint;
  [alphaButton setTitle:@"透明化" forState:UIControlStateNormal];
  [alphaButton addTarget:self
             action:@selector(alphaDidPush)
    forControlEvents:UIControlEventTouchUpInside];

  [self.view addSubview:alphaButton];
}

- (void)alphaDidPush {
  // 标签的 alpha 值以 0.1 为单位递减
  // 0.0 时恢复为 1.0
  if ( label_.alpha < 0.09 ) {
    label_.alpha = 1.0;
  } else {
    label_.alpha -= 0.1;
  }
}
- (void)touchesEnded:(NSSet*)touches withEvent:(UIEvent*)event {
  [self.navigationController setNavigationBarHidden:NO animated:YES];
}
@end
```

执行后的效果如图 2-19 所示。每当单击一次"透明化"逐渐实现透明效果。

图 2-19 执行效果

实例 016 定位屏幕中的图片

实例 016	定位屏幕中的图片
源码路径	\daima\012\

实例说明

在 iOS 应用中，UIView 类使用一个点播绘制模型来展示内容。当一个视图第一次出现在屏幕前，系统会要求它绘制自己的内容。在该流程中，系统会创建一个快照，这个快照是出现在屏幕中的视图内容的可见部分。如果从来没有改变视图的内容，这个视图的绘制代码可能永远不会再被调用。这个快照图像在大部分涉及视图的操作中被重用。如果确实改变了视图内容，也不会直接重新绘制视图内容。相反，使用 setNeedsDisplay 或者 setNeedsDisplayInRect:方法废止该视图，同时让系统在稍后重画内容。系统等待当前运行循环结束，然后开始绘制操作。这个延迟给了开发人员一个机会来废止多个视图，从层次中增加或者删除视图，隐藏、重设大小和重定位视图。所有的改变会稍候在同一时间反应。

UIView 类的 contentMode 属性决定了改变几何结构应该如何解释。大部分内容模式在视图的边界内拉伸或者重定位了已有快照，它不会重新创建一个新的快照。要获取更多关于内容模式如何影响视图的绘制周期，可查看 content modes，当绘制视图内容时，真正的绘制流程会根据视图及其配置改变。系统视图通常会实现私有的绘制方法来解释它们的视图（那些相同的系统视图经常开发接口，好让开发人员可以用来配置视图的真正表现）。对于定制的 UIView 子类，通常可以覆盖 drawRect:方法并使用该方法来绘制视图内容。也有其他方法来提供视图内容，如直接在底部的层设置内容，但是覆盖 drawRect:是最通用的技术。

UIViewContentMode 包含如下所示的模式。

```
typedef enum {
UIViewContentModeScaleToFill,
UIViewContentModeScaleAspectFit,
UIViewContentModeScaleAspectFil,
UIViewContentModeRedraw,
UIViewContentModeCenter,
UIViewContentModeTop,
UIViewContentModeBottom,
 UIViewContentModeLeft,
 UIViewContentModeRight,
UIViewContentModeTopLeft,
UIViewContentModeTopRight,
UIViewContentModeBottomLeft,
UIViewContentModeBottomRight,
} UIViewContentMode;
```

本实例的功能是，使用 UIView 的属性 UIViewContentMode 来改变图片内容在屏幕中的位置。首先在屏幕上方设置显示 UIImageView 图片，在 Label 区域显示当前使用的模式，在下方设置了 1 个"模式修改"按钮。每当单击一次"模式修改"按钮，会调用- (void)changeLabelCaption 方法逐一显示不同的模式。

具体实现

实例文件 UIkitPrjContentMode.h 的实现代码如下所示。

```
#import "SampleBaseController.h"
@interface UIKitPrjContentMode : SampleBaseController
{
@private
 UIImageView* imageView_;
 UILabel* label_;
}
@end
```

实例文件 UIkitPrjContentMode.m 的实现代码如下所示。

```objc
#import "UIKitPrjContentMode.h"
#pragma mark ----- Private Methods Definition -----
@interface UIKitPrjContentMode ()
- (void)modeDidPush;
- (void)changeLabelCaption;
@end
#pragma mark ----- Start Implementation For Methods -----
@implementation UIKitPrjContentMode
// finalize
- (void)dealloc {
    [imageView_ release];
    [label_ release];
    [super dealloc];
}
- (void)viewDidLoad {
    [super viewDidLoad];
    // 背景设置为黑
    self.view.backgroundColor = [UIColor blackColor];
    // 追加图像View
    NSString* path = [NSString stringWithFormat:@"%@/%@", [[NSBundle mainBundle] resourcePath], @"narrow_dog.jpg"];
    UIImage* image = [[[UIImage alloc] initWithContentsOfFile:path] autorelease];
    imageView_ = [[UIImageView alloc] initWithImage:image];
    imageView_.frame = CGRectMake( 0, 0, 320, 320 );
    [self.view addSubview:imageView_];
    // 追加标签
    label_ = [[UILabel alloc] initWithFrame:CGRectMake( 0, 0, 320, 30 )];
    CGPoint newPoint = imageView_.center;
    newPoint.y += 200;
    label_.center = newPoint;
    label_.textAlignment = UITextAlignmentCenter;
    [self.view addSubview:label_];
    [self changeLabelCaption];
    // 追加模式修改按钮
    UIButton* modeButton = [UIButton buttonWithType:UIButtonTypeRoundedRect];
    modeButton.frame = CGRectMake( 0, 0, 100, 40 );
    newPoint = self.view.center;
    newPoint.y = self.view.frame.size.height - 40;
    modeButton.center = newPoint;
    [modeButton setTitle:@"模式修改" forState:UIControlStateNormal];
    [modeButton addTarget:self
                action:@selector(modeDidPush)
      forControlEvents:UIControlEventTouchUpInside];
    [self.view addSubview:modeButton];
}
#pragma mark ----- Private Methods -----
- (void)modeDidPush {
    imageView_.contentMode++;
    [self changeLabelCaption];
}
- (void)changeLabelCaption {
    if ( UIViewContentModeScaleToFill == imageView_.contentMode ) {
        label_.text = @"UIViewContentModeScaleToFill";
    } else if ( UIViewContentModeScaleAspectFit == imageView_.contentMode ) {
        label_.text = @"UIViewContentModeScaleAspectFit";
    } else if ( UIViewContentModeScaleAspectFill == imageView_.contentMode ) {
        label_.text = @"UIViewContentModeScaleAspectFill";
    } else if ( UIViewContentModeRedraw == imageView_.contentMode ) {
        label_.text = @"UIViewContentModeRedraw";
    } else if ( UIViewContentModeCenter == imageView_.contentMode ) {
        label_.text = @"UIViewContentModeCenter";
    } else if ( UIViewContentModeTop == imageView_.contentMode ) {
        label_.text = @"UIViewContentModeTop";
    } else if ( UIViewContentModeBottom == imageView_.contentMode ) {
        label_.text = @"UIViewContentModeBottom";
    } else if ( UIViewContentModeLeft == imageView_.contentMode ) {
        label_.text = @"UIViewContentModeLeft";
    } else if ( UIViewContentModeRight == imageView_.contentMode ) {
```

```
            label_.text = @"UIViewContentModeRight";
        } else if ( UIViewContentModeTopLeft == imageView_.contentMode ) {
            label_.text = @"UIViewContentModeTopLeft";
        } else if ( UIViewContentModeTopRight == imageView_.contentMode ) {
            label_.text = @"UIViewContentModeTopRight";
        } else if ( UIViewContentModeBottomLeft == imageView_.contentMode ) {
            label_.text = @"UIViewContentModeBottomLeft";
        } else if ( UIViewContentModeBottomRight == imageView_.contentMode ) {
            label_.text = @"UIViewContentModeBottomRight";
        } else {
            label_.text = @"else";
        }
}
- (void)touchesEnded:(NSSet*)touches withEvent:(UIEvent*)event {
    [self.navigationController setNavigationBarHidden:NO animated:YES];
}
@end
```

执行后的效果如图 2-20 所示。每当单击一次"模式修改"按钮,会逐一显示不同模式的内容。

图 2-20　执行效果

实例 017　旋转和缩放视图

实例 017	旋转和缩放视图
源码路径	\daima\012\

实例说明

本实例的功能是,使用 UIView 的属性 transform 来翻转或者放缩视图。首先在屏幕上方设置了 UIImageView 区域,在此区域显示一幅图片。在下方设置了 4 个按钮,分别是旋转、扩大、缩小、反转,这 4 个按钮对应如下所示的 4 个方法。

- (void)rotateDidPush:以 90 度为单位旋转。

- (void)bigDidPush:以 0.1 为单位扩大。

- (void)smallDidPush:以 0.1 为单位缩小。

- (void)invertDidPush:左右反转。

具体实现

实例文件 UIkitPrjTransform.h 的实现代码如下所示。

```
#import "SampleBaseController.h"
@interface UIKitPrjTransform : SampleBaseController
{
 @private
    UIImageView* imageView_;
    CGFloat rotate_;
    CGFloat scale_;
    bool    needFlip_;
}
@end
```

实例文件 UIkitPrjTransform.m 的实现代码如下所示。

```objc
#import "UIKitPrjTransform.h"
#pragma mark ----- Private Methods Definition -----
@interface UIKitPrjTransform ()
- (void)rotateDidPush;
- (void)bigDidPush;
- (void)smallDidPush;
- (void)invertDidPush;
- (void)transformWithAnimation;
@end
#pragma mark ----- Start Implementation For Methods -----
@implementation UIKitPrjTransform
// finalize
- (void)dealloc {
  [imageView_ release];
  [super dealloc];
}
- (void)viewDidLoad {
  [super viewDidLoad];
  rotate_ = 0.0;
  scale_ = 1.0;
  needFlip_ = NO;
  // 背景设为黑
  self.view.backgroundColor = [UIColor blackColor];
  // 追加图像 View
  NSString* path = [NSString stringWithFormat:@"%@/%@", [[NSBundle mainBundle] resourcePath], @"dog.jpg"];
  UIImage* image = [[[UIImage alloc] initWithContentsOfFile:path] autorelease];
  imageView_ = [[UIImageView alloc] initWithImage:image];
  CGPoint newPoint = self.view.center;
  newPoint.y = self.view.center.y - 60;
  imageView_.center = newPoint;
  [self.view addSubview:imageView_];
  // 追加 "旋转" 按钮
  UIButton* rotateButton = [UIButton buttonWithType:UIButtonTypeRoundedRect];
  rotateButton.frame = CGRectMake( 0, 0, 50, 40 );
  newPoint = self.view.center;
  newPoint.x -= 75;
  newPoint.y = self.view.frame.size.height - 70;
  rotateButton.center = newPoint;
  [rotateButton setTitle:@"旋转" forState:UIControlStateNormal];
  [rotateButton addTarget:self
            action:@selector(rotateDidPush)
      forControlEvents:UIControlEventTouchUpInside];
  // 追加 "扩大" 按钮
  UIButton* bigButton = [UIButton buttonWithType:UIButtonTypeRoundedRect];
  bigButton.frame = rotateButton.frame;
  newPoint.x += 50;
  bigButton.center = newPoint;
  [bigButton setTitle:@"扩大" forState:UIControlStateNormal];
  [bigButton addTarget:self
            action:@selector(bigDidPush)
      forControlEvents:UIControlEventTouchUpInside];
  // 追加 "缩小" 按钮
  UIButton* smallButton = [UIButton buttonWithType:UIButtonTypeRoundedRect];
  smallButton.frame = rotateButton.frame;
  newPoint.x += 50;
  smallButton.center = newPoint;
  [smallButton setTitle:@"缩小" forState:UIControlStateNormal];
  [smallButton addTarget:self
            action:@selector(smallDidPush)
      forControlEvents:UIControlEventTouchUpInside];
  // 追加 "反转" 按钮
  UIButton* invertButton = [UIButton buttonWithType:UIButtonTypeRoundedRect];
  invertButton.frame = rotateButton.frame;
  newPoint.x += 50;
  invertButton.center = newPoint;
  [invertButton setTitle:@"反转" forState:UIControlStateNormal];
```

实例 017 旋转和缩放视图

```objc
[invertButton addTarget:self
            action:@selector(invertDidPush)
      forControlEvents:UIControlEventTouchUpInside];
[self.view addSubview:rotateButton];
[self.view addSubview:bigButton];
[self.view addSubview:smallButton];
[self.view addSubview:invertButton];
}
#pragma mark ----- Private Methods -----
- (void)rotateDidPush {
    // 以 90 度为单位旋转
    rotate_ += 90.0;
    if ( 359.0 < rotate_ ) {
        rotate_ = 0.0;
    }
    [self transformWithAnimation];
}
- (void)bigDidPush {
    // 以 0.1 为单位扩大
    scale_ += 0.1;
    [self transformWithAnimation];
}
- (void)smallDidPush {
    // 以 0.1 为单位缩小
    scale_ -= 0.1;
    [self transformWithAnimation];
}

- (void)invertDidPush {
    // 左右反转
    needFlip_ = !needFlip_;
    [self transformWithAnimation];
}
- (void)transformWithAnimation {
    [UIView beginAnimations:nil context:NULL];

    CGAffineTransform transformRotate =
      CGAffineTransformMakeRotation( rotate_ * ( M_PI / 180.0 ) );
    CGAffineTransform transformScale =
      CGAffineTransformMakeScale( scale_, scale_ );
    CGAffineTransform transformAll =
      CGAffineTransformConcat( transformRotate, transformScale );
    if ( needFlip_ ) {
        transformAll = CGAffineTransformScale( transformAll, -1.0, 1.0 );
    }
    imageView_.transform = transformAll;

    [UIView commitAnimations];
}
- (void)touchesEnded:(NSSet*)touches withEvent:(UIEvent*)event {
    [self.navigationController setNavigationBarHidden:NO animated:YES];
}
@end
```

执行后的效果如图 2-21 所示。

图 2-21 执行效果

实例 018 伸缩屏幕中的视图

实例 018	伸缩屏幕中的视图
源码路径	\daima\012\

实例说明

本实例的功能是，使用 UIView 的属性 contentStretch 可以设置并改变视图的内容如何拉伸。首先在屏幕上方设置了 UIImageView 区域，在此区域显示一幅图片。在屏幕中间用 x 坐标设置了视图的宽度，用 y 坐标设置了视图的高度。在下方设置了"origin"和"size"两个按钮。

具体实现

实例文件 UIkitPrjContentStretch.h 的实现代码如下所示。

```
#import "SampleBaseController.h"
@interface UIKitPrjContentStretch : SampleBaseController
{
 @private
  UIImageView* imageView_;
  CGRect rect_;
  UILabel* label_;
}
@end
```

实例文件 UIkitPrjContentStretch.m 的实现代码如下所示。

```
#import "UIKitPrjContentStretch.h"
#pragma mark ----- Private Methods Definition -----
@interface UIKitPrjContentStretch ()
- (void)originDidPush;
- (void)sizeDidPush;
- (void)changeLabelCaption;
@end
#pragma mark ----- Start Implementation For Methods -----
@implementation UIKitPrjContentStretch
// finalize
- (void)dealloc {
  [imageView_ release];
  [label_ release];
  [super dealloc];
}
- (void)viewDidLoad {
  [super viewDidLoad];
  rect_ = CGRectMake( 0.0, 0.0, 1.0, 1.0 );
  // 背景设置成黑色
  self.view.backgroundColor = [UIColor blackColor];
  // 追加图像 View
  NSString* path = [NSString stringWithFormat:@"%@/%@", [[NSBundle mainBundle] resourcePath],
@"dog.jpg"];
  UIImage* image = [[[UIImage alloc] initWithContentsOfFile:path] autorelease];
  imageView_ = [[UIImageView alloc] initWithImage:image];
  imageView_.frame = CGRectMake( 0, 0, 320, 320 );
  imageView_.contentMode = UIViewContentModeScaleAspectFit;
  imageView_.contentStretch = rect_;
  [self.view addSubview:imageView_];
  // 追加标签
  label_ = [[UILabel alloc] initWithFrame:CGRectMake( 0, 0, 320, 60 )];
  CGPoint newPoint = imageView_.center;
  newPoint.y += 190;
```

```objc
    label_.center = newPoint;
    label_.textAlignment = UITextAlignmentCenter;
    [self.view addSubview:label_];
    [self changeLabelCaption];
    // 追加 origin 按钮
    UIButton* originChange = [UIButton buttonWithType:UIButtonTypeRoundedRect];
    originChange.frame = CGRectMake( 0, 0, 100, 40 );
    newPoint = self.view.center;
    newPoint.x -= 50;
    newPoint.y = self.view.frame.size.height - 40;
    originChange.center = newPoint;
    [originChange setTitle:@"origin" forState:UIControlStateNormal];
    [originChange addTarget:self
                action:@selector(originDidPush)
        forControlEvents:UIControlEventTouchUpInside];
    // 追加 size 修改按钮
    UIButton* sizeChange = [UIButton buttonWithType:UIButtonTypeRoundedRect];
    sizeChange.frame = CGRectMake( 0, 0, 100, 40 );
    newPoint = originChange.center;
    newPoint.x += 100;
    sizeChange.center = newPoint;
    [sizeChange setTitle:@"size" forState:UIControlStateNormal];
    [sizeChange addTarget:self
              action:@selector(sizeDidPush)
        forControlEvents:UIControlEventTouchUpInside];
    [self.view addSubview:originChange];
    [self.view addSubview:sizeChange];
}
#pragma mark ----- Private Methods -----
- (void)originDidPush {
    if ( 0.99 < rect_.origin.x ) {
        rect_.origin.x = 0.0;
        rect_.origin.y = 0.0;
    } else {
        rect_.origin.x += 0.1;
        rect_.origin.y += 0.1;
    }
    imageView_.contentStretch = rect_;
    [self changeLabelCaption];
}
- (void)sizeDidPush {
    if ( 0.09 > rect_.size.width ) {
        rect_.size.width = 1.0;
        rect_.size.height = 1.0;
    } else {
        rect_.size.width -= 0.1;
        rect_.size.height -= 0.1;
    }
    imageView_.contentStretch = rect_;
    [self changeLabelCaption];
}
- (void)changeLabelCaption {
    label_.numberOfLines = 2;
    label_.text = [NSString stringWithFormat:@"x = %f, y = %f, \r\nw = %f, h = %f",
                      rect_.origin.x, rect_.origin.y, rect_.size.width, rect_.size.height];
}
- (void)touchesEnded:(NSSet*)touches withEvent:(UIEvent*)event {
    [self.navigationController setNavigationBarHidden:NO animated:YES];
}
@end
```

执行后的效果如图 2-22 所示。

第 2 章　界面布局实战

图 2-22　执行效果

实例 019　实现视图的大小自适应

实例 019	实现视图的大小自适应
源码路径	\daima\012\

实例说明

在 iOS 应用中，使用 SizeToFit 方法可以自动将图标图像缩放到合适的大小。例如，在 UILabel 中使用 SizeToFit 方法后，可以根据 UILabel 中保持文本字符串的大小自动调整 UILabel 的大小。如果在 UIButton 中使用了此方法，则会根据按钮的标题大小自动调整按钮的大小。在本实例中，演示了使用 SizeToFit 方法实现 UILabel 文本自适应的方法。

具体实现

实例文件 UIkitPrjFit.h 的实现代码如下所示。

```
#import "SampleBaseController.h"
@interface UIKitPrjFit : SampleBaseController
{
 @private
}
@end
@interface DoubleLabel : UIView
{
 @private
 UILabel* label1_;
 UILabel* label2_;
}
@end
```

实例文件 UIkitPrjFit.m 的实现代码如下所示。

```
#import "UIKitPrjFit.h"
@implementation UIKitPrjFit
- (void)viewDidLoad {
 [super viewDidLoad];
 self.view.backgroundColor = [UIColor blackColor];
// 追加短标签
 UILabel* label1 = [[[UILabel alloc] initWithFrame:CGRectZero] autorelease];
 label1.backgroundColor = [UIColor blueColor];
 label1.textColor = [UIColor whiteColor];
 label1.text = @"短字符串";
 [label1 sizeToFit];
 [self.view addSubview:label1];
 // 追加长标签
 UILabel* label2 = [[[UILabel alloc] initWithFrame:CGRectZero] autorelease];
 label2.backgroundColor = [UIColor blueColor];
```

实例 019　实现视图的大小自适应

```objectivec
    label2.textColor = [UIColor whiteColor];
    label2.text = @"长--------------------字符串";
    [label2 sizeToFit];
    CGPoint newPoint = label2.center;
    newPoint.y += 50;
    label2.center = newPoint;
    [self.view addSubview:label2];
    // 追加 DoubleLabel
    DoubleLabel* doubleLabel = [[[DoubleLabel alloc] init] autorelease];
    doubleLabel.frame = CGRectMake( 0, 100, 320, 200 );
    [doubleLabel sizeToFit];
    [self.view addSubview:doubleLabel];
}

- (void)touchesEnded:(NSSet*)touches withEvent:(UIEvent*)event {
    [self.navigationController setNavigationBarHidden:NO animated:YES];
}
@end
@implementation DoubleLabel
- (void)dealloc {
    [label1_ release];
    [label2_ release];
    [super dealloc];
}
- (id)init {
    if ( (self = [super init]) ) {
        self.backgroundColor = [UIColor blackColor];
        label1_ = [[UILabel alloc] initWithFrame:CGRectZero];
        label1_.text = @"ABC";
        [label1_ sizeToFit];
        label2_ = [[UILabel alloc] initWithFrame:CGRectZero];
        label2_.text = @"ABCDEFGHIJKLMN";
        [label2_ sizeToFit];
        CGPoint newPoint = label2_.center;
        newPoint.x += 100;
        newPoint.y += 50;
        label2_.center = newPoint;
        [self addSubview:label1_];
        [self addSubview:label2_];
    }
    return self;
}
//此类为 UIView 的子类
//其中包含了 label1_ 与 label2_ 两个 UILabel
- (CGSize)sizeThatFits:(CGSize)size {
    CGFloat x1, x2, y1, y2;
    // 将 label1_与 label2_中延伸的一方左侧的坐标设置成 x1
    if ( label1_.frame.origin.x < label2_.frame.origin.x ) {
        x1 = label1_.frame.origin.x;
    } else {
        x1 = label2_.frame.origin.x;
    }
    // 将 label1_与 label2_中延伸的一方右侧的坐标设置成 x2
    if ( label1_.frame.origin.x + label1_.frame.size.width >
         label2_.frame.origin.x + label2_.frame.size.width )
    {
        x2 = label1_.frame.origin.x + label1_.frame.size.width;
    } else {
        x2 = label2_.frame.origin.x + label2_.frame.size.width;
    }
    // 将 label1_与 label2_中向上延伸的一方左侧的坐标设置成 y1
    if ( label1_.frame.origin.y < label2_.frame.origin.y ) {
        y1 = label1_.frame.origin.y;
    } else {
        y1 = label2_.frame.origin.y;
    }
    // 将 label1_与 label2_中向下延伸的一方右侧的坐标设置成 y2
    if ( label1_.frame.origin.y + label1_.frame.size.height >
         label2_.frame.origin.y + label2_.frame.size.height )
```

```
    {
      y2 = label1_.frame.origin.y + label1_.frame.size.height;
    } else {
      y2 = label2_.frame.origin.y + label2_.frame.size.height;
    }
    // 新尺寸的设置
    size.width = x2 - x1;
    size.height = y2 - y1;
    return size;
}
@end
```

执行后的效果如图 2-23 所示。

图 2-23 执行效果

实例 020　实现视图嵌套

实例 020	实现视图嵌套
源码路径	\daima\012\

实例说明

在 iOS 应用中,在一个 UIView 里面可以包含许多的 Subview(其他的 UIView),而这些 Subview 彼此之间有所谓的阶层关系,这有点类似于绘图软件中图层的概念。例如,在下面的代码中,演示了几个在管理图层（Subview）上常用的方法。

（1）新增和移除 Subview。

```
//将 Subview 从当前的 UIView 中移除
[Subview removeFromSuperview];
//替 UIView 增加一个 Subview
[UIView addSubview:Subview];
```

（2）在 UIView 中将 Subview 往前或者往后移动一个图层,往前移动会覆盖住较后层的 Subview,而往后移动则会被较上层的 Subview 所覆盖。

```
//将 Subview 往前移动一个图层（与它的前一个图层对调位置）
[UIView bringSubviewToFront:Subview];
//将 Subview 往后移动一个图层（与它的后一个图层对调位置）
[UIView sendSubviewToBack:Subview];
```

（3）在 UIView 中使用索引 Index 交换两个 Subview 彼此的图层层级。

```
//交换两个图层
[UIView exchangeSubviewAtIndex:indexA withSubviewAtIndex:indexB];
```

（4）使用 Subview 的变数名称取得它在 UIView 中的索引值（Index）。

```
//取得 Index
NSInteger index = [[UIView subviews] indexOfObject:Subview名称];
```

在本实例中,演示了使用 addSubview 方法添加视图,以实现视图嵌套的具体过程。

具体实现

实例文件 UIkitPrjSubviews.h 的实现代码如下所示。

```
#import "SampleBaseController.h"
@interface UIKitPrjSubviews : SampleBaseController
{
 @private
}
@end
```

实例文件 UIkitPrjSubviews.m 的实现代码如下所示。

```
#import "UIKitPrjSubviews.h"
#pragma mark ----- Private Methods Definition -----
@interface UIKitPrjSubviews ()
- (void)button11DidPush:(id)sender;
- (void)button111DidPush:(id)sender;
- (void)button112DidPush:(id)sender;
- (void)button1121DidPush:(id)sender;
- (void)button1122DidPush:(id)sender;
- (void)alertMessage:(UIButton*)button;
@end
#pragma mark ----- Start Implementation For Methods -----
@implementation UIKitPrjSubviews
- (void)viewDidLoad {
 [super viewDidLoad];
 self.view.backgroundColor = [UIColor blackColor];
 // 追加 1-1 按钮
 UIButton* button11 = [UIButton buttonWithType:UIButtonTypeRoundedRect];
 button11.frame = CGRectMake( 10, 10, 300, 300 );
 [button11 setTitle:@"1-1" forState:UIControlStateNormal];
 [button11 addTarget:self
           action:@selector(button11DidPush:)
     forControlEvents:UIControlEventTouchUpInside];
 [self.view addSubview:button11];
 // 追加 1-1-1 按钮
 UIButton* button111 = [UIButton buttonWithType:UIButtonTypeRoundedRect];
 button111.frame = CGRectMake( 20, 20, 260, 100 );
 [button111 setTitle:@"1-1-1" forState:UIControlStateNormal];
 [button111 addTarget:self
            action:@selector(button111DidPush:)
      forControlEvents:UIControlEventTouchUpInside];
 [button11 addSubview:button111];
 // 追加 1-1-2 按钮
 UIButton* button112 = [UIButton buttonWithType:UIButtonTypeRoundedRect];
 button112.frame = CGRectMake( 20, 180, 260, 100 );
 [button112 setTitle:@"1-1-2" forState:UIControlStateNormal];
 [button112 addTarget:self
            action:@selector(button112DidPush:)
      forControlEvents:UIControlEventTouchUpInside];
 [button11 addSubview:button112];
 // 追加 1-1-2-1 按钮
 UIButton* button1121 = [UIButton buttonWithType:UIButtonTypeRoundedRect];
 button1121.frame = CGRectMake( 10, 10, 95, 80 );
 [button1121 setTitle:@"1-1-2-1" forState:UIControlStateNormal];
 [button1121 addTarget:self
             action:@selector(button1121DidPush:)
       forControlEvents:UIControlEventTouchUpInside];
 [button112 addSubview:button1121];
 // 追加 1-1-2-2 按钮
 UIButton* button1122 = [UIButton buttonWithType:UIButtonTypeRoundedRect];
 button1122.frame = CGRectMake( 155, 10, 95, 80 );
 [button1122 setTitle:@"1-1-2-2" forState:UIControlStateNormal];
 [button1122 addTarget:self
             action:@selector(button1122DidPush:)
       forControlEvents:UIControlEventTouchUpInside];
 [button112 addSubview:button1122];
}
#pragma mark ----- Private Methods -----

- (void)button11DidPush:(id)sender {
 [self alertMessage:sender];
}
```

第 2 章　界面布局实战

```objc
- (void)button111DidPush:(id)sender {
  [self alertMessage:sender];
}
- (void)button112DidPush:(id)sender {
  [self alertMessage:sender];
}
- (void)button1121DidPush:(id)sender {
  [self alertMessage:sender];
}
- (void)button1122DidPush:(id)sender {
  [self alertMessage:sender];
}
- (void)alertMessage:(UIButton*)button {
  //显示 self 的标题作为警告框的标题
  NSString* title = [NSString stringWithFormat:@"self = %@", button.titleLabel.text];
  //取得 superview 的标题
  //当 superview 非 UIButton 的情况下，以"UIViewController"替代
  NSString* superViewName;
  if ( [button.superview isKindOfClass:[UIButton class]] ) {
    superViewName = ((UIButton*)button.superview).titleLabel.text;
  } else {
    superViewName = @"UIViewController";
  }
  //取得 subviews 的标题
  NSMutableString* subviews = [[[NSMutableString alloc] initWithCapacity:64] autorelease];
  [subviews setString:@""];
  for ( id view in button.subviews ) {
    NSString* addString;
    if ( [view isKindOfClass:[UIButton class]] ) {
        //如果子元素为 UIButton 时，取 titleLabel 的 text 属性值
      addString = ((UIButton*)view).titleLabel.text;
    } else if ( [view isKindOfClass:[UILabel class]] ) {
        //如果为 UILabel 时取其 text 属性值
      addString = ((UILabel*)view).text;
    } else {
        //上述以外的情况
      addString = [view description];
    }
    if ( [subviews length] > 0 ) {
      [subviews appendString:@", "];
    }
    [subviews appendString:addString];
  }
  NSString* message = [NSString stringWithFormat:@"superview = %@\r\nsubviews = %@", superViewName, subviews];
  UIAlertView* alert = [[[UIAlertView alloc] initWithTitle:title
                                       message:message
                                       delegate:nil
                                cancelButtonTitle:nil
                                otherButtonTitles:@"OK", nil ] autorelease];
  [alert show];
}
- (void)touchesEnded:(NSSet*)touches withEvent:(UIEvent*)event {
  [self.navigationController setNavigationBarHidden:NO animated:YES];
}
@end
```

执行后的效果如图 2-24 所示。

图 2-24　执行效果

实例 021　插入或删除视图中的子元素

实例 021	插入或删除视图中的子元素
源码路径	\daima\012\

实例说明

在 iOS 应用中，除了使用 addSubview 方法添加视图外，还可以使用 insertSubview 方法、insertSubview:aboveSubview 方法和 insertSubview:belowSubview 方法实现。

在本实例中，我们在视图中先追加了父标签 parent 和子标签 child3；然后在上一个标签 CHILD 3 下插入 CHILD 1，在 CHILD 1 上追加 CHILD 2；接下来让 CHILD 1 与 CHILD 3 交换；最后删除 CHILD 3。

具体实现

实例文件 UIkitPrjInsert.h 的实现代码如下所示。

```
#import "SampleBaseController.h"
@interface UIKitPrjInsert : SampleBaseController
{
 @private
  UILabel* parent_;
}
@end
```

实例文件 UIkitPrjInsert.m 的实现代码如下所示。

```
#import "UIKitPrjInsert.h"
#pragma mark ----- Private Methods Definition -----
@interface UIKitPrjInsert ()
- (void)subviewsDidPush;
@end
#pragma mark ----- Start Implementation For Methods -----
@implementation UIKitPrjInsert
// finalize
- (void)dealloc {
  [parent_ release];
  [super dealloc];
}
- (void)viewDidLoad {
  [super viewDidLoad];
  // 将背景设置成黑色
  self.view.backgroundColor = [UIColor blackColor];
  // 追加父标签
  parent_ = [[UILabel alloc] initWithFrame:CGRectMake( 0, 0, 320, 320 )];
  parent_.textAlignment = UITextAlignmentCenter;
  parent_.text = @"PARENT";
  [self.view addSubview:parent_];
  // 追加 1 个子标签
  UILabel* child3 = [[[UILabel alloc] initWithFrame:CGRectZero] autorelease];
  child3.text = @"CHILD 3";
  [child3 sizeToFit];
  [parent_ insertSubview:child3 atIndex:0];
  // 在上一个标签 CHILD 3 下插入 CHILD 1
  UILabel* child1 = [[[UILabel alloc] initWithFrame:CGRectZero] autorelease];
  child1.text = @"CHILD 1";
  [child1 sizeToFit];
  [parent_ insertSubview:child1 belowSubview:child3];
  // 在 CHILD 1 上追加 CHILD 2
  UILabel* child2 = [[[UILabel alloc] initWithFrame:CGRectZero] autorelease];
  child2.text = @"CHILD 2";
  [child2 sizeToFit];
  [parent_ insertSubview:child2 aboveSubview:child1];
  // 让 CHILD 1 与 CHILD 3 交换
  [parent_ exchangeSubviewAtIndex:0 withSubviewAtIndex:2];
  // 如果 CHILD 3 为 PARENT 子元素的话
```

```objc
    if ( [child3 isDescendantOfView:parent_] ) {
      // 删除 CHILD 3
      [child3 removeFromSuperview];
    }
    // 追加 subviews 按钮
    UIButton* subviewsButton = [UIButton buttonWithType:UIButtonTypeRoundedRect];
    subviewsButton.frame = CGRectMake( 0, 0, 150, 40 );
    CGPoint newPoint = self.view.center;
    newPoint.y = self.view.frame.size.height - 40;
    subviewsButton.center = newPoint;
    [subviewsButton setTitle:@"显示 subviews" forState:UIControlStateNormal];
    [subviewsButton addTarget:self
                   action:@selector(subviewsDidPush)
         forControlEvents:UIControlEventTouchUpInside];
    [self.view addSubview:subviewsButton];
}
#pragma mark ----- Private Methods -----
- (void)subviewsDidPush {
    NSMutableString* subviews = [[[NSMutableString alloc] initWithCapacity:64] autorelease];
    [subviews setString:@""];
    // 在 subviews 的 text 后附加字符串
    for ( id view in parent_.subviews ) {
      NSString* addString;
      if ( [view isKindOfClass:[UILabel class]] ) {
        addString = ((UILabel*)view).text;
      } else {
        addString = [view description];
      }
      if ( [subviews length] > 0 ) {
        [subviews appendString:@", "];
      }
      [subviews appendString:addString];
    }
    UIAlertView* alert = [[[UIAlertView alloc] initWithTitle:@"subviews"
                                              message:subviews
                                              delegate:nil
                                       cancelButtonTitle:nil
                                       otherButtonTitles:@"OK", nil ] autorelease];
    [alert show];
}
- (void)touchesEnded:(NSSet*)touches withEvent:(UIEvent*)event {
    [self.navigationController setNavigationBarHidden:NO animated:YES];
}
@end
```

执行效果如图 2-25 所示。

图 2-25 执行效果

实例 022 设置视图位置互换显示（1）

实例 022	设置视图位置互换显示（1）
源码路径	\daima\012\

实例 022 设置视图位置互换显示（1）

实例说明

在 iOS 应用中，可以使用 sendSubviewToFront 方法将某个视图放在前方显示，使用 sendSubviewToBack 方法将某个视图放在后方显示。

本实例的功能是，设置两个重叠的视图，可以设置其中的一个视图靠前显示还是靠后隐藏显示。本实例首先设置了两个标签代表两块视图区域，然后在下方设置了"bringFront"按钮和"sendBack"按钮。单击这两个按钮后，会实现指定标签块的前置和靠后显示效果。

具体实现

实例文件 UIkitPrjSiblings.h 的实现代码如下所示。

```objc
#import "SampleBaseController.h"
@interface UIKitPrjSiblings : SampleBaseController
{
@private
  UILabel* labelA_;
  UILabel* labelB_;
}
@end
```

实例文件 UIkitPrjSiblings.m 的实现代码如下所示。

```objc
#import "UIKitPrjSiblings.h"
#pragma mark ----- Private Methods Definition -----
@interface UIKitPrjSiblings ()
- (void)bringFrontDidPush;
- (void)sendBackDidPush;
@end
#pragma mark ----- Start Implementation For Methods -----
@implementation UIKitPrjSiblings
// finalize
- (void)dealloc {
  [labelA_ release];
  [labelB_ release];
  [super dealloc];
}
- (void)viewDidLoad {
  [super viewDidLoad];
  self.view.backgroundColor = [UIColor whiteColor];
  // 追加标签 A
  labelA_ = [[UILabel alloc] initWithFrame:CGRectMake( 10, 10, 150, 150 )];
  labelA_.backgroundColor = [UIColor redColor];
  labelA_.textAlignment = UITextAlignmentCenter;
  labelA_.text = @"A";
  [self.view addSubview:labelA_];
  // 追加标签 B
  labelB_ = [[UILabel alloc] initWithFrame:CGRectMake( 100, 100, 150, 150 )];
  labelB_.backgroundColor = [UIColor blueColor];
  labelB_.textAlignment = UITextAlignmentCenter;
  labelB_.text = @"B";
  [self.view addSubview:labelB_];
  // 追加 bringFront 按钮
  UIButton* bringFrontButton = [UIButton buttonWithType:UIButtonTypeRoundedRect];
  bringFrontButton.frame = CGRectMake( 0, 0, 100, 40 );
  CGPoint newPoint = self.view.center;
  newPoint.x -= 50;
  newPoint.y = self.view.frame.size.height - 100;
  bringFrontButton.center = newPoint;
  [bringFrontButton setTitle:@"bringFront" forState:UIControlStateNormal];
  [bringFrontButton addTarget:self
                   action:@selector(bringFrontDidPush)
         forControlEvents:UIControlEventTouchUpInside]
  // 追加 sendBack 按钮
  UIButton* sendBackButton = [UIButton buttonWithType:UIButtonTypeRoundedRect];
  sendBackButton.frame = CGRectMake( 0, 0, 100, 40 );
  newPoint.x += 100;
```

```
    sendBackButton.center = newPoint;
    [sendBackButton setTitle:@"sendBack" forState:UIControlStateNormal];
    [sendBackButton addTarget:self
                      action:@selector(sendBackDidPush)
            forControlEvents:UIControlEventTouchUpInside];

    [self.view addSubview:bringFrontButton];
    [self.view addSubview:sendBackButton];
}
#pragma mark ----- Private Methods -----
- (void)bringFrontDidPush {
    [self.view bringSubviewToFront:labelA_];
}

- (void)sendBackDidPush {
    [self.view sendSubviewToBack:labelA_];
}
- (void)touchesEnded:(NSSet*)touches withEvent:(UIEvent*)event {
    [self.navigationController setNavigationBarHidden:NO animated:YES];
}
@end
```

执行后的效果如图 2-26 所示。

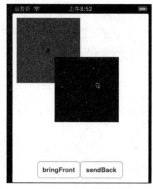

图 2-26　执行效果

实例 023　设置视图位置互换显示（2）

实例 023	设置视图位置互换显示（2）
源码路径	\daima\012\

实例说明

在 iOS 应用中，可以使用 UIView 的 tag 属性为某个视图设置一个标签，这样在具体操作时可以直接调用这个 tag，就相当于调用了这个视图。例如，在程序里有一个顶层容器 UIView，我们可以在它上面进行多个子 UIImageView 的添加和删除操作。在添加的时候，可以为每个子 UIImageView 添加一个 tag（大于 0 的整数）。接下来如何进行删除视图的操作，可以先通过顶层 UIView 的 subviews 方法，得到所有子 UIImageView，然后根据指定的 tag 删除 UIView 上对应的子 UIImageView。由此可见，tag 起了一个标签的作用。

本实例的功能是设置了 10 个视图，并为每个视图设置了对应的 tag。其中将第 8 个视图的 tag 设置为 999。在下方设置了一个 "search 999" 按钮，当单击此按钮后会检索显示出值为 999 的视图。

具体实现

实例文件 UIkitPrjTag.h 的实现代码如下所示。

```objc
#import "SampleBaseController.h"
@interface UIKitPrjTag : SampleBaseController
{
 @private
  UILabel* parent_;
}
@end
```

实例文件 UIkitPrjTag.m 的实现代码如下所示。

```objc
#import "UIKitPrjTag.h"
#pragma mark ----- Private Methods Definition -----
@interface UIKitPrjTag ()
- (void)searchDidPush;
@end
#pragma mark ----- Start Implementation For Methods -----
@implementation UIKitPrjTag
// finalize
- (void)dealloc {
  [parent_ release];
  [super dealloc];
}
- (void)viewDidLoad {
  [super viewDidLoad];
  // 背景设置成黑色
  self.view.backgroundColor = [UIColor blackColor];
  // 追加父标签
  parent_ = [[UILabel alloc] initWithFrame:CGRectMake( 0, 0, 320, 320 )];
  parent_.textAlignment = UITextAlignmentCenter;
  parent_.text = @"PARENT";
  [self.view addSubview:parent_];
  // 追加 10 个子标签
  for ( int i = 1; i <= 10; ++i ){
    UILabel* child = [[[UILabel alloc] initWithFrame:CGRectZero] autorelease];
    child.text = [NSString stringWithFormat:@"CHILD %d", i];
    [child sizeToFit];
    CGPoint newPoint = child.center;
    newPoint.y += 30 * i;
    child.center = newPoint;
    [parent_ addSubview:child];
    // 将第 8 个标签的 tag 设置成 999
    if ( 8 == i ) {
      child.tag = 999;
    }
  }
  // 追加 search 按钮
  UIButton* searchButton = [UIButton buttonWithType:UIButtonTypeRoundedRect];
  searchButton.frame = CGRectMake( 0, 0, 150, 40 );
  CGPoint newPoint = self.view.center;
  newPoint.y = self.view.frame.size.height - 40;
  searchButton.center = newPoint;
  [searchButton setTitle:@"search 999" forState:UIControlStateNormal];
  [searchButton addTarget:self
              action:@selector(searchDidPush)
      forControlEvents:UIControlEventTouchUpInside];
  [self.view addSubview:searchButton];
}
#pragma mark ----- Private Methods -----
- (void)searchDidPush {
  NSString* message;
  //从 parent_ 的子元素中检索 tag 为 999 的元素,找到后显示警告框
  UILabel* label = (UILabel*)[parent_ viewWithTag:999];
  if ( label ) {
    message = label.text;
  } else {
    message = @"nothing";
  }
  UIAlertView* alert = [[[UIAlertView alloc] initWithTitle:@"search 999"
                                      message:message
                                      delegate:nil
                              cancelButtonTitle:nil
                              otherButtonTitles:@"OK", nil ] autorelease];

  [alert show];
```

```
- (void)touchesEnded:(NSSet*)touches withEvent:(UIEvent*)event {
    [self.navigationController setNavigationBarHidden:NO animated:YES];
}
@end
```

执行后的效果如图 2-27 所示。

图 2-27 执行效果

实例 024 获得屏幕内视图的坐标

实例 024	获得屏幕内视图的坐标
源码路径	\daima\012\

实例说明

在 iOS 应用中，可以使用如下 4 个方法在不同的坐标系间转换。
- convertPoint:fromView：从 fromView 的本地坐标系变换到自己的本地坐标系。
- convertPoint:toView：从自己的本地坐标系变换到 toView 的本地坐标系。
- convertRect:fromView：从 fromView 的本地坐标系变换到自己的本地坐标系。
- convertRect:toView：从自己的本地坐标系变换到 toView 的本地坐标系。

在本实例中，设置了"CHILD 1"和"CHILD 2"两个标签，分别以"CHILD 1"和"CHILD 2"本地坐标系转换为对方的坐标。在具体实现时，首先将"CHILD 1"本地坐标系中的"CHILD 1"的右下方的（100，100）的坐标和"CHILD 1"的矩形（0，0，100，100）变换到"CHILD 2"的本地坐标系。

具体实现

实例文件 UIkitPrjConvertPoint.h 的实现代码如下所示。

```
#import "SampleBaseController.h"
@interface UIKitPrjConvertPoint : SampleBaseController
{
 @private
  UILabel* parent_;
  UILabel* child1_;
  UILabel* child2_;
}
@end
```

实例文件 UIkitPrjConvertPoint.m 的实现代码如下所示。

```
#import "UIKitPrjConvertPoint.h"
#pragma mark ----- Private Methods Definition -----
@interface UIKitPrjConvertPoint ()
- (void)toViewDidPush;
```

```objc
- (void)fromViewDidPush;
@end
#pragma mark ----- Start Implementation For Methods -----
@implementation UIKitPrjConvertPoint
// finalize
- (void)dealloc {
  [parent_ release];
  [child1_ release];
  [child2_ release];
  [super dealloc];
}
- (void)viewDidLoad {
  [super viewDidLoad];
  // 背景设置成黑色
  self.view.backgroundColor = [UIColor blackColor];
  // 追加父标签
  parent_ = [[UILabel alloc] initWithFrame:CGRectMake( 0, 0, 320, 320 )];
  parent_.textAlignment = UITextAlignmentCenter;
  [self.view addSubview:parent_];
  // 追加两个子标签
  child1_ = [[UILabel alloc] initWithFrame:CGRectMake( 10, 10, 100, 100 )];
  child1_.text = @"CHILD 1";
  child1_.textAlignment = UITextAlignmentCenter;
  child1_.backgroundColor = [UIColor blackColor];
  child1_.textColor = [UIColor whiteColor];
  child2_ = [[UILabel alloc] initWithFrame:CGRectMake( 110, 110, 100, 100 )];
  child2_.text = @"CHILD 2";
  child2_.textAlignment = UITextAlignmentCenter;
  child2_.backgroundColor = [UIColor blackColor];
  child2_.textColor = [UIColor whiteColor];
  [parent_ addSubview:child1_];
  [parent_ addSubview:child2_];
  // 追加 toView 按钮
  UIButton* toViewButton = [UIButton buttonWithType:UIButtonTypeRoundedRect];
  toViewButton.frame = CGRectMake( 0, 0, 100, 40 );
  CGPoint newPoint = self.view.center;
  newPoint.x -= 50;
  newPoint.y = self.view.frame.size.height - 40;
  toViewButton.center = newPoint;
  [toViewButton setTitle:@"toView" forState:UIControlStateNormal];
  [toViewButton addTarget:self
              action:@selector(toViewDidPush)
       forControlEvents:UIControlEventTouchUpInside];
  // 追加 fromView 按钮
  UIButton* fromViewButton = [UIButton buttonWithType:UIButtonTypeRoundedRect];
  fromViewButton.frame = CGRectMake( 0, 0, 100, 40 );
  newPoint.x += 100;
  fromViewButton.center = newPoint;
  [fromViewButton setTitle:@"fromView" forState:UIControlStateNormal];
  [fromViewButton addTarget:self
              action:@selector(fromViewDidPush)
       forControlEvents:UIControlEventTouchUpInside];
  [self.view addSubview:toViewButton];
  [self.view addSubview:fromViewButton];
}
#pragma mark ----- Private Methods -----
- (void)toViewDidPush {
  CGPoint newPoint = [child1_ convertPoint:CGPointMake( 100, 100 ) toView:child2_];
  CGRect newFrame = [child1_ convertRect:CGRectMake( 0, 0, 100, 100 ) toView:child2_];
  NSString* title = [NSString stringWithFormat:@"( %f, %f )", newPoint.x, newPoint.y];
  NSString* message = [NSString stringWithFormat:@"( %f, %f, %f, %f )",
                          newFrame.origin.x, newFrame.origin.y, newFrame.size.width,
newFrame.size.height];
  UIAlertView* alert = [[[UIAlertView alloc] initWithTitle:title
                                            message:message
                                            delegate:nil
                                  cancelButtonTitle:nil
                                  otherButtonTitles:@"OK", nil ] autorelease];
  [alert show];
}
```

第 2 章 界面布局实战

```
- (void)fromViewDidPush {
    CGPoint newPoint = [child1_ convertPoint:CGPointMake( 0, 0 ) fromView:child2_];
    CGRect newFrame = [child1_ convertRect:CGRectMake( 0, 0, 100, 100 ) fromView:child2_];
    NSString* title = [NSString stringWithFormat:@"( %f, %f )", newPoint.x, newPoint.y];
    NSString* message = [NSString stringWithFormat:@"( %f, %f, %f, %f )",
                          newFrame.origin.x, newFrame.origin.y, newFrame.size.width, newFrame.size.height];
    UIAlertView* alert = [[[UIAlertView alloc] initWithTitle:title
                                      message:message
                                      delegate:nil
                              cancelButtonTitle:nil
                              otherButtonTitles:@"OK", nil ] autorelease];
    [alert show];
}
- (void)touchesEnded:(NSSet*)touches withEvent:(UIEvent*)event {
    [self.navigationController setNavigationBarHidden:NO animated:YES];
}
@end
```

执行后的效果如图 2-28 所示。

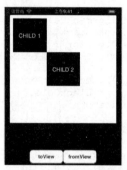

图 2-28 执行效果

实例 025 实现视图外观的自动调整

实例 025	实现视图外观的自动调整
源码路径	\daima\012\

实例说明

在 iOS 应用中，UIView 有两个重要的方法：setNeedsDisplay 和 setNeedsLayout。这两个方法都是异步执行的，而 setNeedsDisplay 会自动调用 drawRect 方法，这样可以拿到 UIGraphicsGetCurrent-Context，就可以画画了。而方法 setNeedsLayout 会默认调用 layoutSubViews，这样就可以处理子视图中的一些数据。由此可见，方法 setNeedsDisplay 方便绘图，而方法 layoutSubViews 方便处理数据。

在本实例中，使用方法 layoutSubView 定制了视图外观。我们可以在自己的定制视图中重载 layoutSubViews 方法，以便调整子视图的尺寸和位置。假如一个视图具有很大的滚动区域，就需要使用几个子视图来"平铺"，而不是创建一个内存很可能装不下的大视图。在这个方法的实现中，视图可以隐藏所有不需显示在屏幕上的子视图，或者在重新定位之后将它们用于显示新的内容。作为这个过程的一部分，视图也可以将用于"平铺"的子视图标识为需要重画。

具体实现

实例文件 UIkitPrjLayoutSubviews.h 的实现代码如下所示。

```
#import "SampleBaseController.h"
#pragma mark ----- LayoutTest -----
//定义 UIView 的子类
//子类中拥有 child1_ 以及 child2_ 两个标签子元素
@interface LayoutTest : UILabel
{
 @private
  UILabel* child1_;
  UILabel* child2_;
}
@end
#pragma mark ----- UIKitPrjLayoutSubviews -----
@interface UIKitPrjLayoutSubviews : SampleBaseController
{
 @private
  LayoutTest* label_;
}
@end
```

实例文件 UIkitPrjLayoutSubviews.m 的实现代码如下所示。

```
#import "UIKitPrjLayoutSubviews.h"
#pragma mark ----- Private Methods Definition -----
@interface UIKitPrjLayoutSubviews ()
- (void)setNeedsDidPush;
- (void)resizeDidPush;
@end
#pragma mark ----- Start Implementation For Methods -----
@implementation UIKitPrjLayoutSubviews
// finalize
- (void)dealloc {
  [label_ release];
  [super dealloc];
}
- (void)viewDidLoad {
  [super viewDidLoad];
  self.view.backgroundColor = [UIColor grayColor];

  label_ = [[LayoutTest alloc] init];
  [self.view addSubview:label_];
  // 追加 setNeedsLayout 按钮
  UIButton* setNeedsButton = [UIButton buttonWithType:UIButtonTypeRoundedRect];
  setNeedsButton.frame = CGRectMake( 0, 0, 150, 40 );
  CGPoint newPoint = self.view.center;
  newPoint.x -= 75;
  newPoint.y = self.view.frame.size.height - 40;
  setNeedsButton.center = newPoint;
  [setNeedsButton setTitle:@"setNeedsLayout" forState:UIControlStateNormal];
  [setNeedsButton addTarget:self
                action:@selector(setNeedsDidPush)
      forControlEvents:UIControlEventTouchUpInside];
  [self.view addSubview:setNeedsButton];
  // 追加 layoutIfNeeded 按钮
  UIButton* layoutButton = [UIButton buttonWithType:UIButtonTypeRoundedRect];
  layoutButton.frame = CGRectMake( 0, 0, 150, 40 );
  newPoint.x += 150;
  layoutButton.center = newPoint;
  [layoutButton setTitle:@"layoutIfNeeded" forState:UIControlStateNormal];
  [layoutButton addTarget:self
               action:@selector(resizeDidPush)
     forControlEvents:UIControlEventTouchUpInside];

  [self.view addSubview:layoutButton];
}
#pragma mark ----- Private Methods -----
```

```objc
- (void)setNeedsDidPush {
  // 取得 child1_ 并使之尺寸变大
  UIView* child1 = (UIView*)[label_.subviews objectAtIndex:0];
  child1.frame = CGRectMake( 0, 0, 160, 160 );
  child1.center = label_.center;
}
- (void)resizeDidPush {
  //此后，调用 setNeedsLayout 方法
  //layoutSubviews 方法将会被自动调用
  [label_ setNeedsLayout];
  [label_ layoutIfNeeded];
}
- (void)touchesEnded:(NSSet*)touches withEvent:(UIEvent*)event {
    [self.navigationController setNavigationBarHidden:NO animated:YES];
}
@end
#pragma mark ----- Implementation For LayoutTest -----
//实现 LayoutTest 类
@implementation LayoutTest

// 释放处理
- (void)dealloc {
  [child1_ release];
  [child2_ release];
  [super dealloc];
}
// 初始化处理
- (id)init {
  if ( (self = [super init]) ) {
      //调整自己的大小，追加两个标签
    self.frame = CGRectMake( 0, 0, 320, 320 );
    child1_ = [[UILabel alloc] initWithFrame:CGRectZero];
    child1_.text = @"CHILD 1";
    [child1_ sizeToFit];
    child1_.backgroundColor = [UIColor redColor];
    child1_.textColor = [UIColor whiteColor];
    child2_ = [[UILabel alloc] initWithFrame:CGRectZero];
    child2_.text = @"CHILD 2";
    [child2_ sizeToFit];
    child2_.backgroundColor = [UIColor blueColor];
    child2_.textColor = [UIColor whiteColor];
    child2_.center = CGPointMake( child2_.center.x, child2_.center.y + 30 );
    [self addSubview:child1_];
    [self addSubview:child2_];
  }
  return self;
}
//外观调整方法
- (void)layoutSubviews {
  // 调用父类中的相同方法
  [super layoutSubviews];
  // 再显示子元素
  // 将 child1_ 移动到左下
  CGRect newRect = child1_.frame;
  newRect.origin.x = 0;
  newRect.origin.y = self.frame.size.height - child1_.frame.size.height;
  child1_.frame = newRect;
  // 将 child2_ 移动到右上
  newRect = child2_.frame;
  newRect.origin.x = self.frame.size.width - child2_.frame.size.width;
  newRect.origin.y = 0;
  child2_.frame = newRect;
}
@end
```

执行效果如图 2-29 所示。

实例 026　自动调整视图中的子元素

图 2-29　执行效果

实例 026　自动调整视图中的子元素

实例 026	自动调整视图中的子元素
源码路径	\daima\012\

实例说明

在 iOS 应用中，在 UIView 中有一个名为 autoresizingMask 的属性，此属性的功能是自动调整子控件与父控件中间的位置和宽高。属性 autoresizingMask 对应如下所示的枚举值。

```
enum{
UIViewAutoresizingNone                =0,
UIViewAutoresizingFlexibleLeftMargin  = 1 << 0,
UIViewAutoresizingFlexibleWidth       =1<<1,
UIViewAutoresizingFlexibleRightMargin =1<<2,
UIViewAutoresizingFlexibleTopMargin   =1<<3,
UIViewAutoresizingFlexibleHeight =1<<4,
UIViewAutoresizingFlexibleBottomMargin=1<<5
};
typedef NSUInteger UIViewAutoresizing;
```

对上述枚举值的具体说明如下所示。

- **UIViewAutoresizingNone**：表示不自动调整。
- **UIViewAutoresizingFlexibleLeftMargin**：表示自动调整与 superView 左边的距离，也就是说，与 superView 右边的距离不变。
- **UIViewAutoresizingFlexibleRightMargin**：表示自动调整与 superView 的右边距离，也就是说，与 superView 左边的距离不变。
- **UIViewAutoresizingFlexibleTopMargin**：表示自动调整与 superView 顶部的距离，也就是说，与 superView 底部的距离不变。
- **UIViewAutoresizingFlexibleBottomMargin**：表示自动调整与 superView 底部的距离，也就是说，与 superView 顶部的距离不变。
- **UIViewAutoresizingFlexibleWidth**：表示自动调整子元素的宽度，效果居中，子元素的尺寸发生变化。
- **UIViewAutoresizingFlexibleHeight**：表示自动调整子元素的高度，效果居中，子元素的尺寸发生变化。

在具体使用时，我们可以同时设置多个枚举，例如下面的代码：

```
subView.autoresizingMask=UIViewAutoresizingFlexibleLeftMargin |UIViewAutoresizing
FlexibleRightMargin;
```

第2章 界面布局实战

如果有多个枚举,在枚举值之间用"|"分开界限。除此之外,还有一个属性——autoresizesSubviews,此属性设置是否可以让其 subviews 自动进行调整,默认状态是 YES,就是允许;如果设置成 NO,那么 subView 的 autoresizingMask 属性失效。

在本实例中,设置了两个 UILabel 区域,分别表示父视图和子视图。在中部设置了 bottomDidSwitch 选项按钮,在下方设置了 "resize" 按钮,当单击此按钮时会根据中间 bottomDidSwitch 选择的选项在顶部显示对应的视图。

具体实现

实例文件 UIkitPrjAutoResize.m 的实现代码如下所示。

```objc
#import "UIKitPrjAutoResize.h"
#pragma mark ----- Private Methods Definition -----
@interface UIKitPrjAutoResize ()
- (void)resizeDidPush;
- (void)leftDidSwitch:(id)sender;
- (void)rightDidSwitch:(id)sender;
- (void)topDidSwitch:(id)sender;
- (void)bottomDidSwitch:(id)sender;
- (void)widthDidSwitch:(id)sender;
- (void)heightDidSwitch:(id)sender;
- (void)addSwitch:(NSString*)caption frame:(CGRect)frame action:(SEL)action;
@end

#pragma mark ----- Start Implementation For Methods -----
@implementation UIKitPrjAutoResize
// finalize
- (void)dealloc {
  [child_ release];
  [parent_ release];
  [super dealloc];
}
- (void)viewDidLoad {
  [super viewDidLoad];
  // 背景设置成黑色
  self.view.backgroundColor = [UIColor blackColor];
  // 追加父标签
  parent_ = [[UILabel alloc] initWithFrame:CGRectMake( 0, 0, 160, 160 )];
  parent_.backgroundColor = [UIColor whiteColor];
  [self.view addSubview:parent_];
  // 追加一个子元素标签
  child_ = [[UILabel alloc] initWithFrame:CGRectInset( parent_.bounds, 30, 30 )];
  child_.text = @"CHILD 1";
  child_.backgroundColor = [UIColor redColor];
  child_.textColor = [UIColor blackColor];
  [parent_ addSubview:child_];
  // 追加 Resize 按钮
  UIButton* resizeButton = [UIButton buttonWithType:UIButtonTypeRoundedRect];
  resizeButton.frame = CGRectMake( 0, 0, 150, 40 );
  CGPoint newPoint = self.view.center;
  newPoint.y = self.view.frame.size.height - 40;
  resizeButton.center = newPoint;
  [resizeButton setTitle:@"Resize" forState:UIControlStateNormal];
  [resizeButton addTarget:self
              action:@selector(resizeDidPush)
       forControlEvents:UIControlEventTouchUpInside];
  [self.view addSubview:resizeButton];
  // 追加 Switch
  CGRect labelRect = CGRectMake( 5, 201, 310, 30 );
  [self addSwitch:@"FlexibleLeftMargin" frame:labelRect action:@selector(leftDidSwitch:)];
  labelRect.origin = CGPointMake( 5, labelRect.origin.y + 30 );
  [self addSwitch:@"FlexibleRightMargin" frame:labelRect action:@selector(rightDidSwitch:)];
  labelRect.origin = CGPointMake( 5, labelRect.origin.y + 30 );
  [self addSwitch:@"FlexibleTopMargin" frame:labelRect action:@selector(topDidSwitch:)];
  labelRect.origin = CGPointMake( 5, labelRect.origin.y + 30 );
```

```objc
    [self addSwitch:@"FlexibleBottomMargin" frame:labelRect action:@selector(bottomDidSwitch:)];
    labelRect.origin = CGPointMake( 5, labelRect.origin.y + 30 );
    [self addSwitch:@"FlexibleWidth" frame:labelRect action:@selector(widthDidSwitch:)];
    labelRect.origin = CGPointMake( 5, labelRect.origin.y + 30 );
    [self addSwitch:@"FlexibleHeight" frame:labelRect action:@selector(heightDidSwitch:)];
}
#pragma mark ----- Private Methods -----
- (void)resizeDidPush {
    if ( 200 == parent_.frame.size.width ) {
        parent_.frame = CGRectMake( 0, 0, 160, 160 );
    } else {
        parent_.frame = CGRectMake( 0, 0, 200, 200 );
    }
}
- (void)leftDidSwitch:(id)sender {
    if ( [sender isOn] ) {
        child_.autoresizingMask |= UIViewAutoresizingFlexibleLeftMargin;
    } else {
        child_.autoresizingMask &= ~UIViewAutoresizingFlexibleLeftMargin;
    }
}
- (void)rightDidSwitch:(id)sender {
    if ( [sender isOn] ) {
        child_.autoresizingMask |= UIViewAutoresizingFlexibleRightMargin;
    } else {
        child_.autoresizingMask &= ~UIViewAutoresizingFlexibleRightMargin;
    }
}
- (void)topDidSwitch:(id)sender {
    if ( [sender isOn] ) {
        child_.autoresizingMask |= UIViewAutoresizingFlexibleTopMargin;
    } else {
        child_.autoresizingMask &= ~UIViewAutoresizingFlexibleTopMargin;
    }
}
- (void)bottomDidSwitch:(id)sender {
    if ( [sender isOn] ) {
        child_.autoresizingMask |= UIViewAutoresizingFlexibleBottomMargin;
    } else {
        child_.autoresizingMask &= ~UIViewAutoresizingFlexibleBottomMargin;
    }
}
- (void)widthDidSwitch:(id)sender {
    if ( [sender isOn] ) {
        child_.autoresizingMask |= UIViewAutoresizingFlexibleWidth;
    } else {
        child_.autoresizingMask &= ~UIViewAutoresizingFlexibleWidth;
    }
}
- (void)heightDidSwitch:(id)sender {
    if ( [sender isOn] ) {
        child_.autoresizingMask |= UIViewAutoresizingFlexibleHeight;
    } else {
        child_.autoresizingMask &= ~UIViewAutoresizingFlexibleHeight;
    }
}
- (void)addSwitch:(NSString*)caption frame:(CGRect)frame action:(SEL)action {
    UILabel* label = [[[UILabel alloc] initWithFrame:frame] autorelease];
    label.text = caption;
    UISwitch* newSwitch = [[[UISwitch alloc] initWithFrame:CGRectZero] autorelease];
    CGPoint newPoint = label.center;
    newPoint.x += 100;
    newSwitch.center = newPoint;
    [newSwitch addTarget:self action:action forControlEvents:UIControlEventValueChanged];
    [self.view addSubview:label];
    [self.view addSubview:newSwitch];
}
- (void)touchesEnded:(NSSet*)touches withEvent:(UIEvent*)event {
    [self.navigationController setNavigationBarHidden:NO animated:YES];
}
@end
```

执行效果如图 2-30 所示。

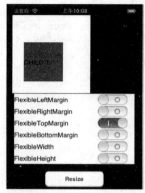

图 2-30　执行效果

实例 027　实现不同界面之间的跳转处理

实例 027	实现不同界面之间的跳转处理
源码路径	\daima\027\

实例说明

在 iOS 应用中，UIViewController 类提供了一个显示用的 View 界面，同时包含 View 加载、卸载事件的重定义功能。需要注意的是在自定义其子类实现时，必须在 Interface Builder 中手动关联 view 属性。UIViewController 是 iOS 开发中最常见也最重要的部件之一，可以说绝大多数的 App 都用到了 UIViewController 来管理页面的 View。UIViewController 是 MVC 的核心结构和桥梁构成，是绝大多数开发者所花时间最多的部分。一个 ViewController 应该且只应该管理一个 view hierarchy，而通常来说一个完整的 View hierarchy 指的是整整占满的一个屏幕。

在本实例中，通过使用 UIViewController 类实现了两个不同界面之间的切换。其中，第一个界面显示文本"Hello，world!"和一个"界面跳转"按钮。单击此按钮后会进入第二个界面，第二个显示文本"你好、世界！"和一个"界面跳转"按钮，单击此按钮后会返回到第一个界面。

具体实现

实例文件 ViewController1.m 的具体实现代码如下所示。

```objc
#import "ViewController1.h"
@implementation ViewController1
- (void)viewDidLoad {
 [super viewDidLoad];
 // 追加 "Hello, world!" 标签
 // 背景为白色、文字为黑色
 UILabel* label = [[[UILabel alloc] initWithFrame:self.view.bounds] autorelease];
 label.text = @"Hello, world!";
 label.textAlignment = UITextAlignmentCenter;
 label.backgroundColor = [UIColor whiteColor];
 label.textColor = [UIColor blackColor];
 label.autoresizingMask = UIViewAutoresizingFlexibleWidth | UIViewAutoresizingFlexibleHeight;
 [self.view addSubview:label];
 // 追加按钮
 // 点击按钮后跳转到其他画面
 UIButton* button = [UIButton buttonWithType:UIButtonTypeRoundedRect];
 [button setTitle:@"画面跳转" forState:UIControlStateNormal];
 [button sizeToFit];
 CGPoint newPoint = self.view.center;
 newPoint.y += 50;
 button.center = newPoint;
```

```
  button.autoresizingMask =
    UIViewAutoresizingFlexibleTopMargin | UIViewAutoresizingFlexibleBottomMargin;
  [button addTarget:self
            action:@selector(buttonDidPush)
   forControlEvents:UIControlEventTouchUpInside];
  [self.view addSubview:button];
}
- (void)buttonDidPush {
   // 自己移向背面
   // 结果是 ViewController2 显示在前
   [self.view.window sendSubviewToBack:self.view];
}
@end
```

实例文件 ViewController2.m 的具体实现代码如下所示。

```
#import "ViewController2.h"
@implementation ViewController2
- (void)viewDidLoad {
  [super viewDidLoad];
  // 追加"您好、世界！"标签
  // 背景为黑色、文字为白色
  UILabel* label = [[[UILabel alloc] initWithFrame:self.view.bounds] autorelease];
  label.text = @"您好、世界！";
  label.textAlignment = UITextAlignmentCenter;
  label.backgroundColor = [UIColor blackColor];
  label.textColor = [UIColor whiteColor];
  label.autoresizingMask = UIViewAutoresizingFlexibleWidth | UIViewAutoresizingFlexibleHeight;
  [self.view addSubview:label];
  // 追加按钮
  // 点击按钮后画面跳转
  UIButton* button = [UIButton buttonWithType:UIButtonTypeRoundedRect];
  [button setTitle:@"画面跳转" forState:UIControlStateNormal];
  [button sizeToFit];
  CGPoint newPoint = self.view.center;
  newPoint.y += 50;
  button.center = newPoint;
  button.autoresizingMask =
    UIViewAutoresizingFlexibleTopMargin | UIViewAutoresizingFlexibleBottomMargin;
  [button addTarget:self
            action:@selector(buttonDidPush)
   forControlEvents:UIControlEventTouchUpInside];
  [self.view addSubview:button];
}
- (void)buttonDidPush {
     // 自己移向背面
     // 结果是 ViewController1 显示在前
   [self.view.window sendSubviewToBack:self.view];
}
@end
```

执行后的效果如图 2-31 所示，单击"页面跳转"按钮后会进入第二个界面，如图 2-32 所示。

图 2-31　第一个界面

图 2-32　第二个界面

实例 028 通过列表实现不同界面之间的跳转

实例 028	通过列表实现不同界面之间的跳转
源码路径	\daima\028\

实例说明

在 iOS 应用中,可以使用 UINavigationController 实现导航列表样式的跳转处理。UINavigation-Controller 可以翻译为导航控制器,由 Navigation bar、Navigation View、Navigation toolbar 等组成。

在本实例中,通过使用 UIViewController 类实现了两个不同界面之间的切换。和前一个实例不同,本实例更加人性化,使用了 UINavigationController 实现导航转换,并且在界面的左上角专门设置了一个"主菜单"按钮,通过此按钮可以随时返回到主界面。

具体实现

实例文件 ViewController1.m 的具体实现代码如下所示。

```
#import "ViewController1.h"
@implementation ViewController1
// initialize
- (id)init {
  if ( (self = [super init]) ) {
    // 设置 tabBar 的标题
    self.title = @"Hello";
    UIImage* icon = [UIImage imageNamed:@"ball1.png"];
    self.tabBarItem =
      [[[UITabBarItem alloc] initWithTitle:@"Hello" image:icon tag:0] autorelease];
  }
  return self;
}
- (void)viewDidLoad {
  [super viewDidLoad];
  // 追加 "Hello, world!" 标签
  // 背景为白色,文字为黑色
  UILabel* label = [[[UILabel alloc] initWithFrame:self.view.bounds] autorelease];
  label.text = @"Hello, world!";
  label.textAlignment = UITextAlignmentCenter;
  label.backgroundColor = [UIColor whiteColor];
  label.textColor = [UIColor blackColor];
  label.autoresizingMask = UIViewAutoresizingFlexibleWidth | UIViewAutoresizingFlexibleHeight;
  [self.view addSubview:label];
}
@end
```

实例文件 ViewController2.m 的具体实现代码如下所示。

```
#import "ViewController2.h"
@implementation ViewController2
// initialize
- (id)init {
  if ( (self = [super init]) ) {
    // 设置 tabBar 的标题
    self.title = @"您好";
    UIImage* icon = [UIImage imageNamed:@"ball2.png"];
    self.tabBarItem =
      [[[UITabBarItem alloc] initWithTitle:@"您好" image:icon tag:0] autorelease];
  }
  return self;
}
```

```
- (void)viewDidLoad {
  [super viewDidLoad];
  // 追加 "您好、世界" 标签
  // 背景为黑色，文字为白色
  UILabel* label = [[[UILabel alloc] initWithFrame:self.view.bounds] autorelease];
  label.text = @"您好、世界！";
  label.textAlignment = UITextAlignmentCenter;
  label.backgroundColor = [UIColor blackColor];
  label.textColor = [UIColor whiteColor];
  label.autoresizingMask = UIViewAutoresizingFlexibleWidth | UIViewAutoresizingFlexibleHeight;
  [self.view addSubview:label];
}
@end
```

和上一个实例相比，本实例新增了文件 TopMenuController.m，此文件用于实现屏幕左上角的"主菜单"处理事件功能。文件 TopMenuController.m 的具体实现代码如下所示。

```
#import "TopMenuController.h"
@implementation TopMenuController
- (void)dealloc {
  [items_ release];
  [super dealloc];
}
- (id)init {
  if ( (self = [super initWithStyle:UITableViewStylePlain]) ) {
    self.title = @"主菜单";
    // 初始化显示用的数组
    items_ = [[NSMutableArray alloc] initWithObjects:
              @"ViewController1",
              @"ViewController2",
              nil ];
  }
  return self;
}
#pragma mark ----- UITableViewDataSource Methods -----
- (NSInteger)tableView:(UITableView*)tableView
numberOfRowsInSection:(NSInteger)section {
  return [items_ count];
}
- (UITableViewCell*)tableView:(UITableView*)tableView
cellForRowAtIndexPath:(NSIndexPath*)indexPath {
  // 检测单元类型是否已经登记
  UITableViewCell* cell = [tableView dequeueReusableCellWithIdentifier:@"simple-cell"];
  if ( !cell ) {
    // 如果该单元类型没有登记，则新创建
    cell = [[[UITableViewCell alloc] initWithFrame:CGRectZero reuseIdentifier:@"simple-cell"] autorelease];
  }
  // 设置单元标签中显示的文字
  cell.textLabel.text = [items_ objectAtIndex:indexPath.row];
  return cell;
}
#pragma mark ----- UITableViewDelegate Methods -----
- (void)tableView:(UITableView*)tableView
didSelectRowAtIndexPath:(NSIndexPath*)indexPath {
  Class class = NSClassFromString( [items_ objectAtIndex:indexPath.row] );
  id viewController = [[[class alloc] init] autorelease];
  if ( viewController ) {
    [self.navigationController pushViewController:viewController animated:YES];
  }
}
@end
```

执行效果如图 2-33 所示，单击列表中的某个选项会跳转到对应的界面，如图 2-34 所示。

第 2 章 界面布局实战

图 2-33　主界面　　　　图 2-34　跳转到 ViewController1 界面

单击图 2-34 所示中的"主菜单"按钮会返回到图 2-33 所示的主界面。

实例 029　通过 UITabBarController 选项卡实现不同界面之间的跳转

实例 029	通过 UITabBarController 选项卡实现不同界面之间的跳转
源码路径	\daima\029\

实例说明

在 iOS 应用中，和 UINavigationController 功能类似的是 UITabBarController，它也可以用来控制多个页面导航，用户可以在多个视图控制器之间移动，并可以定制屏幕底部的选项卡栏。UITabBarController 能够在手机屏幕中实现一个选项卡效果，单击一个选项卡可以来到一个指定的视图界面。

在本实例中，通过使用 UIViewController 类实现了两个不同的视图界面 ViewController1 和 ViewController2。然后在屏幕下方创建了两个 UITabBarController 选项卡"Hello"和"您好"，单击某个选项卡会进入对应的 UIViewController 视图界面。

具体实现

实例文件 ViewController1.m 的具体实现代码如下所示。

```objc
#import "ViewController1.h"
@implementation ViewController1
- (id)init {
  if ( (self = [super init]) ) {
    // 设置 tabBar 的标题
    self.title = @"Hello";
    UIImage* icon = [UIImage imageNamed:@"ball1.png"];
    self.tabBarItem =
      [[[UITabBarItem alloc] initWithTitle:@"Hello" image:icon tag:0] autorelease];
  }
  return self;
}
- (void)viewDidLoad {
  [super viewDidLoad];
  // 追加 "Hello, world!" 标签
  // 背景为白色，文字为黑色
  UILabel* label = [[[UILabel alloc] initWithFrame:self.view.bounds] autorelease];
  label.text = @"Hello, world!";
  label.textAlignment = UITextAlignmentCenter;
  label.backgroundColor = [UIColor whiteColor];
  label.textColor = [UIColor blackColor];
  label.autoresizingMask = UIViewAutoresizingFlexibleWidth | UIViewAutoresizingFlexibleHeight;
```

```objc
  [self.view addSubview:label];
}
#pragma mark ----- Responder -----
- (void)touchesEnded:(NSSet*)touches withEvent:(UIEvent*)event {
  NSLog( @"parentViewController = %x", self.parentViewController );
  NSLog( @"tabBarController = %x", self.tabBarController );
}
@end
```

实例文件 ViewController2.m 的具体实现代码如下所示。

```objc
#import "ViewController2.h"
@implementation ViewController2
- (id)init {
  if ( ( self = [super init]) ) {
    // 设置 tabBar 的标题
    self.title = @"您好";
    UIImage* icon = [UIImage imageNamed:@"ball2.png"];
    self.tabBarItem =
      [[[UITabBarItem alloc] initWithTitle:@"您好" image:icon tag:0] autorelease];
  }
  return self;
}
- (void)viewDidLoad {
  [super viewDidLoad];
  // 追加 "您好、世界！" 标签
  // 背景为黑色，文字为白色
  UILabel* label = [[[UILabel alloc] initWithFrame:self.view.bounds] autorelease];
  label.text = @"您好、世界！";
  label.textAlignment = UITextAlignmentCenter;
  label.backgroundColor = [UIColor blackColor];
  label.textColor = [UIColor whiteColor];
  label.autoresizingMask = UIViewAutoresizingFlexibleWidth | UIViewAutoresizingFlexibleHeight;
  [self.view addSubview:label];
}
@end
```

实例文件 HelloWorldAppDelegate.m 的具体实现代码如下所示。

```objc
#import "HelloWorldAppDelegate.h"
#import "ViewController1.h"
#import "ViewController2.h"
@implementation HelloWorldAppDelegate
@synthesize window = window_;
- (void)applicationDidFinishLaunching:(UIApplication *)application {
  // 初始化 window 对象
  CGRect bounds = [[UIScreen mainScreen] bounds];
  window_ = [[UIWindow alloc] initWithFrame:bounds];
  // 创建根 Controller
  rootController_ = [[UITabBarController alloc] init];
  // 创建两个画面的 ViewController1 及 ViewController2
  ViewController1* tab1 = [[[ViewController1 alloc] init] autorelease];
  ViewController2* tab2 = [[[ViewController2 alloc] init] autorelease];
  // 将创建完成的 ViewController 追加到数组 Controller 中
  NSArray* controllers = [NSArray arrayWithObjects:tab1, tab2, nil];
  [(UITabBarController*)rootController_ setViewControllers:controllers animated:NO];
  // 将 Controller 的 view 追加到 window 中
  [window_ addSubview:rootController_.view];
  [window_ makeKeyAndVisible];
}
- (void)dealloc {
  [rootController_ release];
  [window_ release];
  [super dealloc];
}
@end
```

执行效果如图 2-35 所示，单击下方的选项卡会进入对应的视图界面。

图 2-35 执行效果

实例 030 在布局中实现一个模态对话框

实例 030	在布局中实现一个模态对话框
源码路径	\daima\030\

实例说明

在 Windows 应用程序中，经常使用模态（Model）对话框来和用户进行简单的交互，比如登录框。在 iOS 应用中有时我们也希望做同样的事情，此时可以使用 UIModalTransition 属性来实现。

在本实例中，在主视图界面设置了一个"调出模态对话框"按钮，当单击此按钮时会以动画的效果显示一个模态对话框。在这个模态对话框中，在上方显示了文本"您好，我是模态画面"，在下方显示一个"Good-bye"按钮，单击此按钮会返回到主视图。

具体实现

实例文件 UIKitPrjModal.m 的具体实现代码如下所示。

```
#import "UIKitPrjModal.h"
@implementation UIKitPrjModal
- (void)viewDidLoad {
  [super viewDidLoad];
  self.view.backgroundColor = [UIColor whiteColor];
  // 追加调用模态画面的按钮
  UIButton* modalButton = [UIButton buttonWithType:UIButtonTypeRoundedRect];
  [modalButton setTitle:@"调出模态对话框" forState:UIControlStateNormal];
  [modalButton sizeToFit];
  modalButton.center = self.view.center;
  [modalButton addTarget:self
              action:@selector(modalDidPush)
      forControlEvents:UIControlEventTouchUpInside];
  [self.view addSubview:modalButton];
}
- (void)modalDidPush {
  // 显示模态对话框
  ModalDialog* dialog = [[[ModalDialog alloc] init] autorelease];
  dialog.modalTransitionStyle = UIModalTransitionStyleFlipHorizontal;
  [self presentModalViewController:dialog animated:YES];
}
@end
@implementation ModalDialog
- (void)viewDidLoad {
  [super viewDidLoad];
  // 追加1个标签
  UILabel* label = [[[UILabel alloc] initWithFrame:self.view.bounds] autorelease];
  label.backgroundColor = [UIColor blackColor];
```

```
        label.textColor = [UIColor whiteColor];
        label.textAlignment = UITextAlignmentCenter;
        label.text = @"你好,我是模态画面。";
        [self.view addSubview:label];
        // 追加关闭按钮
        UIButton* goodbyeButton = [UIButton buttonWithType:UIButtonTypeRoundedRect];
        [goodbyeButton setTitle:@"Good-bye" forState:UIControlStateNormal];
        [goodbyeButton sizeToFit];
        CGPoint newPoint = self.view.center;
        newPoint.y += 80;
        goodbyeButton.center = newPoint;
        [goodbyeButton addTarget:self
                    action:@selector(goodbyeDidPush)
            forControlEvents:UIControlEventTouchUpInside];
        [self.view addSubview:goodbyeButton];
}
- (void)goodbyeDidPush {
    // 关闭模态对话框
    [self dismissModalViewControllerAnimated:YES];
}
@end
```

执行效果如图 2-36 所示,单击"调出模态对话框"按钮后会以动画的效果显示一个模态对话框,如图 2-37 所示。

图 2-36 执行效果

图 2-37 模态对话框视图界面

实例 031 实现仿 iPhone 的底部选项卡

实例 031	实现仿 iPhone 的底部选项卡
源码路径	\daima\030\

实例说明

在 Mac 的 iPhone 产品中,屏幕底部的选项卡界面十分美观大方。在 iOS 开发应用中,我们可以使用 UITabBar 实现类似的功能。在 UIKit 中 UITabbar 代表了标签栏,而 UITabBarController 对其进行了封装,令多个不同的视图管理与切换变得更加轻松。在构建一个标签栏控制器时,首先需要为每个按钮准备一个单独的页,每一页都应被创建为 UIViewController 对象。

在 iOS 应用中,可以使用 UIKit 中提供的系统标签项目作为标签条的项目。针对特定的功能(或画面),系统标签项目提供与此功能相匹配的图标图片以及标题的组合。这样不仅能简单地设置标签项目,而且因为使用了这些用户已经习惯的标签项目,将方便用户更直观地操作,应用程序的界面也将更加人性化。

可以在 UITabBarItem 初始化时调用 initWithTabBarSystemItem:tag:方法来设置系统标签项目,例如下面的代码。

```
// self 为 UITabBarController 中设置的 UIViewController 的子类实例
self.tarBarItem = [[[UITabBarItem alloc] initWithTabBarSystemItem:UITabBarSystemItemFeatured
tag:0] autorelease];
```

正如上述代码中所示，通过在 initWithTabBarSystemItem:tag: 方法的第一个参数中指定 UITabBarSystemItem 类型的常量来设置各种系统标签项目。不需要 tag 的话可将其设置成 0。这样就将创建的 UITabBarItem 实例设置到 UITabBarController 中追加的各画面的 tabBarItem 属性中了。UITabBarSystemItem 设置了选项卡的显示图标，各个值的具体说明如下所示。

- UITabBarSystemItemMore：显示其他未显示图标。
- UITabBarSystemItemFavorites：显示用户自己的个人收藏内容。
- UITabBarSystemItemFeatured：显示应用程序的推荐内容。
- UITabBarSystemItemTopRated：显示用户评价高的内容。
- UITabBarSystemItemRecents：显示用户最近访问的内容。
- UITabBarSystemItemContacts：显示通信录画面。
- UITabBarSystemItemHistory：显示用户操作的历史记录。
- UITabBarSystemItemBookmarks：显示应用程序的书签画面。
- UITabBarSystemItemSearch：显示搜索画面。
- UITabBarSystemItemDownloads：显示下载中/下载完成的内容。
- UITabBarSystemItemMostRecent：显示最新内容。
- UITabBarSystemItemMostViewed：显示观看最多最有人气的内容。

在本实例中，通过使用 UITabBarSystemItem 设置在屏幕底部显示了 5 种不同的选项卡。

具体实现

实例文件 UIKitPrjTabBar.m 的具体实现代码如下所示。

```
#import "UIKitPrjTabBar.h"
@implementation UIKitPrjTabBar
- (void)viewDidLoad {
 [super viewDidLoad];
 self.title = @"UITabBarController";
 id scene1 = [[[SampleScene alloc] initWithSystemItem:UITabBarSystemItemFeatured
badge:nil] autorelease];
 id scene2 = [[[SampleScene alloc] initWithSystemItem:UITabBarSystemItemMostViewed
badge:@"1"] autorelease];
 id scene3 = [[[SampleScene alloc] initWithSystemItem:UITabBarSystemItemSearch
badge:nil] autorelease];
 id scene4 = [[[SampleScene alloc] initWithSystemItem:UITabBarSystemItemBookmarks
badge:nil] autorelease];
 id scene5 = [[[SampleScene alloc] initWithSystemItem:UITabBarSystemItemMostRecent
badge:@"2"] autorelease];
 id scene6 = [[[SampleScene alloc] initWithSystemItem:UITabBarSystemItemTopRated
badge:nil] autorelease];
 id scene7 = [[[SampleScene alloc] initWithSystemItem:UITabBarSystemItemHistory
badge:nil] autorelease];
 id scene8 = [[[SampleScene alloc] initWithSystemItem:UITabBarSystemItemDownloads
badge:nil] autorelease];
 id scene9 = [[[SampleScene alloc] initWithSystemItem:UITabBarSystemItemContacts
badge:nil] autorelease];
id scene10 = [[[SampleScene alloc] initWithSystemItem:UITabBarSystemItemFavorites
badge:nil] autorelease];
 id scene11 = [[[SampleScene alloc] initWithSystemItem:UITabBarSystemItemRecents
badge:nil] autorelease];
 id scene12 = [[[SampleScene alloc] initWithFileName:@"smile.png" title:@" 微 笑 "]
autorelease];
 NSArray* scenes = [NSArray arrayWithObjects:
                        scene1,
                        scene2,
                        scene3,
```

实例 031　实现仿 iPhone 的底部选项卡

```
                         scene4,
                         scene5,
                         scene6,
                         scene7,
                         scene8,
                         scene9,
                         scene10,
                         scene11,
                         scene12,
                         nil ];
    [self setViewControllers:scenes animated:NO];
    self.customizableViewControllers =
      [NSArray arrayWithObjects:scene4, scene5, scene6, scene7, scene8, scene9, scene12,
nil];
}
- (void)viewWillAppear:(BOOL)animated {
    [super viewWillAppear:YES];
    [self.navigationController setNavigationBarHidden:YES animated:NO];
    [self.navigationController setToolbarHidden:YES animated:NO];
}
@end
@implementation SampleScene
- (id)initWithSystemItem:(UITabBarSystemItem)item badge:(NSString*)badge {
  if ( (self = [super init]) ) {
    self.tabBarItem =
      [[[UITabBarItem alloc] initWithTabBarSystemItem:item tag:0] autorelease];
    self.tabBarItem.badgeValue = badge;
  }
  return self;
}
- (void)viewDidLoad {
  [super viewDidLoad];
  UILabel* label = [[[UILabel alloc] init] autorelease];
  label.frame = self.view.bounds;
  label.autoresizingMask =
    UIViewAutoresizingFlexibleWidth | UIViewAutoresizingFlexibleHeight;
  label.backgroundColor = [UIColor blackColor];
  [self.view addSubview:label];
}
- (id)initWithFileName:(NSString*)fileName title:(NSString*)title {
  if ( (self = [super init]) ) {
    UIImage* icon = [UIImage imageNamed:fileName];
    self.tabBarItem =
      [[[UITabBarItem alloc] initWithTitle:title image:icon tag:0] autorelease];
  }
  return self;
}
#pragma mark ----- Responder -----

- (void)touchesEnded:(NSSet*)touches withEvent:(UIEvent*)event {
    [self.navigationController setNavigationBarHidden:NO animated:YES];
}
@end
```

执行效果如图 2-38 所示。

图 2-38　执行效果

实例 032　实现导航条效果

实例 032	实现导航条效果
源码路径	\daima\030\

实例说明

在 iOS 开发应用中，我们可以使用 UINavigationController 在屏幕中实现导航条效果。iOS 中的导航控制器是作为栈来实现的，它控制了一个视图控制器栈，遵循先进后出原则。在设计导航控制器的时候，需要指定应用程序运行的第一个视图，在程序的整个视图层次中，这个视图位于最底层，被称作根控制器。

在本实例中，通过使用 UINavigationController 在屏幕顶部定制了一个导航条。其中在导航条上方显示一段文本"第一行信息"，在文本下方从左到右依次显示一幅图片、文本标题和一个图片样式的按钮。

具体实现

实例文件 UIKitPrjNavigation.m 的具体实现代码如下所示。

```
#import "UIKitPrjNavigation.h"
@implementation UIKitPrjNavigation
- (void)viewDidLoad {
  [super viewDidLoad];
  // 第1行信息的追加
  self.navigationItem.prompt = @"第1行信息";
  // 设置标题
  self.navigationItem.title = @"标题";
  // 在右侧追加按钮
  UIBarButtonItem* rightItem =
    [[[UIBarButtonItem alloc] initWithBarButtonSystemItem:UIBarButtonSystemItemCompose
                                          target:nil
                                          action:nil ] autorelease];
  self.navigationItem.rightBarButtonItem = rightItem;
  // 在左侧追加 UIImageView
  UIImage* image = [UIImage imageNamed:@"face.jpg"];
  UIImageView* imageView = [[[UIImageView alloc] initWithImage:image] autorelease];
  UIBarButtonItem* icon =
    [[[UIBarButtonItem alloc] initWithCustomView:imageView] autorelease];
  self.navigationItem.leftBarButtonItem = icon;

  UILabel* label = [[[UILabel alloc] init] autorelease];
  label.frame = self.view.bounds;
  label.autoresizingMask = UIViewAutoresizingFlexibleWidth | UIViewAutoresizingFlexibleHeight;
  label.backgroundColor = [UIColor blackColor];
  [self.view addSubview:label];
}
- (void)viewWillAppear:(BOOL)animated {
  [super viewWillAppear:animated];
  [self.navigationController setNavigationBarHidden:NO animated:NO];
  [self.navigationController setToolbarHidden:NO animated:NO];
}
#pragma mark ----- Responder -----
- (void)touchesEnded:(NSSet*)touches withEvent:(UIEvent*)event {
  self.navigationItem.leftBarButtonItem = nil;
}
@end
```

执行后的效果如图 2-39 所示。

实例 033　在导航条中添加一个滑动条

图 2-39　执行效果

实例 033　在导航条中添加一个滑动条

实例 033	在导航条中添加一个滑动条
源码路径	\daima\030\

实例说明

在 iOS 开发应用中，当使用 UINavigationController 在屏幕中实现导航条效果时，可以使用其 titleView 属性设置任意的 UIView 子类，并且在设置 UIView 子类时没有任何限制。

在本实例中，首先使用 UINavigationController 在屏幕顶部定制了一个导航条，然后使用 titleView 属性在导航条中设置了一个 UISlider 滑块。将此 UISlider 滑块的属性 minimumValue 设置为 0.0，将其属性 maximumValue 设置为 1.0。当滑块滑动时调用(void)sliderDidChange 方法改动屏幕的颜色。

具体实现

实例文件 UIKitPrjTitleView.m 的具体实现代码如下所示。

```
#import "UIKitPrjTitleView.h"
//声明私有方法
@interface UIKitPrjTitleView ()
- (void)sliderDidChange;
@end
@implementation UIKitPrjTitleView
// finalize
- (void)dealloc {
  [slider_ release];
  [label_ release];
  [super dealloc];
}
- (void)viewDidLoad {
  [super viewDidLoad];
  //顶部信息设置
  self.navigationItem.prompt = @"移动滑块后将改变画面颜色";
  //创建 UISlider 实例，滑块变化时调用 sliderDidChange 方法
  slider_ = [[UISlider alloc] init];
  slider_.frame = self.navigationController.navigationBar.bounds;
  slider_.minimumValue = 0.0;
  slider_.maximumValue = 1.0;
  slider_.value = slider_.maximumValue / 2.0;
  [slider_ addTarget:self
            action:@selector(sliderDidChange)
   forControlEvents:UIControlEventValueChanged];
  self.navigationItem.titleView = slider_;
  //创建标签，并根据滑块的值改变标签的颜色
  label_ = [[UILabel alloc] init];
```

```
    label_.frame = CGRectInset( self.view.bounds, 10, 10 );
    label_.autoresizingMask = UIViewAutoresizingFlexibleWidth | UIViewAutoresizingFlexible
Height;
    label_.backgroundColor = [UIColor blackColor];
    [self.view addSubview:label_];
    //调用 sliderDidChange 方法设置滑块初始值
    [self sliderDidChange];
}
//确保显示导航条以及工具条
- (void)viewWillAppear:(BOOL)animated {
    [super viewWillAppear:animated];
    [self.navigationController setNavigationBarHidden:NO animated:NO];
    [self.navigationController setToolbarHidden:NO animated:NO];
    [self.navigationItem setHidesBackButton:YES animated:NO];
}
#pragma mark ----- Private Methods -----
//画面显示时隐藏返回按钮,触摸画面后显示返回按钮
- (void)touchesEnded:(NSSet*)touches withEvent:(UIEvent*)event {
    [self.navigationItem setHidesBackButton:NO animated:YES];
}
//实现滑块移动时调用的方法,改变标签的颜色
- (void)sliderDidChange {
    UIColor* color = [[[UIColor alloc] initWithRed:slider_.value
                                             green:slider_.value
                                              blue:slider_.value
                                             alpha:1.0 ] autorelease];
    label_.backgroundColor = color;
}
@end
```

执行后的效果如图 2-40 所示。

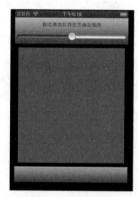

图 2-40 执行效果

实例 034 在屏幕中显示一个工具条

实例 034	在屏幕中显示一个工具条
源码路径	\daima\030\

实例说明

在 iOS 开发应用中,使用 UIViewController 的 setToolbarItems:animated 方法可以在画面中添加一个工具条。如果要显示工具条,需要将 UINavigationController 的 toolbarHidden 属性设置为 NO。但是这样只会显示一个空的工具条。所以在本实例中,通过使用 UIViewController 的 setToolbarItems: animated 方法在工具条中设置了 3 个 UIBarButtonItem 按钮选项。

具体实现

实例文件 UIKitPrjToolbar.m 的具体实现代码如下所示。

```objc
#import "UIKitPrjToolbar.h"
@interface NextViewController ()
- (void)nextDidPush;
@end
@implementation NextViewController
- (void)viewDidLoad {
  [super viewDidLoad];
  self.title = @"Next";
  self.view.backgroundColor = [UIColor whiteColor];
  // 设置跳转到下一画面的按钮
  UIButton* button = [UIButton buttonWithType:UIButtonTypeRoundedRect];
  [button setTitle:@"下一画面" forState:UIControlStateNormal];
  [button sizeToFit];
  button.center = self.view.center;
  button.autoresizingMask =
    UIViewAutoresizingFlexibleTopMargin | UIViewAutoresizingFlexibleBottomMargin;
  [button addTarget:self
           action:@selector(nextDidPush)
   forControlEvents:UIControlEventTouchUpInside];
  [self.view addSubview:button];
}
- (void)nextDidPush {
  UIViewController* nextViewController = [[[NextViewController alloc] init] autorelease];
  // 开始时先将 hidesBottomBarWhenPushed 设置成 YES
  static BOOL nowFirst = YES;
  if ( nowFirst ) {
    nextViewController.hidesBottomBarWhenPushed = YES;
    nowFirst = FALSE;
    NSLog( @"nowFirst" );
  }
  [self.navigationController pushViewController:nextViewController animated:YES];
}
@end
#pragma mark ----- Private Methods Definition -----
@interface UIKitPrjToolbar ()
- (void)buttonDidPush;
@end
#pragma mark ----- Start Implementation For Methods -----
@implementation UIKitPrjToolbar
- (void)viewDidLoad {
  [super viewDidLoad];
  self.title = @"UIKitPrjToolbar";
  // 工具条左侧显示的按钮
  UIBarButtonItem* button1 =
    [[[UIBarButtonItem alloc] initWithBarButtonSystemItem:UIBarButtonSystemItemRefresh
                                                  target:self
                                                  action:@selector(buttonDidPush) ] autorelease];
  // 自动伸缩按钮以及按钮间的空白
  UIBarButtonItem* spacer =
    [[[UIBarButtonItem alloc] initWithBarButtonSystemItem:UIBarButtonSystemItemFlexibleSpace
                                                  target:nil
                                                  action:nil ] autorelease];
  // 工具条右侧显示的按钮
  UIBarButtonItem* button2 =
    [[[UIBarButtonItem alloc] initWithBarButtonSystemItem:UIBarButtonSystemItemUndo
                                                  target:self
                                                  action:@selector(buttonDidPush) ] autorelease];
  // 全部保存到 NSArray 中
  NSArray* buttons = [NSArray arrayWithObjects:button1, spacer, button2, nil];
  // 将准备好的 NSArray 作为工具条的项目设置进去
  [self setToolbarItems:buttons animated:YES];
}
- (void)viewWillAppear:(BOOL)animated {
  [super viewWillAppear:animated];
  [self.navigationController setNavigationBarHidden:NO animated:YES];
  [self.navigationController setToolbarHidden:NO animated:YES];
}
#pragma mark ----- Private Methods -----
- (void)buttonDidPush {
```

第2章 界面布局实战

```
        UIAlertView* alert = [[[UIAlertView alloc] initWithTitle:@"INFORMATION"
                                                  message:@"buttonDidPush"
                                                  delegate:nil
                                                  cancelButtonTitle:nil
                                                  otherButtonTitles:@"OK", nil ] autorelease];
        [alert show];
}
@end
```

执行后的效果如图 2-41 所示。

图 2-41 执行效果

实例 035 在工具条中添加系统按钮

实例 035	在工具条中添加系统按钮
源码路径	\daima\030\

实例说明

在 iOS 开发应用中,有很多个内置的按钮样式,我们通过设置 **UIBarButtonSystemItem** 的值可以实现需要的按钮样式。在本实例中定义了 **arrayWithObjects** 数组,通过 **UIBarButtonSystemItem** 在工具条中设置了 21 种样式的按钮。当单击屏幕时会触发(void)touchesEnded 方法,将工具条中的按钮向左移动。

具体实现

实例文件 UIKitPrjToolbarSystemItem.m 的具体实现代码如下所示。

```
#import "UIKitPrjToolbarSystemItem.h"
#pragma mark ----- Private Methods Definition -----
@interface UIKitPrjToolbarSystemItem ()
- (UIBarButtonItem*)barButtonSystemItem:(UIBarButtonSystemItem)systemItem;
@end
#pragma mark ----- Start Implementation For Methods -----
@implementation UIKitPrjToolbarSystemItem
- (void)viewDidLoad {
  [super viewDidLoad];
  self.title = @"UIBarButtonSystemItem列表";

  [self setToolbarItems:[NSArray arrayWithObjects:
        [self barButtonSystemItem:UIBarButtonSystemItemDone],
        [self barButtonSystemItem:UIBarButtonSystemItemCancel],
        [self barButtonSystemItem:UIBarButtonSystemItemEdit],
        [self barButtonSystemItem:UIBarButtonSystemItemSave],
        [self barButtonSystemItem:UIBarButtonSystemItemAdd],
        [self barButtonSystemItem:UIBarButtonSystemItemCompose],
        [self barButtonSystemItem:UIBarButtonSystemItemReply],
        [self barButtonSystemItem:UIBarButtonSystemItemAction],
```

```objc
            [self barButtonSystemItem:UIBarButtonSystemItemOrganize],
            [self barButtonSystemItem:UIBarButtonSystemItemBookmarks],
            [self barButtonSystemItem:UIBarButtonSystemItemSearch],
            [self barButtonSystemItem:UIBarButtonSystemItemRefresh],
            [self barButtonSystemItem:UIBarButtonSystemItemStop],
            [self barButtonSystemItem:UIBarButtonSystemItemCamera],
            [self barButtonSystemItem:UIBarButtonSystemItemTrash],
            [self barButtonSystemItem:UIBarButtonSystemItemPlay],
            [self barButtonSystemItem:UIBarButtonSystemItemPause],
            [self barButtonSystemItem:UIBarButtonSystemItemRewind],
            [self barButtonSystemItem:UIBarButtonSystemItemFastForward],
            [self barButtonSystemItem:UIBarButtonSystemItemUndo],
            [self barButtonSystemItem:UIBarButtonSystemItemRedo],
            nil]];
    // 追加标签
    UILabel* label = [[[UILabel alloc] init] autorelease];
    label.frame = self.view.bounds;
    label.autoresizingMask = UIViewAutoresizingFlexibleWidth | UIViewAutoresizingFlexibleHeight;
    label.numberOfLines = 3;
    label.textAlignment = UITextAlignmentCenter;
    label.text = @"点击画面后工具条的按钮将移向左侧。";
    [self.view addSubview:label];
}
- (void)viewWillAppear:(BOOL)animated {
    [super viewWillAppear:animated];
    [self.navigationController setToolbarHidden:NO animated:NO];
}
-
(BOOL)shouldAutorotateToInterfaceOrientation:(UIInterfaceOrientation)interfaceOrient
ation {
    return YES;
}
#pragma mark ----- Responder -----
- (void)touchesEnded:(NSSet*)touches withEvent:(UIEvent*)event {
    // 准备移动后容纳工具条按钮的数组
    NSMutableArray* newItems =
        [[[NSMutableArray alloc] initWithCapacity:self.toolbarItems.count] autorelease];
    // 首先抽取左手第 2 个以后（包括第 2）的按钮并追加
    NSRange range = NSMakeRange( 1, self.toolbarItems.count - 1 );
    [newItems addObjectsFromArray:[self.toolbarItems subarrayWithRange:range]];
    // 接着将左侧按钮追加到最后
    [newItems addObject:[self.toolbarItems objectAtIndex:0]];
    // 将移动后的工具条按钮集合设置到工具条中
    [self setToolbarItems:newItems animated:YES];
}
#pragma mark ----- Private Methods -----
- (UIBarButtonItem*)barButtonSystemItem:(UIBarButtonSystemItem)systemItem {
    UIBarButtonItem* button =
        [[[UIBarButtonItem alloc] initWithBarButtonSystemItem:systemItem
                                        target:nil
                                        action:nil] autorelease];
    return button;
}
@end
```

执行效果如图 2-42 所示。

图 2-42 执行效果

实例 036　在工具条中自定义按钮（1）

实例 036	在工具条中自定义按钮（1）
源码路径	\daima\030\

实例说明

在 iOS 开发应用中，除了可以使用 UIKit 中的系统按钮外，还可以在导航条和工具条中添加自定义按钮。在本实例中，使用 initWithTitle:style:target:action: 方法创建了文本按钮，使用 initWithImage:style:target:action: 方法创建了图标按钮。

具体实现

实例文件 UIKitPrjBarButtonWithTitle.m 的具体实现代码如下所示。

```objc
#import "UIKitPrjBarButtonWithTitle.h"
#pragma mark ----- Private Methods Definition -----
@interface UIKitPrjBarButtonWithTitle ()
- (UIBarButtonItem*)barButtonItemWithStyle:(UIBarButtonItemStyle)style;
- (void)buttonDidPush:(id)sender;
@end
#pragma mark ----- Start Implementation For Methods -----
@implementation UIKitPrjBarButtonWithTitle
- (void)viewDidLoad {
  [super viewDidLoad];
  self.title = @"UIBarButtonItem";
  self.navigationItem.rightBarButtonItem =
    [self barButtonItemWithStyle:UIBarButtonItemStyleBordered];
  UIImage* image = [UIImage imageNamed:@"smile.png"];
  UIBarButtonItem* icon = [[[UIBarButtonItem alloc] initWithImage:image
                                    style:UIBarButtonItemStylePlain
                                    target:self
                                    action:@selector(buttonDidPush:)] autorelease];
        [self setToolbarItems:[NSArray arrayWithObjects:
        [self barButtonItemWithStyle:UIBarButtonItemStylePlain],
        [self barButtonItemWithStyle:UIBarButtonItemStyleBordered],
        [self barButtonItemWithStyle:UIBarButtonItemStyleDone],
        icon,
        nil]];
}
- (void)viewWillAppear:(BOOL)animated {
  [super viewWillAppear:animated];
  [self.navigationController setNavigationBarHidden:NO animated:NO];
  [self.navigationController setToolbarHidden:NO animated:NO];
}
#pragma mark ----- Private Methods -----
- (UIBarButtonItem*)barButtonItemWithStyle:(UIBarButtonItemStyle)style {
  NSString* title;
  switch ( style ) {
    case UIBarButtonItemStylePlain:
      title = @"Plain";
      break;
    case UIBarButtonItemStyleBordered:
      title = @"Bordered";
      break;
    default: //< UIBarButtonItemStyleDone
      title = @"Done";
      break;
  }
  UIBarButtonItem* button = [[[UIBarButtonItem alloc] initWithTitle:title
                                    style:style
                                    target:nil
                                    action:nil] autorelease];
  return button;
}
```

```
- (void)buttonDidPush:(id)sender {
  if ( [sender isKindOfClass:[UIBarButtonItem class]] ) {
    UIBarButtonItem* item = sender;
    switch ( item.style ) {
      case UIBarButtonItemStylePlain:
        item.style = UIBarButtonItemStyleBordered;
        break;
      case UIBarButtonItemStyleBordered:
        item.style = UIBarButtonItemStyleDone;
        break;
      default: //< UIBarButtonItemStyleDone
        item.style = UIBarButtonItemStylePlain;
        break;
    }
  }
}
@end
```

执行后的效果如图 2-43 所示。

图 2-43 执行效果

实例 037　在工具条中自定义按钮（2）

实例 037	在工具条中自定义按钮（2）
源码路径	\daima\030\

实例说明

在 iOS 开发应用中，在 UIBarButtonItem 中除了可以使用文本和图片创建自定义按钮外，还可以使用任意的 UIView 子类来创建。在本实例中，首先在导航条中添加了 UIImageView，然后分别向工具条中添加了 UISwitch 和 UISegmentedControl。

具体实现

实例文件 UIKitPrjCustomBarButton.m 的具体实现代码如下所示。

```
#import "UIKitPrjCustomBarButton.h"
@implementation UIKitPrjCustomBarButton
- (void)viewDidLoad {
  [super viewDidLoad];
  self.title = @"CustomBarButton";
// 在导航条中追加 UIImageView
  UIImage* image = [UIImage imageNamed:@"face.jpg"];
  UIImageView* imageView = [[[UIImageView alloc] initWithImage:image] autorelease];
  UIBarButtonItem* icon =
    [[[UIBarButtonItem alloc] initWithCustomView:imageView] autorelease];
  self.navigationItem.rightBarButtonItem = icon;
// 向工具条中追加 UISwitch
  UISwitch* theSwitch = [[[UISwitch alloc] init] autorelease];
```

```
    theSwitch.on = YES;
    UIBarButtonItem* switchBarButton =
      [[[UIBarButtonItem alloc] initWithCustomView:theSwitch] autorelease];
// 向工具条中追加 UISegmentedControl
    NSArray* segments = [NSArray arrayWithObjects:@"1", @"2", @"3", nil];
    UISegmentedControl* segmentedControl =
      [[[UISegmentedControl alloc] initWithItems:segments] autorelease];
    segmentedControl.selectedSegmentIndex = 1;
    segmentedControl.frame = CGRectMake( 0, 0, 100, 30 );
    UIBarButtonItem* segmentedBarButton =
    [[[UIBarButtonItem alloc] initWithCustomView:segmentedControl] autorelease];
    [self setToolbarItems:[NSArray arrayWithObjects:
        switchBarButton,
        segmentedBarButton,
        nil]];
}
- (void)viewWillAppear:(BOOL)animated {
    [super viewWillAppear:animated];
    [self.navigationController setNavigationBarHidden:NO animated:NO];
    [self.navigationController setToolbarHidden:NO animated:NO];
}
@end
```

执行后的效果如图 2-44 所示。

图 2-44　执行效果

实例 038　改变状态栏的颜色

实例 038	改变状态栏的颜色
源码路径	\daima\030\

实例说明

在 iOS 开发应用中，在 UIViewController 中定义了设置状态栏的相关方法，例如 viewDidLoad、viewWillAppear、viewDidAppear 等。在本实例的屏幕中，在中间设置了一个 UILabel 控件来显示文本 "触摸画面后，切换状态条颜色"。当用户触摸屏幕时会触发 touchesEnded 方法，此方法会改变状态栏的颜色，并隐藏当前被调用的状态栏。

具体实现

实例文件 UIKitPrjStatusBarColor.m 的具体实现代码如下所示。

```
#import "UIKitPrjStatusBarColor.h"
@implementation UIKitPrjStatusBarColor
- (void)viewDidLoad {
  [super viewDidLoad];
  UILabel* label = [[[UILabel alloc] init] autorelease];
```

实例 038　改变状态栏的颜色

```
    label.frame = self.view.bounds;
    label.autoresizingMask = UIViewAutoresizingFlexibleWidth | UIViewAutoresizingFlexibleHeight;
    label.text = @"触摸画面后，切换状态条颜色";
    label.numberOfLines = 2;
    [self.view addSubview:label];
    self.wantsFullScreenLayout = YES;
}
- (void)viewWillAppear:(BOOL)animated {
    [super viewWillAppear:YES];
    [self.navigationController setNavigationBarHidden:NO animated:NO];
    [self.navigationController setToolbarHidden:YES animated:NO];
}
#pragma mark ----- Responder -----
- (void)touchesEnded:(NSSet*)touches withEvent:(UIEvent*)event {
    UIApplication* app = [UIApplication sharedApplication];
    if ( UIStatusBarStyleDefault == app.statusBarStyle ) {
      app.statusBarStyle = UIStatusBarStyleBlackOpaque;
      [self.navigationController setNavigationBarHidden:NO animated:NO];
    } else if ( UIStatusBarStyleBlackOpaque == app.statusBarStyle ) {
      app.statusBarStyle = UIStatusBarStyleBlackTranslucent;
      [self.navigationController setNavigationBarHidden:YES animated:NO];
    } else {
      app.statusBarStyle = UIStatusBarStyleDefault;
      [self.navigationController setNavigationBarHidden:NO animated:NO];
    }
}
@end
```

执行后的效果如图 2-45 所示。

图 2-45　执行效果

第 3 章　iOS 控件应用实战

作为智能设备开发项目来说，几乎所有的应用功能都需要用控件来实现。控件就如同在 Web 开发中的一个模块，通过调用这些控件能够实现对应的功能效果。其实在本书第 2 章的内容中，所有的实例都是基于控件实现的，无论是 Button 还是 TextView，都是控件。之所以将它们作为单独的一章来讲解，是因为屏幕布局的重要性。在本章的内容中，将通过具体的实例来讲解 iOS 系统中各个常用控件的基本用法。

实例 039　使用文本、键盘和按钮（1）

实例 039	使用文本、键盘和按钮
源码路径	\daima\039\

实例说明

在 iOS 应用中，标签（UILabel）用于在视图中显示字符串，这是通过设置其 text 属性实现的。可以控制标签中文本的属性有很多，例如字体和字号、对齐方式以及颜色。标签可以在视图中显示静态文本，也可显示在代码中生成的动态输出。在本书前面的实例中，其实已经多次用到了 UILabel 标签控件。

具体实现

（1）打开 Xcode，新建一个名为 "UILabelDemo" 的 "Single View Applicatiom" 项目，如图 3-1 所示。
（2）设置新建项目的工程名，然后设置设备为 "iPhone"，如图 3-2 所示。
（3）设置一个界面，整个界面为空，效果如图 3-3 所示。

图 3-1　新建 Xcode 项目

图 3-2　设置设备

图 3-3　空界面

(4)编写文件 ViewController.m,在此创建了一个 UILabel 对象,并分别设置了显示文本的字体、颜色、背景颜色和水平位置等。并且在此文件中使用了自定义控件 UILabelEx,此控件可以设置文本的垂直方向位置。文件 ViewController.m 的实现代码如下所示。

```
- (void)viewDidLoad
{
    [superviewDidLoad];
#if 0
//载入视图
- (void)viewDidLoad
{
    [superviewDidLoad];

#if 0
//创建 UIlabel 对象
UILabel* label = [[UILabel alloc] initWithFrame:self.view.bounds];
    //设置显示文本
    label.text = @"This is a UILabel Demo,";
    //设置文本字体
    label.font = [UIFont fontWithName:@"Arial" size:35];
    //设置文本颜色
    label.textColor = [UIColor yellowColor];
    //设置文本水平显示位置
   label.textAlignment = UITextAlignmentCenter;
    //设置背景颜色
    label.backgroundColor = [UIColor blueColor];
    //设置单词折行方式
    label.lineBreakMode = UILineBreakModeWordWrap;
    //设置 label 是否可以显示多行,0 则显示多行
    label.numberOfLines = 0;
    //根据内容大小,动态设置 UILabel 的高度
  CGSize size = [label.text sizeWithFont:label.font constrainedToSize:self.view.bounds.size lineBreakMode:label.lineBreakMode];
    CGRect rect = label.frame;
    rect.size.height = size.height;
    label.frame = rect;
#endif
#if 1
//使用自定义控件 UILabelEx,此控件可以设置文本的垂直方向位置
UILabelEx* label = [[UILabelExalloc] initWithFrame:self.view.bounds];
    label.text = @"This is a UILabel Demo,";
    label.font = [UIFontfontWithName:@"Arial"size:35];
    label.textColor = [UIColoryellowColor];
    label.textAlignment = UITextAlignmentCenter;
    label.backgroundColor = [UIColorblueColor];
    label.lineBreakMode = UILineBreakModeWordWrap;
    label.numberOfLines = 0;
    label.verticalAlignment = VerticalAlignmentTop;//设置文本垂直方向顶部对齐

#endif
    //将 label 对象添加到 view 中,这样才可以显示
    [self.view addSubview:label];
    [label release];
}
```

(5)接下来开始看自定义控件 UILabelEx 的实现过程。首先在文件 UILabelEx.h 中定义一个枚举类型,在里面分别设置顶了部、居中和底部对齐 3 种类型。具体代码如下所示。

```
#import <UIKit/UIKit.h>
//定义一个枚举类型,包含顶部、居中、底部对齐三种类型
typedef enum {
    VerticalAlignmentTop,
    VerticalAlignmentMiddle,
    VerticalAlignmentBottom,
} VerticalAlignment;
@interface UILabelEx : UILabel
{
```

```
    VerticalAlignment _verticalAlignment;
}
@property (nonatomic, assign) VerticalAlignment verticalAlignment;
@end
```

然后看文件 UILabelEx.m，在此设置了文本显示类型，并重写了两个父类。具体代码如下所示。

@implementation UILabelEx

@synthesize verticalAlignment = _verticalAlignment;

```
-(id) initWithFrame:(CGRect)frame
{
    if (self = [super initWithFrame:frame]) {
        self.verticalAlignment = VerticalAlignmentMiddle;
    }
    return  self;
}
//设置文本显示类型
-(void) setVerticalAlignment:(VerticalAlignment)verticalAlignment
{
    _verticalAlignment = verticalAlignment;
    [selfsetNeedsDisplay];
}
//重写父类(CGRect) textRectForBounds:(CGRect)bounds limitedToNumberOfLines:(NSInteger)numberOfLines
-(CGRect) textRectForBounds:(CGRect)bounds limitedToNumberOfLines:(NSInteger)numberOfLines
{
    CGRect textRect = [supertextRectForBounds:bounds limitedToNumberOfLines:numberOfLines];
    switch (self.verticalAlignment) {
        caseVerticalAlignmentTop:
            textRect.origin.y = bounds.origin.y;
            break;

        caseVerticalAlignmentBottom:
            textRect.origin.y = bounds.origin.y + bounds.size.height - textRect.size.height;
            break;

        caseVerticalAlignmentMiddle:
        default:
            textRect.origin.y = bounds.origin.y + (bounds.size.height - textRect.size.height) / 2.0;
    }
    return  textRect;
}
//重写父类 -(void) drawTextInRect:(CGRect)rect
-(void) drawTextInRect:(CGRect)rect
{
    CGRect realRect = [selftextRectForBounds:rect limitedToNumberOfLines:self.numberOfLines];
    [super drawTextInRect:realRect];
}
@end
```

这样整个实例就讲解完毕，执行后的效果如图 3-4 所示。

图 3-4 执行效果

实例 040 使用文本、键盘和按钮（2）

实例 040	使用文本、键盘和按钮（2）
源码路径	\daima\040\

实例说明

在 iOS 应用中，最常见的与用户交互的方式是检测用户轻按按钮（UIButton）并对此做出反应。按钮在 iOS 中是一个视图元素，用于响应用户在界面中触发的事件。按钮通常用 Touch Up Inside 事件来体现，能够抓取用户用手指按下按钮并在该按钮上松开发生的事件。当检测到事件后，便可能触发相应视图控件中的操作（IBAction）。

在 iOS 应用中，文本框也是一种常见的信息输入机制。文本框（UITextField）能够让我们在应用程序中输入一行喜欢的任何信息，这类似于 Web 表单中的表单字段。当我们在文本框中输入数据时，可以使用各种 iOS 键盘将其输入限制为数字或文本。和按钮一样，文本框也能响应事件，但是通常将其实现为被动（passive）界面元素，这意味着视图控制器可随时通过 text 属性读取其内容。文本视图（UITextView）与文本框类似，差别在于文本视图可显示一个可滚动和编辑的文本块，供用户阅读或修改。仅当需要的输入很多时，才应使用文本视图。

无论是文本框还是按钮，在本书前面的实例中都已经多次用到过。在本实例中将创建一个 Mad-Libs 式故事生成器，让用户通过 3 个文本框（UITextField）输入一个名词（地点）、一个动词和一个数字；用户还可输入或修改一个模板，该模板包含将生成的故事概要。由于模板可能有多行，因此将使用一个文本视图（UITextView）来显示这些信息。当用户按下按钮（UIButton）时将触发一个操作，该操作生成故事并将其输出到另一个文本视图中。

具体实现

实现文件 ViewController.h 的实现代码如下所示。

```
#import <UIKit/UIKit.h>
@interface ViewController : UIViewController
@property (strong, nonatomic) IBOutlet UITextField *thePlace;
@property (strong, nonatomic) IBOutlet UITextField *theVerb;
@property (strong, nonatomic) IBOutlet UITextField *theNumber;
@property (strong, nonatomic) IBOutlet UITextView *theTemplate;
@property (strong, nonatomic) IBOutlet UITextView *theStory;
@property (strong, nonatomic) IBOutlet UIButton *theButton;
- (IBAction)createStory:(id)sender;
- (IBAction)hideKeyboard:(id)sender;
@end
```

实现文件 ViewController.h 的实现代码如下所示。

```
#import "ViewController.h"
@implementation ViewController
@synthesize thePlace;
@synthesize theVerb;
@synthesize theNumber;
@synthesize theTemplate;
@synthesize theStory;
@synthesize theButton;
- (void)didReceiveMemoryWarning
{
    [super didReceiveMemoryWarning];
}
#pragma mark - View lifecycle
- (void)viewDidLoad
{
```

```objc
    UIImage *normalImage = [[UIImage imageNamed:@"whiteButton.png"]
                            stretchableImageWithLeftCapWidth:12.0
                            topCapHeight:0.0];
    UIImage *pressedImage = [[UIImage imageNamed:@"blueButton.png"]
                             stretchableImageWithLeftCapWidth:12.0
                             topCapHeight:0.0];
    [self.theButton setBackgroundImage:normalImage
                              forState:UIControlStateNormal];
    [self.theButton setBackgroundImage:pressedImage
                              forState:UIControlStateHighlighted];
    [super viewDidLoad];
}
- (void)viewDidUnload
{
    [self setThePlace:nil];
    [self setTheVerb:nil];
    [self setTheNumber:nil];
    [self setTheTemplate:nil];
    [self setTheStory:nil];
    [self setTheButton:nil];
    [super viewDidUnload];
    // Release any retained subviews of the main view.
    // e.g. self.myOutlet = nil;
}
- (void)viewWillAppear:(BOOL)animated
{
    [super viewWillAppear:animated];
}

- (void)viewDidAppear:(BOOL)animated
{
    [super viewDidAppear:animated];
}
- (void)viewWillDisappear:(BOOL)animated
{
  [super viewWillDisappear:animated];
}
- (void)viewDidDisappear:(BOOL)animated
{
  [super viewDidDisappear:animated];
}
-
(BOOL)shouldAutorotateToInterfaceOrientation:(UIInterfaceOrientation)interfaceOrientation
{
    return (interfaceOrientation != UIInterfaceOrientationPortraitUpsideDown);
}
/*  1:*/- (IBAction)createStory:(id)sender {
    /*  2:*/    self.theStory.text=[self.theTemplate.text
    /*  3:*/                        stringByReplacingOccurrencesOfString:@"<place>"
    /*  4:*/                        withString:self.thePlace.text];
    /*  5:*/    self.theStory.text=[self.theStory.text
    /*  6:*/                        stringByReplacingOccurrencesOfString:@"<verb>"
    /*  7:*/                        withString:self.theVerb.text];
    /*  8:*/    self.theStory.text=[self.theStory.text
    /*  9:*/                        stringByReplacingOccurrencesOfString:@"<number>"
    /* 10:*/                        withString:self.theNumber.text];
    /* 11:*/}
- (IBAction)hideKeyboard:(id)sender {
    [self.thePlace resignFirstResponder];
    [self.theVerb resignFirstResponder];
    [self.theNumber resignFirstResponder];
    [self.theTemplate resignFirstResponder];
}
@end
```

单击 Xcode 工具栏中的 Run 按钮。最终的执行效果如图 3-5 所示。在文本框中输入信息,单击"构造"按钮后的效果如图 3-6 所示。

实例 041　在屏幕中显示一个指定的文本

图 3-5　初始执行效果　　　　图 3-6　单击按钮后的效果

实例 041　在屏幕中显示一个指定的文本

实例 041	在屏幕中显示一个指定的文本
源码路径	\daima\041\

实例说明

在 iOS 应用中，使用 UILabel 控件可以在屏幕中显示文本。在本实例中，使用 UILabel 控件的 font 属性设置了显示文本的字体，并使用其 size 属性设置了文本的大小。

具体实现

实例文件 UIKitPrjSimple.m 的具体实现代码如下所示。

```
#import "UIKitPrjSimple.h"
@implementation UIKitPrjSimple
- (void)viewDidLoad {
  [super viewDidLoad];
  UILabel* label = [[[UILabel alloc] init] autorelease];
  label.frame = self.view.bounds;
  label.autoresizingMask =
    UIViewAutoresizingFlexibleWidth | UIViewAutoresizingFlexibleHeight;
  label.text = @"good";
  label.textAlignment = UITextAlignmentCenter;
  label.backgroundColor = [UIColor blackColor];
  label.textColor = [UIColor whiteColor];
  label.font = [UIFont fontWithName:@"Zapfino" size:48];
  [self.view addSubview:label];
}
@end
```

执行后的效果如图 3-7 所示。

图 3-7　执行效果

实例 042　设置屏幕中文本的对齐方式

实例 042	设置屏幕中文本的对齐方式
源码路径	\daima\041\

实例说明

本实例还是使用了 UILabel 控件，首先在屏幕中显示了 3 段文本。然后使用 backgroundColor 设置了背景颜色，并分别使用如下 3 个对齐属性设置了文本的对齐方式。

- UITextAlignmentLeft：左对齐。
- UITextAlignmentCenter：居中对齐。
- UITextAlignmentRight：右对齐。

具体实现

实例文件 UIKitPrjAlignment.m 的具体实现代码如下所示。

```
#import "UIKitPrjAlignment.h"
@implementation UIKitPrjAlignment
- (void)viewDidLoad {
  [super viewDidLoad];
  self.title = @"UITextAlignment";
  self.view.backgroundColor = [UIColor blackColor];
  UILabel* label1 = [[[UILabel alloc] initWithFrame:CGRectMake( 0, 10, 320, 30 )] autorelease];
  UILabel* label2 = [[[UILabel alloc] initWithFrame:CGRectMake( 0, 50, 320, 30 )] autorelease];
  UILabel* label3 = [[[UILabel alloc] initWithFrame:CGRectMake( 0, 90, 320, 30 )] autorelease];
  label1.textAlignment = UITextAlignmentLeft;
  label2.textAlignment = UITextAlignmentCenter;
  label3.textAlignment = UITextAlignmentRight;
  label1.text = @"UITextAlignmentLeft";
  label2.text = @"UITextAlignmentCenter";
  label3.text = @"UITextAlignmentRight";
  [self.view addSubview:label1];
  [self.view addSubview:label2];
  [self.view addSubview:label3];
}
@end
```

执行后的效果如图 3-8 所示。

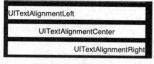

图 3-8　执行效果

实例 043　设置屏幕中标签的颜色和文本的颜色

实例 043	设置屏幕中标签的颜色和文本的颜色
源码路径	\daima\041\

实例说明

本实例还是使用了 UILabel 控件，首先在屏幕中通过 for 循环的方式显示了 8 行文本。然后使用 backgroundColor 设置了背景颜色，使用 for 循环设置了这 8 行文本的颜色。在本实例中，还使用属性 backgroundColor 设置了标签的颜色，使用属性 textColor 设置了 8 行文本的颜色。

具体实现

实例文件 UIKitPrjColor.m 的具体实现代码如下所示。

```
#import "UIKitPrjColor.h"
@implementation UIKitPrjColor
- (void)viewDidLoad {
  [super viewDidLoad];
```

```
    self.title = @"textColor";
    self.view.backgroundColor = [UIColor blackColor];
    NSInteger y = 0;
    for ( int r = 0; r < 2; ++r ){
        for ( int g = 0; g < 2; ++g ){
          for ( int b = 0; b < 2; ++b ){
            UIColor* color = [[[UIColor alloc] initWithRed:r*0.5
                                                    green:g*0.5
                                                     blue:b*0.5
                                                    alpha:1.0] autorelease];
            UILabel* label =
              [[[UILabel alloc] initWithFrame:CGRectMake( 0, y, 320, 30 )] autorelease];
            label.textColor = color;
            label.text =
              [NSString stringWithFormat:@"  R:%.1f  G:%.1f  B:%.1f", r*0.5, g*0.5, b*0.5];
            [self.view addSubview:label];
            y += 40;
          }
        }
    }
}
@end
```

执行后的效果如图 3-9 所示。

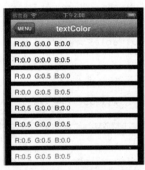

图 3-9　执行效果

实例 044　设置屏幕中显示不同字体的文本

实例 044	设置屏幕中显示不同字体的文本
源码路径	\daima\041\

实例说明

其实在本章前面的实例 041 中已经演示了设置 UILabel 控件中文本字体的方法，设置字体功能是通过如下代码实现的。

```
label.font = [UIFont fontWithName:@"Zapfino" size:48];
```

在本实例中，首先定义了 tableView 表格，然后定义了 NSString* familyName 获取当前系统内字体格式的名字，并最终在 tableView 表格行中用对应的字体显示对应的字体名。

具体实现

实例文件 UIKitPrjFont.m 的具体实现代码如下所示。

```
#import "UIKitPrjFont.h"
@implementation UIKitPrjFont
- (void)dealloc {
  [familyNames_ release];
  [fontNames_ release];
```

```objc
    [super dealloc];
}
- (id)init {
  if ( (self = [super initWithStyle:UITableViewStyleGrouped]) ) {
    self.title = @"MENU";
    if ( !fontNames_ ) {
      fontNames_ = [[NSMutableDictionary alloc] initWithCapacity:32];
      familyNames_ = [[[UIFont familyNames] sortedArrayUsingSelector:@selector(compare:)] retain];
      for ( id familyName in familyNames_ ) {
        NSArray* fonts = [UIFont fontNamesForFamilyName:familyName];
        NSLog( [fonts description] );
        [fontNames_ setObject:fonts forKey:familyName];
      }
    }
  }
  return self;
}
#pragma mark UITableView methods
- (NSInteger)numberOfSectionsInTableView:(UITableView*)tableView {
  return [fontNames_ count];
}
- (NSInteger)tableView:(UITableView*)tableView
  numberOfRowsInSection:(NSInteger)section
{
  NSString* familyName = [familyNames_ objectAtIndex:section];
  return [[fontNames_ objectForKey:familyName] count];
}
- (UITableViewCell*)tableView:(UITableView*)tableView
  cellForRowAtIndexPath:(NSIndexPath*)indexPath
{
  static NSString *CellIdentifier = @"Cell";

  UITableViewCell *cell = [tableView dequeueReusableCellWithIdentifier:CellIdentifier];
  if (cell == nil) {
    cell = [[[UITableViewCell alloc] initWithStyle:UITableViewCellStyleDefault reuseIdentifier:CellIdentifier] autorelease];
  }
  NSString* familyName = [familyNames_ objectAtIndex:indexPath.section];
  NSArray* values = [fontNames_ objectForKey:familyName];
  NSString* fontName = [values objectAtIndex:indexPath.row];
  cell.textLabel.text = fontName;
  UIFont* font = [UIFont fontWithName:fontName size:[UIFont labelFontSize]];
  cell.textLabel.font = font;
  return cell;
}
- (NSString*)tableView:(UITableView*)tableView
 titleForHeaderInSection:(NSInteger)section
{
  return [familyNames_ objectAtIndex:section];
}
@end
```

执行后的效果如图 3-10 所示。

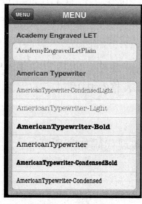

图 3-10 执行效果

实例 045 自动调整屏幕中的文本大小

实例 045	自动调整屏幕中的文本大小
源码路径	\daima\041\

实例说明

在 iOS 应用中,可以使用 UILabel 控件的属性 adjustsFontSizeToFitWidth 设置文本文字自适应大小,此时需要将 adjustsFontSizeToFitWidth 的值设置为 YES。而在下面的代码中,使用值 baselineAdjustment 控制文本的基线位置,只有文本行数为 1 时有效。

```
label1.baselineAdjustment = UIBaselineAdjustmentAlignCenters;
```

在本实例中,首先使用 UILabel 控件显示了 6 行文本,然后通过属性 adjustsFontSizeToFitWidth 设置了各行文本的自适应属性。

具体实现

实例文件 UIKitPrjAdjust.m 的具体实现代码如下所示。

```objc
#import "UIKitPrjAdjust.h"
@implementation UIKitPrjAdjust
- (void)viewDidLoad {
    [super viewDidLoad];
    self.title = @"Adjust";
    self.view.backgroundColor = [UIColor blackColor];
    UILabel* label0 = [[[UILabel alloc] initWithFrame:CGRectMake( 0, 10, 320, 40 )] autorelease];
    UILabel* label1 = [[[UILabel alloc] initWithFrame:CGRectMake( 0, 60, 320, 40 )] autorelease];
    UILabel* label2 = [[[UILabel alloc] initWithFrame:CGRectMake( 0, 110, 320, 40 )] autorelease];
    UILabel* label3 = [[[UILabel alloc] initWithFrame:CGRectMake( 0, 160, 320, 40 )] autorelease];
    UILabel* label4 = [[[UILabel alloc] initWithFrame:CGRectMake( 0, 210, 320, 40 )] autorelease];
    UILabel* label5 = [[[UILabel alloc] initWithFrame:CGRectMake( 0, 260, 320, 40 )] autorelease];
    label0.text = @"标签中放入了很长的文本字符串。今后要注意不要这么长。";
    label1.text = @"标签中放入了很长的文本字符串。今后要注意不要这么长。";
    label2.text = @"标签中放入了很长的文本字符串。今后要注意不要这么长。";
    label3.text = @"UIBaselineAdjustmentAlignBaselines 放入很长的文本字符串。";
    label4.text = @"UIBaselineAdjustmentAlignCenters 放入很长的文本字符串。";
    label5.text = @"UIBaselineAdjustmentNone 放入很长的文本字符串。";
    label1.adjustsFontSizeToFitWidth = YES;
    label2.adjustsFontSizeToFitWidth = YES;
    label2.minimumFontSize = 14;
    label3.adjustsFontSizeToFitWidth = YES;
    label3.baselineAdjustment = UIBaselineAdjustmentAlignBaselines;
    label4.adjustsFontSizeToFitWidth = YES;
    label4.baselineAdjustment = UIBaselineAdjustmentAlignCenters;
    label5.adjustsFontSizeToFitWidth = YES;
    label5.baselineAdjustment = UIBaselineAdjustmentNone;
    [self.view addSubview:label0];
    [self.view addSubview:label1];
    [self.view addSubview:label2];
    [self.view addSubview:label3];
    [self.view addSubview:label4];
    [self.view addSubview:label5];
}
@end
```

执行后的效果如图 3-11 所示。

图 3-11　执行效果

实例 046　在一个 UILabel 控件中显示多行文本

实例 046	在一个 UILabel 控件中显示多行文本
源码路径	\daima\041\

实例说明

在 iOS 应用中，一个 UILabel 控件默认只显示一行文本。其实通过其 numberOfLines 属性可设置显示多行文本。在本实例中，首先使用 UILabel 控件显示了 6 行文本，然后通过属性 adjustsFontSizeToFitWidth 设置了各行文本的自适用属性。

具体实现

实例文件 UIKitPrjMultiline.m 的具体实现代码如下所示。

```
#import "UIKitPrjMultiline.h"
@implementation UIKitPrjMultiline
- (void)viewDidLoad {
  [super viewDidLoad];
  self.title = @"Multiline";
  self.view.backgroundColor = [UIColor blackColor];
  UILabel* label0 = [[[UILabel alloc] initWithFrame:CGRectMake( 10,  10, 300, 60 )] autorelease];
  UILabel* label1 = [[[UILabel alloc] initWithFrame:CGRectMake( 10,  90, 300, 60 )] autorelease];
  UILabel* label2 = [[[UILabel alloc] initWithFrame:CGRectMake( 10, 170, 300, 60 )] autorelease];
  NSString* longText =
    @"放入很长的文本字符串。今后要注意不要这么长。"
     "现在已经进入到第 3 行了。";
  label0.text = label1.text = label2.text = longText;
  label0.numberOfLines = 1;
  label1.numberOfLines = 2;
  label2.numberOfLines = 3;
  [self.view addSubview:label0];
  [self.view addSubview:label1];
  [self.view addSubview:label2];
}
@end
```

执行后的效果如图 3-12 所示。

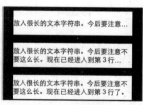

图 3-12　执行效果

实例 047　设置文本的换行和省略模式

实例 047	设置文本的换行和省略模式
源码路径	\daima\041\

实例说明

在 iOS 应用中，一个 UILabel 控件默认只显示一行文本。当一行中的文本过多时，我们可以设置换行显示或用省略方式显示，这一功能是通过 UILabel 的 UILineBreakMode 属性实现的。例如 UILineBreakModeClip 表示截去多余部分，UILineBreakModeHeadTruncation 表示截去头部，UILineBreakModeTailTruncation 表示截去尾部，UILineBreakModeMiddleTruncation 表示截去中间部分。

在本实例中，首先使用 UILabel 控件显示了 6 行 UILabel 文本，然后通过属性 numberOfLines 设置了每个 UILabel 中的文本多行显示，最后使用 UILineBreakMode 属性设置了每个 UILabel 文本的换行和省略方式。

具体实现

实例文件 UIKitPrjLineBreak.m 的具体实现代码如下所示。

```
#import "UIKitPrjLineBreak.h"
@implementation UIKitPrjLineBreak
- (void)viewDidLoad {
 [super viewDidLoad];
 self.title = @"LineBreak";
 self.view.backgroundColor = [UIColor blackColor];
 UILabel* label0 = [[[UILabel alloc] initWithFrame:CGRectMake( 0, 10, 320, 40 )] autorelease];
 UILabel* label1 = [[[UILabel alloc] initWithFrame:CGRectMake( 0, 60, 320, 40 )] autorelease];
 UILabel* label2 = [[[UILabel alloc] initWithFrame:CGRectMake( 0, 110, 320, 40 )] autorelease];
 UILabel* label3 = [[[UILabel alloc] initWithFrame:CGRectMake( 0, 160, 320, 40 )] autorelease];
 UILabel* label4 = [[[UILabel alloc] initWithFrame:CGRectMake( 0, 210, 320, 40 )] autorelease];
 UILabel* label5 = [[[UILabel alloc] initWithFrame:CGRectMake( 0, 260, 320, 40 )] autorelease];
 label0.text   = @"UILineBreakModeWordWrap       test/test  test/test  test/test abcdefghijklmnopqrstuwxyz12345678901234567890";
 label1.text   = @"UILineBreakModeCharacterWrap___ test/test  test/test  test/test abcdefghijklmnopqrstuwxyz12345678901234567890";
 label2.text   = @"UILineBreakModeClip        test/test  test/test  test/test abcdefghijklmnopqrstuwxyz12345678901234567890";
 label3.text   = @"UILineBreakModeHeadTruncation__ test/test  test/test  test/test abcdefghijklmnopqrstuwxyz12345678901234567890";
 label4.text   = @"UILineBreakModeTailTruncation__ test/test  test/test  test/test abcdefghijklmnopqrstuwxyz12345678901234567890";
 label5.text   = @"UILineBreakModeMiddleTruncation test/test  test/test  test/test abcdefghijklmnopqrstuwxyz12345678901234567890";
 label0.numberOfLines = 2;
 label1.numberOfLines = 2;
 label2.numberOfLines = 2;
 label3.numberOfLines = 2;
 label4.numberOfLines = 2;
 label5.numberOfLines = 2;
 label0.lineBreakMode = UILineBreakModeWordWrap;
 label1.lineBreakMode = UILineBreakModeCharacterWrap;
 label2.lineBreakMode = UILineBreakModeClip;
 label3.lineBreakMode = UILineBreakModeHeadTruncation;
 label4.lineBreakMode = UILineBreakModeTailTruncation;
 label5.lineBreakMode = UILineBreakModeMiddleTruncation;
 [self.view addSubview:label0];
 [self.view addSubview:label1];
 [self.view addSubview:label2];
 [self.view addSubview:label3];
 [self.view addSubview:label4];
 [self.view addSubview:label5];
}
@end
```

执行后的效果如图 3-13 所示。

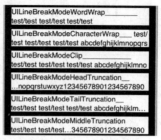

图 3-13 执行效果

实例 048　实现文本的阴影效果

实例 048	实现文本的阴影效果
源码路径	\daima\041\

实例说明

在 iOS 应用中，可以设置 UILabel 控件的文本实现阴影效果。在具体实现时，可以使用属性 shadowColor 设置阴影的颜色，使用属性 shadowOffset 设置阴影的位置。

在本实例中，首先使用 UILabel 控件显示了 4 行 UILabel 文本，然后通过属性 textAlignment 和 shadowColor 设置了阴影效果。

具体实现

实例文件 UIKitPrjShadow.m 的具体实现代码如下所示。

```
#import "UIKitPrjShadow.h"
@implementation UIKitPrjShadow
- (void)viewDidLoad {
    [super viewDidLoad];
    self.title = @"shadowOffset";
    self.view.backgroundColor = [UIColor blackColor];
    UILabel* label0 = [[[UILabel alloc] initWithFrame:CGRectMake( 0, 10, 320, 40 )] autorelease];
    UILabel* label1 = [[[UILabel alloc] initWithFrame:CGRectMake( 0, 60, 320, 40 )] autorelease];
    UILabel* label2 = [[[UILabel alloc] initWithFrame:CGRectMake( 0, 110, 320, 40 )] autorelease];
    UILabel* label3 = [[[UILabel alloc] initWithFrame:CGRectMake( 0, 160, 320, 40 )] autorelease];
    label0.textAlignment = UITextAlignmentCenter;
    label1.textAlignment = UITextAlignmentCenter;
    label2.textAlignment = UITextAlignmentCenter;
    label3.textAlignment = UITextAlignmentCenter;
    label1.shadowColor = [UIColor grayColor];
    label2.shadowColor = [UIColor grayColor];
    label3.shadowColor = [UIColor grayColor];
    label2.shadowOffset = CGSizeMake( 1, 1 );
    label3.shadowOffset = CGSizeMake( 3, 0 );
    label0.text = @"无阴影";
    label1.text = @"默认的 shadowOffset";
    label2.text = @"shadowOffset = CGSizeMake( 1, 1 )";
    label3.text = @"shadowOffset = CGSizeMake( 3, 0 )";
    [self.view addSubview:label0];
    [self.view addSubview:label1];
    [self.view addSubview:label2];
    [self.view addSubview:label3];
}
@end
```

执行后的效果如图 3-14 所示。

图 3-14 执行效果

实例 049　高亮显示屏幕中的文本

实例 049	高亮显示屏幕中的文本
源码路径	\daima\041\

实例说明

在 iOS 应用中，可以设置 UILabel 控件的文本实现高亮显示效果。在具体实现时，只需将其 highlighted 属性设置为 YES 即可，并且可以通过 highlightedTextColor 设置高亮文本的颜色。在本实例中，首先使用 UILabel 控件显示了 3 行 UILabel 文本，然后通过属性 highlightedTextColor 设置了 3 行文本的 3 种不同的颜色。

具体实现

实例文件 UIKitPrjHighlighted.m 的具体实现代码如下所示。

```
#import "UIKitPrjHighlighted.h"
#pragma mark ----- Private Methods Definition -----
@interface UIKitPrjHighlighted ()
- (void)buttonDidPush;
@end
#pragma mark ----- Start Implementation For Methods -----

@implementation UIKitPrjHighlighted
- (void)dealloc {
  [items_ release];
  [super dealloc];
}
- (id)init {
  if ( (self = [super initWithStyle:UITableViewStyleGrouped]) ) {
    self.title = @"MENU";
    if ( !items_ ) {
      items_ = [[NSArray alloc] initWithObjects:@"redColor",
                                                @"blueColor",
                                                @"greenColor",
                                                nil ];
    }
  }
  return self;
}
- (void)viewDidLoad {
  [super viewDidLoad];
  self.title = @"Highlighted";
  self.view.backgroundColor = [UIColor blackColor];
  // 按钮追加
  UIButton* button = [UIButton buttonWithType:UIButtonTypeRoundedRect];
  [button setTitle:@"higlighted 切替" forState:UIControlStateNormal];
  [button sizeToFit];
  CGPoint newPoint = self.view.center;
  newPoint.y = self.view.frame.size.height - 120;
  button.center = newPoint;
  [button addTarget:self
             action:@selector(buttonDidPush)
   forControlEvents:UIControlEventTouchUpInside];
  [self.view addSubview:button];
}
#pragma mark ----- Private Methods -----
```

```objc
- (void)buttonDidPush {
  NSInteger rows = [self.tableView numberOfRowsInSection:0];
  for ( int i = 0; i < rows; ++i ){
    NSIndexPath* indexPath = [NSIndexPath indexPathForRow:i inSection:0];
    UITableViewCell* cell = [self.tableView cellForRowAtIndexPath:indexPath];
    cell.textLabel.highlighted = !cell.textLabel.highlighted;
  }
}
#pragma mark UITableView methods
- (NSInteger)tableView:(UITableView*)tableView
  numberOfRowsInSection:(NSInteger)section
{
  return [items_ count];
}
- (UITableViewCell*)tableView:(UITableView*)tableView
  cellForRowAtIndexPath:(NSIndexPath*)indexPath
{
  static NSString *CellIdentifier = @"Cell";

  UITableViewCell *cell = [tableView dequeueReusableCellWithIdentifier:CellIdentifier];
  if (cell == nil) {
    cell = [[[UITableViewCell alloc] initWithStyle:UITableViewCellStyleDefault reuseIdentifier:CellIdentifier] autorelease];
  }
  cell.textLabel.text = [items_ objectAtIndex:indexPath.row];
  SEL selector = NSSelectorFromString( cell.textLabel.text );
  UIColor* color = [UIColor performSelector:selector];
  cell.textLabel.highlightedTextColor = color;
  return cell;
}
- (NSIndexPath*)tableView:(UITableView*)tableView willSelectRowAtIndexPath:(NSIndexPath*)indexPath {
  UITableViewCell* cell = [tableView cellForRowAtIndexPath:indexPath];
  NSLog( @"%d", cell.textLabel.highlighted );
  return indexPath;
}
- (void)tableView:(UITableView*)tableView didSelectRowAtIndexPath:(NSIndexPath*)indexPath {
  UITableViewCell* cell = [tableView cellForRowAtIndexPath:indexPath];
  NSLog( @"%d", cell.textLabel.highlighted );
}
@end
```

执行后的效果如图 3-15 所示。单击"highlighted 切替"按钮会呈现高亮效果显示,如图 3-16 所示。

图 3-15 执行效果

图 3-16 高亮显示效果

实例 050 定制一个文本绘制方法

实例 050	定制一个文本绘制方法
源码路径	\daima\041\

实例说明

在 iOS 应用中,虽然 UILabel 控件的属性和方法能够满足我们大多数需要。但是有时候需要实现一些特殊的效果，此时我们可以自行定制绘制文本的方法。在本实例中，定义了一个子类 ThickTextLabel，通过此类绘制了两行等价于重合效果的文本。

具体实现

实例文件 UIKitPrjDrawTextOverride.h 的具体实现代码如下所示。

```
#import "UIKitPrjDrawTextOverride.h"
@implementation ThickTextLabel
- (void)drawTextInRect:(CGRect)rect {
  for ( int y = -2; y <= 4; ++y ){
    for ( int x = -2; x <= 4; ++x ){
      CGRect drawRect = CGRectMake( rect.origin.x + x,
                                    rect.origin.y + y,
                                    rect.size.width,
                                    rect.size.height );
      [self.text drawInRect:drawRect withFont:self.font
                          lineBreakMode:self.lineBreakMode
                              alignment:self.textAlignment];
    }
  }
}

@end
@implementation UIKitPrjDrawTextOverride
- (void)viewDidLoad {
  [super viewDidLoad];
  // 普通的 UILabel
  UILabel* normalLabel = [[[UILabel alloc] init] autorelease];
  normalLabel.frame = CGRectMake( 0, 50, 320, 60 );
  normalLabel.textAlignment = UITextAlignmentCenter;
  normalLabel.font = [UIFont systemFontOfSize:48];
  normalLabel.text = @"普通文本";
  [self.view addSubview:normalLabel];
  // 使用定制 UILabel 子类
  ThickTextLabel* newLabel = [[[ThickTextLabel alloc] init] autorelease];
  newLabel.frame = CGRectMake( 0, 130, 320, 60 );
  newLabel.textAlignment = UITextAlignmentCenter;
  newLabel.font = [UIFont systemFontOfSize:48];
  newLabel.text = @"普通文本";
  [self.view addSubview:newLabel];
}
@end
```

执行后的效果如图 3-17 所示。

图 3-17　执行效果

实例 051　按下按钮后触发一个事件

实例 051	按下按钮后触发一个事件
源码路径	\daima\051\

实例说明

在 iOS 应用中，使用 Button 控件可以实现一个按钮效果。Button 控件的最大好处是通过触摸触发事件处理程序，最终实现交互处理功能。在本实例中，设置了一个"危险!请勿触摸！"按钮，按下按钮后会执行 buttonDidPush 方法，弹出一个对话框，在对话框中显示"哈哈，这是笑话!!"。

具体实现

实例文件 UIKitPrjButtonTap.m 的具体实现代码如下所示。

```objc
#import "UIKitPrjButtonTap.h"
@implementation UIKitPrjButtonTap
- (void)viewDidLoad {
  [super viewDidLoad];
  UIButton* button = [UIButton buttonWithType:UIButtonTypeRoundedRect];
  [button setTitle:@"危险!请勿触摸!" forState:UIControlStateNormal];
  [button sizeToFit];
  [button addTarget:self
            action:@selector(buttonDidPush)
   forControlEvents:UIControlEventTouchUpInside];
  button.center = self.view.center;
  button.autoresizingMask = UIViewAutoresizingFlexibleLeftMargin |
                            UIViewAutoresizingFlexibleRightMargin |
                            UIViewAutoresizingFlexibleTopMargin |
                            UIViewAutoresizingFlexibleBottomMargin;
  [self.view addSubview:button];
}
- (void)buttonDidPush {
  UIAlertView* alert = [[[UIAlertView alloc] init] autorelease];
  alert.message = @"哈哈，这是笑话!!";
  [alert addButtonWithTitle:@"OK"];
  [alert show];
}
@end
```

执行后的效果如图 3-18 所示。

图 3-18　执行效果

实例 052　在屏幕中显示不同的按钮

实例 052	在屏幕中显示不同的按钮
源码路径	\daima\051\

实例说明

在 iOS 应用中，使用 Button 控件可以实现不同样式的按钮效果。通过使用方法 ButtonWithType 可以指定几种不同的 UIButtonType 的类型常量，使用不同的常量显示不同外观样式的按钮。UIButtonType 指定一个按钮的风格，有如下几种常用的外观风格。

- UIButtonTypeCustom：无按钮的样式。
- UIButtonTypeRoundedRect：一个圆角矩形样式的按钮。
- UIButtonTypeDetailDisclosure：一个详细披露按钮。
- UIButtonTypeInfoLight：一个信息按钮，有一个浅色背景。
- UIButtonTypeInfoDark：一个信息按钮，有一个黑暗的背景。
- UIButtonTypeContactAdd：一个联系人添加按钮。

实例 052 在屏幕中显示不同的按钮

在本实例中,在屏幕中演示了各种不同外观样式按钮。

具体实现

实例文件 **UIKitPrjButtonWithType.m** 的具体实现代码如下所示。

```objc
#import "UIKitPrjButtonWithType.h"
static const CGFloat kRowHeight = 80.0;
#pragma mark ----- Private Methods Definition -----
@interface UIKitPrjButtonWithType ()
- (UIButton*)buttonForThisSampleWithType:(UIButtonType)type;
@end
#pragma mark ----- Start Implementation For Methods -----
@implementation UIKitPrjButtonWithType
- (void)dealloc {
    [dataSource_ release];
    [buttons_ release];
    [super dealloc];
}
- (void)viewDidLoad {
    [super viewDidLoad];
    self.tableView.rowHeight = kRowHeight;
    dataSource_ = [[NSArray alloc] initWithObjects:
                   @"Custom",
                   @"RoundedRect",
                   @"DetailDisclosure",
                   @"InfoLight",
                   @"InfoDark",
                   @"ContactAdd",
                   nil ];
    UIButton* customButton = [self buttonForThisSampleWithType:UIButtonTypeCustom];
    UIImage* image = [UIImage imageNamed:@"frame.png"];
    UIImage* stretchableImage = [image stretchableImageWithLeftCapWidth:20 topCapHeight:20];
    [customButton setBackgroundImage:stretchableImage forState:UIControlStateNormal];
    customButton.frame = CGRectMake( 0, 0, 200, 60 );
        //self.tableView.backgroundColor = [UIColor lightGrayColor];
    buttons_ = [[NSArray alloc] initWithObjects:
                customButton,
                [self buttonForThisSampleWithType:UIButtonTypeRoundedRect],
                [self buttonForThisSampleWithType:UIButtonTypeDetailDisclosure],
                [self buttonForThisSampleWithType:UIButtonTypeInfoLight],
                [self buttonForThisSampleWithType:UIButtonTypeInfoDark],
                [self buttonForThisSampleWithType:UIButtonTypeContactAdd],
                nil ];
}
- (void)viewDidUnload {
    [dataSource_ release];
    [super viewDidUnload];
}
- (NSInteger)tableView:(UITableView*)tableView
 numberOfRowsInSection:(NSInteger)section
{
    return [dataSource_ count];
}
- (UITableViewCell*)tableView:(UITableView*)tableView
        cellForRowAtIndexPath:(NSIndexPath*)indexPath
{
    static NSString* CellIdentifier = @"CellStyleDefault";
    UITableViewCell* cell = [tableView dequeueReusableCellWithIdentifier:CellIdentifier];
    if ( nil == cell ) {
        cell = [[[UITableViewCell alloc] initWithStyle:UITableViewCellStyleDefault reuseIdentifier:CellIdentifier] autorelease];
    }
    cell.textLabel.text = [dataSource_ objectAtIndex:indexPath.row];
    UIButton* button = [buttons_ objectAtIndex:indexPath.row];
    button.frame = CGRectMake( cell.contentView.bounds.size.width - button.bounds.size.width - 20,
                              ( kRowHeight - button.bounds.size.height ) / 2,
                              button.bounds.size.width,
                              button.bounds.size.height );
    [cell.contentView addSubview:button];
    return cell;
```

```
}
#pragma mark ----- Private Methods -----
- (UIButton*)buttonForThisSampleWithType:(UIButtonType)type {
  UIButton* button = [UIButton buttonWithType:type];
  [button setTitle:@"UIButton" forState:UIControlStateNormal];
  [button setTitleColor:[UIColor blackColor] forState:UIControlStateNormal];
  [button sizeToFit];
  return button;
}
@end
```

执行后的效果如图 3-19 所示。

图 3-19 执行效果

实例 053 点击按钮后改变按钮的文字

实例 053	点击按钮后改变按钮的文字
源码路径	\daima\051\

实例说明

在 iOS 应用中，使用 Button 控件可以实现一个按钮效果。通过设置 Button 控件的 setTitle:forState: 方法可以设置按钮的状态变化时标题字符串的变化形式。例如，setTitleColor:forState:方法可以设置标题颜色的变化形式，setTitleShadowColor:forState:方法可以设置标题阴影的变化形式。

在本实例中，首先在屏幕中设置了一个"点我"按钮。当点击此按钮时，会使用方法 setTitleColor:forState: 和 setTitleShadowColor:forState:设置标题的颜色和阴影变化效果，并改变了按钮上的显示文本为"变样子了"。

具体实现

实例文件 UIKitPrjChangeTextByState.m 的具体实现代码如下所示。

```
#import "UIKitPrjChangeTextByState.h"
@implementation UIKitPrjChangeTextByState
- (void)dealloc {
  [button_ release];
  [super dealloc];
}
- (void)viewDidLoad {
  [super viewDidLoad];
  button_ = [[UIButton buttonWithType:UIButtonTypeRoundedRect] retain];
  button_.frame = CGRectMake( 0, 0, 200, 60 );
  button_.center = self.view.center;
  button_.autoresizingMask = UIViewAutoresizingFlexibleLeftMargin |
                             UIViewAutoresizingFlexibleRightMargin |
                             UIViewAutoresizingFlexibleTopMargin |
                             UIViewAutoresizingFlexibleBottomMargin;
  [self.view addSubview:button_];
  //设置按钮标题字体及阴影颜色
```

```
button_.titleLabel.font = [UIFont boldSystemFontOfSize:24];
button_.titleLabel.shadowOffset = CGSizeMake( 1.0, 1.0 );
//设置通常状态下的显示特征
[button_ setTitle:@"点我" forState:UIControlStateNormal];
[button_ setTitleColor:[UIColor blackColor] forState:UIControlStateNormal];
[button_ setTitleShadowColor:[UIColor grayColor] forState:UIControlStateNormal];
//设置按钮被触摸时即高亮状态下的显示特征
[button_ setTitle:@"变样子了" forState:UIControlStateHighlighted];
[button_ setTitleColor:[UIColor blueColor] forState:UIControlStateHighlighted];
[button_ setTitleShadowColor:[UIColor whiteColor] forState:UIControlStateHighlighted];
//设置按钮非活性（无效）状态下的显示特征
[button_ setTitle:@"StateDisable" forState:UIControlStateDisabled];
[button_ setTitleColor:[UIColor grayColor] forState:UIControlStateDisabled];
[button_ setTitleShadowColor:[UIColor blackColor] forState:UIControlStateDisabled];
UIBarButtonItem* disableButton =
    [[[UIBarButtonItem alloc] initWithTitle:@"DISABLE"
                            style:UIBarButtonItemStyleBordered
                            target:self
                            action:@selector(disableDidPush)] autorelease];
NSArray* barButtons = [NSArray arrayWithObjects:disableButton, nil];
[self setToolbarItems:barButtons animated:YES];
}
- (void)disableDidPush {
    button_.enabled = !button_.enabled;
}
@end
```

执行后的效果如图 3-20 所示。点击"点我"按钮后会改变按钮上的文本，单击"DISABLE"按钮后会改变按钮文本的阴影效果。

图 3-20　执行效果

实例 054　点击按钮后实现阴影反转

实例 054	点击按钮后实现阴影反转
源码路径	\daima\051\

实例说明

在 iOS 应用中，使用 Button 控件的 reversesTitleShadowWhenHighlighted 属性，可以确定按钮高亮时是否改变阴影的 Bool 值，其默认值是 NO。当设置为 YES 时，阴影在雕刻与浮雕感之间变化。在本实例中，点击屏幕中按钮后会实现阴影反转效果。

具体实现

实例文件 UIKitPrjReversesTitleShadow.m 的具体代码如下所示。

```
#import "UIKitPrjReversesTitleShadow.h"
#pragma mark ----- Private Methods Definition -----
```

```objectivec
@interface UIKitPrjReversesTitleShadow ()
- (UIButton*)buttonForThisSample;
@end
#pragma mark ----- Start Implementation For Methods -----
@implementation UIKitPrjReversesTitleShadow
- (void)viewDidLoad {
  [super viewDidLoad];
  UIButton* button1 = [self buttonForThisSample];
  button1.frame = CGRectMake( 0, 0, 200, 60 );
  button1.center = self.view.center;
  button1.reversesTitleShadowWhenHighlighted = YES;
  [self.view addSubview:button1];
  UIButton* button2 = [self buttonForThisSample];
  button2.frame = button1.frame;
  CGPoint newPoint = button1.center;
  newPoint.y += 100;
  button2.center = newPoint;
  [self.view addSubview:button2];
}
#pragma mark ----- Private Methods -----
- (UIButton*)buttonForThisSample {
  UIButton* button = [UIButton buttonWithType:UIButtonTypeRoundedRect];
  button.autoresizingMask = UIViewAutoresizingFlexibleLeftMargin |
                            UIViewAutoresizingFlexibleRightMargin |
                            UIViewAutoresizingFlexibleTopMargin |
                            UIViewAutoresizingFlexibleBottomMargin;
  button.titleLabel.font = [UIFont boldSystemFontOfSize:24];
  button.titleLabel.shadowOffset = CGSizeMake( 3.0, 3.0 );
  [button setTitle:@"按钮" forState:UIControlStateNormal];
  [button setTitleColor:[UIColor blackColor] forState:UIControlStateNormal];
  [button setTitleShadowColor:[UIColor grayColor] forState:UIControlStateNormal];
  return button;
}
@end
```

执行后的效果如图 3-21 所示。

图 3-21　执行效果

实例 055　点击按钮时实现闪烁效果

实例 055	点击按钮时实现闪烁效果
源码路径	\daima\051\

实例说明

在 iOS 应用中，使用 Button 控件的 showsTouchWhenHighlighted 属性可以控制当按钮按下时是否闪光。此属性是一个 BOOL 值，默认为 NO，当为 YES 时表示按下时会有白色光点，其中的图片和按钮事件的不会因闪光改变。在本实例中，使用"showsTouchWhenHighlighted=YES"设置屏幕中的按钮被按下时呈现闪烁效果。

具体实现

实例文件 UIKitPrjShowsTouch.m 的具体实现代码如下所示。

```objc
#import "UIKitPrjShowsTouch.h"
#pragma mark ----- Private Methods Definition -----
@interface UIKitPrjShowsTouch ()
- (UIButton*)buttonForThisSample;
@end
#pragma mark ----- Start Implementation For Methods -----
@implementation UIKitPrjShowsTouch
- (void)viewDidLoad {
  [super viewDidLoad];
  UIButton* button1 = [self buttonForThisSample];
  button1.frame = CGRectMake( 0, 0, 200, 60 );
  button1.center = self.view.center;
  button1.showsTouchWhenHighlighted = YES;
  [self.view addSubview:button1];
  UIButton* button2 = [self buttonForThisSample];
  button2.frame = button1.frame;
  CGPoint newPoint = button1.center;
  newPoint.y += 100;
  button2.center = newPoint;
  [self.view addSubview:button2];
}
#pragma mark ----- Private Methods -----
- (UIButton*)buttonForThisSample {
  UIButton* button = [UIButton buttonWithType:UIButtonTypeRoundedRect];
  button.autoresizingMask = UIViewAutoresizingFlexibleLeftMargin |
                   UIViewAutoresizingFlexibleRightMargin |
                   UIViewAutoresizingFlexibleTopMargin |
                   UIViewAutoresizingFlexibleBottomMargin;
  button.titleLabel.font = [UIFont boldSystemFontOfSize:24];
  [button setTitle:@"UIButton" forState:UIControlStateNormal];
  return button;
}
@end
```

执行后的效果如图 3-22 所示。

图 3-22 执行效果

实例 056 在按钮中添加图像

实例 056	在按钮中添加图像
源码路径	\daima\051\

实例说明

在 iOS 应用中,在 Button 控件中可以添加指定的图像,这个功能是通过 setImage:forState:方法实现的。使用此方法的格式如下所示:

setImage：(UIImage 的*)图像 forState:（UIControlState）状态

在上述格式中，参数"图像"指图像使用指定的状态，"状态"指状态使用指定的标题。在 UIControlState 值的描述。如果没有指定一个属性，则默认使用 UIControlStateNormal 的值。如果 UIControlStateNormal 值未设置，则属性默认为一个系统的价值，因此至少应该设置为正常状态的值。

另外，也可以使用 setBackgroundImage:forState:方法为按钮设置背景图片，此方法的用法和 setImage:forState:的类似。

在本实例中，使用 Button 控件设置了 5 个按钮，并且使用 setImage:forState:方法在按钮中添加了指定的图片，使用 setBackgroundImage:forState:方法为按钮设置了指定的背景图片。

具体实现

实例文件 UIKitPrjButtonWithImage.m 的具体实现代码如下所示。

```objc
#import "UIKitPrjButtonWithImage.h"
#pragma mark ----- Private Methods Definition -----
@interface UIKitPrjButtonWithImage()
- (UIButton*)buttonForThisSample;
@end
#pragma mark ----- Start Implementation For Methods -----
@implementation UIKitPrjButtonWithImage
- (void)viewDidLoad {
  [super viewDidLoad];
  UIButton* button1 = [self buttonForThisSample];
  button1.frame = CGRectMake( 0, 0, 150, 40 );
  button1.center = self.view.center;
  CGPoint newPoint = button1.center;
  newPoint.y -= 120;
  button1.center = newPoint;
  UIImage* image1 = [UIImage imageNamed:@"Dog.png"];
  [button1 setImage:image1 forState:UIControlStateNormal];
  UIImage* image2 = [UIImage imageNamed:@"DogHighlight.png"];
  [button1 setImage:image2 forState:UIControlStateHighlighted];
  UIImage* image3 = [UIImage imageNamed:@"DogDisable.png"];
  [button1 setImage:image3 forState:UIControlStateDisabled];
  [self.view addSubview:button1];

  UIButton* button2 = [self buttonForThisSample];
  button2.frame = button1.frame;
  newPoint = button1.center;
  newPoint.y += 60;
  button2.center = newPoint;
  [button2 setImage:image1 forState:UIControlStateNormal];
  button2.adjustsImageWhenHighlighted = NO;
  [self.view addSubview:button2];

  UIButton* button3 = [self buttonForThisSample];
  button3.frame = button2.frame;
  newPoint = button2.center;
  newPoint.y += 60;
  button3.center = newPoint;
  [button3 setImage:image1 forState:UIControlStateNormal];
  [self.view addSubview:button3];

  UIButton* button4 = [self buttonForThisSample];
  button4.frame = button3.frame;
  newPoint = button3.center;
  newPoint.y += 60;
  button4.center = newPoint;
  [button4 setImage:image1 forState:UIControlStateNormal];
  button4.enabled = NO;
  button4.adjustsImageWhenDisabled = YES;
  [self.view addSubview:button4];

  UIButton* button5 = [self buttonForThisSample];
  button5.frame = CGRectMake( 0, 0, 180, 60 );
  newPoint = button4.center;
  newPoint.y += 80;
```

```
    button5.center = newPoint;
//  [button5 setImage:image1 forState:UIControlStateNormal];
    UIImage* backImage = [UIImage imageNamed:@"frame.png"];
    UIImage* stretchableImage = [backImage stretchableImageWithLeftCapWidth:20 topCapHeight:20];
    [button5 setBackgroundImage:stretchableImage forState:UIControlStateNormal];
    [self.view addSubview:button5];
}
#pragma mark ----- Private Methods -----
- (UIButton*)buttonForThisSample {
    UIButton* button = [UIButton buttonWithType:UIButtonTypeRoundedRect];
    button.autoresizingMask = UIViewAutoresizingFlexibleLeftMargin |
                              UIViewAutoresizingFlexibleRightMargin |
                              UIViewAutoresizingFlexibleTopMargin |
                              UIViewAutoresizingFlexibleBottomMargin;
    button.titleLabel.font = [UIFont boldSystemFontOfSize:24];
    [button setTitle:@"按钮" forState:UIControlStateNormal];
    return button;
}
@end
```

执行后的效果如图 3-23 所示。

图 3-23 执行效果

实例 057 调整屏幕中按钮的边间距

实例 057	调整屏幕中按钮的边间距
源码路径	\daima\051\

实例说明

在 iOS 应用中，可以设置 Button 控件按钮的边间距，这个功能是通过 contentEdgeInsets 属性实现的。此属性设置了按钮内容的内外线边缘绘制区域，其中按钮内容包含按钮图片和标题。

在本实例中，使用 Button 控件设置了 4 个按钮，并通过 contentEdgeInsets 属性设置了按钮的边间距。

具体实现

实例文件 UIKitPrjEdgeInsets.m 的具体实现代码如下所示。

```
#import "UIKitPrjEdgeInsets.h"
#pragma mark ----- Private Methods Definition -----
@interface UIKitPrjEdgeInsets ()
- (UIButton*)buttonForThisSample;
@end
#pragma mark ----- Start Implementation For Methods -----
@implementation UIKitPrjEdgeInsets
- (void)viewDidLoad {
    [super viewDidLoad];
```

```objc
    UIButton* button1 = [self buttonForThisSample];
    button1.frame = CGRectMake( 0, 0, 150, 40 );
    button1.center = self.view.center;
    CGPoint newPoint = button1.center;
    newPoint.y -= 180;
    button1.center = newPoint;
    [self.view addSubview:button1];

    UIButton* button2 = [self buttonForThisSample];
    button2.frame = button1.frame;
    newPoint = button1.center;
    newPoint.y += 70;
    button2.center = newPoint;
    UIEdgeInsets insets;
    insets.top = insets.bottom = insets.left = insets.right = 10;
    button2.contentEdgeInsets = insets;
    [self.view addSubview:button2];

    UIButton* button3 = [self buttonForThisSample];
    button3.frame = button1.frame;
    newPoint = button2.center;
    newPoint.y += 70;
    button3.center = newPoint;
    insets.top = insets.bottom = insets.left = insets.right = 10;
    button3.titleEdgeInsets = insets;
    [self.view addSubview:button3];

    UIButton* button4 = [self buttonForThisSample];
    button4.frame = button1.frame;
    newPoint = button3.center;
    newPoint.y += 70;
    button4.center = newPoint;
    insets.top = insets.bottom = insets.left = insets.right = 10;
    button4.imageEdgeInsets = insets;
    [self.view addSubview:button4];
}

#pragma mark ----- Private Methods -----

- (UIButton*)buttonForThisSample {
    UIButton* button = [UIButton buttonWithType:UIButtonTypeRoundedRect];
    button.autoresizingMask = UIViewAutoresizingFlexibleLeftMargin |
                    UIViewAutoresizingFlexibleRightMargin |
                    UIViewAutoresizingFlexibleTopMargin |
                    UIViewAutoresizingFlexibleBottomMargin;
    button.titleLabel.font = [UIFont boldSystemFontOfSize:24];
    [button setTitle:@"U按钮" forState:UIControlStateNormal];
    UIImage* image = [UIImage imageNamed:@"Dog.png"];
    [button setImage:image forState:UIControlStateNormal];
    return button;
}
@end
```

执行后的效果如图 3-24 所示。

图 3-24　执行效果

实例 058 设置按钮中文本的换行和省略格式

实例 058	设置按钮中文本的换行和省略格式
源码路径	\daima\051\

实例说明

在 iOS 应用中,可以设置 UILineBreakMode 常量实现换行或省略模式。UIButton 的 titleLabel 属性也是 UILabel 的实例,可以设置其 LineBreakMode 模式实现换行显示或省略模式显示样式。

在本实例中,使用 Button 控件设置了 2 个按钮,其中在第一个按钮中实现了换行模式,在第二个按钮中实现了省略模式。

具体实现

实例文件 UIKitPrjLineBreakMode.m 的具体实现代码如下所示。

```
#import "UIKitPrjLineBreakMode.h"
@implementation UIKitPrjLineBreakMode
- (void)viewDidLoad {
  [super viewDidLoad];
  UIButton* button = [UIButton buttonWithType:UIButtonTypeRoundedRect];
  button.frame = CGRectMake( 80, 50, 160, 50 );
  [button setTitle:@"故意放了很长的字符串,超出了按钮区域。" forState:UIControlStateNormal];
  button.titleLabel.lineBreakMode = UILineBreakModeWordWrap;
  [self.view addSubview:button];
  UIButton* button1 = [UIButton buttonWithType:UIButtonTypeRoundedRect];
  button1.frame = CGRectMake( 80, 120, 150, 40 );
  [button1 setTitle:@"特地放入很长的字符串、超出按钮标题区域。" forState:UIControlStateNormal];
  //button1.titleLabel.lineBreakMode = UILineBreakMode;
  [self.view addSubview:button1];
}
@end
```

执行后的效果如图 3-25 所示。

图 3-25 执行效果

实例 059 在屏幕中显示一个文本输入框

实例 059	在屏幕中显示一个文本输入框
源码路径	\daima\059\

实例说明

在 iOS 应用中,可以使用控件 UITextField 在屏幕中显示一个文本输入框。UITextField 通常用于外部数据输入,以实现人机交互。在本实例中,使用 UITextField 控件设置了 1 个文本输入框,并设置在框中显示提示文本"请输入信息"。

具体实现

实例文件 UIKitPrjPlaceholder.m 的具体实现代码如下所示。

```objc
#import "UIKitPrjPlaceholder.h"
@implementation UIKitPrjPlaceholder
- (void)viewDidLoad {
 [super viewDidLoad];
 self.view.backgroundColor = [UIColor whiteColor];
 UITextField* textField = [[[UITextField alloc] init] autorelease];
 textField.frame = CGRectMake( 20, 100, 280, 30 );
 textField.borderStyle = UITextBorderStyleRoundedRect;
 textField.contentVerticalAlignment = UIControlContentVertical AlignmentCenter;
 textField.placeholder = @"请输入信息";
 [self.view addSubview:textField];
}
@end
```

执行后的效果如图 3-26 所示。

图 3-26 执行效果

实例 060 设置文本输入框的边框线样式

实例 060	设置文本输入框的边框线样式
源码路径	\daima\059\

实例说明

在 iOS 应用中，当使用控件 UITextField 在屏幕中设置一个文本输入框后，可以继续使用其 borderStyle 属性设置输入框的边框线样式。在本实例中，首先使用 UITextField 控件设置了 4 个文本输入框，然后使用 borderStyle 属性为这 4 个输入框设置了不同的边框线样式。

具体实现

实例文件 UIKitPrjSimple.m 的具体实现代码如下所示。

```objc
#import "UIKitPrjSimple.h"
@implementation UIKitPrjSimple
- (void)dealloc {
 [textFields_ release];
 [super dealloc];
}
- (void)viewDidLoad {
 [super viewDidLoad];
 self.view.backgroundColor = [UIColor whiteColor];

 UITextField* textField1 = [[[UITextField alloc] init] autorelease];
 textField1.delegate = self;
 textField1.frame = CGRectMake( 20, 20, 280, 30 );
 textField1.borderStyle = UITextBorderStyleLine;
 textField1.text = @"aaaaaaaaa";
```

```
    textField1.returnKeyType = UIReturnKeyNext;
    [self.view addSubview:textField1];

    UITextField* textField2 = [[[UITextField alloc] init] autorelease];
    textField2.delegate = self;
    textField2.frame = CGRectMake( 20, 60, 280, 30 );
    textField2.borderStyle = UITextBorderStyleBezel;
    textField2.text = @"bbbbbbbbb";
    textField2.returnKeyType = UIReturnKeyNext;
    [self.view addSubview:textField2];

    UITextField* textField3 = [[[UITextField alloc] init] autorelease];
    textField3.delegate = self;
    textField3.frame = CGRectMake( 20, 100, 280, 30 );
    textField3.borderStyle = UITextBorderStyleRoundedRect;
    textField3.text = @"cccccccccc";
    textField3.returnKeyType = UIReturnKeyNext;
    [self.view addSubview:textField3];

    UITextField* textField4 = [[[UITextField alloc] init] autorelease];
    textField4.delegate = self;
    textField4.frame = CGRectMake( 20, 140, 280, 30 );
    textField4.borderStyle = UITextBorderStyleNone;
    textField4.text = @"dddddddddd";
    textField4.returnKeyType = UIReturnKeyNext;
    [self.view addSubview:textField4];

    textFields_ = [[NSArray alloc] initWithObjects:textField1, textField2, textField3, textField4, nil];
}
- (void)textFieldDidBeginEditing:(UITextField*)textField {
    currentFieldIndex_ = [textFields_ indexOfObject:textField];
}
- (BOOL)textFieldShouldReturn:(UITextField*)textField {
    if ( textFields_.count <= ++currentFieldIndex_ ) {
        currentFieldIndex_ = 0;
    }
    UITextField* newField = [textFields_ objectAtIndex:currentFieldIndex_];
    if ( [newField canBecomeFirstResponder] ) {
        [newField becomeFirstResponder];
    }
    return YES;
}
@end
```

执行后的效果如图 3-27 所示。

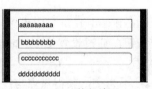

图 3-27　执行效果

实例 061　设置文本输入框的字体和颜色

实例 061	设置文本输入框的字体和颜色
源码路径	\daima\059\

实例说明

在 iOS 应用中，当使用控件 UITextField 在屏幕中设置一个文本输入框后，可以继续使用其 backgroundColor 属性设置输入框的背景颜色，使用其 font 属性设置字体。在本实例中，首先使用

UITextField 控件设置了 1 个文本输入框，然后设置了输入框中默认显示的文本为"看我的字体和颜色"，并设置了文本的字体和整个输入框的背景颜色。

具体实现

实例文件 UIKitPrjChangeColorAndFont.m 的具体实现代码如下所示。

```
#import "UIKitPrjChangeColorAndFont.h"
@implementation UIKitPrjChangeColorAndFont
- (void)viewDidLoad {
    [super viewDidLoad];
    self.view.backgroundColor = [UIColor whiteColor];
    UITextField* textField = [[[UITextField alloc] init] autorelease];
    textField.frame = CGRectMake( 20, 100, 280, 50 );
    textField.borderStyle = UITextBorderStyleBezel;
    textField.backgroundColor = [UIColor blackColor];//设置背景色
    textField.textColor = [UIColor redColor];//设置文本颜色
    textField.textAlignment = UITextAlignmentCenter;
    textField.font = [UIFont systemFontOfSize:36];//设置字体大小
    textField.text = @"看我的字体和颜色";
    [self.view addSubview:textField];
}
@end
```

执行后的效果如图 3-28 所示。

图 3-28 执行效果

实例 062 在文本输入框中设置一个清空按钮

实例 062	在文本输入框中设置一个清空按钮
源码路径	\daima\059\

实例说明

在 iOS 应用中，当使用控件 UITextField 在屏幕中设置一个文本输入框后，可以继续使用其 ClearButtonMode 属性设置一个清空按钮，通过设置 ClearButtonMode 可以指定是否以及何时显示清除按钮。此属性主要有如下几种类型。

- UITextFieldViewModeAlways：不为空，获得焦点与没有获得焦点都显示清空按钮。
- UITextFieldViewModeNever：不显示清空按钮。
- UITextFieldViewModeWhileEditing：不为空，且在编辑状态时（及获得焦点）显示清空按钮。
- UITextFieldViewModeUnlessEditing：不为空，且不在编译状态时（焦点不在输入框上）显示清空按钮。

在本实例中，首先使用 UITextField 控件设置了 4 个文本输入框，然后分别为这 4 个输入框设置了不同类型的清空按钮。

具体实现

实例文件 UIKitPrjClearButtonMode.m 的具体实现代码如下所示。

```
#import "UIKitPrjClearButtonMode.h"
@implementation UIKitPrjClearButtonMode
```

```objc
- (void)dealloc {
  [textFields_ release];
  [super dealloc];
}
- (void)viewDidLoad {
  [super viewDidLoad];
  self.view.backgroundColor = [UIColor whiteColor];

  UITextField* textField1 = [[[UITextField alloc] init] autorelease];
  textField1.delegate = self;
  textField1.clearsOnBeginEditing = YES;
  textField1.frame = CGRectMake( 20, 20, 280, 30 );
  textField1.borderStyle = UITextBorderStyleRoundedRect;
  textField1.clearButtonMode = UITextFieldViewModeNever;
  textField1.text = @"UITextFieldViewModeNever";
  [self.view addSubview:textField1];

  UITextField* textField2 = [[[UITextField alloc] init] autorelease];
  textField2.delegate = self;
  textField2.frame = CGRectMake( 20, 60, 280, 30 );
  textField2.borderStyle = UITextBorderStyleRoundedRect;
  textField2.clearButtonMode = UITextFieldViewModeWhileEditing;
  textField2.text = @"UITextFieldViewModeWhileEditing";
  [self.view addSubview:textField2];

  UITextField* textField3 = [[[UITextField alloc] init] autorelease];
  textField3.delegate = self;
  textField3.frame = CGRectMake( 20, 100, 280, 30 );
  textField3.borderStyle = UITextBorderStyleRoundedRect;
  textField3.clearButtonMode = UITextFieldViewModeUnlessEditing;
  textField3.text = @"UITextFieldViewModeUnlessEditing";
  [self.view addSubview:textField3];

  UITextField* textField4 = [[[UITextField alloc] init] autorelease];
  textField4.delegate = self;
  textField4.frame = CGRectMake( 20, 140, 280, 30 );
  textField4.borderStyle = UITextBorderStyleRoundedRect;
  textField4.clearButtonMode = UITextFieldViewModeAlways;
  textField4.text = @"UITextFieldViewModeAlways";
  [self.view addSubview:textField4];

  textFields_ = [[NSArray alloc] initWithObjects:textField1, textField2, textField3, textField4, nil];
}

- (BOOL)textFieldShouldClear:(UITextField*)textField {
  NSLog( @"textFieldShouldClear:%@", textField.text );
  return YES;
}
@end
```

执行后的效果如图 3-29 所示。

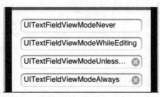

图 3-29 执行效果

实例 063 为文本输入框设置背景图片

实例 063	为文本输入框设置背景图片
源码路径	\daima\059\

第 3 章 iOS 控件应用实战

实例说明

在 iOS 应用中，当使用控件 UITextField 在屏幕中设置一个文本输入框后，可以继续使用其 Background 属性设置一个背景图片。在本实例中，首先用 UITextField 在屏幕中设置一个文本输入框，然后使用 "textField.background = stretchableWhitePaper" 语句为输入框设置了背景图片。

具体实现

实例文件 UIKitPrjBackground.m 的具体实现代码如下所示。

```
#import "UIKitPrjBackground.h"
@implementation UIKitPrjBackground
- (void)viewDidLoad {
  [super viewDidLoad];
  self.view.backgroundColor = [UIColor whiteColor];
  //导入背景图片，并设置成自动伸缩
  UIImage* imageWhitePaper = [UIImage imageNamed:@"paper.png"];
  UIImage* stretchableWhitePaper = [imageWhitePaper stretchableImageWithLeftCapWidth:20 topCapHeight:20];
  UIImage* imageGrayPaper = [UIImage imageNamed:@"paperGray.png"];
  UIImage* stretchableGrayPaper = [imageGrayPaper stretchableImageWithLeftCapWidth:20 topCapHeight:20];
  //创建 UITextField 实例
  UITextField* textField = [[[UITextField alloc] init] autorelease];
  textField.delegate = self;
  textField.frame = CGRectMake( 20, 100, 280, 50 );
  textField.background = stretchableWhitePaper;//设置背景图片
  textField.disabledBackground = stretchableGrayPaper;
  textField.text = @"有图片";
  textField.textAlignment = UITextAlignmentCenter;
  textField.contentVerticalAlignment = UIControlContentHorizontalAlignmentCenter;
  [self.view addSubview:textField];
}
- (BOOL)textFieldShouldReturn:(UITextField*)textField {
  textField.enabled = NO;
  return YES;
}
@end
```

执行后的效果如图 3-30 所示。

图 3-30 执行效果

实例 064 在文本输入框中添加 UIView 元素

实例 064	在文本输入框中添加 UIView 元素
源码路径	\daima\059\

实例说明

在 iOS 应用中，当使用控件 UITextField 在屏幕中设置一个文本输入框后，我们可以继续在输入框中添加其他 UIView 元素。在本实例中，首先用 UITextField 在屏幕中设置了 2 个文本输入框，然后使用 UIImageView 在第一个输入框的左侧和右侧分别添加了图片，在第二个输入框的右侧添加了一个显示详情按钮。

具体实现

实例文件 UIKitPrjAddView.m 的具体实现代码如下所示。

```
#import "UIKitPrjAddView.h"
```

```
@implementation UIKitPrjAddView
- (void)viewDidLoad {
 [super viewDidLoad];
 self.view.backgroundColor = [UIColor whiteColor];
 UIImage* imageForLeft = [UIImage imageNamed:@"leftDog.png"];
 UIImageView* imageViewForLeft = [[[UIImageView alloc] initWithImage:imageForLeft]
autorelease];
 UIImage* imageForRight = [UIImage imageNamed:@"rightDog.png"];
 UIImageView* imageViewForRight = [[[UIImageView alloc] initWithImage:imageForRight]
autorelease];
 UITextField* textField1 = [[[UITextField alloc] init] autorelease];
 textField1.borderStyle = UITextBorderStyleRoundedRect;
 textField1.frame = CGRectMake( 20, 30, 280, 50 );
 textField1.text = @"一直在左右显示图片";
 textField1.textAlignment = UITextAlignmentCenter;
 textField1.contentVerticalAlignment = UIControlContentHorizontalAlignmentCenter;
 textField1.leftView = imageViewForLeft;//输入框左侧追加UIImageView
 textField1.rightView = imageViewForRight;//输入框右侧追加UIImageView
 textField1.leftViewMode = UITextFieldViewModeAlways;//让左侧UIView一直显示
 textField1.rightViewMode = UITextFieldViewModeAlways;//让右侧UIView一直显示
 [self.view addSubview:textField1];
 UITextField* textField2 = [[[UITextField alloc] init] autorelease];;
 textField2.borderStyle = UITextBorderStyleRoundedRect;
 textField2.frame = CGRectMake( 20, 100, 280, 50 );
 textField2.text = @"非编辑状态时右侧显示详细按钮";
 textField2.contentVerticalAlignment = UIControlContentHorizontalAlignmentCenter;
 UIButton* button = [UIButton buttonWithType:UIButtonTypeDetailDisclosure];
 textField2.rightView = button;//输入框的右侧追加详细按钮
 textField2.rightViewMode = UITextFieldViewModeUnlessEditing;//只在非编辑模式下才显示
 [self.view addSubview:textField2];
}
@end
```

执行后的效果如图 3-31 所示。

图 3-31 执行效果

实例 065 监视文本输入框的状态

实例 065	监视文本输入框的状态
源码路径	\daima\059\

实例说明

在 iOS 应用中，可以向控件 UITextField 的 delegate 属性中设置一个继承了的 UITextField Delegate 协议的委托类，来监视 UITextField 文本输入框的各种状态，我们可以继续在输入框中添加其他 UIView 元素。本实例演示了使用 UITextField Delegate 协议的委托类监视文本输入框状态的方法。

具体实现

实例文件 UIKitPrjObserving.m 的具体实现代码如下所示。

```
#import "UIKitPrjObserving.h"
@implementation UIKitPrjObserving
- (void)viewDidLoad {
 [super viewDidLoad];
 self.view.backgroundColor = [UIColor whiteColor];
 UITextField* textField1 = [[[UITextField alloc] init] autorelease];
```

```
textField1.delegate = self;
textField1.frame = CGRectMake( 20, 50, 280, 40 );
textField1.borderStyle = UITextBorderStyleRoundedRect;
textField1.text = @"通过NSLog确认delegate的设置";
textField1.clearButtonMode = UITextFieldViewModeAlways;
textField1.contentVerticalAlignment = UIControlContentHorizontalAlignmentCenter;
[self.view addSubview:textField1];
}

- (BOOL)textFieldShouldBeginEditing:(UITextField*)textField {
    NSLog( @"textFieldShouldBeginEditing %@", textField.text );
    return TRUE;
}
- (void)textFieldDidBeginEditing:(UITextField*)textField {
    NSLog( @"textFieldDidBeginEditing %@", textField.text );
}
- (BOOL)textFieldShouldEndEditing:(UITextField*)textField {
    NSLog( @"textFieldShouldEndEditing %@", textField.text );
    return YES;
}
- (void)textFieldDidEndEditing:(UITextField*)textField {
    NSLog( @"textFieldDidEndEditing %@", textField.text );
}
- (BOOL)textField:(UITextField*)textField
shouldChangeCharactersInRange:(NSRange)range replacementString:(NSString*)string
{
    NSLog( @"shouldChangeCharactersInRange %@", string );
    return YES;
}
- (BOOL)textFieldShouldClear:(UITextField*)textField {
    NSLog( @"textFieldShouldClear %@", textField.text );
    return YES;
}
- (BOOL)textFieldShouldReturn:(UITextField*)textField {
    NSLog( @"textFieldShouldReturn %@", textField.text );
    return YES;
}
@end
```

执行后的效果如图 3-32 所示。

图 3-32　执行效果

实例 066　实现一个开关效果

实例 066	实现一个开关效果
源码路径	\daima\066\

实例说明

在大多数传统桌面应用程序中，通过复选框和单选按钮来实现开关功能。在 iOS 中，Apple 放弃了这些界面元素，取而代之的是开关和分段控件。在 iOS 应用中，使用开关控件 UISwitch 来实现"开/关" UI 元素，它类似于传统的物理开关，如图 3-33 所示。开关的可配置选项很少，应将其用于处理布尔值。

图 3-33　开关控件向用户提供了开和关两个选项

为了利用开关，我们将使用其 Value Changed 事件来检测开关切换，并通过属性 on 或实例方法 isOn 来获取当前值。检查开关时将返回一个布尔值，这意味着可将其与 TRUE 或 FALSE（YES/NO）进行比较以确定其状态，还可直接在条件语句中判断结果。例如，要检查开关 mySwitch 是否是开的，可使用类似于下面的代码。

```
if([mySwitch isOn]){<switch is on>}else{<switch is off>}
```
在本实例中,通过 Switch 控件实现了一个开关效果。

具体实现

实例文件 RootViewController.m 的具体实现代码如下所示。

```objc
#import "RootViewController.h"
@implementation RootViewController
- (void)dealloc {
  [items_ release];
  [super dealloc];
}
#pragma mark UIViewController methods
- (void)viewDidLoad {
  [super viewDidLoad];

  self.title = @"MENU";
  if ( !items_ ) {
    items_ = [[NSArray alloc] initWithObjects:
                            @"UIKitPrjSwitch",
                            nil ];
  }
}
- (void)viewWillAppear:(BOOL)animated {
  [super viewWillAppear:animated];

  [self.navigationController setNavigationBarHidden:NO animated:NO];
  [self.navigationController setToolbarHidden:NO animated:NO];
  // 恢复条的颜色
  [UIApplication sharedApplication].statusBarStyle = UIStatusBarStyleDefault;
  self.navigationController.navigationBar.barStyle = UIBarStyleDefault;
  self.navigationController.navigationBar.translucent = NO;
  self.navigationController.navigationBar.tintColor = nil;
  self.navigationController.toolbar.barStyle = UIBarStyleDefault;
  self.navigationController.toolbar.translucent = NO;
  self.navigationController.toolbar.tintColor = nil;
}

#pragma mark UITableView methods
- (NSInteger)tableView:(UITableView*)tableView
  numberOfRowsInSection:(NSInteger)section
{
  return [items_ count];
}
- (UITableViewCell*)tableView:(UITableView*)tableView
  cellForRowAtIndexPath:(NSIndexPath*)indexPath
{
  static NSString *CellIdentifier = @"Cell";
  UITableViewCell *cell = [tableView dequeueReusableCellWithIdentifier:CellIdentifier];
  if (cell == nil) {
    cell  =  [[[UITableViewCell  alloc]  initWithStyle:UITableViewCellStyleDefault
reuseIdentifier:CellIdentifier] autorelease];
  }
  NSString* title = [items_ objectAtIndex:indexPath.row];
  cell.textLabel.text  =   [title  stringByReplacingOccurrencesOfString:@"UIKitPrj"
withString:@""];

  return cell;
}
- (void)tableView:(UITableView*)tableView
  didSelectRowAtIndexPath:(NSIndexPath*)indexPath
{
  NSString* className = [items_ objectAtIndex:indexPath.row];
  Class class = NSClassFromString( className );
  UIViewController* viewController = [[[class alloc] init] autorelease];
  if ( !viewController ) {
    NSLog( @"%@ was not found.", className );
    return;
  }
```

```
    [self.navigationController pushViewController:viewController animated:YES];
}
@end
```

实例文件 UIKitPrjSwitch.m 的具体实现代码如下所示。

```
#import "UIKitPrjSwitch.h"
@implementation UIKitPrjSwitch
- (void)dealloc {
  [switch1_ release];
  [switch2_ release];
  [super dealloc];
}
- (void)viewDidLoad {
  [super viewDidLoad];
  self.view.backgroundColor = [UIColor whiteColor];
  //UISwitch 对象的创建与初始化
  switch1_ = [[UISwitch alloc] init];
  switch1_.center = CGPointMake( 100, 50 );
  //默认设置为 ON
  switch1_.on = YES;
  //开关变化时，调用 switchDidChange 方法
  [switch1_ addTarget:self
               action:@selector(switchDidChange)
     forControlEvents:UIControlEventValueChanged];
  [self.view addSubview:switch1_];
  //创建第二个 UISwitch 对象
  switch2_ = [[UISwitch alloc] init];
  switch2_.center = CGPointMake( 100, 100 );
  //默认设置为 OFF
  switch2_.on = NO;
  [self.view addSubview:switch2_];
}
- (void)switchDidChange {
  //switch1_ 变化后 switch2_ 也一起改变
  [switch2_ setOn:!switch2_.on animated:YES];
}
@end
```

执行后的效果如图 3-34 所示。单击图中的 Switch 后，在新界面会显示两个开关，如图 3-35 所示。

图 3-34　执行效果

图 3-35　开关界面

实例 067　改变 UISWitch 文本和颜色

实例 067	改变 UISWitch 文本和颜色
源码路径	\daima\067\

实例说明

我们知道，iOS 中的 Switch 控件默认的文本为 ON 和 OFF 两种，不同的语言显示不同，颜色均为蓝色和亮灰色。如果想改变上面的 ON 和 OFF 文本，我们必须重新从 UISwitch 继承一个新类，然后在新的 Switch 类中修改替换原有的 Views。在本实例中，我们根据上述原理改变了 UISwitch 的文本和颜色。

具体实现

本实例具体的实现代码如下所示。

实例 067　改变 UISWitch 文本和颜色

```
#import <UIKit/UIKit.h>
//该方法是 SDK 文档中没有的，添加一个 category
@interface UISwitch (extended)
- (void) setAlternateColors:(BOOL) boolean;
@end
//自定义 Slider 类
@interface _UISwitchSlider : UIView
@end
 @interface UICustomSwitch : UISwitch {
}
- (void) setLeftLabelText:(NSString *)labelText
                     font:(UIFont*)labelFont
                    color: (UIColor *)labelColor;
- (void) setRightLabelText:(NSString *)labelText
                      font:(UIFont*)labelFont
                     color:(UIColor *)labelColor;
- (UILabel*) createLabelWithText:(NSString*)labelText
                            font:(UIFont*)labelFont
                           color:(UIColor*)labelColor;
@end
```

这样在上述代码中添加了一个名为 "extended" 的 category，主要作用是声明一下 UISwitch 的 setAlternateColors 消息，否则在使用的时候会出现找不到该消息的警告。其实，setAlternateColors 已经在 UISwitch 中实现，只是没有在头文件中公开而已，所以在此只是做一个声明。当调用 setAlternateColors:YES 时，UISwitch 的状态为 "on" 时会显示为橙色，否则为亮蓝色。对应的文件 UICustomSwitch.m 的实现代码如下所示。

```
#import "UICustomSwitch.h"
 @implementation UICustomSwitch
- (id)initWithFrame:(CGRect)frame {
    if (self = [super initWithFrame:frame]) {
        // Initialization code
    }
    return self;
}
- (void)drawRect:(CGRect)rect {
    // Drawing code
}
- (void)dealloc {
    [super dealloc];
}
- (_UISwitchSlider *) slider {
    return [[self subviews] lastObject];
}
- (UIView *) textHolder {
    return [[[self slider] subviews] objectAtIndex:2];
}
- (UILabel *) leftLabel {
    return [[[self textHolder] subviews] objectAtIndex:0];
}
- (UILabel *) rightLabel {
    return [[[self textHolder] subviews] objectAtIndex:1];
}

// 创建文本标签
- (UILabel*) createLabelWithText:(NSString*)labelText
                            font:(UIFont*)labelFont
                           color:(UIColor*)labelColor{
    CGRect rect = CGRectMake(-25.0f, -10.0f, 50.0f, 20.0f);
    UILabel *label = [[UILabel alloc] initWithFrame: rect];
    label.text = labelText;
    label.font = labelFont;
    label.textColor = labelColor;
    label.textAlignment = UITextAlignmentCenter;
    label.backgroundColor = [UIColor clearColor];
    return label;
}
// 重新设定左边的文本标签
- (void) setLeftLabelText:(NSString *)labelText
                     font:(UIFont*)labelFont
```

```
                                 color:(UIColor *)labelColor
{
    @try {
        //
        [[self leftLabel] setText:labelText];
        [[self leftLabel] setFont:labelFont];
        [[self leftLabel] setTextColor:labelColor];
    } @catch (NSException *ex) {
        //
        UIImageView* leftImage = (UIImageView*)[self leftLabel];
        leftImage.image = nil;
        leftImage.frame = CGRectMake(0.0f, 0.0f, 0.0f, 0.0f);
        [leftImage addSubview: [[self createLabelWithText:labelText
                                                    font:labelFont
                                                   color:labelColor] autorelease]];
    }
}

// 重新设定右边的文本
- (void) setRightLabelText:(NSString *)labelText font:(UIFont*)labelFont color:(UIColor *)labelColor {
    @try {
        //
        [[self rightLabel] setText:labelText];
        [[self rightLabel] setFont:labelFont];
        [[self rightLabel] setTextColor:labelColor];
    } @catch (NSException *ex) {
        //
        UIImageView* rightImage = (UIImageView*)[self rightLabel];
        rightImage.image = nil;
        rightImage.frame = CGRectMake(0.0f, 0.0f, 0.0f, 0.0f);
        [rightImage addSubview: [[self createLabelWithText:labelText
                                                     font:labelFont
                                                    color:labelColor] autorelease]];
    }
}
@end
```

由此可见，具体的实现过程就是替换原有的标签 view 以及 slider。使用方法非常简单，只需设置一下左右文本以及颜色即可，比如下面的代码：

```
switchCtl = [[UICustomSwitch alloc] initWithFrame:frame];
//    [switchCtl setAlternateColors:YES];
[switchCtl setLeftLabelText:@"Yes"
                       font:[UIFont boldSystemFontOfSize: 17.0f]
                      color:[UIColor whiteColor]];
[switchCtl setRightLabelText:@"No"
                        font:[UIFont boldSystemFontOfSize: 17.0f]
                       color:[UIColor grayColor]];
```

这样上面的代码将显示 Yes、No 两个选项，如图 3-36 所示。

图 3-36　显示效果

实例 068　显示具有开关状态的开关

实例 068	显示具有开关状态的开关
源码路径	\daima\068\

实例说明

本实例简单地演示了 UISwitch 控件的基本用法。首先通过方法 - (IBAction)switch Changed:(id)sender 获取了开关的状态，然后通过 setOn:setting 设置了开关的显示状态。

实例 068 显示具有开关状态的开关

具体实现

（1）打开 Xcode，创建一个名为 "UIswitch" 的工程。

（2）文件 UIswitchViewController.h 的实现代码如下所示。

```objc
#import <UIKit/UIKit.h>
@interface UIswitchViewController : UIViewController
{
    UISwitch* leftSwitch;
    UISwitch* rightSwitch;
}
@property(nonatomic,retain)UISwitch*leftSwitch;
@property(nonatomic,retain)UISwitch*rightSwitch;
@end
```

（3）文件 UIswitchViewController.m 的实现代码如下所示。

```objc
#import "UIswitchViewController.h"
@interface UIswitchViewController ()
@end
@implementation UIswitchViewController
@synthesize leftSwitch,rightSwitch;
- (id)initWithNibName:(NSString *)nibNameOrNil bundle:(NSBundle *)nibBundleOrNil
{
    self = [super initWithNibName:nibNameOrNil bundle:nibBundleOrNil];
    if (self) {
        // Custom initialization
    }
    return self;
}
- (void)viewDidLoad
{
    [super viewDidLoad];
    leftSwitch=[[UISwitch alloc]initWithFrame:CGRectMake(0, 0, 40, 20)];
    rightSwitch=[[UISwitch alloc] initWithFrame:CGRectMake(0,240, 40, 20)];
    [leftSwitch addTarget:self action:@selector(switchChanged:) forControlEvents:UIControlEventValueChanged];

    [self.view addSubview:leftSwitch];
    [rightSwitch addTarget:self action:@selector(switchChanged:) forControlEvents:UIControlEventValueChanged];
    [self.view addSubview:rightSwitch];
 // Do any additional setup after loading the view.
}
- (IBAction)switchChanged:(id)sender {
    UISwitch *mySwitch = (UISwitch *)sender;
    BOOL setting = mySwitch.isOn;     //获得开关状态
    if(setting)
    {
       NSLog(@"YES");
    }else {
       NSLog(@"NO");
    }
    [leftSwitch setOn:setting animated:YES];   //设置开关状态
    [rightSwitch setOn:setting animated:YES];
}
- (void)viewDidUnload
{
    [super viewDidUnload];
    // Release any retained subviews of the main view.
}
- (BOOL)shouldAutorotateToInterfaceOrientation:(UIInterfaceOrientation)interfaceOrientation
{
    return (interfaceOrientation == UIInterfaceOrientationPortrait);
}
@end
```

执行后的效果如图 3-37 所示。

图 3-37　执行效果

实例 069　在屏幕中显示一个分段选项

实例 069	在屏幕中显示一个分段选项
源码路径	\daima\069\

实例说明

在 iOS 应用中，当用户输入的不仅仅是布尔值时，可使用分段控件 UISegmentedControl 实现我们需要的功能。分段控件提供一栏按钮（有时称为按钮栏），但只能激活其中一个按钮，如图 3-38 所示。

图 3-38　分段控件

如果我们按 Apple 指南使用 UISegmentedControl，分段控件会导致用户在屏幕上看到的内容发生变化。它们常用于在不同类别的信息之间选择，或在不同的应用程序屏幕——如配置屏幕和结果屏幕之间切换。

在本实例中，实现了一个基本的分段分段卡效果。

具体实现过程如下。

（1）打开 Xcode，创建一个名为 "UISegmentedControlDemo" 的工程。

（2）文件 ViewController.h 的实现代码如下所示。

```
#import <UIKit/UIKit.h>
@interface ViewController : UIViewController{
}
@end
```

（3）文件 ViewController.m 的实现代码如下所示。

```
#import "ViewController.h"
@implementation ViewController
- (void)didReceiveMemoryWarning
{
    [super didReceiveMemoryWarning];
    // Release any cached data, images, etc that aren't in use.
}
#pragma mark - View lifecycle
-(void)selected:(id)sender{
    UISegmentedControl* control = (UISegmentedControl*)sender;
    switch (control.selectedSegmentIndex) {
        case 0:
```

```objc
            //
            break;
        case 1:
            //
            break;
        case 2:
            //
            break;
        default:
            break;
    }
}
- (void)viewDidLoad{
    [super viewDidLoad];
    UISegmentedControl* mySegmentedControl = [[UISegmentedControl alloc]initWithItems:nil];
    mySegmentedControl.segmentedControlStyle = UISegmentedControlStyleBezeled;
    UIColor *myTint = [[ UIColor alloc]initWithRed:0.66 green:1.0 blue:0.77 alpha:1.0];
    mySegmentedControl.tintColor = myTint;
    mySegmentedControl.momentary = YES;
    [mySegmentedControl insertSegmentWithTitle:@"First" atIndex:0 animated:YES];
    [mySegmentedControl insertSegmentWithTitle:@"Second" atIndex:2 animated:YES];
    [mySegmentedControl insertSegmentWithImage:[UIImage imageNamed:@"pic"] atIndex:3 animated:YES];
    //[mySegmentedControl removeSegmentAtIndex:0 animated:YES];//删除一个片段
    //[mySegmentedControl removeAllSegments];//删除所有片段
    [mySegmentedControl setTitle:@"ZERO" forSegmentAtIndex:0];//设置标题
    NSString* myTitle = [mySegmentedControl titleForSegmentAtIndex:1];//读取标题
    NSLog(@"myTitle:%@",myTitle);
    //[mySegmentedControl setImage:[UIImage imageNamed:@"pic"] forSegmentAtIndex:1];//设置
    UIImage* myImage = [mySegmentedControl imageForSegmentAtIndex:2];//读取
    [mySegmentedControl setWidth:100 forSegmentAtIndex:0];//设置 Item 的宽度
    [mySegmentedControl addTarget:self action:@selector(selected:) forControlEvents:UIControlEventValueChanged];
    //[self.view addSubview:mySegmentedControl];//添加到父视图
    self.navigationItem.titleView = mySegmentedControl;//添加到导航栏
}
- (void)viewDidUnload{
    [super viewDidUnload];
    // Release any retained subviews of the main view.
    // e.g. self.myOutlet = nil;
}
- (void)viewWillAppear:(BOOL)animated
{
    [super viewWillAppear:animated];
}
- (void)viewDidAppear:(BOOL)animated
{
    [super viewDidAppear:animated];
}
- (void)viewWillDisappear:(BOOL)animated
{
    [super viewWillDisappear:animated];
}
- (void)viewDidDisappear:(BOOL)animated
{
    [super viewDidDisappear:animated];
}
- (BOOL)shouldAutorotateToInterfaceOrientation:(UIInterfaceOrientation)interfaceOrientation
{
    // Return YES for supported orientations
    return (interfaceOrientation != UIInterfaceOrientationPortraitUpsideDown);
}
@end
```

执行后的效果如图 3-39 所示。

第 3 章 iOS 控件应用实战

图 3-39 执行效果

实例 070　选择一个分段卡后可以改变屏幕的背景颜色

实例 070	选择一个分段卡后可以改变屏幕的背景颜色
源码路径	\daima\070\

实例说明

在 iOS 应用中，分段控件 UISegmentedControl 的功能是供我们选择一个选项，选择后可以实现不同的功能。在本实例中提供了"Black"和"White"两个选项，当选择"Black"分段卡后屏幕的背景颜色变为黑色，当选择"White"分段卡后屏幕的背景颜色变为白色。

具体实现

实例文件 UIKitPrjSegmentedControl.m 的具体实现代码如下所示。

```
#import "UIKitPrjSegmentedControl.h"
@implementation UIKitPrjSegmentedControl
- (void)viewDidLoad {
  [super viewDidLoad];
  self.view.backgroundColor = [UIColor blackColor];
  //创建两个选项的字符串数组
  NSArray* items = [NSArray arrayWithObjects:@"Black", @"White", nil];
  //以 NSArray 为参数初始化选择控件
  UISegmentedControl* segment =
    [[[UISegmentedControl alloc] initWithItems:items] autorelease];
  //左侧第一选项默认被选择
  segment.selectedSegmentIndex = 0;
  segment.frame = CGRectMake( 0, 0, 130, 30 );
  //注册选项被选择时调用方法
  [segment addTarget:self
            action:@selector(segmentDidChange:)
    forControlEvents:UIControlEventValueChanged];
  //将选择控件对象追加到导航条的右侧
  UIBarButtonItem* barButton =
    [[[UIBarButtonItem alloc] initWithCustomView:segment] autorelease];
  self.navigationItem.rightBarButtonItem = barButton;
}
//选项选择发生变化时调用此方法
- (void)segmentDidChange:(id)sender {
  if ( [sender isKindOfClass:[UISegmentedControl class]] ) {
    UISegmentedControl* segment = sender;
    if ( 0 == segment.selectedSegmentIndex ) {
        //第一个选项被选择后将画面背景设置成黑色
      self.view.backgroundColor = [UIColor blackColor];
    } else {
```

```objc
        //第二个选项被选择后将画面背景设置成白色
        self.view.backgroundColor = [UIColor whiteColor];
    }
  }
}
@end
```

执行后的效果如图 3-40 所示。

图 3-40 执行效果

实例 071 设置分段卡的显示样式

实例 071	设置分段卡的显示样式
源码路径	\daima\070\

实例说明

在 iOS 应用中,使用分段控件 UISegmentedControl 在屏幕中设置一个分段卡后,可以继续使用其 segmentedControlStyle 属性设置分段卡的显示样式。一共有如下 4 种样式。

- UISegmentedControlStylePlain:
- UISegmentedControlStyleBordered:
- UISegmentedControlStyleBar:
- UISegmentedControlStyleBezeled:

在本实例中,使用 segmentedControlStyle 属性设置了显示样式是 UISegmentedControlStylePlain。

具体实现

实例文件 UIKitPrjSegmentedControlStyle.m 的具体实现代码如下所示。

```objc
#import "UIKitPrjSegmentedControlStyle.h"
@implementation UIKitPrjSegmentedControlStyle
- (void)viewDidLoad {
  [super viewDidLoad];
  self.view.backgroundColor = [UIColor whiteColor];
  NSArray* items = [NSArray arrayWithObjects:@"Plain", @"Borderd", @"Bar", nil];
  UISegmentedControl* segment =
    [[[UISegmentedControl alloc] initWithItems:items] autorelease];
  segment.segmentedControlStyle = UISegmentedControlStylePlain;
  segment.selectedSegmentIndex = 0;
  segment.frame = CGRectMake( 10, 50, 300, 30 );
  //segment.momentary = YES;
  [segment addTarget:self
          action:@selector(segmentDidChange:)
    forControlEvents:UIControlEventValueChanged];

  [self.view addSubview:segment];
}
```

```
- (void)segmentDidChange:(id)sender {
  if ( [sender isKindOfClass:[UISegmentedControl class]] ) {
    UISegmentedControl* segment = sender;
    switch ( segment.selectedSegmentIndex ) {
      case 0: segment.segmentedControlStyle = UISegmentedControlStylePlain; break;
      case 1: segment.segmentedControlStyle = UISegmentedControlStyleBordered; break;
      default: segment.segmentedControlStyle = UISegmentedControlStyleBar; break;
    }
  }
}
@end
```

执行效果如图 3-41 所示。

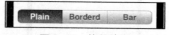

图 3-41 执行效果

实例 072　设置不显示分段卡的选择状态

实例 072	设置不显示分段卡的选择状态
源码路径	\daima\070\

实例说明

在 iOS 应用中，当选择分段卡中的某一个选项后，被选择的选项会默认呈现蓝色高亮样式显示。其实将其属性 momentary 设置为 YES 后，会取消这个特殊样式效果，即选择后不显示分段卡的选择状态。在本实例中，通过使用 momentary = YES，取消了屏幕中分段卡的选择状态样式。

具体实现

实例文件 UIKitPrjMomentary.m 的具体实现代码如下所示。

```
#import "UIKitPrjMomentary.h"
@implementation UIKitPrjMomentary
- (void)viewDidLoad {
  [super viewDidLoad];
  self.view.backgroundColor = [UIColor blackColor];
  NSArray* items = [NSArray arrayWithObjects:@"Black", @"White", nil];
  UISegmentedControl* segment =
    [[[UISegmentedControl alloc] initWithItems:items] autorelease];
  segment.momentary = YES;
  segment.frame = CGRectMake( 0, 0, 130, 30 );
  [segment addTarget:self
            action:@selector(segmentDidChange:)
    forControlEvents:UIControlEventValueChanged];
  UIBarButtonItem* barButton = [[[UIBarButtonItem alloc] initWithCustomView:segment] autorelease];
  self.navigationItem.rightBarButtonItem = barButton;
}
- (void)segmentDidChange:(id)sender {
  if ( [sender isKindOfClass:[UISegmentedControl class]] ) {
    UISegmentedControl* segment = sender;
    if ( 0 == segment.selectedSegmentIndex ) {
      self.view.backgroundColor = [UIColor blackColor];
    } else {
      self.view.backgroundColor = [UIColor whiteColor];
    }
  }
}
@end
```

执行效果如图 3-42 所示。

实例073 改变分段卡的显示颜色

图 3-42 执行效果

实例 073 改变分段卡的显示颜色

实例 073	改变分段卡的显示颜色
源码路径	\daima\070\

实例说明

在 iOS 应用中，当使用分段控件 UISegmentedControl 在屏幕中设置一个分段卡后，可以将其属性 segmentedControlStyle 设置为 UISegmentedControlStyleBezeled，这样分段卡便以不同的颜色样式显示。在本实例中，通过如下代码设置了分段卡以不同的颜色样式呈现出来。

```
segment.segmentedControlStyle = UISegmentedControlStylePlain
```

具体实现

实例文件 UIKitPrjTintColor.m 的具体实现代码如下所示。

```
#import "UIKitPrjTintColor.h"
@implementation UIKitPrjTintColor
- (void)viewDidLoad {
  [super viewDidLoad];
  self.view.backgroundColor = [UIColor whiteColor];
  NSArray* items = [NSArray arrayWithObjects:@"Plain", @"Borderd", @"Bar", nil];
  UISegmentedControl* segment =
    [[[UISegmentedControl alloc] initWithItems:items] autorelease];
  segment.segmentedControlStyle = UISegmentedControlStylePlain;
  segment.tintColor = [UIColor redColor];
  segment.selectedSegmentIndex = 0;
  segment.frame = CGRectMake( 10, 50, 300, 30 );
  [segment addTarget:self
       action:@selector(segmentDidChange:)
       forControlEvents:UIControlEventValueChanged];
  [self.view addSubview:segment];
}
- (void)segmentDidChange:(id)sender {
  if ( [sender isKindOfClass:[UISegmentedControl class]] ) {
    UISegmentedControl* segment = sender;
    switch ( segment.selectedSegmentIndex ) {
      case 0: segment.segmentedControlStyle = UISegmentedControlStylePlain; break;
      case 1: segment.segmentedControlStyle = UISegmentedControlStyleBordered; break;
      default: segment.segmentedControlStyle = UISegmentedControlStyleBar; break;
    }
  }
}
@end
```

执行后的效果如图 3-43 所示。

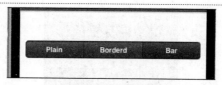

图 3-43 执行效果

实例 074 选择某个选项时在此分段卡中显示一幅图片

实例 074	选择某个选项时在此分段卡中显示一幅图片
源码路径	\daima\070\

实例说明

在 iOS 应用中，当使用分段控件 UISegmentedControl 在屏幕中设置一个分段卡后，可以使用方法 initWithItems 来初始化 UISegmentedControl 的内容，可以使用方法 setTitle:forSegmentAtIndex 设置在上面显示的文本，使用方法 setImage:forSegmentAtIndex 设置在上面显示的图像。在本实例中，使用方法 setImage: forSegmentAtIndex 和 setTitle:forSegmentAtIndex 为分段卡设置了显示的文本和图像。

具体实现

实例文件 UIKitPrjTitleAndImage.m 的具体实现代码如下所示。

```
#import "UIKitPrjTitleAndImage.h"
@implementation UIKitPrjTitleAndImage
- (void)dealloc {
  [titles_ release];
  [images_ release];
  [super dealloc];
}
- (void)viewDidLoad {
  [super viewDidLoad];
  self.view.backgroundColor = [UIColor whiteColor];
  //读入图标图片
  UIImage* image1 = [UIImage imageNamed:@"Elephant.png"];
  UIImage* image2 = [UIImage imageNamed:@"Lion.png"];
  UIImage* image3 = [UIImage imageNamed:@"Dog.png"];
  //创建选项显示用的图标、文本标题的NSArray类型数组
  images_ = [[NSArray alloc] initWithObjects:image1, image2, image3, nil];
  titles_ = [[NSArray alloc] initWithObjects:@"Elephant", @"Lion", @"Dog", nil];
  //选项初期为文本型
  UISegmentedControl* segment =
    [[[UISegmentedControl alloc] initWithItems:titles_] autorelease];
  segment.segmentedControlStyle = UISegmentedControlStyleBordered;
  segment.frame = CGRectMake( 10, 50, 300, 30 );
  [segment addTarget:self
            action:@selector(segmentDidChange:)
      forControlEvents:UIControlEventValueChanged];

  [self.view addSubview:segment];
}
//选项被选中后，将其中的文本型标题换成图标
- (void)segmentDidChange:(id)sender {
  if ( [sender isKindOfClass:[UISegmentedControl class]] ) {
    UISegmentedControl* segment = sender;
    for ( int i = 0; i < segment.numberOfSegments; ++i ){
      if ( i == segment.selectedSegmentIndex ) {
        [segment setImage:[images_ objectAtIndex:i] forSegmentAtIndex:i];
      } else {
        [segment setTitle:[titles_ objectAtIndex:i] forSegmentAtIndex:i];
      }
    }
  }
}
@end
```

执行后的效果如图 3-44 所示。

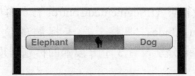

图 3-44 执行效果

实例 075 设置指定图片作为分段卡的选项

实例 075	设置指定图片作为分段卡的选项
源码路径	\daima\070\

实例说明

在 iOS 应用中，当使用分段控件 UISegmentedControl 在屏幕中设置一个分段卡后，可以使用方法 initWithItems:items 设置使用指定的图片作为分段卡的选项。在本实例中，使用方法 initWithItems:items 指定了 3 幅图片作为选项的选项。

具体实现

实例文件 UIKitPrjSegmentedControlWithImage.m 的具体实现代码如下所示。

```
#import "UIKitPrjSegmentedControlWithImage.h"
@implementation UIKitPrjSegmentedControlWithImage
- (void)viewDidLoad {
  [super viewDidLoad];
  self.view.backgroundColor = [UIColor whiteColor];
  UIImage* image1 = [UIImage imageNamed:@"Elephant.png"];
  UIImage* image2 = [UIImage imageNamed:@"Lion.png"];
  UIImage* image3 = [UIImage imageNamed:@"Dog.png"];
  NSArray* items = [NSArray arrayWithObjects:image1, image2, image3, nil];

  UISegmentedControl* segment =
    [[[UISegmentedControl alloc] initWithItems:items] autorelease];
  segment.segmentedControlStyle = UISegmentedControlStyleBar;
  segment.selectedSegmentIndex = 0;
  segment.frame = CGRectMake( 60, 50, 200, 40 );
[self.view addSubview:segment];
}
@end
```

执行后的效果如图 3-45 所示。

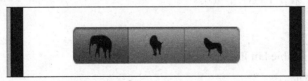

图 3-45 执行效果

实例 076 修改分段卡标题的位置

实例 076	修改分段卡标题的位置
源码路径	\daima\070\

实例说明

在 iOS 应用中，当使用分段控件 UISegmentedControl 在屏幕中设置一个分段卡后，可以使用方法 setContentOffset:forSegmentAtIndex 调整分段卡标题的位置。在本实例中，使用方法 setContentOffset:forSegmentAtIndex 分别设置 3 个分段卡的标题居顶、居中和居低显示。

具体实现

实例文件 UIKitPrjContentOffset.m 的具体实现代码如下所示。

```
#import "UIKitPrjContentOffset.h"
@implementation UIKitPrjContentOffset
- (void)viewDidLoad {
  [super viewDidLoad];
  self.view.backgroundColor = [UIColor whiteColor];
  NSArray* items = [NSArray arrayWithObjects:@"aaaaa", @"bbbbb", @"ccccc", nil];
  UISegmentedControl* segment =
    [[[UISegmentedControl alloc] initWithItems:items] autorelease];
  segment.segmentedControlStyle = UISegmentedControlStyleBar;
  segment.selectedSegmentIndex = 0;
  segment.frame = CGRectMake( 10, 50, 300, 40 );
  [segment setContentOffset:CGSizeMake( 0, -7 ) forSegmentAtIndex:0];
  [segment setContentOffset:CGSizeMake( 0,  7 ) forSegmentAtIndex:2];
  [self.view addSubview:segment];
}
@end
```

执行后的效果如图 3-46 所示。

图 3-46　执行效果

实例 077　设置某个选项不可用

实例 077	设置某个选项不可用
源码路径	\daima\070\

实例说明

在 iOS 应用中，当使用分段控件 UISegmentedControl 在屏幕中设置一个分段卡后，可以使用方法 setEnabled:forSegmentAtIndex 设置某个选项不可用。在本实例中，使用方法 setEnabled:forSegmentAtIndex 设置了第 2 个选项不可用。

具体实现

实例文件 UIKitPrjEnabled.m 的具体实现代码如下所示。

```
#import "UIKitPrjEnabled.h"
@implementation UIKitPrjEnabled
- (void)viewDidLoad {
  [super viewDidLoad];
  self.view.backgroundColor = [UIColor whiteColor];
  NSArray* items = [NSArray arrayWithObjects:@"Enabled", @"Disabled", @"Enabled", nil];
  UISegmentedControl* segment =
    [[[UISegmentedControl alloc] initWithItems:items] autorelease];
  segment.segmentedControlStyle = UISegmentedControlStyleBordered;
  segment.selectedSegmentIndex = 0;
  segment.frame = CGRectMake( 10, 50, 300, 40 );
  [segment setEnabled:NO forSegmentAtIndex:1];
```

```
    [self.view addSubview:segment];
}
@end
```

执行后的效果如图 3-47 所示。

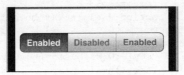

图 3-47　执行效果

实例 078　插入\删除分段卡中的选项（1）

实例 078	插入\删除分段卡中的选项（1）
源码路径	\daima\070\

实例说明

在 iOS 应用中，当使用分段控件 UISegmentedControl 在屏幕中设置一个分段卡后，可以使用方法 initWithItems 设置选项的个数。使用如下代码在指定索引插入一个选项并设置题目：

```
[segmentedControl insertSegmentWithTitle:@"insert" atIndex:3 animated:NO];
```

使用如下代码在指定索引插入一个选项并设置图片：

```
[segmentedControl insertSegmentWithImage:[UIImage imageNamed:@"mei.png"] atIndex:2 animated:NO];
```

使用如下代码移除指定索引的选项：

```
[segmentedControl removeSegmentAtIndex:0 animated:NO];
```

在本实例中，使用分段控件 UISegmentedControl 在屏幕中设置一个有 3 个选项的分段卡，然后分别设置了"插入"、"删除"和"全部删除"3 个按钮，通过这 3 个按钮可以控制分段卡中选项的个数。

具体实现

实例文件 UIKitPrjInsertAndRemove.m 的具体实现代码如下所示。

```
#import "UIKitPrjInsertAndRemove.h"
@implementation UIKitPrjInsertAndRemove
- (void)dealloc {
  [segment_ release];
  [super dealloc];
}
- (void)viewDidLoad {
  [super viewDidLoad];
  self.view.backgroundColor = [UIColor whiteColor];
  //初始化选择控件
  segment_ = [[UISegmentedControl alloc] init];
  segment_.segmentedControlStyle = UISegmentedControlStyleBordered;
  segment_.frame = CGRectMake( 10, 50, 300, 30 );
  [self.view addSubview:segment_];
  //向其中追加三个选项
  [segment_ insertSegmentWithTitle:@"3" atIndex:0 animated:NO];
  [segment_ insertSegmentWithTitle:@"2" atIndex:0 animated:NO];
  [segment_ insertSegmentWithTitle:@"1" atIndex:0 animated:NO];
  //在工具条上追加 Insert、Remove、RemoveAll 三个按钮，并注册三个按钮的响应方法
  UIBarButtonItem* insertButton =
    [[[UIBarButtonItem alloc] initWithTitle:@"插入"
                       style:UIBarButtonItemStyleBordered
```

```objc
                                    target:self
                                    action:@selector(insertDidPush)] autorelease];
    UIBarButtonItem* removeButton =
      [[[UIBarButtonItem alloc] initWithTitle:@"删除"
                                    style:UIBarButtonItemStyleBordered
                                    target:self
                                    action:@selector(removeDidPush)] autorelease];
    UIBarButtonItem* removeAllButton =
      [[[UIBarButtonItem alloc] initWithTitle:@"全部删除"
                                    style:UIBarButtonItemStyleBordered
                                    target:self
                                    action:@selector(removeAllDidPush)] autorelease];
    NSArray* items = [NSArray arrayWithObjects:insertButton, removeButton, removeAllButton, nil];
    [self setToolbarItems:items animated:YES];
}
//Insert 按钮的响应方法
- (void)insertDidPush {
    NSNumber* number = [NSNumber numberWithInteger:segment_.numberOfSegments + 1];
    [segment_ insertSegmentWithTitle:[number stringValue]
                             atIndex:[number integerValue]
                            animated:YES];
}
//Remove 按钮的响应方法
- (void)removeDidPush {
    [segment_ removeSegmentAtIndex:segment_.numberOfSegments - 1 animated:YES];
}
//RemoveAll 按钮的响应方法
- (void)removeAllDidPush {
    [segment_ removeAllSegments];
}
@end
```

执行后的效果如图 3-48 所示。

图 3-48　执行效果

实例 079　插入\删除分段卡中的选项（2）

实例 079	插入\删除分段卡中的选项（2）
源码路径	\daima\079\

实例说明

对 Apple 和 iOS 设备来说，提供有趣、平滑和美妙的用户体验是其成功的关键，而程序员的任务就是开发能够实现这样体验的程序。iOS SDK 的界面选项可以用有趣而独特的方式呈现应用程序的功能。

滑块（UISlider）是常用的界面组件，能够让用户可以用可视化方式设置指定范围内的值。假设我们想让用户提高或降低速度，采取让用户输入值的方式并不合理，相反，可提供一个如图 3-49 所示的滑块，让用户能够轻按并来回拖曳。在幕后，这将设置一个 value 属性，应用程序可使用它来设置速度。这不要求用户理解幕后的细节，也不需要用户执行除使用手指拖曳之外的其他操作。

图 3-49 使用滑块收集特定范围内的值

和按钮一样，滑块也能响应事件，还可像文本框一样被读取。如果希望用户对滑块的调整立刻影响应用程序，则需要让它触发操作。

滑块为用户提供了一种可见的做范围调整的方法，用户可以通过拖动一个滑动条改变它的值，并且可以对其配置以适合不同值域。可以设置滑块值的范围，也可以在两端加上图片，以及进行各种调整让它更美观。滑块非常适合用于表示在很大范围（但不精确）的数值中进行选择，比如音量设置、灵敏度控制等诸如此类的用途。

在本实例中，演示了 3 种不同样式的滑块控件。

具体实现

（1）打开 Xcode，新建一个名为 "test_project" 的工程，如图 3-50 所示。

（2）准备一幅名为 "circularSliderThumbImage.png" 的图片作为素材，如图 3-51 所示。

（3）设计 UI 界面，在界面中设置了如下 3 个控件。

- UISlider：放在界面的顶部，用于实现滑块功能。

图 3-50 新建 Xcode 工程

图 3-51 素材图片

- UIProgressView：这是一个进度条控件，放在界面中间，能够实现进度条效果。
- UICircularSlider：这是一个自定义滑块控件，放在界面底部，能够实现圆环状的滑块效果。

最终的 UI 界面效果如图 3-52 所示。

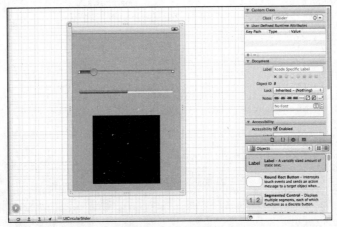

图 3-52 UI 界面

（4）看文件 UICircularSlider.m 的源码，此文件是 UICircularSlider Library 的一部分，这里的 UICircularProgressView 是一款自由软件，读者们可以免费获取这个软件，并且可以重新分配和/或修改使用，读者可以从网络中免费获取 UICircularProgressView。此文件的最终代码如下所示。

```objc
#import "UICircularSlider.h"
@interface UICircularSlider()
@property (nonatomic) CGPoint thumbCenterPoint;
#pragma mark - Init and Setup methods
- (void)setup;
#pragma mark - Thumb management methods
- (BOOL)isPointInThumb:(CGPoint)point;
#pragma mark - Drawing methods
- (CGFloat)sliderRadius;
- (void)drawThumbAtPoint:(CGPoint)sliderButtonCenterPoint inContext:(CGContextRef)context;
- (CGPoint)drawCircularTrack:(float)track atPoint:(CGPoint)point withRadius:(CGFloat)radius inContext:(CGContextRef)context;
- (CGPoint)drawPieTrack:(float)track atPoint:(CGPoint)point withRadius:(CGFloat)radius inContext:(CGContextRef)context;
@end
#pragma mark -
@implementation UICircularSlider
@synthesize value = _value;
- (void)setValue:(float)value {
    if (value != _value) {
        if (value > self.maximumValue) { value = self.maximumValue; }
        if (value < self.minimumValue) { value = self.minimumValue; }
        _value = value;
        [self setNeedsDisplay];
        [self sendActionsForControlEvents:UIControlEventValueChanged];
    }
}
@synthesize minimumValue = _minimumValue;
- (void)setMinimumValue:(float)minimumValue {
    if (minimumValue != _minimumValue) {
        _minimumValue = minimumValue;
        if (self.maximumValue < self.minimumValue)  { self.maximumValue = self.minimumValue; }
        if (self.value < self.minimumValue) { self.value = self.minimumValue; }
    }
}
@synthesize maximumValue = _maximumValue;
- (void)setMaximumValue:(float)maximumValue {
    if (maximumValue != _maximumValue) {
        _maximumValue = maximumValue;
        if (self.minimumValue > self.maximumValue)  { self.minimumValue = self.maximumValue; }
        if (self.value > self.maximumValue)         { self.value = self.maximumValue; }
    }
}
@synthesize minimumTrackTintColor = _minimumTrackTintColor;
- (void)setMinimumTrackTintColor:(UIColor *)minimumTrackTintColor {
    if (![minimumTrackTintColor isEqual:_minimumTrackTintColor]) {
        _minimumTrackTintColor = minimumTrackTintColor;
        [self setNeedsDisplay];
    }
}
@synthesize maximumTrackTintColor = _maximumTrackTintColor;
- (void)setMaximumTrackTintColor:(UIColor *)maximumTrackTintColor {
    if (![maximumTrackTintColor isEqual:_maximumTrackTintColor]) {
        _maximumTrackTintColor = maximumTrackTintColor;
        [self setNeedsDisplay];
    }
}
@synthesize thumbTintColor = _thumbTintColor;
- (void)setThumbTintColor:(UIColor *)thumbTintColor {
    if (![thumbTintColor isEqual:_thumbTintColor]) {
        _thumbTintColor = thumbTintColor;
        [self setNeedsDisplay];
    }
}
@synthesize continuous = _continuous;
```

```objc
@synthesize sliderStyle = _sliderStyle;
- (void)setSliderStyle:(UICircularSliderStyle)sliderStyle {
    if (sliderStyle != _sliderStyle) {
        _sliderStyle = sliderStyle;
        [self setNeedsDisplay];
    }
}
@synthesize thumbCenterPoint = _thumbCenterPoint;

/** @name Init and Setup methods */
#pragma mark - Init and Setup methods
- (id)initWithFrame:(CGRect)frame {
    self = [super initWithFrame:frame];
    if (self) {
        [self setup];
    }
    return self;
}
- (void)awakeFromNib {
    [self setup];
}

- (void)setup {
    self.value = 0.0;
    self.minimumValue = 0.0;
    self.maximumValue = 1.0;
    self.minimumTrackTintColor = [UIColor blueColor];
    self.maximumTrackTintColor = [UIColor whiteColor];
    self.thumbTintColor = [UIColor darkGrayColor];
    self.continuous = YES;
    self.thumbCenterPoint = CGPointZero;
    
    UITapGestureRecognizer *tapGestureRecognizer = [[UITapGestureRecognizer alloc] initWithTarget:self action:@selector(tapGestureHappened:)];
    [self addGestureRecognizer:tapGestureRecognizer];
    
    UIPanGestureRecognizer *panGestureRecognizer = [[UIPanGestureRecognizer alloc] initWithTarget:self action:@selector(panGestureHappened:)];
    panGestureRecognizer.maximumNumberOfTouches = panGestureRecognizer.minimumNumberOfTouches;
    [self addGestureRecognizer:panGestureRecognizer];
}

/** @name Drawing methods */
#pragma mark - Drawing methods
#define kLineWidth 5.0
#define kThumbRadius 12.0
- (CGFloat)sliderRadius {
    CGFloat radius = MIN(self.bounds.size.width/2, self.bounds.size.height/2);
    radius -= MAX(kLineWidth, kThumbRadius);
    return radius;
}
- (void)drawThumbAtPoint:(CGPoint)sliderButtonCenterPoint inContext:(CGContextRef)context {
    UIGraphicsPushContext(context);
    CGContextBeginPath(context);
    
    CGContextMoveToPoint(context, sliderButtonCenterPoint.x, sliderButtonCenterPoint.y);
    CGContextAddArc(context, sliderButtonCenterPoint.x, sliderButtonCenterPoint.y, kThumbRadius, 0.0, 2*M_PI, NO);
    
    CGContextFillPath(context);
    UIGraphicsPopContext();
}

- (CGPoint)drawCircularTrack:(float)track atPoint:(CGPoint)center withRadius:(CGFloat)radius inContext:(CGContextRef)context {
    UIGraphicsPushContext(context);
    CGContextBeginPath(context);
    
    float angleFromTrack = translateValueFromSourceIntervalToDestinationInterval(track, self.minimumValue, self.maximumValue, 0, 2*M_PI);
```

```objc
        CGFloat startAngle = -M_PI_2;
        CGFloat endAngle = startAngle + angleFromTrack;
        CGContextAddArc(context, center.x, center.y, radius, startAngle, endAngle, NO);

        CGPoint arcEndPoint = CGContextGetPathCurrentPoint(context);

        CGContextStrokePath(context);
        UIGraphicsPopContext();

        return arcEndPoint;
}

- (CGPoint)drawPieTrack:(float)track atPoint:(CGPoint)center withRadius:(CGFloat)radius inContext:(CGContextRef)context {
        UIGraphicsPushContext(context);

        float  angleFromTrack = translateValueFromSourceIntervalToDestinationInterval(track, self.minimumValue, self.maximumValue, 0, 2*M_PI);

        CGFloat startAngle = -M_PI_2;
        CGFloat endAngle = startAngle + angleFromTrack;
        CGContextMoveToPoint(context, center.x, center.y);
        CGContextAddArc(context, center.x, center.y, radius, startAngle, endAngle, NO);

        CGPoint arcEndPoint = CGContextGetPathCurrentPoint(context);

        CGContextClosePath(context);
        CGContextFillPath(context);
        UIGraphicsPopContext();

        return arcEndPoint;
}

- (void)drawRect:(CGRect)rect {
    CGContextRef context = UIGraphicsGetCurrentContext();

        CGPoint middlePoint;
        middlePoint.x = self.bounds.origin.x + self.bounds.size.width/2;
        middlePoint.y = self.bounds.origin.y + self.bounds.size.height/2;

        CGContextSetLineWidth(context, kLineWidth);

        CGFloat radius = [self sliderRadius];
        switch (self.sliderStyle) {
            case UICircularSliderStylePie:
                [self.maximumTrackTintColor setFill];
                [self drawPieTrack:self.maximumValue atPoint:middlePoint withRadius:radius inContext:context];
                [self.minimumTrackTintColor setStroke];
                [self drawCircularTrack:self.maximumValue  atPoint:middlePoint  withRadius:radius inContext:context];
                [self.minimumTrackTintColor setFill];
                self.thumbCenterPoint = [self drawPieTrack:self.value atPoint:middlePoint withRadius:radius inContext:context];
                break;
            case UICircularSliderStyleCircle:
            default:
                [self.maximumTrackTintColor setStroke];
                [self drawCircularTrack:self.maximumValue  atPoint:middlePoint with Radius:radius inContext:context];
                [self.minimumTrackTintColor setStroke];
                self.thumbCenterPoint = [self drawCircularTrack:self.value atPoint:middlePoint withRadius:radius inContext:context];
                break;
        }

        [self.thumbTintColor setFill];
        [self drawThumbAtPoint:self.thumbCenterPoint inContext:context];
}

/** @name Thumb management methods */
#pragma mark - Thumb management methods
- (BOOL)isPointInThumb:(CGPoint)point {
```

```objc
        CGRect thumbTouchRect = CGRectMake(self.thumbCenterPoint.x - kThumbRadius, self.
thumbCenterPoint.y - kThumbRadius, kThumbRadius*2, kThumbRadius*2);
        return CGRectContainsPoint(thumbTouchRect, point);
}

/** @name UIGestureRecognizer management methods */
#pragma mark - UIGestureRecognizer management methods
- (void)panGestureHappened:(UIPanGestureRecognizer *)panGestureRecognizer {
    CGPoint tapLocation = [panGestureRecognizer locationInView:self];
    switch (panGestureRecognizer.state) {
        case UIGestureRecognizerStateChanged: {
            CGFloat radius = [self sliderRadius];
            CGPoint sliderCenter = CGPointMake(self.bounds.size.width/2, self.bounds.
            size.height/2);
            CGPoint sliderStartPoint = CGPointMake(sliderCenter.x, sliderCenter.y -
            radius);
            CGFloat angle = angleBetweenThreePoints(sliderCenter, sliderStartPoint,
            tapLocation);
            
            if (angle < 0) {
                angle = -angle;
            }
            else {
                angle = 2*M_PI - angle;
            }
            
            self.value = translateValueFromSourceIntervalToDestinationInterval(angle, 0,
2*M_PI, self.minimumValue, self.maximumValue);
            break;
        }
        default:
            break;
    }
}
- (void)tapGestureHappened:(UITapGestureRecognizer *)tapGestureRecognizer {
    if (tapGestureRecognizer.state == UIGestureRecognizerStateEnded) {
        CGPoint tapLocation = [tapGestureRecognizer locationInView:self];
        if ([self isPointInThumb:tapLocation]) {
        }
        else {
        }
    }
}
@end
/** @name Utility Functions */
#pragma mark - Utility Functions
float translateValueFromSourceIntervalToDestinationInterval(float sourceValue, float
sourceIntervalMinimum, float sourceIntervalMaximum, float destinationIntervalMinimum,
float destinationIntervalMaximum) {
    float a, b, destinationValue;
    
    a = (destinationIntervalMaximum - destinationIntervalMinimum) / (sourceInterval
Maximum - sourceIntervalMinimum);
    b = destinationIntervalMaximum - a*sourceIntervalMaximum;
    
    destinationValue = a*sourceValue + b;
    
    return destinationValue;
}

CGFloat angleBetweenThreePoints(CGPoint centerPoint, CGPoint p1, CGPoint p2) {
    CGPoint v1 = CGPointMake(p1.x - centerPoint.x, p1.y - centerPoint.y);
    CGPoint v2 = CGPointMake(p2.x - centerPoint.x, p2.y - centerPoint.y);
    
    CGFloat angle = atan2f(v2.x*v1.y - v1.x*v2.y, v1.x*v2.x + v1.y*v2.y);
    
    return angle;
}
```

（5）再看文件 UICircularSliderViewController.m，此文件也是借助了自由软件 UICircular ProgressView，读者们可以免费获取这个软件，并且可以重新分配或修改使用，读者可以从网络中免费获取 UICircularProgressView。此文件的最终代码如下所示。

```objc
#import "UICircularSliderViewController.h"
#import "UICircularSlider.h"
@interface UICircularSliderViewController ()
@property (unsafe_unretained, nonatomic) IBOutlet UISlider *slider;
@property (unsafe_unretained, nonatomic) IBOutlet UIProgressView *progressView;
@property (unsafe_unretained, nonatomic) IBOutlet UICircularSlider *circularSlider;
@end
@implementation UICircularSliderViewController
@synthesize slider = _slider;
@synthesize progressView = _progressView;
@synthesize circularSlider = _circularSlider;
- (void)viewDidLoad {
    [super viewDidLoad];
    [self.circularSlider addTarget:self action:@selector(updateProgress:) forControlEvents:UIControlEventValueChanged];
    [self.circularSlider setMinimumValue:self.slider.minimumValue];
    [self.circularSlider setMaximumValue:self.slider.maximumValue];
}

- (void)viewDidUnload {
    [self setProgressView:nil];
    [self setCircularSlider:nil];
    [self setSlider:nil];
    [super viewDidUnload];
}

- (BOOL)shouldAutorotateToInterfaceOrientation:(UIInterfaceOrientation)interfaceOrientation {
    return YES;
}

- (IBAction)updateProgress:(UISlider *)sender {
    float progress = translateValueFromSourceIntervalToDestinationInterval(sender.value, sender.minimumValue, sender.maximumValue, 0.0, 1.0);
    [self.progressView setProgress:progress];
    [self.circularSlider setValue:sender.value];
    [self.slider setValue:sender.value];
}
@end
```

这样整个实例就介绍完毕了，执行后的效果如图 3-53 所示。

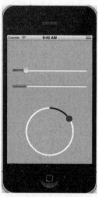

图 3-53 执行效果

实例 080 滑动滑块时显示对应的值

实例 080	滑动滑块时显示对应的值
源码路径	\daima\080\

实例说明

在 iOS 应用中，当使用 UISlider 控件设置一个滑块后，可以使用其属性设置 minimumValue 最小值，使用其属性设置 maximumValue 最小值。在本实例中，设置了滑块的最大值是 1.0，最小值是 0.0，初始值是 0.5，然后通过方法 sliderDidChange:(id)sender 设置滑动时改变滑块的值。

具体实现

实例文件 UIKitPrjSlider.m 的具体实现代码如下所示。

```objc
#import "UIKitPrjSlider.h"
@implementation UIKitPrjSlider
- (void)dealloc {
    [label_ release];
    [super dealloc];
}
- (void)viewDidLoad {
    [super viewDidLoad];
    label_ = [[UILabel alloc] init];
    label_.frame = self.view.bounds;
    label_.autoresizingMask =
      UIViewAutoresizingFlexibleWidth | UIViewAutoresizingFlexibleHeight;
    label_.text = @"0.5";
    label_.font = [UIFont boldSystemFontOfSize:36];
    label_.textAlignment = UITextAlignmentCenter;
    [self.view addSubview:label_];
    
    UISlider* slider = [[[UISlider alloc] init] autorelease];
    slider.frame = CGRectMake( 0, 0, 250, 50 );
    slider.minimumValue = 0.0;
    slider.maximumValue = 1.0;
    slider.value = 0.5;  //设置初始值
    slider.center = self.view.center;
    [slider addTarget:self
            action:@selector(sliderDidChange:)
     forControlEvents:UIControlEventValueChanged];
    [self.view addSubview:slider];
}
- (void)sliderDidChange:(id)sender {
    if ( [sender isKindOfClass:[UISlider class]] ) {
        UISlider* slider = sender;
        label_.text = [NSString stringWithFormat:@"%0.1f", slider.value];
    }
}
@end
```

执行后的效果如图 3-54 所示。

图 3-54 执行效果

实例 081　滑动滑块控制文字的大小

实例 081	滑动滑块控制文字的大小
源码路径	\daima\080\

实例说明

在 iOS 应用中，当使用 UISlider 控件设置一个滑块后，可以使用 UIImage 为滑块设置表示放大和缩小的图像素材。在本实例中，设置了滑块的最大值是 1.0，最小值是 0.0，初始值是 0.5，然后为滑块左侧图标设置了 minimumValueImage 属性，为滑块右侧图标设置了 maximumValueImage 属性。最后定义了方法 - (void)sliderDidChange:(id)sender，当滑块变化时会响应此方法，通过此方法设置标签的文字字体放大或缩小，缩放级别是 96。

具体实现

实例文件 UIKitPrjSliderWithImage.m 的具体实现代码如下所示。

```
#import "UIKitPrjSliderWithImage.h"
@implementation UIKitPrjSliderWithImage
- (void)dealloc {
  [label_ release];
  [super dealloc];
}
- (void)viewDidLoad {
  [super viewDidLoad];
  //追加标签，将通过滑块控制标签文字大小
  label_ = [[UILabel alloc] init];
  label_.frame = self.view.bounds;
  label_.autoresizingMask =
    UIViewAutoresizingFlexibleWidth | UIViewAutoresizingFlexibleHeight;
  label_.text = @"标题";
  label_.font = [UIFont boldSystemFontOfSize:48];
  label_.textAlignment = UITextAlignmentCenter;
  [self.view addSubview:label_];
  //创建并初始化滑块对象
  UISlider* slider = [[[UISlider alloc] init] autorelease];
  slider.frame = CGRectMake( 0, 0, 250, 50 );
  slider.minimumValue = 0.0;
  slider.maximumValue = 1.0;
  slider.value = 0.5; //初期值的设置
  slider.center = self.view.center;
  //读入左侧及右侧用的图标图片，并设置到 minimumValueImage 及 maximumValueImage 属性中
  UIImage* imageForMin = [UIImage imageNamed:@"roope_small.png"];
  UIImage* imageForMax = [UIImage imageNamed:@"roope_big.png"];
  slider.minimumValueImage = imageForMin;
  slider.maximumValueImage = imageForMax;
  [self.view addSubview:slider];
  //注册滑块变化时的响应方法
  [slider addTarget:self
          action:@selector(sliderDidChange:)
   forControlEvents:UIControlEventValueChanged];
}
//滑块变化时的响应方法，其中设置标签的文字字体
- (void)sliderDidChange:(id)sender {
  if ( [sender isKindOfClass:[UISlider class]] ) {
    UISlider* slider = sender;
    label_.font = [UIFont boldSystemFontOfSize:( 96 * slider.value )];
  }
}
@end
```

执行后的效果如图 3-55 所示。

图 3-55 执行效果

实例 082　自定义一个滑块

实例 082	自定义一个滑块
源码路径	\daima\080\

实例说明

在 iOS 应用中，我们可以自行定义一个指定样式的滑块。在本实例中，使用方法 setThumbImage:forState 将一幅图片定义为滑块的中间部分，使用方法 setMinimumTrackImage:forState 定义了滑块的左侧部分，使用方法 setMaximumTrackImage:forState 定义了滑块的右侧部分。

具体实现

实例文件 UIKitPrjSetThumbImage.m 的具体实现代码如下所示。

```
#import "UIKitPrjSetThumbImage.h"
@implementation UIKitPrjSetThumbImage
- (void)viewDidLoad {
    [super viewDidLoad];
    self.view.backgroundColor = [UIColor whiteColor];
    //创建并初始化滑块对象
    UISlider* slider = [[[UISlider alloc] init] autorelease];
    slider.frame = CGRectMake( 0, 0, 250, 50 );
    slider.minimumValue = 0.0;
    slider.maximumValue = 1.0;
    slider.value = 0.5; //设置初期值
    slider.center = self.view.center;
    UIImage* imageForThumb = [UIImage imageNamed:@"thumb.png"];
    UIImage* imageMinBase = [UIImage imageNamed:@"left.png"];
    UIImage* imageForMin = [imageMinBase stretchableImageWithLeftCapWidth:4 topCapHeight:0];
    UIImage* imageMaxBase = [UIImage imageNamed:@"right.png"];
    UIImage* imageForMax = [imageMaxBase stretchableImageWithLeftCapWidth:4 topCapHeight:0];
    [slider setThumbImage:imageForThumb forState:UIControlStateNormal];
    [slider setThumbImage:imageForThumb forState:UIControlStateHighlighted];
    [slider setMinimumTrackImage:imageForMin forState:UIControlStateNormal];
    [slider setMaximumTrackImage:imageForMax forState:UIControlStateNormal];

    [self.view addSubview:slider];
}
@end
```

执行后的效果如图 3-56 所示。

图 3-56　执行效果

实例 083　实现一个日期选择器

实例 083	实现一个日期选择器
源码路径	\daima\083\

实例说明

在 iOS 应用中，选择器是 iOS 的一种独特功能，它们通过转轮界面效果的提供一系列多值选项。选择器的每个组件显示数行可供用户选择的值，而不是水果或数字。在桌面应用程序中，与选择器最接近的组件是下拉列表。图 3-57 显示了标准的日期选择器（UIDatePicker）。

图 3-57　选择器提供了一些值供选择

当用户需要选择多个（通常相关）值时应使用选择器。它们通常用于设置日期和事件，但是可以对其进行定制以处理开发人员能想到的任何选择方式。Apple 认为，在选择日期和时间方面，选择器是一种不错的界面元素，所以特意提供了如下两种形式的选择器。

- 日期选择器：这种方式易于实现，且专门用于处理日期和时间。
- 自定义选择器视图：可以根据需要配置成显示任意数量的组件。

在本实例中，使用 UIDatePicker 控件实现了一个日期选择器,该选择器通过模态切换方式显示。

具体实现

（1）创建项目。

使用模板 Single View Application 新建一个项目，并将其命名为"Date"。模板创建的初始场景/视图控制器将包含日期计算逻辑，但我们还需添加一个场景和视图控制器，它们将用于显示日期选择器界面。

为了使用日期选择器显示日期并在用户选择日期时作出响应，需要在项目中添加一个 DateChooserViewController 类。为此单击项目导航器左下角的"+"按钮，在弹出的对话框中选择 iOS Cocoa Touch 和图标 UIViewController subclass，再单击 Next 按钮。在下一个屏幕中输入名称"DateChooserViewController"。在最后一个设置屏幕中，从 Group 下拉列表中选择项目代码编组，然后再单击 Create 按钮。

在 Interface Builder 编辑器中打开文件 MainStoryboard.storyboard，使用快捷键"Control+Option+Command+3"打开对象库，并将一个视图控制器拖曳到 Interface Builder 编辑器的空白区域（或文档大纲区域）。此时项目将包含两个场景，为了将新增的视图控制器关联到 DateChooserViewController 类，在文档大纲区域中选择第二个场景的 View Controller 图标，按下快捷键"Option+Command+3"打开 Identity Inspector，并从 Class 下拉列表中选择 DateChooserViewController。

选择第一个场景的 View Controller 图标，并确保仍显示了 Identity Inspector。在 Identity 部分，将视图控制器标签设置为 Initial。对第二个场景重复上述操作，将其视图控制器标签设置为 Date Chooser。此时的文档大纲区域将显示 Initial Scene 和 Date Chooser Scene。

在初始场景中，将包含一个用于显示输出的标签——outputLabel。还将添加一个操作——showDateChooser，用于显示日期选择场景。另外，初始场景的视图控制器类 ViewController 需要包含一个属性（dateChooserVisible），用于跟踪日期选择场景是否可见；它还需要一个方法（calculateDateDifference），用于计算当前日期和选定日期相差多少天。

而日期选择场景不需要输出口，而只需两个操作——setDatetime 和 dismissDateChooser。其中，前者在用户通过日期选择器时被调用，而后者用于 iPhone 视图，当用户触摸日期场景中的按钮时关闭该场景。在 DateChooserViewController 中，还将添加一个非常重要的属性（delegate），它存储了指向初始场景的视图控制器的引用。我们将利用这个属性访问 ViewController 类的属性和方法。

（2）设计界面。

打开文件 MainStoryboard.storyboard，并滚动到在编辑器中能够看到初始场景。打开对象库

（Control+Option+Command+3），并拖曳一个工具栏到该视图底部。默认情况下，工具栏只包含一个名为"item"的按钮。双击 item，并将其改为 Choose a Date。接下来，从对象库拖曳两个灵活间距栏按钮项（Flexible Space Bar Button Item）到工具栏中，并将它们分别放在按钮 Choose a Date 两边。这将让按钮"选择日期"位于工具栏中央。

接下来，在视图中央添加一个标签。使用 Attributes Inspector（Option+Command+4）增大标签的字体，让文本居中显示，并将标签扩大到至少能够容纳 5 行文本。将文本改为"没有选择"。最终的视图如图 3-58 所示。

图 3-58　设计的 UI 图

然后选择该场景的视图，并将其背景色设置为 Scroll View Texted Background Color。拖曳一个日选择器到视图顶部。如果您创建的是该应用程序的 iPad 版，该视图最终将显示为弹出框，因此只有左上角部分可见。

然后在日期选择器下方，放置一个标签，并将其文本改为"选择日期"。最后，如果创建的是该应用程序的 iPhone 版，拖曳一个按钮到视图底部，它将用于关闭日期选择场景。将该按钮的标签设置为"确定"。图 3-59 显示了设计的日期选择界面。

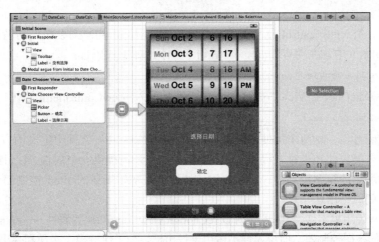

图 3-59　日期选择场景

（3）创建切换。

按住"Ctrl"键，从初始场景的视图控制器拖曳到日期选择场景的视图控制器，我们可以直接在文档大纲区域中这样做，也可以使用 Interface Builder 编辑器中场景的可视化表示。在 Xcode 中要求我们指定故事板切换类型时，选择 Modal（iPhone）或 Popover（iPad）。在文档大纲区域，初始场景中将新增一行,其内容为 Segue from UIViewController to DateChooseViewController。选择这一行并打开 Attributes Inspector(Option+Command+4),以配置该切换。

给切换指定标识符 toDateChooser。我们将手工触发该切换,因此将标识符设置为前面所说的值很重要,这样代码才能正确运行。

(4)创建并连接输出口和操作。

本演示程序的每个场景都需要建立两个连接——初始场景是一个操作和一个输出口,而日期选择场景是两个操作。这些输出口和操作如下所述。

- outputLabel(UILabel):该标签在初始场景中显示日期计算的结果。
- showDateChooser:这是一个操作方法,由初始场景中的栏按钮项触发。
- dismissDateChooser:这是一个操作方法,由日期选择场景中的"Done"按钮触发。
- setDateTime:这是一个操作方法,在日期选择器的值发生变化时触发。

接下来切换到助手编辑器,并首先连接初始场景的输出口。首先添加输出口。选择初始场景中的输出标签,按住"Control"键并从该标签拖曳到文件 ViewController.h 中编译指令@interface 下方。在 Xcode 提示时,创建一个名为"outputLabel"的新输出口,然后添加操作。在这个项目中,除了一个连接是输出口外,其他连接都是操作。在初始场景中,按住"Control"键并从按钮"选择日期"拖曳到文件 ViewController.h 中属性定义的下方。在 Xcode 提示时,添加一个名为"showDateChooser"的新操作。

切换到第二个场景(日期选择场景),按住"Control"键,并从日期选择器拖曳到文件 DateChooserViewController.h 中编译指令@interface 下方。在 Xcode 提示时,新建一个名为"setDateTime"的操作,并将触发事件指定为 Value Changed。如果开发的是该应用程序的 iPad 版,至此创建并连接操作和输出口的工作就完成了,用户将触摸弹出框的外面来关闭弹出框。如果创建的是 iPhone 版,还需按住"Control"键,并从按钮"Done"拖曳到文件 DateChooserView Controller.h 中,以创建由该按钮触发的操作 dismissDateChooser。

(5)实现场景切换逻辑。

在应用程序逻辑中,需要处理两项主要任务。首先,需要处理初始场景的视图控制器和日期选择场景的视图控制器之间的交互;其次,需要计算并显示两个日期相差多少天。首先来处理视图控制器之间的通信。

在本实例中,类 ViewController 和类 DateChooserViewController 需要彼此访问对方的属性。在文件 ViewController.h 中,在现有的#import 语句下方添加如下代码行:

```
#import "DateChooserViewController.h"
```

同样在文件 DateChooserViewController.h 中,添加导入 ViewController.h 的代码:

```
#import "ViewController.h"
```

添加这些代码行后,这两个类便可彼此访问对方的接口(.h)文件中定义的方法和属性了。

除了让这两个类彼此知道对方提供的方法和属性外,还需提供一个属性,让日期选择视图控制器能够访问初始场景的视图控制器。它将通过该属性调用初始场景的 iPad 控制器中的日期计算方法,并在自己关闭时指出这一点。

如果该项目只使用模态切换,则可使用 DateChooserViewController 的属性 presentingViewController 来获取初始场景的视图控制器,但该属性不适用于弹出框。为了保持模态实现和弹出框的实现一致,将给类 DateChooserViewController 添加一个 delegate 属性:

```
@property (strong, nonatomic) id delegate;
```

上述代码定义了一个类型为 id 的属性,这意味着它可以指向任何对象,就像 Apple 类内置的 delegate 属性一样。

接下来,修改文件 DateChooserViewController.m,在@implementation 后面添加配套的变异指

令@synthesize：

```
@synthesize delegate;
```

最后执行清理工作，将该实例变量/属性设置为 nil。需要在文件 DateChooserViewController.m 的方法 viewDidUnload 中添加如下代码行：

```
[self setDelegate:nil];
```

要想设置属性 delegate，可以在 ViewController.m 的方法 prepareForSegue:sender 中实现。当初始场景和日期选择场景之间的切换被触发时会调用这个方法。修改文件 ViewController.h，在其中添加该方法，具体代码如下所示。

```
- (void)prepareForSegue:(UIStoryboardSegue *)segue sender:(id)sender {
    ((DateChooserViewController *)segue.destinationViewController).delegate=self;
}
```

通过上述代码，将参数 segue 的属性 destinationViewController 强制转换为一个 DateChooserViewController，并将其 delegate 属性设置为 self，即初始场景的 VewController 类的当前实例。

这样通过导入接口文件，让两个场景的视图控制器能够彼此访问对方的方法和属性。而属性 delegate 提供了一种交换信息的机制。

在这个应用程序中，切换是在视图控制器之间，而不是对象和视图控制器之间创建的。通常将这种切换称为"手工"切换，因为需要在方法 showDateChooser 中使用代码来触发它。

在触发场景时，首先需要检查当前是否显示了日期选择器，这是通过一个布尔属性（dateChooserVisible）进行判断的。因此需要在 ViewController 类中添加该属性。为此，修改文件 ViewController.h，在其中包含该属性的定义：

```
@property (nonatomic) Boolean dateChooserVisible;
```

布尔值不是对象，因此声明这种类型的属性/变量时，不需要使用关键字 strong，也无需在使用完后将其设置为 nil。然而确实需要在文件 ViewController.m 中添加配套的编译指令@synthesize：

```
@synthesize dateChooserVisible;
```

接下来实现方法 showDateChooser，使其首先核实属性 dateChooserVisible 不为 YES，再调用 performSegueWithIdentifier:sender 启动到日期选择场景的切换，然后将属性 dateChooserVisible 设置为 YES，以便我们知道当前显示了日期选择场景。这个功能是通过文件 ViewController.m 中的方法 showDateChooser 实现的，具体代码如下所示。

```
- (IBAction)showDateChooser:(id)sender {
    if (self.dateChooserVisible!=YES) {
        [self performSegueWithIdentifier:@"toDateChooser" sender:sender];
        self.dateChooserVisible=YES;
    }
}
```

此时可以运行该应用程序，并触摸"选择日期"按钮显示日期选择场景。但是用户将无法关闭模态的日期选择场景，因为还没有给"确定"按钮触发的操作编写代码。开始实现当用户单击日期选择场景中的 Done 时关闭该场景。前面已经建立了到操作 dismissDateChooser 的连接，因此只需在该方法中调用 dismissViewControllerAnimated:completion 即可。这一功能是通过文件 DateChooserViewController.m 中的方法 dismissDateChooser 实现的，其实现代码如下所示。

```
- (IBAction)dismissDateChooser:(id)sender {
    [self dismissViewControllerAnimated:YES completion:nil];
}
```

（6）实现日期计算逻辑。

为了实现日期选择器，最核心的工作是编写 calculateDateDifference 的代码。为了实现我们制定的目标（显示当前日期与选择器中的日期相差多少天），需要完成如下 3 个工作。

- 获取当前的日期。

- 显示日期和时间。
- 计算这两个日期之间相差多少天。

在具体编写代码之前，先来看看完成这些任务所需的方法和数据类型。为了获取当前的日期并将其存储在一个 NSDate 对象中，只需使用 date 方法初始化一个 NSDate。在初始化这种对象时，它默认存储当前日期。这意味着完成第一项任务只需一行代码即可实现。

```
todaysDate=[NSDate date];
```

显示日期和时间比获取当前日期要复杂。由于将在标签（UILabel）中显示输出，并且知道它将如何显示在屏幕上，因此真正的问题是如何根据 NSDate 对象获得一个字符串并设置其格式。

有趣的是，有一个类可以为我们处理这项工作！我们将创建并初始化一个 NSDateFormatter 对象；然后使用该对象的 setDateFormat 和一个模式字符串创建一种自定义格式。最后调用 NSDateFormatter 的另一个方法 stringFromDate 将这种格式应用于日期，这个方法接受一个 NSDate 作为参数，并以指定格式返回一个字符串。

假如已经将一个 NDDate 存储在变量 todaysDate 中，并要以"月份，日，年，小时：分：秒"（AM 或 PM）的格式输出，则可使用如下代码。

```
dateFormat= [[NSDateFormatter alloc] init];
[dateFormat setDateFormat:@"MMMM d,yyyy hh:mm:ssa"];
todaysDateString=[dateFormat stringFromDate:todaysDate];
```

首先，分配并初始化了一个 NSDateFormatter 对象，再将其存储到 dateFormat 中；然后将字符串@"MMMMd, yyyy hh: mm:ssa"用作格式化字符串以设置格式；最后使用 dateFormat 对象的实例方法 stringFromDate 生成一个新的字符串，并将其存储在 todaysDateString 中。

要想计算两个日期相差多少天，可以使用 NSDate 对象的实例方法 timeIntervalSinceDate 实现，而无需进行复杂的计算。这个方法返回两个日期相差多少秒，假如有两个 NSDate 对象：todaysDate 和 futureDate，可以使用如下代码计算它们之间相差多少秒。

```
NSTimeInterval difference;
    difference=[todaysDate timeIntervalSinceDate:futureDate];
```

注意，到这里将结果存储到了一个类型为 NSTimeInterval 的变量中。这种类型并非对象，在内部，它只是一个双精度浮点数。通常可以使用 C 语言数据类型 double 来声明这种变量，但 Apple 使用新类型 NSTimeInterval 进行了抽象，让我们知道日期差异计算的结果是什么。

为了计算两个日期相差多少天并显示结果,可在 ViewController.m 中实现方法 calculateDateDifference，它接受一个参数（chosenDate）。编写该方法后，可在日期选择视图控制器中编写调用该方法的代码，而这些代码将在用户使用日期选择器时被执行。

首先，在文件 ViewController.h 中，添加日期计算方法的原型：

```
- (void) calculateDateDifference: (NSDate *)chosenDate;
```

接下来在文件 ViewController.m 中添加方法 calculateDateDifference，其实现代码如下所示。

```
- (void)calculateDateDifference:(NSDate *)chosenDate {
  NSDate *todaysDate;
    NSString *differenceOutput;
    NSString *todaysDateString;
  NSString *chosenDateString;
    NSDateFormatter *dateFormat;
    NSTimeInterval difference;

    todaysDate=[NSDate date];
    difference = [todaysDate timeIntervalSinceDate:chosenDate] / 86400;

    dateFormat = [[NSDateFormatter alloc] init];
    [dateFormat setDateFormat:@"MMMM d, yyyy hh:mm:ssa"];
    todaysDateString = [dateFormat stringFromDate:todaysDate];
  chosenDateString = [dateFormat stringFromDate:chosenDate];
```

```
    differenceOutput=[[NSString alloc] initWithFormat:
                        @"选择的日期 (%@) 和今天 (%@) 相差: %1.2f 天",
                        chosenDateString,todaysDateString,fabs(difference)];
    self.outputLabel.text=differenceOutput;
}
```

上述代码的具体实现流程如下所示。

第 1 步：声明将要使用的 todaysDate 存储当前日期，differenceOutput 是最终要显示给用户的经过格式化的字符串；todaysDateString 包含当前日期的格式化版本；chosenDateString 将存储传递给这个方法的日期的格式化版本；dateFormat 是日期格式化对象，而 difference 是一个双精度浮点数变量，用于存储两个日期相差的秒数。

第 2 步：给 todaysDate 分配内存，并将其初始化为一个新的 NSDate 对象。这将自动把当前日期和时间存储到这个对象中。

第 3 步：使用 timeIntervalSinceDate 计算 todaysDate 和[sender date]之间相差多少秒。sender 将是日期选择器对象，而 date 方法命令 UIDatePicker 实例以 NSDate 对象的方式返回其日期和时间，这给我们要实现的方法提供了所需的一切。将结果除以 86400 并存储到变量 difference 中。86400 是一天的秒数，这样便能够显示两个日期相差的天数而不是秒数。

第 4 步：创建一个新的日期格式器（NSDateFormatter）对象，再使用它来格式化 todaysDate 和 chosenDate，并将结果存储到变量 todaysDateString 和 chosenDateString 中。

第 5 步：设置最终输出字符串的格式。分配一个新的字符串变量（differenceOutput）并使用 initWithFormat 对其进行初始化。提供的格式字符串包含要向用户显示的消息以及占位符%@和%1.2f，这分别表示字符串以及带一个前导零和两位小数的浮点数。这些占位符将替换为 todayDateString、chosenDateString 以及两个日期相差的天数的绝对值（fabs（defference））。

第 6 步：对我们加入到视图中的标签 differenceResult 进行更新，使其显示 differenceOutput 存储的值。

为了完成该项目，需要添加调用 calculateDateDifference 的代码，以便在用户选择日期时更新输出。实际上需要在两个地方调用 calculateDateDifference——用户选择日期时以及显示日期选择场景时。在第二种情况下，用户还未选择日期，且日期选择器显示的是当前日期。

首先考虑最重要的用例：对用户的操作做出响应，即在方法 setDateTime 被调用时，计算两个日期相差的天数，此方法在日期选择器的值发生变化时被触发。在文件 DateChooserViewController.m 中，设置方法 setDateTime 的实现代码如下所示。

```
- (IBAction)setDateTime:(id)sender {
    [(ViewController *)delegate calculateDateDifference:((UIDatePicker *)sender).date];
}
```

这样通过属性 delegate 来访问 ViewController.m 中的方法 calculateDateDifference，并将日期选择器的 date 属性传递给这个方法。需要注意的是，如果用户在没有显式选择日期的情况下退出选择器，将不会进行日期计算。

在这种情况下，可假定用户选择的是当前日期。为处理这种隐式选择，可以在文件 DateChooserViewController.m 中，设置方法 viewDidAppear 的实现代码，如下所示。

```
-(void)viewDidAppear:(BOOL)animated {
    [(ViewController *)self.delegate calculateDateDifference:[NSDate date]];
}
```

上述的代码与方法 setDateTime 相同，但是传递的是包含当前日期的 NSDate 对象，而不是日期选择器返回的日期。这确保即使用户马上关闭模态场景或弹出框，也将显示计算得到的结果。

（7）生成应用程序。

到此为止，本章的日期选择器实例编写完毕。执行后的效果如图 3-60 所示。

第 3 章　iOS 控件应用实战

图 3-60　执行效果

实例 084　获取当前的时间

实例 084	获取当前的时间
源码路径	\daima\084\

实例说明

在 iOS 应用中，UIDatePicker 是一个控制器类，封装了 UIPickerView，但是它是 UIControl 的子类，专门用于接受日期、时间和持续时长的输入。日期选取器的各列会按照指定的风格进行自动配置，这样就让开发者不必关心如何配置表盘这样的底层操作。开发者也可以对其进行定制，令其使用任何范围的日期。

在本实例中，使用 UIDatePicker 控件设置了一个日期选择器。当单击下方的"显示日期"按钮后，会调用方法- (void)buttonDidPush，使用"yyyy/MM/dd HH:mm"格式在对话框中显示当前的日期。

具体实现

实例文件 UIKitPrjDatePicker.m 的具体实现代码如下所示。

```
#import "UIKitPrjDatePicker.h"
@implementation UIKitPrjDatePicker
- (void)dealloc {
  [datePicker_ release];
  [super dealloc];
}
- (void)viewDidLoad {
  [super viewDidLoad];
  datePicker_ = [[UIDatePicker alloc] init];
  [self.view addSubview:datePicker_];
  [datePicker_ addTarget:self
             action:@selector(pickerDidChange:)
     forControlEvents:UIControlEventValueChanged];
  UIButton* button = [UIButton buttonWithType:UIButtonTypeRoundedRect];
  [button setTitle:@"显示日期" forState:UIControlStateNormal];
  [button sizeToFit];
  CGPoint newPoint = self.view.center;
  newPoint.y += 50;
  button.center = newPoint;
  [button addTarget:self
          action:@selector(buttonDidPush)
   forControlEvents:UIControlEventTouchUpInside];
  [self.view addSubview:button];
}
- (void)buttonDidPush {
  NSDateFormatter* formatter = [[[NSDateFormatter alloc] init] autorelease];
  [formatter setDateFormat:@"yyyy/MM/dd HH:mm"];
  NSString* dateString = [formatter stringFromDate:datePicker_.date];
```

```
    UIAlertView* alert = [[[UIAlertView alloc] init] autorelease];
    alert.message = dateString;
    [alert addButtonWithTitle:@"好"];
    [alert show];
}
- (void)pickerDidChange:(id)sender {
    if ( [sender isKindOfClass:[UIDatePicker class]] ) {
        UIDatePicker* picker = sender;
        NSLog( @"%@", [picker.date description] );
    }
}
@end
```

执行后的效果如图 3-61 所示。

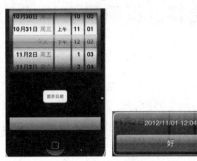

图 3-61　执行效果

实例 085　设置日期选择器中的时间间隔

实例 085	设置日期选择器中的时间间隔
源码路径	\daima\084\

实例说明

在 iOS 应用中，当使用控件 UIDatePicker 添加一个日期选择器后，可以使用其属性 minuteInterval 设置时间间隔，具体格式如下所示：

```
datePicker.minuteInterval = 30;
```

其中 30 便是设置的时间间隔。在本实例中，首先使用 UIDatePicker 控件设置了一个日期选择器，然后设置时间间隔为 30 分钟。

具体实现

实例文件 UIKitPrjMinuteInterval.m 的具体实现代码如下所示。

```
#import "UIKitPrjMinuteInterval.h"
@implementation UIKitPrjMinuteInterval
- (void)viewDidLoad {
    [super viewDidLoad];
    UIDatePicker* datePicker = [[[UIDatePicker alloc] init] autorelease];
    datePicker.minuteInterval = 30;
    datePicker.minimumDate = [NSDate date];
    datePicker.maximumDate = [NSDate dateWithTimeIntervalSinceNow:60*60*24*31];
    [self.view addSubview:datePicker];
}
@end
```

执行后的效果如图 3-62 所示。

图 3-62 执行效果

实例 086 设置日期选择器框的显示样式（1）

实例 086	设置日期选择器框的显示样式（1）
源码路径	\daima\084\

实例说明

在 iOS 应用中，当使用控件 UIDatePicker 添加一个日期选择器后，可以使用其属性 UIDatePickerMode 设置选择器框的显示样式，支持的样式如下所示。

- UIDatePickerModeTime：显示小时/分钟和可选的上午/下午指定不同的区域设置（例如 6 | 53 | 下午）。
- UIDatePickerModeDate：显示一天/月/年，取决于区域设置（例如十一月| 15 | 2007）。
- UIDatePickerModeDateAndTime：显示日期/小时/分和可选的上午/下午指定不同的区域设置（例如 11 月 15 | 6 | 53 | 下午）。
- UIDatePickerModeCountDownTimer：显示小时和分钟（如 1 | 53）。

在本实例中，首先使用 UIDatePicker 控件设置了一个日期选择器，然后设置了一个"模式切换"按钮，当单击此按钮时会切换成不同的显示样式。

具体实现

实例文件 UIKitPrjDatePickerMode.m 的具体实现代码如下所示。

```
#import "UIKitPrjDatePickerMode.h"
@implementation UIKitPrjDatePickerMode
- (void)dealloc {
  [datePicker_ release];
  [super dealloc];
}
- (void)viewDidLoad {
  [super viewDidLoad];

  datePicker_ = [[UIDatePicker alloc] init];
  datePicker_.datePickerMode = UIDatePickerModeCountDownTimer;
  [self.view addSubview:datePicker_];
  UIButton* button = [UIButton buttonWithType:UIButtonTypeRoundedRect];
  [button setTitle:@"模式切换" forState:UIControlStateNormal];
  [button sizeToFit];
  CGPoint newPoint = self.view.center;
  newPoint.y += 50;
  button.center = newPoint;
  [button addTarget:self
             action:@selector(buttonDidPush)
    forControlEvents:UIControlEventTouchUpInside];
  [self.view addSubview:button];
}
- (void)buttonDidPush {
  if ( UIDatePickerModeCountDownTimer < ++datePicker_.datePickerMode ) {
```

```
    datePicker_.datePickerMode = UIDatePickerModeTime;
  }
}
@end
```

执行后的效果如图 3-63 所示。

图 3-63　执行效果

实例 087　设置日期选择器框的显示样式（2）

实例 087	设置日期选择器框的显示样式（2）
源码路径	\daima\084\

实例说明

在 iOS 应用中，当使用控件 UIDatePicker 添加一个日期选择器后，可以使用方法+ (id)dateWithTimeIntervalSinceNow 返回当前的时间。在本实例中，首先使用 UIDatePicker 控件设置了一个日期选择器，然后设置了一个"返回来到今日"按钮，当单击此按钮时会返回到当前今日时间。

具体实现

实例文件 UIKitPrjSetDate.m 的具体实现代码如下所示。

```
#import "UIKitPrjSetDate.h"
@implementation UIKitPrjSetDate
- (void)dealloc {
  [datePicker_ release];
  [super dealloc];
}
- (void)viewDidLoad {
  [super viewDidLoad];
  datePicker_ = [[UIDatePicker alloc] init];
  datePicker_.date = [NSDate dateWithTimeIntervalSinceNow:-1*60*60*24*1];
  [self.view addSubview:datePicker_];
  UIButton* button = [UIButton buttonWithType:UIButtonTypeRoundedRect];
  [button setTitle:@"返回来到今日" forState:UIControlStateNormal];
  [button sizeToFit];
  CGPoint newPoint = self.view.center;
  newPoint.y += 50;
  button.center = newPoint;
  [button addTarget:self
             action:@selector(buttonDidPush)
   forControlEvents:UIControlEventTouchUpInside];
  [self.view addSubview:button];
}
- (void)buttonDidPush {
  [datePicker_ setDate:[NSDate date]];
}
@end
```

执行后的效果如图 3-64 所示。

第 3 章 iOS 控件应用实战

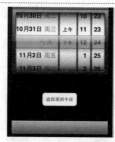

图 3-64 执行效果

实例 088　实现自动倒计时功能

实例 088	实现自动倒计时功能
源码路径	\daima\084\

实例说明

在 iOS 应用中，我们可以自己定制一个日期选择器框。例如在本实例中，将属性 datePickerMode 设置为 UIDatePickerModeCountDownTimer，这个选择框变成了一个倒计时牌，然后通过 NSTimer 定期更新选择框的 countDownDuration 属性，这样就实现了自动倒计时功能。

具体实现

实例文件 UIKitPrjCountDownTimer.m 的具体实现代码如下所示。

```
#import "UIKitPrjCountDownTimer.h"
@implementation UIKitPrjCountDownTimer
- (void)dealloc {
  [timer_ release];
  [datePicker_ release];
  [super dealloc];
}
- (void)viewDidLoad {
  [super viewDidLoad];
  // 倒计时牌的创建与初始化
  datePicker_ = [[UIDatePicker alloc] init];
  datePicker_.datePickerMode = UIDatePickerModeCountDownTimer;
  // 初始显示为 5 min
  datePicker_.countDownDuration = 60*5;
  [self.view addSubview:datePicker_];
  // 创建每隔 1 秒重复调用的 Timer
  timer_ = [[NSTimer timerWithTimeInterval:60.0
                     target:self
                     selector:@selector(timerFireMethod:)
                     userInfo:nil
                     repeats:YES] retain];
  // Timer 运行开始
  NSRunLoop* runLoop = [NSRunLoop currentRunLoop];
  [runLoop addTimer:timer_ forMode:NSDefaultRunLoopMode];
}
- (void)viewWillDisappear:(BOOL)animated {
  [super viewWillDisappear:animated];
  // 画面隐藏时停止 Timer
  if ( [timer_ isValid] ) {
    [timer_ invalidate];
  }
}
- (void)timerFireMethod:(NSTimer*)theTimer {
  // Timer 会在每隔 1 秒时调用此方法
  // 至计数为 0.0 为止，每间隔 1 分钟进行减算
  NSTimeInterval now = datePicker_.countDownDuration;
```

```
        if ( 0.0 < now ) {
          datePicker_.countDownDuration = now - 60.0;
        }
      }
@end
```

执行效果如图 3-65 所示。

图 3-65 执行效果

实例 089 使用选择器视图

实例 089	使用选择器视图
源码路径	\daima\089\

实例说明

从外观上说，选择器视图（UIPickerView）类似于日期选择器，但其实现几乎完全不同。在选择器视图中，只定义了整体行为和外观，选择器视图包含的组件数以及每个组件的内容都将由开发者定义。图 3-66 所示的选择器视图包含两个组件，它们分别显示文本和图像。

要想在应用程序中添加选择器视图，可以使用 Interface Builder 编辑器从对象库拖曳选择器视图到视图中。但是不能在 Connections Inspector 中配置选择器视图的外观，而需要编写遵守两个协议的代码，其中一个协议提供选择器的布局（数据源协议），另一个协议提供选择器将包含的信息（委托）。可使用 Connections Inspector 将委托和数据源输出口连接到一个类，也可以使用代码设置这些属性。

图 3-66 可以对选择器视图进行配置

在本实例中，实现了两个 UIPickerView 控件之间的数据依赖。

具体实现

（1）首先在工程中创建一个 songInfo.plist 文件用以储存数据，如图 3-67 所示。
添加的内容如图 3-68 所示。
（2）在 ViewController 中设置一个选取器 pickerView 对象，两个数组存放选取器数据和一个字典，读取 plist 文件。具体代码如下所示。

```
#import <UIKit/UIKit.h>
@interface ViewController : UIViewController<UIPickerViewDelegate,UIPickerViewDataSource>
{
//定义滑轮组件
    UIPickerView *pickerView;
//    储存第一个选取器的的数据
    NSArray *singerData;
//    储存第二个选取器
    NSArray *singData;
```

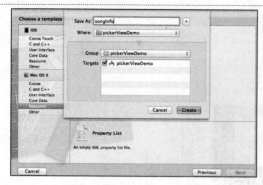
图 3-67　新建 songInfo.plist 文件

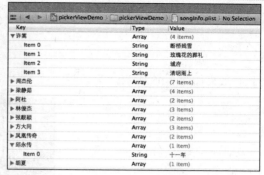
图 3-68　添加的数据

```
//    读取 plist 文件数据
    NSDictionary *pickerDictionary;
}
-(void)buttonPressed:(id)sender;
@end
```

（3）在 ViewController.m 文件的 ViewDidLoad 完成初始化。首先定义如下两个宏定义：

```
#define singerPickerView 0
#define singPickerView 1
```

分别表示两个选取器的索引序号值，放在 #import "ViewController.h"后面。

```
- (void)viewDidLoad
{
    [super viewDidLoad];
  // Do any additional setup after loading the view, typically from a nib.

    pickerView = [[UIPickerView alloc] initWithFrame:CGRectMake(0, 0, 320, 216)];
//    指定 Delegate
    pickerView.delegate=self;
    pickerView.dataSource=self;
//    显示选中框
    pickerView.showsSelectionIndicator=YES;
    [self.view addSubview:pickerView];
//    获取 mainBundle
    NSBundle *bundle = [NSBundle mainBundle];
//    获取 songInfo.plist 文件路径
    NSURL *songInfo = [bundle URLForResource:@"songInfo" withExtension:@"plist"];
//    把 plist 文件里的内容存入数组
    NSDictionary *dic = [NSDictionary dictionaryWithContentsOfURL:songInfo];
    pickerDictionary=dic;
//    将字典里面的内容取出放到数组中
    NSArray *components = [pickerDictionary allKeys];
//选取出第一个滚轮中的值
    NSArray *sorted = [components sortedArrayUsingSelector:@selector(compare:)];
    singerData = sorted;
//    根据第一个滚轮中的值，选取第二个滚轮中的值
    NSString *selectedState = [singerData objectAtIndex:0];
    NSArray *array = [pickerDictionary objectForKey:selectedState];
    singData=array;
//    添加按钮
    CGRect frame = CGRectMake(120, 250, 80, 40);
    UIButton *selectButton = [UIButton buttonWithType:UIButtonTypeRoundedRect];
    selectButton.frame=frame;
    [selectButton setTitle:@"SELECT" forState:UIControlStateNormal];

    [selectButton addTarget:self action:@selector(buttonPressed:) forControlEvents:
UIControlEventTouchUpInside];
    [self.view addSubview:selectButton];
}
```

实现按钮事件的代码如下所示。

```
-(void) buttonPressed:(id)sender
```

```objc
{
//获取选取器某一行索引值
    NSInteger singerrow =[pickerView selectedRowInComponent:singerPickerView];
    NSInteger singrow = [pickerView selectedRowInComponent:singPickerView];
//将 singerData 数组中的值取出
    NSString *selectedsinger = [singerData objectAtIndex:singerrow];
    NSString *selectedsing = [singData objectAtIndex:singrow];
    NSString *message = [[NSString alloc] initWithFormat:@"你选择了 %@ 的 %@",selectedsinger,selectedsing];

    UIAlertView *alert = [[UIAlertView alloc] initWithTitle:@"提示"
                                                    message:message
                                                   delegate:self
                                          cancelButtonTitle:@"OK"
                                          otherButtonTitles: nil];
    [alert show];
}
```

(4) 关于两个协议的代理方法的实现代码如下所示。

```objc
#pragma mark -
#pragma mark Picker Date Source Methods
//返回显示的列数
-(NSInteger)numberOfComponentsInPickerView:(UIPickerView *)pickerView
{
//返回几就有几个选取器
    return 2;
}
//返回当前列显示的行数
-(NSInteger)pickerView:(UIPickerView *)pickerView numberOfRowsInComponent:(NSInteger)component
{
    if (component==singerPickerView) {
        return [singerData count];
    }
    return [singData count];
}
#pragma mark Picker Delegate Methods
//返回当前行的内容,此处是将数组中数值添加到滚动的那个显示栏上
-(NSString*)pickerView:(UIPickerView *)pickerView titleForRow:(NSInteger)row forComponent:(NSInteger)component
{
    if (component==singerPickerView) {
        return [singerData objectAtIndex:row];
    }
    return [singData objectAtIndex:row];
}
-(void)pickerView:(UIPickerView *)pickerViewt didSelectRow:(NSInteger)row inComponent:(NSInteger)component
{
//如果选取的是第一个选取器
    if (component == singerPickerView) {
//得到第一个选取器的当前行
        NSString *selectedState =[singerData objectAtIndex:row];
//根据从 pickerDictionary 字典中取出的值,选择对应的值
        NSArray *array = [pickerDictionary objectForKey:selectedState];
        singData=array;
        [pickerView selectRow:0 inComponent:singPickerView animated:YES];
//重新装载第二个滚轮中的值
        [pickerView reloadComponent:singPickerView];
    }
}
//设置滚轮的宽度
-(CGFloat)pickerView:(UIPickerView *)pickerView widthForComponent:(NSInteger)component
{
    if (component == singerPickerView) {
        return 120;
    }
    return 200;
}
```

在这个方法中，-(void)pickerView:(UIPickerView *)pickerViewt didSelectRow:(NSInteger)row inComponent:(NSInteger)component，把(UIPickerView *)pickerView 参数改成了(UIPickerView *) pickerViewt，因为定义的 pickerView 对象和参数发生冲突，所以将参数进行了修改。

这样整个实例介绍完毕，执行后的效果如图 3-69 所示。

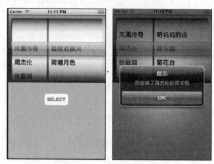

图 3-69　执行效果

实例 090　自定义一个选择器

实例 090	自定义一个选择器
源码路径	\daima\090\

实例说明

在本实例中将创建一个自定义选择器，它包含两个组件，一个显示动物图像，另一个显示动物声音。当用户在自定义选择器视图中选择动物图像或声音时，在输出标签中将显示出用户所做的选择。在实现本实例时，必须让一个类遵守选择器委托协议和选择器数据源协议，但是这个应用程序的很多核心处理都与前一个应用程序相同。初始场景包含一个输出标签，还有一个只包含一个按钮的工具栏。触摸该按钮将切换到自定义选择器场景。在该场景中，用户可以操纵自定义选择器，并通过触摸"好"按钮返回到初始场景。另外，在本实例中添加了防止出现多个弹出框的逻辑，逻辑和实现几乎相同。

具体实现

（1）打开 Xcode，使用模板 Single View Application 新建一个项目，并将其命名为"CustomPicker"，如图 3-70 所示。建议选择设备是"iPad"。

图 3-70　创建 Xcode 项目

（2）添加图片资源。

为了让自定义选择器显示动物照片，需要在项目中添加一些图像。为此，将文件夹 Images 拖曳到码编组中，在 Xcode 询问时选择复制文件并创建编组。

打开项目中的 Images 编组，核实其中有 7 幅图像：bear.png、cat.png、dog.png、goose.png、mouse.pmg、pig.png 和 snake.png。

（3）添加 Animal ChooserViewController 类。

AnimalChooserViewController 类将负责处理这样的场景，即其中有一个包含动物和声音的自定义选择器。为此，单击项目导航器左下角的+按钮，新建一个 UIViewController 子类，并将其命名为"AnimalChooserViewController"，将这个新类放到项目代码编组中。

（4）添加动物选场景并关联视图控制器。

打开文件 MainStoryboard.storyboard 和对象库（Control+Option+Command+3），将一个视图控制器拖曳到 Interface Builder 编辑器的空白区域（或文档大纲区域）。选择新场景的视图控制器图标，按 Option+Command+3 打开 Identity Inspector，并从 Class 下拉列表中选择 AnimalChooserViewController。使用 Identity Inspector 将第一个场景的视图控制器标签设置为 Initial，将第二个场景的视图控制器标签设置为 Animal Chooser。这些修改将立即在文档大纲中反映出来。

（5）规划变量和连接。

本项目需要的输出口和操作与前一个项目相同，但有一个例外。在前一个项目中，当日期选择器的值发生变化时，需要执行一个方法，但在这个项目中，我们将实现选择器协议，其中包含的一个方法将在用户使用选择器时自动被调用。

在初始场景中，将包含一个输出标签（outputLabel），还有一个用于显示动物选择场景的操作（showAnimalChooser）。该场景的视图控制器类 ViewController 将通过属性 animalChooserVisible 跟踪动物选择场景是否可见，还有一个显示用户选择的动物和声音的方法——displayAnimal: WithSound: FromComponent。

（6）添加表示自定义选择器组件的常量。

在创建自义选择器时，必须实现各种协议方法，而在这些方法中需要使用数字来引用组件。为了简化自定义选择器实现，可以定义一些常量，这样就可使用符号来引用组件了。

在本实例项目中，组件 0 表示动物组件，而组件 1 为声音组件。通过在实现文件开头定义几个常量，可以通过名称来引用组件。为此，在文件 AnimalChooserView.m 中，在#import 代码后面添加下面的代码。

```
#define kComponentCount 2
#define kAnimalComponent 0
#define kSoundComponent 1
```

第一个常量 kComponent Count 是要在选择器中显示的组件数，而其他两个常量——kAnimal Component 和 kSoundComponent 可用于引用选择器中不同的组件，而无需借助于它们的实际编号。

（7）设计界面。

打开文件 MainStoryboard.storyboard，滚动到在编辑器中能够看到初始场景。打开对象库（Control+Option+ Command+3），并拖曳一个工具栏到该视图底部。修改默认栏按钮项的文本，将其改为"选择图片和文字"。使用两个灵活间距栏按钮项（Flexible Space Bar Button Item）让该按钮位于工具栏中央。

然后在视图中央添加一个标签，将其文本改为 Nothing Selected。使用 Attributes Inspector，让文本居中、增大标签的字体并将标签扩大到至少能够容纳 5 行文本。下面的图 3-71 显示了初始视图的布局。

使用和前面配置日期选择场景一样的方法配置动物选择场景：设置背景色，添加一个文本为"请选择图像和文字"的标签，但拖曳一个选择器视图对象到场景顶部。因为我们创建的是 iPad 版，所以该视图最终将显示为弹出框，因此只有左上角部分可见。下面的图 3-72 是设计的图像选择界面。

接下来开始设置选择器视图的数据源和委托。在这个项目中，让类 AnimalChooserViewController 承担双重职责，即充当选择器视图的数据源和委托。换句话说，AnimalChooserViewController 类将负责实现让自定义选择器能够正常运行所需的所有方法。

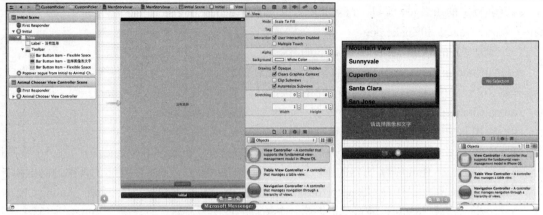

图 3-71　初始场景　　　　　　　　　图 3-72　图像选择场景

要为选择器视图设置数据源和委托，可以在动物选择场景或文档大纲区域选择，再打开 Connections Inspector（Option+Command+6）。从输出口 dataSource 拖曳到文档大纲中的视图控制器图标 Animal Chooser，对输出口 delegate 做相同的处理。完成这些处理后，Connection Inspector 如图 3-73 所示，这样将选择器视图的输出口 dataSource 和 delegate 连接到视图控制器对象 Animal Chooser。

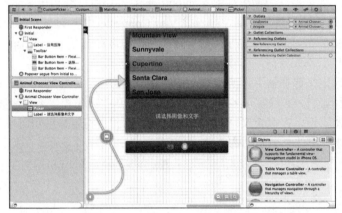

图 3-73　Connection Inspector 界面

（8）创建切换。

现在在场景之间创建切换。按住"Ctrl"键，从初始场景的视图控制器拖曳到图像选择场景的视图控制器，创建一个模态切换（iPhone）或弹出切换（iPad）。创建切换后，将新增内容为 Segue from UIViewController to AnimalChooseViewController 的一行。打开 Attributes　Inspector（Option+Command+4）以配置该切换。

给切换指定标识符 toAnimalChooser，在实现代码中我们将使用这个 ID 来触发切换。

在该应用程序的 iPad 版中，需要设置弹出框的锚。为此打开 Attributes Inspector，并从文本框 Anchor 拖曳到初始场景中工具栏上的 Choose an Animal and Sound 按钮。接下来，选择图像并选择场景的视图对象，并打开 Size Inspector。将宽度和高度都设置为大约 320 点，调整该视图的内容，使其间距比较合适。

接下来，选择日期选择场景的视图对象，并打开 Size Inspector。将宽度和高度都设置为大约 320 点，调整该视图的内容，使其刚好居中。

（9）创建并连接输出口和操作。

本项目一共需要建立两个连接，分别是初始场景的一个操作接口和一个输出口。具体说明如下所示。

- outputLabel（UILabel）：该标签在初始场景中显示用户与选择器视图交互的结果。
- showAnimalChooser：这是一个操作方法，由初始场景中的栏按钮项"选择图像和文字"触发。

切换到助手编辑器并建立连接。选择初始场景中的输出标签，按住 Control 并从该标签拖曳到文件 ViewController.h 编译指令@interface 下方。在 Xcode 提示时，创建一个名为"outputLabel"的新输出口。

在初始场景中按住"Ctrl"键，并从按钮"选择图像和文字"拖曳到文件 ViewController.h 中属性定义的下方。在 Xcode 提示时，添加一个名为"showAnimalChooser"的新操作。

（10）实现场景切换逻辑。

在自定义选择器视图的实现时，需要确保 iPad 版不会显示多个相互堆叠的动物选择场景，为此将采取 DateCalc 采取的方式。

修改两个视图控制器类的接口文件，让它们彼此导入对方的接口文件。为此在文件 ViewController.h 中，在现有的#import 语句下方添加如下代码行：

```
#import "AnimalChooserViewController.h"
```

在文件 AnimalChooserViewController.h 中，添加导入 ViewController.h 的代码：

```
#import"ViewController.h"
```

使用属性 delegate 来访问初始场景的视图控制器，在文件 AnimalChooserViewController.h 中，在编译指令@interface 后面添加如下代码行：

```
@property (strong, nonatomic) id delegate;
```

接下来修改文件 AnimalChooserViewController.m，在@implementation 后面添加配套的编译指令@synthesize：

```
@synthesize delegate;
```

开始执行清理工作，将该实例"变量/属性"设置为 nil。为此，在文件 AnimalChooserViewController.m 的方法 viewDidUnload 中添加如下代码：

```
[self setDelegate:nil];
```

为了设置属性 delegate，修改文件 ViewController.m，在其中添加如下所示的代码。

```
- (void)prepareForSegue:(UIStoryboardSegue *)segue sender:(id)sender {
    ((AnimalChooserViewController *)segue.destinationViewController).delegate=self;
}
```

在本项目中，我们使用一个属性（animalChooserVisible）来存储动物选择场景的当前可见性。修改文件 ViewController.h，在其中包含该属性的定义：

```
@property (nonatomic) Boolean animalChooserVisible;
```

在文件 ViewController.m 中添加配套的编译指令@synthesize：

```
@synthesize animalChooserVisible;
```

实现方法 showAnimalChooser，使其在标记 animalChooserVisible 为 NO 时调用 performSegueWithIdentifier:sender。下面的代码显示了在文件 ViewController.m 中实现的方法 showAnimalChooser。

```
- (IBAction)showAnimalChooser:(id)sender {
    if (self.animalChooserVisible!=YES) {
        [self performSegueWithIdentifier:@"toAnimalChooser" sender:sender];
        self.animalChooserVisible=YES;
    }
}
```

为了在图像选择场景关闭时将标记 animalChooserVisible 设置为 NO，可在文件 AnimalChooserViewController.m 的方法 viewWillDisappear 中使用如下所示的代码。

```
- (void)viewWillDisappear:(BOOL)animated
{
    [super viewWillDisappear:animated];
}
```

（11）实现自定义选择器视图。

在这个示例项目中，将创建一个自定义选择器视图并选择它，在两个组件中分别显示图像和文本。

要显示选择器，需要给它提供数据。我们已经将图像资源加入到项目中，但要将这些图像提供给选择器，需要通过名称引用它们。另外，还需要在动物图像和动物名之间进行转换，即如果用户选择了小猪图像，我们希望应用程序显示 Pig，而不是 pig.png。为此，我们将创建一个动物图像数组（animalImages）和一个动物名数组（animalName）；在这两个数组中，同一种动物的图像和名称的索引相同。例如，如果用户选择的动物图像对应于数组 animal Images 的第三个元素，则可从数组 animalNames 的第三个元素获取动物名。我们还需要表示动物声音的数据，它们显示在选择器视图的第二个组件中。因此还需创建第三个数组：animalSounds。

在文件 AnimalChooserViewController.h 中，通过如下代码将这 3 个数组声明为属性：

```
@property (strong, nonatomic) NSArray *animalNames;
@property (strong, nonatomic) NSArray *animalSounds;
@property (strong, nonatomic) NSArray *animalImages;
```

然后，在文件 AnimalChooserViewController.m 中，添加配套的编译指令@synthesize：

```
@synthesize animalNames;
@synthesize animalSounds;
@synthesize animalImages;
```

再在方法 viewDidUnload 中清理这些属性：

```
[self setAnimalNames:nill;
[self setAnimalImages:nil];
[self setAnimalSounds:nil];
```

现在需要分配并初始化每个数组。对于名称和声音数组，只需在其中存储字符串即可。

然而对于图像数组来说，需要在其中存储 UIImageView。在文件 AnimalChooserViewController.m 中方法 viewDidLoad 的实现代码如下所示。

```
- (void)viewDidLoad
{
    self.animalNames=[[NSArray alloc]initWithObjects:
                    @"Mouse",@"Goose",@"Cat",@"Dog",@"Snake",@"Bear",@"Pig",nil];
    self.animalSounds=[[NSArray alloc]initWithObjects:
@"Oink",@"Rawr",@"Ssss",@"Roof",@"Meow",@"Honk",@"Squeak",nil];
    self.animalImages=[[NSArray alloc]initWithObjects:
                    [[UIImageView alloc] initWithImage:[UIImage imageNamed:@"mouse.png"]],
                    [[UIImageView alloc] initWithImage:[UIImage imageNamed:@"goose.png"]],
                    [[UIImageView alloc] initWithImage:[UIImage imageNamed:@"cat.png"]],
                    [[UIImageView alloc] initWithImage:[UIImage imageNamed:@"dog.png"]],
                    [[UIImageView alloc] initWithImage:[UIImage imageNamed:@"snake.png"]],
                    [[UIImageView alloc] initWithImage:[UIImage imageNamed:@"bear.png"]],
                    [[UIImageView alloc] initWithImage:[UIImage imageNamed:@"pig.png"]],
                    nil
                    ];
    [super viewDidLoad];
}
```

在文件 AnimalChooserViewController.h 中，将@interface 行设置为如下格式：

```
@interface AnimalChooserViewController:
UIViewController <UIPickerViewDataSource>
```

这样将这个类声明为遵守协议 UIPickerViewDataSource。

接下来，实现方法 numberOfComponentsInPickerView，此方法返回选择器将显示多少个组件。由于我们为此定义了一个常量（kComponentCount），因此只需返回该常量即可，具体代码如下所示。

```
- (NSInteger)numberOfComponentsInPickerView:(UIPickerView *)pickerView {
    return kComponentCount;
}
```

必须实现的另一个数据源方法是 pickerView:numberOfRowsInComponent，功能是根据编号返回相应组件将显示的元素数。为简化确定组件的方式，可以使用常量 kAnimalComponent 和 kSound-

Component，并使用类 NArray 的方法 count 来获取数组包含的元素数。pickerView:number OfRowsIn-Component 的实现代码如下所示。

```
- (NSInteger)pickerView:(UIPickerView *)pickerView
numberOfRowsInComponent:(NSInteger)component {
    if (component==kAnimalComponent) {
        return [self.animalNames count];
    } else {
        return [self.animalSounds count];
    }
}
```

在上述代码中，第 3 行检查查询的组件是否是动物组件。如果是，第 4 行返回数组 animalNames 包含的元素数（也可以返回图像数组包含的元素数）。如果查询的不是动物组件，便可认为查询的是声音组件（第 5 行），因此返回数组 Sounds 包含的元素数。

这就是实现数据源协议需要做的全部工作。其他与选择器视图相关的工作由选择器视图委托协议（UIPickerViewDelegate）处理。

选择器视图委托协议负责定制选择器的显示方式，以及在用户在选择器中选择时做出反应。在文件 AnimalChooserViewController.h 中，指出我们要遵守委托协议：

```
@interface AnimalChooserViewController:UIViewController
<UIPickerViewDataSource, UIPickerViewDelegate>
```

要生成我们所需的选择器，需要实现多个委托方法，但其中最重要的是 pickerView:viewForRow:forComponent:reusingView。这个方法接受组件和行号作为参数，并返回要在选择器相应位置显示的自定义视图。

在我们的实现中，需要给第一个组件返回动物图像，并给第二个组件返回标签，其中包含对动物声音的描述。在项目中，通过如下代码实现这个方法。

```
- (UIView *)pickerView:(UIPickerView *)pickerView viewForRow:(NSInteger)row
        forComponent:(NSInteger)component reusingView:(UIView *)view {
    if (component==kAnimalComponent) {
        return [self.animalImages objectAtIndex:row];
    } else {
        UILabel *soundLabel;
        soundLabel=[[UILabel alloc] initWithFrame:CGRectMake(0,0,100,32)];
//      [soundLabel autorelease];
        soundLabel.backgroundColor=[UIColor clearColor];
        soundLabel.text=[self.animalSounds objectAtIndex:row];
        return soundLabel;
    }
}
```

如果现在运行该应用程序，将看到自定义选择器，但显得比较拥挤。为调整选择器视图的组件大小，可以实现另外两个委托方法——pickerView：rowHeightForComponent 和 pickerView:widthForComponent。

对于这个示例应用程序，通过试错确定动物组件的宽度应为 75 点，而声音组件在宽度大约为 150 点时看起来最佳。这两个组件都应使用固定的行高-55 点。上述功能是在文件 AnimalChooserViewController.m 中实现的，具体代码如下所示：

```
- (CGFloat)pickerView:(UIPickerView *)pickerView
rowHeightForComponent:(NSInteger)component {
    return 55.0;
}
- (CGFloat)pickerView:(UIPickerView *)pickerView widthForComponent:(NSInteger)component
{
    if (component==kAnimalComponent) {
        return 75.0;
    } else {
        return 150.0;
    }
}
```

在日期选择器示例中，将选择器连接到一个操作方法，并使用事件 Value Changed 来捕获用户修改选择器的操作。不幸的是，自定义选择器的工作原理不是这样的。要获取用户在自定义选择器中所做的选择，必须实现另一个委托方法：pickerView:didSelectRowIncomponent。给这个方法提供的参数为用户选择的组件和行号。这个委托方法给我们提供了用户选择的组件和行号，但是并没有指出其他组件的状态。要获取其他组件的值，我们必须使用选择器的实例方法 selectedRowInComponent，并将我们要获取其值的组件作为参数提供给它。

在这个项目中，当用户做出选择时，将调用方法 displayAnimal:withSound:fromComponent，将选择情况显示在初始场景的输出标签中。我们还没有实现这个方法，现在就要这样做。在文件 ViewController.h 中，添加这个方法的原型：

```
- (void)displayAnimal:(NSString *)chosenAnimal
withSound: (NSString *)chosenSound
fromComponent: (NSString*)chosenComponent;
```

在文件 ViewControler.m 中实现这个方法，它应将传入的字符串参数显示在输出标签中。具体代码如下所示：

```
- (void)displayAnimal:(NSString *)chosenAnimal withSound:(NSString *)chosenSound
fromComponent:(NSString *)chosenComponent {
    NSString *animalSoundString;
    animalSoundString=[[NSString alloc]
                initWithFormat:@"你改变 %@ (%@ 和声音文字 %@)",chosenComponent,
chosen Animal,chosenSound];
    self.outputLabel.text=animalSoundString;
}
```

根据字符串参数 chosenComponent、chosenAnimal 和 chosenSound 的内容，创建了一个字符串：animalSoundString，然后设置输出标签的内容，以显示这个字符串。

有了用于显示用户选择情况的机制后，需要在用户选择时做出响应了。在文件 AnimalChooserViewController.m 中，实现方法 pickerView:didSelectRow:inComponent，具体代码如下所示。

```
- (void)pickerView:(UIPickerView *)pickerView didSelectRow:(NSInteger)row
        inComponent:(NSInteger)component {

    ViewController *initialView;
    initialView=(ViewController *)self.delegate;

    if (component==kAnimalComponent) {
        int chosenSound=[pickerView selectedRowInComponent:kSoundComponent];
        [initialView displayAnimal:[self.animalNames objectAtIndex:row]
                    withSound:[self.animalSounds objectAtIndex:chosenSound]
                fromComponent:@"动物图像"];
    } else {
        int chosenAnimal=[pickerView selectedRowInComponent:kAnimalComponent];
        [initialView displayAnimal:[self.animalNames objectAtIndex:chosenAnimal]
                    withSound:[self.animalSounds objectAtIndex:row]
                fromComponent:@"声音"];
    }
}
```

与日期选择器一样，用户显示自定义选择器后，也可能在不做任何选择的情况下关闭它。在这种情况下，我们应假定用户想选择默认的动物和声音。为实现这一点，我们可在动物选择场景显示后，立即更新初始场景中的输出标签，让其显示默认的动物名和声音以及一条消息，让消息指出用户没有做任何选择（nothing yet...）。

与日期选择器一样，可以在文件 AnimalChooserViewController.m 的方法 viewDidAppear 中实现，具体代码如下所示：

```
-(void)viewDidAppear:(BOOL)animated {
    ViewController *initialView;
    initialView=(ViewController *)self.delegate;
    [initialView displayAnimal:[self.animalNames objectAtIndex:0]
                withSound:[self.animalSounds objectAtIndex:0]
            fromComponent:@"还没有..."];
}
```

通过调用方法 displayAnimal:withSound:fromComponent，并将动物名数组和声音数组的第一个元素传递给它，因为它们是选择器默认显示的元素。对于参数 fromComponent，则将其设置为一个字符串，指出用户还未做出选择。

运行后，当用户在选择器视图（显示在一个弹出框中）做出选择后，输出标签将立即更新。执行效果如图 3-74 所示。

图 3-74　执行效果

实例 091　实现一个数字选择器

实例 091	实现一个数字选择器
源码路径	\daima\091\

实例说明

本实例也是通过 UIPickerView 控件实现的，首先实现了 UIPickerView 控件的创建与初始化操作，然后使用 numberOfComponentsInPickerView 方法将选择列数设置为 3 列，使用 numberOfRowsInComponent 方法设置各列拥有 10 行。

具体实现

实例文件 UIKitPrjPickerView.m 的具体代码如下所示。

```
#import "UIKitPrjPickerView.h"
@implementation UIKitPrjPickerView
- (void)viewDidLoad {
  [super viewDidLoad];
  //UIPickerView 的创建与初始化
  UIPickerView* picker = [[[UIPickerView alloc] init] autorelease];
  picker.delegate = self;//将 delegate 设置成自己
  picker.dataSource = self;//将 dataSource 也设置为自己
  [self.view addSubview:picker];
}
- (NSInteger)numberOfComponentsInPickerView:(UIPickerView*)pickerView {
  return 3;//选择列数为 3 列
}
- (NSInteger)pickerView:(UIPickerView*)pickerView numberOfRowsInComponent:(NSInteger)component {
  return 10;//各列拥有 10 行
}
- (NSString*)pickerView:(UIPickerView*)pickerView
  titleForRow:(NSInteger)row forComponent:(NSInteger)component
{
  //显示各选项的内容
  return [NSString stringWithFormat:@"%d-%d", row, component];
}
@end
```

执行后的效果如图 3-75 所示。

图 3-75 执行效果

实例 092 突出显示选择器中的某一行

实例 092	突出显示选择器中的某一行
源码路径	\daima\091\

实例说明

本实例也是通过 UIPickerView 控件实现的，首先使用 UIPickerView 控件创建了一个选择器，然后将属性 showsSelectionIndicator 设置为 YES，这表示明确地标示出选择器中的某一行。然后在下方设置了一个"信息显示"按钮，当单击此按钮时，调用方法 selectedRowInComponent 在弹出框中显示当前选中的行。

具体实现

实例文件 UIKitPrjSelectedRow.m 的具体代码如下所示。

```
#import "UIKitPrjSelectedRow.h"
@implementation UIKitPrjSelectedRow
- (void)dealloc {
  [picker_ release];
  [super dealloc];
}
- (void)viewDidLoad {
  [super viewDidLoad];
  picker_ = [[UIPickerView alloc] init];
  picker_.delegate = self;
  picker_.showsSelectionIndicator = YES;
  [self.view addSubview:picker_];
  UIButton* button = [UIButton buttonWithType:UIButtonTypeRoundedRect];
  [button setTitle:@"信息显示" forState:UIControlStateNormal];
  [button sizeToFit];
  CGPoint newPoint = self.view.center;
  newPoint.y += 50;
  button.center = newPoint;
  [button addTarget:self
            action:@selector(buttonDidPush)
   forControlEvents:UIControlEventTouchUpInside];
  [self.view addSubview:button];
}
- (NSInteger)numberOfComponentsInPickerView:(UIPickerView*)pickerView {
  return 3;
}
- (NSInteger)pickerView:(UIPickerView*)pickerView numberOfRowsInComponent:(NSInteger)component {
  return 10;
}
- (NSString*)pickerView:(UIPickerView*)pickerView
  titleForRow:(NSInteger)row forComponent:(NSInteger)component
```

```
{
  return [NSString stringWithFormat:@"%d", row];
}
- (void)buttonDidPush {
  NSInteger component0 = [picker_ selectedRowInComponent:0];
  NSInteger component1 = [picker_ selectedRowInComponent:1];
  NSInteger component2 = [picker_ selectedRowInComponent:2];
  NSString* message = [NSString stringWithFormat:@"%d - %d - %d", component0, component1, component2];
  UIAlertView* alert = [[[UIAlertView alloc] init] autorelease];
  alert.message = message;
  [alert addButtonWithTitle:@"OK"];
  [alert show];
}
@end
```

执行后的效果如图 3-76 所示。

图 3-76　执行效果

实例 093　向选择器中添加 UIView 子类

实例 093	向选择器中添加 UIView 子类
源码路径	\daima\091\

实例说明

在 iOS 应用中，当通过 UIPickerView 控件在屏幕中实现选择器效果时，可以继续使用 UIView 子类向选择器中添加新的元素。在本实例中，首先导入了选项所使用的图片，然后追加了信息显示按钮，当触摸此按钮时选中的选项图片将显示在工具条上，最后注册按钮点击后的响应方法。使用方法 numberOfComponentsInPickerView 返回列数，使用方法返回行数。当单击下方的"信息显示"按钮时，调用 viewForRow:forComponent:方法取得选择行每列中的 UIView 对象，并以取得的 UIView 对象为基础创建 UIBarButtonItem 对象。

具体实现

实例文件 UIKitPrjViewForRow.m 的具体代码如下所示。

```
#import "UIKitPrjViewForRow.h"
@implementation UIKitPrjViewForRow
- (void)dealloc {
  [dataSource_ release];
  [picker_ release];
  [super dealloc];
}
- (void)viewDidLoad {
  [super viewDidLoad];
  //创建选择框并初始化
  picker_ = [[UIPickerView alloc] init];
  picker_.delegate = self;
```

```objc
    picker_.showsSelectionIndicator = YES;
    [self.view addSubview:picker_];
    //导入选项用的图片
    UIImage* imageDog = [UIImage imageNamed:@"Dog.png"];
    UIImage* imageMonkey = [UIImage imageNamed:@"Monkey.png"];
    UIImage* imageElephant = [UIImage imageNamed:@"Elephant.png"];
    UIImage* imageLion = [UIImage imageNamed:@"Lion.png"];
    NSArray* components1 =
      [NSArray arrayWithObjects:imageDog, imageMonkey, imageElephant, imageLion, nil];
    NSArray* components2 =
      [NSArray arrayWithObjects:imageDog, imageMonkey, imageElephant, imageLion, nil];
    NSArray* components3 =
      [NSArray arrayWithObjects:imageDog, imageMonkey, imageElephant, imageLion, nil];
    dataSource_ = [[NSArray alloc] initWithObjects:components1, components2, components3,
nil];
    //追加信息显示按钮,当触摸此按钮时,选中选项图片将显示在工具条上
    UIButton* button = [UIButton buttonWithType:UIButtonTypeRoundedRect];
    [button setTitle:@"信息显示" forState:UIControlStateNormal];
    [button sizeToFit];
    CGPoint newPoint = self.view.center;
    newPoint.y += 50;
    button.center = newPoint;
    //注册按钮点击后的响应方法
    [button addTarget:self
            action:@selector(buttonDidPush)
     forControlEvents:UIControlEventTouchUpInside];
    [self.view addSubview:button];
}
- (NSInteger)numberOfComponentsInPickerView:(UIPickerView*)pickerView {
    return 3;//返回列数
}
- (NSInteger)pickerView:(UIPickerView*)pickerView numberOfRowsInComponent:(NSInteger)
component {
    return 4;//返回行数
}
//以 UIView 作为选项时必须实现此方法
- (UIView*)pickerView:(UIPickerView*)pickerView
    viewForRow:(NSInteger)row forComponent:(NSInteger)component reusingView:(UIView*)view
{
    //如果参数 view 已经初始化,则直接显示(再利用)
    UIImageView* imageView = (UIImageView*)view;
    if ( !imageView ) {
    //否则创建并初始化 UIImageView
      UIImage* image = [[dataSource_ objectAtIndex:component] objectAtIndex:row];
      imageView = [[[UIImageView alloc] initWithImage:image] autorelease];
    }
    return imageView;
}
//按钮[信息显示]被触摸时的响应方法
- (void)buttonDidPush {
    static const int kNumbersOfComponent = 3;
    NSMutableArray* items = [[NSMutableArray alloc] initWithCapacity:3];
    for ( int i = 0; i < kNumbersOfComponent; ++i ){
    //调用 viewForRow:forComponent:方法取得选择行每列中的 UIView 对象
      UIImageView* imageView =
        (UIImageView*)[picker_ viewForRow:[picker_ selectedRowInComponent:i] forComponent:i];
      UIImageView* newImageView =
        [[[UIImageView alloc] initWithImage:imageView.image] autorelease];
    //以取得的 UIView 对象为基础创建 UIBarButtonItem 对象
      UIBarButtonItem* barButton =
        [[[UIBarButtonItem alloc] initWithCustomView:newImageView] autorelease];
      [items addObject:barButton];
    }
    //将上述 UIBarButtonItem 对象数组设置到工具条中
    [self setToolbarItems:items];
    [items release];
}
@end
```

执行后的效果如图 3-77 所示。

图 3-77 执行效果

实例 094 设置选择器框行和列尺寸

实例 094	设置选择器框行和列尺寸
源码路径	\daima\091\

实例说明

在 iOS 应用中，当通过 UIPickerView 控件在屏幕中实现选择器时，可以继续使用方法 pickerView widthForComponent 和 pickerViewrowHeightForComponent 调整选择器框的行和列的尺寸。在本实例中，就是用这两个方法设置了行和列的宽度。

具体实现

实例文件 UIKitPrjChangeWidthAndHeight.m 的具体代码如下所示。

```
#import "UIKitPrjChangeWidthAndHeight.h"
@implementation UIKitPrjChangeWidthAndHeight
- (void)viewDidLoad {
  [super viewDidLoad];

  UIPickerView* picker = [[[UIPickerView alloc] init] autorelease];
  picker.delegate = self;
  [self.view addSubview:picker];
}
- (NSInteger)numberOfComponentsInPickerView:(UIPickerView*)pickerView {
  return 2;
}
- (NSInteger)pickerView:(UIPickerView*)pickerView numberOfRowsInComponent:(NSInteger)
component {
  return 10;
}
- (NSString*)pickerView:(UIPickerView*)pickerView
  titleForRow:(NSInteger)row forComponent:(NSInteger)component
{
  if ( 0 == component ) {
    // 第 1 列
    return [NSString stringWithFormat:@"%2d", row+1];
  } else {
    // 第 2 列
    return [NSString stringWithFormat:@"比较长的字符串 其中%d", row+1];
  }
}
- (CGFloat)pickerView:(UIPickerView*)pickerView widthForComponent:(NSInteger)component {
  if ( 0 == component ) {
    // 第 1 列变窄
    return 50;
  } else {
    // 第 2 列变宽
    return 250;
```

```
    }
- (CGFloat)pickerView:(UIPickerView*)pickerView rowHeightForComponent:(NSInteger)component {
    if ( 0 == component ) {
        // 第 1 列变短
        return 30;
    } else {
        // 第 2 列变长
        return 60;
    }
}
@end
```

执行后的效果如图 3-78 所示。

图 3-78 执行效果

实例 095 实现一个播放器的活动指示器

实例 095	实现一个播放器的活动指示器
源码路径	\daima\095\

实例说明

在 iOS 应用中，可以使用控件 UIActivityIndicatorView 实现一个活动指示器效果。在开发过程中，可以使用 UIActivityIndicatorView 实例提供轻型视图，这些视图显示一个标准的旋转进度轮。当使用这些视图时，最重要的一个关键词是小。20×20 像素是大多数指示器样式获得最清楚显示效果的大小。只要稍大一点，指示器都会变得模糊。

iOS 提供了几种不同样式的 UIActivityIndicatorView 类。UIActivityIndicatorViewStyleWhite 和 UIActivityIndicatorViewStyleGray 是最简洁的。黑色背景下最适合白色版本的外观，白色背景最适合灰色外观。它非常瘦小，而且采用夏普风格。选择白色还是灰色时要格外注意，全白显示在白色背景下将不能显示任何内容。而 UIActivityIndicatorViewStyleWhiteLarge 只能用于深色背景。它提供最大、最清晰的指示器。

在本实例中，首先在根视图中使用 tableView 实现了一个列表效果，然后在次级视图设置了一个播放界面。当单击播放、暂停和快进按钮时会显示对应的提示效果，这个提示效果是通过 UIActivityIndicatorView 实现的。

具体实现

实例文件 RootViewController.m 用于实现根视图，具体代码如下所示。

```
#import "RootViewController.h"
@implementation RootViewController
- (void)dealloc {
    [items_ release];
    [super dealloc];
}
```

```objc
#pragma mark UIViewController methods
- (void)viewDidLoad {
  [super viewDidLoad];
  self.title = @"MENU";
  if ( !items_ ) {
    items_ = [[NSArray alloc] initWithObjects:
                        @"UIKitPrjActivityIndicator",
                        nil ];
  }
}
- (void)viewWillAppear:(BOOL)animated {
  [super viewWillAppear:animated];
  [self.navigationController setNavigationBarHidden:NO animated:NO];
  [self.navigationController setToolbarHidden:NO animated:NO];
  [UIApplication sharedApplication].statusBarStyle = UIStatusBarStyleDefault;
  self.navigationController.navigationBar.barStyle = UIBarStyleDefault;
  self.navigationController.navigationBar.translucent = NO;
  self.navigationController.navigationBar.tintColor = nil;
  self.navigationController.toolbar.barStyle = UIBarStyleDefault;
  self.navigationController.toolbar.translucent = NO;
  self.navigationController.toolbar.tintColor = nil;
}
#pragma mark UITableView methods
- (NSInteger)tableView:(UITableView*)tableView
  numberOfRowsInSection:(NSInteger)section
{
  return [items_ count];
}


- (UITableViewCell*)tableView:(UITableView*)tableView
  cellForRowAtIndexPath:(NSIndexPath*)indexPath
{
  static NSString *CellIdentifier = @"Cell";

  UITableViewCell *cell = [tableView dequeueReusableCellWithIdentifier:CellIdentifier];
  if (cell == nil) {
    cell = [[[UITableViewCell alloc] initWithStyle:UITableViewCellStyleDefault reuseIdentifier:CellIdentifier] autorelease];
  }
  NSString* title = [items_ objectAtIndex:indexPath.row];
  cell.textLabel.text = [title stringByReplacingOccurrencesOfString:@"UIKitPrj" withString:@""];
return cell;
}
- (void)tableView:(UITableView*)tableView
  didSelectRowAtIndexPath:(NSIndexPath*)indexPath
{
  NSString* className = [items_ objectAtIndex:indexPath.row];
  Class class = NSClassFromString( className );
  UIViewController* viewController = [[[class alloc] init] autorelease];
  if ( !viewController ) {
    NSLog( @"%@ was not found.", className );
    return;
  }
  [self.navigationController pushViewController:viewController animated:YES];
}
@end
```

文件 **UIKitPrjActivityIndicator.m** 实现次级视图，具体实现代码如下所示。

```objc
#import "UIKitPrjActivityIndicator.h"
@implementation UIKitPrjActivityIndicator
- (void)dealloc {
  [indicator_ release];
  [super dealloc];
}
- (void)viewDidLoad {
  [super viewDidLoad];
  self.view.backgroundColor = [UIColor lightGrayColor];
  UIBarButtonItem* playButton =
```

```objc
    [[[UIBarButtonItem alloc] initWithBarButtonSystemItem:UIBarButtonSystemItemPlay
                                                  target:self
                                                  action:@selector(playDidPush)]
   autorelease];
   UIBarButtonItem* pauseButton =
     [[[UIBarButtonItem alloc] initWithBarButtonSystemItem:UIBarButtonSystemItemPause
                                                  target:self
                                                  action:@selector(pauseDidPush)]
   autorelease];
   UIBarButtonItem* changeButton =
     [[[UIBarButtonItem alloc] initWithBarButtonSystemItem:UIBarButtonSystemItemFastForward
                                                  target:self
                                                  action:@selector(changeDidPush)]
   autorelease];
   NSArray* items = [NSArray arrayWithObjects:playButton, pauseButton, changeButton, nil];
   [self setToolbarItems:items animated:YES];

   indicator_ =
     [[UIActivityIndicatorView alloc] initWithActivityIndicatorStyle:UIActivityIndicatorViewStyleWhiteLarge];
   [self.view addSubview:indicator_];
}
- (void)playDidPush {
   if ( UIActivityIndicatorViewStyleWhiteLarge == indicator_.activityIndicatorViewStyle )
{
      indicator_.frame = CGRectMake( 0, 0, 50, 50 );
   } else {
      indicator_.frame = CGRectMake( 0, 0, 20, 20 );
   }
   indicator_.center = self.view.center;
   [indicator_ startAnimating];
}
- (void)pauseDidPush {
   indicator_.hidesWhenStopped = NO;
   [indicator_ stopAnimating];
}
- (void)changeDidPush {
   [self pauseDidPush];
   if ( UIActivityIndicatorViewStyleGray < ++indicator_.activityIndicatorViewStyle ) {
      indicator_.activityIndicatorViewStyle = UIActivityIndicatorViewStyleWhiteLarge;
   }
   [self playDidPush];
}
@end
```

执行后的效果如图3-79所示，次级视图界面如图3-80所示。

图3-79　执行效果

图3-80　次级视图界面

实例096　实现一个蓝色进度条效果

实例096	实现一个蓝色进度条效果
源码路径	\daima\096\

实例 096 实现一个蓝色进度条效果

实例说明

在 iOS 应用中，UIProgressView 与 UIActivityIndicatorView 相似，只不过它提供了一个接口可以显示一个进度条，这样就能让用户知道当前操作完成了多少。在开发过程中，可以使用控件 UIProgressView 实现一个进度条效果。

在本实例中，首先使用方法 initWithProgressViewStyle 创建并初始化了 UIProgressView 对象，然后通过其 center 属性和 frame 属性设置其显示位置，并添加到显示画面中。UIProgressViewStyle 属性可以设置如下两种样式。

- 标准进度条。
- 深灰色进度条，用于工具栏中。

具体实现

实例文件 UIKitPrjProgressView.m 的具体代码如下所示。

```
#import "UIKitPrjProgressView.h"
#pragma mark ----- Private Methods Definition -----
@interface UIKitPrjProgressView ()
- (void)updateProgress:(UIProgressView*)progressView;
@end
#pragma mark ----- Start Implementation For Methods -----
@implementation UIKitPrjProgressView
- (void)dealloc {
  [progressView_ release];
  [super dealloc];
}
- (void)viewDidLoad {
  [super viewDidLoad];
  self.view.backgroundColor = [UIColor whiteColor];
  progressView_ =
    [[UIProgressView alloc] initWithProgressViewStyle:UIProgressViewStyleDefault];
  progressView_.center = self.view.center;
  progressView_.autoresizingMask = UIViewAutoresizingFlexibleTopMargin |
                       UIViewAutoresizingFlexibleBottomMargin;
  [self.view addSubview:progressView_];
}
- (void)viewDidAppear:(BOOL)animated {
  [super viewDidAppear:animated];
  [self updateProgress:progressView_];
}
- (void)viewWillDisappear:(BOOL)animated {
  [super viewWillDisappear:animated];
  progressView_.hidden = YES;
}
- (void)updateProgress:(UIProgressView*)progressView {
  if ( [progressView isHidden] || 1.0 <= progressView.progress ) {
    return;
  }
  progressView.progress += 0.1;
  [self performSelector:@selector(updateProgress:)
          withObject:progressView
          afterDelay:1.0];
}
@end
```

执行后的效果如图 3-81 所示。

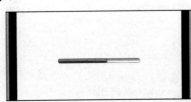

图 3-81 执行效果

实例 097　在进度条中显示进度百分比

实例 097	在进度条中显示进度百分比
源码路径	\daima\096\

实例说明

本实例的功能是，在进度条上方以文字的形式显示进度百分比，下方进度的变化和上方百分比的变化是对应的。首先创建了 UIProgressView 控件与 UILable 控件，并设置了默认值。根据 UIProgressView 及 UILabel 的状态重新绘制了外观。然后创建并初始化 ProgressViewWithLabel 实例，将定制的按钮添加到工具条中，并使其居中显示。最后实时更新下载数目，追加单位是 0.1，逐步递增。

具体实现

实例文件 UIKitPrjProgressViewOnToolbar.m 的具体代码如下所示。

```
#import "UIKitPrjProgressViewOnToolbar.h"
#pragma mark ----- ProgressViewWithLabel -----
@implementation ProgressViewWithLabel
@synthesize progressView = progressView_;
@synthesize textLabel = textLabel_;
- (void)dealloc {
  [progressView_ release];
  [textLabel_ release];
  [super dealloc];
}
//初始化处理
//创建 UIProgressView 与 UILable，并设置默认值
- (id)init {
  if ( (self = [super init]) ) {
    self.opaque = NO;
    progressView_ =
      [[UIProgressView alloc] initWithProgressViewStyle:UIProgressViewStyleBar];
    textLabel_ = [[UILabel alloc] init];
    textLabel_.textAlignment = UITextAlignmentCenter;
    textLabel_.backgroundColor = [UIColor colorWithRed:0.0 green:0.0 blue:0.0 alpha:0.0];
    textLabel_.textColor = [UIColor whiteColor];
    textLabel_.font = [UIFont boldSystemFontOfSize:14];
    textLabel_.shadowColor = [UIColor blackColor];
    [self addSubview:progressView_];
    [self addSubview:textLabel_];
    self.frame = CGRectMake( 0, 0, progressView_.bounds.size.width, progressView_.bounds.size.height * 3 );
  }
  return self;
}
//外观设置
//根据 UIProgressView 及 UILabel 的状态重新绘制外观
- (void)layoutSubviews {
  CGRect newFrame = self.bounds;
  newFrame.size.height -= progressView_.frame.size.height;
  textLabel_.frame = newFrame;
  newFrame = progressView_.frame;
  newFrame.origin.y = self.bounds.size.height - newFrame.size.height;
  progressView_.frame = newFrame;
}
@end
#pragma mark ----- Private Methods Definition -----

@interface UIKitPrjProgressViewOnToolbar ()
- (void)updateProgress:(ProgressViewWithLabel*)component;
@end
#pragma mark ----- Start Implementation For Methods -----
```

```
@implementation UIKitPrjProgressViewOnToolbar
- (void)dealloc {
  [component_ release];
  [super dealloc];
}
- (void)viewDidLoad {
  [super viewDidLoad];
  self.view.backgroundColor = [UIColor whiteColor];
  //创建并初始化 ProgressViewWithLabel 实例
  component_ = [[ProgressViewWithLabel alloc] init];
  //以创建的 ProgressViewWithLabel 实例为参数创建定制按钮
  UIBarButtonItem* barButton =
    [[[UIBarButtonItem alloc] initWithCustomView:component_] autorelease];;
  UIBarButtonItem* flexibleSpace =
    [[[UIBarButtonItem alloc] initWithBarButtonSystemItem:UIBarButtonSystemItemFlexibleSpace
                                      target:nil
                                      action:nil] autorelease];
  //将定制按钮追加到工具条中并使其居中显示
  NSArray* items = [NSArray arrayWithObjects:flexibleSpace, barButton, flexibleSpace, nil];
  [self setToolbarItems:items animated:NO];
}
- (void)viewDidAppear:(BOOL)animated {
  [super viewDidAppear:animated];
  [self updateProgress:component_];
}
- (void)viewWillDisappear:(BOOL)animated {
  [super viewWillDisappear:animated];
  component_.hidden = YES;
}
- (void)updateProgress:(ProgressViewWithLabel*)component {
  if ( [component isHidden] || 1.0 <= component.progressView.progress ) {
    return;
  }
  //实时改变下载件数，追加 0.1，表示完成一件下载
  component.progressView.progress += 0.1;
  component.textLabel.text =
    [NSString stringWithFormat:@"Downloading %.0f of 10", component.progressView.progress * 10];
  [self performSelector:@selector(updateProgress:)
             withObject:component
             afterDelay:1.0];
}
@end
```

执行后的效果如图 3-82 所示。

图 3-82　执行效果

实例 098　在屏幕中实现一个检索框效果

实例 098	在屏幕中实现一个检索框效果
源码路径	\daima\098\

实例说明

在 iOS 应用中，可以使用 UISearchBar 控件实现一个检索框效果。UISearchBar 控件各个属性的具体说明如表 3-1 所示。

表 3-1　　　　　　　　　　　　　UISearchBar 控件的属性

属　性	作　用
UIBarStyle barStyle	控件的样式
id<UISearchBarDelegate> delegate	设置控件的委托
NSString *text	控件上面的显示的文字
NSString *prompt	显示在顶部的单行文字，通常作为一个提示行
NSString *placeholder	半透明的提示文字，输入搜索内容消失
BOOL showsBookmarkButton	是否在控件的右端显示一个书的按钮（没有文字的时候）
BOOL showsCancelButton	是否显示 cancel 按钮
BOOL showsSearchResultsButton	是否在控件的右端显示搜索结果按钮（没有文字的时候）
BOOL searchResultsButtonSelected	搜索结果按钮是否被选中
UIColor *tintColor	bar 的颜色（具有渐变效果）
BOOL translucent	指定控件是否会有透视效果
UITextAutocapitalizationType autocapitalizationType	设置在什么的情况下自动大写
UITextAutocorrectionType autocorrectionType	对于文本对象自动校正风格
UIKeyboardType keyboardType	键盘的样式
NSArray *scopeButtonTitles	搜索栏下部的选择栏，数组里面的内容是按钮的标题
NSInteger selectedScopeButtonIndex	搜索栏下部的选择栏按钮的个数
BOOL showsScopeBar	控制搜索栏下部的选择栏是否显示出来

本实例的功能是，在上方使用 UISearchBar 控件设置了一个检索框，在下方显示了 NSMutableArray 中存储的数据，本实例设置了显示 0～63 共 64 个数字。在检索框中输入数字关键字后，可以检索出对应的结果。

具体实现

实例文件 UIKitPrjSearchBar.m 的具体代码如下所示。

```
#import "UIKitPrjSearchBar.h"
@implementation UIKitPrjSearchBar
- (void)dealloc {
 [searchBar_ release];
 [dataSource_ release];
 [dataBase_ release];
 [super dealloc];
}
- (void)viewDidLoad {
 [super viewDidLoad];
searchBar_ = [[UISearchBar alloc] init];
 searchBar_.frame = CGRectMake( 0, 0, self.tableView.bounds.size.width, 0 );
 searchBar_.delegate = self;
 [searchBar_ sizeToFit];
 self.tableView.tableHeaderView = searchBar_;
dataBase_ = [[NSMutableArray alloc] initWithCapacity:64];
 dataSource_ = [[NSMutableArray alloc] initWithCapacity:64];
 for ( int i = 0; i < 64; ++i ){
   [dataBase_ addObject:[NSString stringWithFormat:@"%d", i]];
   [dataSource_ addObject:[NSString stringWithFormat:@"%d", i]];
 }
}
//此处 datasource_为 NSArray 类型的实例变量
//dataBase_为保持全部数据的 NSArray 类型变量
- (void)searchBarSearchButtonClicked:(UISearchBar*)searchBar {
 //首先删除数据资源中的内容
 [dataSource_ removeAllObjects];
 for ( NSString* data in dataBase_ ) {
  if ( [data hasPrefix:searchBar.text] ) {
  //将满足检索条件的数据追加到数据资源变量中
    [dataSource_ addObject:data];
  }
```

```
    }
    //表格更新
    [self.tableView reloadData];
    //隐藏键盘
    [searchBar resignFirstResponder];
}
#pragma mark UITableView methods

- (NSInteger)tableView:(UITableView*)tableView
  numberOfRowsInSection:(NSInteger)section
{
    return [dataSource_ count];
}
- (UITableViewCell*)tableView:(UITableView*)tableView
  cellForRowAtIndexPath:(NSIndexPath*)indexPath
{
    static NSString *CellIdentifier = @"Cell";

    UITableViewCell *cell = [tableView dequeueReusableCellWithIdentifier:CellIdentifier];
    if (cell == nil) {
        cell = [[[UITableViewCell alloc] initWithStyle:UITableViewCellStyleDefault reuseIdentifier:CellIdentifier] autorelease];
    }
    cell.textLabel.text = [dataSource_ objectAtIndex:indexPath.row];
    return cell;
}
@end
```

执行后的效果如图 3-83 所示。

图 3-83　执行效果

实例 099　实现一个实时显示检索框效果

实例 099	实现一个实时显示检索框效果
源码路径	\daima\098\

实例说明

本实例是在上一个实例的基础上实现的，在实例中使用 searchBarCancelButtonClicked:(UISearchBar*)searchBar 设置"取消"按钮隐藏输入键盘，然后使用方法 searchBar:textDidChange 实现了实时检索功能。

具体实现

实例文件 UIKitPrjRealTimeSearch.m 的具体代码如下所示。

```
#import "UIKitPrjRealTimeSearch.h"
@implementation UIKitPrjRealTimeSearch
- (void)viewDidLoad {
    [super viewDidLoad];
    searchBar_.keyboardType = UIKeyboardTypeNumberPad;
    searchBar_.showsCancelButton = YES;
```

```
}
- (void)searchBar:(UISearchBar*)searchBar textDidChange:(NSString*)searchText {
  if ( 0 == searchText.length ) {
      //检索字符串为空时，显示全部数据
    [dataSource_ release];
    dataSource_ = [[NSMutableArray alloc] initWithArray:dataBase_];
    [self.tableView reloadData];
  } else {
      //检索字符串非空时，进行实时检索
    [dataSource_ removeAllObjects];
    for ( NSString* data in dataBase_ ) {
      if ( [data hasPrefix:searchBar.text] ) {
        [dataSource_ addObject:data];
      }
    }
    [self.tableView reloadData];
  }
  //实时检索时保持键盘为显示状态
}
- (void)searchBarCancelButtonClicked:(UISearchBar*)searchBar {
  searchBar.text = @"";
  [searchBar resignFirstResponder];//隐藏键盘
}
@end
```

执行后的效果如图 3-84 所示。

图 3-84　执行效果

实例 100　设置检索框的背景颜色

实例 100	设置检索框的背景颜色
源码路径	\daima\098\

实例说明

在 iOS 应用中，当使用 UISearchBar 控件实现一个检索框效果后，可以使用其属性 barStyle、translucent 和 tintColor 设置检索框的背景颜色。在本实例中，在屏幕上方设置了一个检索框，在中间设置了拥有 64 个数字的数组，在下方设置了 3 个按钮——black、translucent 和 tintColor，当单击按钮后可以显示不同颜色的样式。

具体实现

实例文件 UIKitPrjBarStyle.m 的具体代码如下所示。

```
#import "UIKitPrjBarStyle.h"
@implementation UIKitPrjBarStyle
- (void)viewDidLoad {
  [super viewDidLoad];
  searchBar_.barStyle = UIBarStyleBlack;
  searchBar_.translucent = YES;
```

```
  searchBar_.tintColor = nil;
  UIBarButtonItem* blackButton =
    [[[UIBarButtonItem alloc] initWithTitle:@"black"
                              style:UIBarButtonItemStyleDone
                              target:self
                              action:@selector(blackDidPush:)] autorelease];
  UIBarButtonItem* translucentButton =
    [[[UIBarButtonItem alloc] initWithTitle:@"translucent"
                              style:UIBarButtonItemStyleDone
                              target:self
                              action:@selector(translucentDidPush:)] autorelease];
  UIBarButtonItem* tintButton =
    [[[UIBarButtonItem alloc] initWithTitle:@"tintColor"
                              style:UIBarButtonItemStyleBordered
                              target:self
                              action:@selector(tintDidPush:)] autorelease];
  NSArray* items = [NSArray arrayWithObjects:blackButton, translucentButton, tintButton,
nil];
  [self setToolbarItems:items animated:NO];
}
- (void)blackDidPush:(UIBarButtonItem*)sender {
  if ( UIBarButtonItemStyleDone == sender.style ) {
    searchBar_.barStyle = UIBarStyleDefault;
    sender.style = UIBarButtonItemStyleBordered;
  } else {
    searchBar_.barStyle = UIBarStyleBlack;
    sender.style = UIBarButtonItemStyleDone;
  }
}
- (void)translucentDidPush:(UIBarButtonItem*)sender {
  if ( UIBarButtonItemStyleDone == sender.style ) {
    searchBar_.translucent = NO;
    sender.style = UIBarButtonItemStyleBordered;
  } else {
    searchBar_.translucent = YES;
    sender.style = UIBarButtonItemStyleDone;
  }
}
- (void)tintDidPush:(UIBarButtonItem*)sender {
  if ( UIBarButtonItemStyleDone == sender.style ) {
    searchBar_.tintColor = nil;
    sender.style = UIBarButtonItemStyleBordered;
  } else {
    searchBar_.tintColor = [UIColor redColor];
    sender.style = UIBarButtonItemStyleDone;
  }
}
@end
```

执行后的效果如图3-85所示。

图3-85 执行效果

实例101 在检索框中添加一个书签按钮

实例101	在检索框中添加一个书签按钮
源码路径	\daima\098\

实例说明

在 iOS 应用中,当使用 UISearchBar 控件实现一个检索框效果后,可以将其属性 showsBookmark Button 设置为 YES,这样便可以在检索框中添加一个书签按钮。在本实例中,当触摸了检索框中的书签按钮后,会触发 searchBarBookmarkButtonClicked 方法,以模态画面的形式显示书签列表。

具体实现

实例文件 UIKitPrjBookmarkButton.m 的具体代码如下所示。

```
#import "UIKitPrjBookmarkButton.h"
@implementation UIKitPrjBookmarkButton
- (void)viewDidLoad {
  [super viewDidLoad];
  //显示书签按钮
  searchBar_.showsBookmarkButton = YES;
}
//触摸书签按钮后,以模态画面的形式显示书签列表
- (void)searchBarBookmarkButtonClicked:(UISearchBar*)searchBar {
  id rootViewController = [[BookmarkDialog alloc] initWithParent:self];
  id navi = [[UINavigationController alloc] initWithRootViewController:rootViewController];
  [rootViewController release];
  [self presentModalViewController:navi animated:YES];
  [navi release];
}
- (void)setCurrentText:(NSString*)text {
  searchBar_.text = text;
  [self searchBarSearchButtonClicked:searchBar_];
}
@end
#pragma mark ----- BookmarkDialog -----
@implementation BookmarkDialog
- (void)dealloc {
  [dataSource_ release];
  [super dealloc];
}
- (id)initWithParent:(UIViewController*)parent {
  if ( (self = [super init]) ) {
    parent_ = parent;
  }
  return self;
}
- (void)viewDidLoad {
  [super viewDidLoad];
  dataSource_ = [[NSArray alloc] initWithObjects:@"11",
                                                 @"22",
                                                 @"33",
                                                 nil ];
  self.title = @"Bookmarks";
  UIBarButtonItem* barButton =
    [[[UIBarButtonItem alloc] initWithBarButtonSystemItem:UIBarButtonSystemItemDone
                                          target:self
                                          action:@selector(doneDidPush)]
autorelease];
  self.navigationItem.rightBarButtonItem = barButton;
}
- (void)doneDidPush {
  [self dismissModalViewControllerAnimated:YES];
}
- (void)tableView:(UITableView*)tableView
  didSelectRowAtIndexPath:(NSIndexPath*)indexPath
{
  [parent_ setCurrentText:[dataSource_ objectAtIndex:indexPath.row]];
  [self dismissModalViewControllerAnimated:YES];
}
- (NSInteger)tableView:(UITableView*)tableView
  numberOfRowsInSection:(NSInteger)section
{
  return [dataSource_ count];
}
```

```
- (UITableViewCell*)tableView:(UITableView*)tableView
  cellForRowAtIndexPath:(NSIndexPath*)indexPath
{
  static NSString *CellIdentifier = @"Cell";
  UITableViewCell *cell = [tableView dequeueReusableCellWithIdentifier:CellIdentifier];
  if (cell == nil) {
    cell = [[[UITableViewCell alloc] initWithStyle:UITableViewCellStyleDefault reuseIdentifier:CellIdentifier] autorelease];
  }
  cell.textLabel.text = [dataSource_ objectAtIndex:indexPath.row];
  return cell;
}
@end
```

执行效果如图 3-86 所示。单击书签按钮后，会在新界面中显示书签中的数据，如图 3-87 所示。书签中存储的是当前设备已经使用过的搜索关键字。

图 3-86　执行效果

图 3-87　书签中的数据

实例 102　在检索框中添加一个范围条

实例 102	在检索框中添加一个范围条
源码路径	\daima\098\

实例说明

在 iOS 应用中，当使用 UISearchBar 控件实现一个检索框效果后，可以添加一个范围条。在本实例中，将属性 showsScopeBar 设置为 YES，这样便在检索框中添加了一个范围条。

具体实现

实例文件 UIKitPrjScopeBar.m 的具体代码如下所示。

```
#import "UIKitPrjScopeBar.h"
@implementation UIKitPrjScopeBar
- (void)viewDidLoad {
  [super viewDidLoad];
  UISearchBar* searchBar = [[[UISearchBar alloc] init] autorelease];
  searchBar.frame = CGRectMake( 0, 0, 320, 0 );
  searchBar.delegate = self;
  searchBar.scopeButtonTitles = [NSArray arrayWithObjects:@"AAA", @"BBB", @"CCC", nil];
  searchBar.showsScopeBar = YES;
  [searchBar sizeToFit];
  [self.view addSubview:searchBar];
}
- (void)searchBar:(UISearchBar*)searchBar
  selectedScopeButtonIndexDidChange:(NSInteger)selectedScope
{
  NSLog( @"selectedScopeButtonIndexDidChange %d", selectedScope );
  NSLog( @"selectedScopeButtonIndex %d", searchBar.selectedScopeButtonIndex );
}
@end
```

执行后的效果如图 3-88 所示。

图 3-88 执行效果

实例 103 添加或删除屏幕中的翻页数目

实例 103	添加或删除屏幕中的翻页数目
源码路径	\daima\103\

实例说明

在 iOS 应用中，当在项目中和 UIScrollView 配合来显示大量数据时，会使用 UIPageControl 控件来控制 UIScrollView 的翻页。在滚动 ScrollView 时可通过 PageControll 中的小白点来观察当前页面的位置，也可通过点击 PageControl 中的小白点来滚动到指定的页面。

在本实例中设置了两个视图界面：RootViewController 和 UIKitPrjPageControl。具体说明如下所示。

（1）在根视图 RootViewController 中设置了一个 UITableView 控件，用于列表显示信息，在本例中只显示了一行文本"RootViewController"，并且在载入视图界面时，设置在屏幕顶部显示一个"主屏幕"按钮。

（2）在子视图 UIKitPrjPageControl 中，使用 UIPageControl 控件实现了翻页效果，其中使用属性 numberOfPages 设置了 5 个分页，使用属性设置 currentPage 默认显示的当前页是第一页。在屏幕底部显示"添加新页"和"删除不喜欢的页"两个按钮，通过这两个按钮可以添加或删除 UIPageControl 控件中的翻页。

具体实现

实例文件 RootViewController.m 的具体代码如下所示。

```objc
#import "RootViewController.h"
@implementation RootViewController
- (void)dealloc {
  [items_ release];
  [super dealloc];
}
#pragma mark UIViewController methods
- (void)viewDidLoad {
  [super viewDidLoad];
  
  self.title = @"主屏幕";
  if ( !items_ ) {
    items_ = [[NSArray alloc] initWithObjects:
                          @"UIKitPrjPageControl",
                          nil ];
  }
```

```objc
}
- (void)viewWillAppear:(BOOL)animated {
  [super viewWillAppear:animated];
  [self.navigationController setNavigationBarHidden:NO animated:NO];
  [self.navigationController setToolbarHidden:NO animated:NO];
  // 恢复条的颜色
  [UIApplication sharedApplication].statusBarStyle = UIStatusBarStyleDefault;
  self.navigationController.navigationBar.barStyle = UIBarStyleDefault;
  self.navigationController.navigationBar.translucent = NO;
  self.navigationController.navigationBar.tintColor = nil;
  self.navigationController.toolbar.barStyle = UIBarStyleDefault;
  self.navigationController.toolbar.translucent = NO;
  self.navigationController.toolbar.tintColor = nil;
}
#pragma mark UITableView methods
- (NSInteger)tableView:(UITableView*)tableView
  numberOfRowsInSection:(NSInteger)section
{
  return [items_ count];
}
- (UITableViewCell*)tableView:(UITableView*)tableView
  cellForRowAtIndexPath:(NSIndexPath*)indexPath
{
  static NSString *CellIdentifier = @"Cell";
  UITableViewCell *cell = [tableView dequeueReusableCellWithIdentifier:CellIdentifier];
  if (cell == nil) {
    cell = [[[UITableViewCell alloc] initWithStyle:UITableViewCellStyleDefault reuseIdentifier:CellIdentifier] autorelease];
  }
  NSString* title = [items_ objectAtIndex:indexPath.row];
  cell.textLabel.text = [title stringByReplacingOccurrencesOfString:@"UIKitPrj" withString:@""];

  return cell;
}
- (void)tableView:(UITableView*)tableView
  didSelectRowAtIndexPath:(NSIndexPath*)indexPath
{
  NSString* className = [items_ objectAtIndex:indexPath.row];
  Class class = NSClassFromString( className );
  UIViewController* viewController = [[[class alloc] init] autorelease];
  if ( !viewController ) {
    NSLog( @"%@ 没有了.", className );
    return;
  }
  [self.navigationController pushViewController:viewController animated:YES];
}
@end
```

实例文件 UIKitPrjPageControl.m 的具体代码如下所示。

```objc
#import "UIKitPrjPageControl.h"
@implementation UIKitPrjPageControl
- (void)dealloc {
  [pageControl_ release];
  [super dealloc];
}
- (void)viewDidLoad {
  [super viewDidLoad];
pageControl_ = [[UIPageControl alloc] init];
  pageControl_.frame = CGRectMake( 0, self.view.bounds.size.height - 30, 320, 30 );
  pageControl_.backgroundColor = [UIColor blackColor]; //必须改变背景色
  [pageControl_ addTarget:self
             action:@selector(pageControlDidChange:)
        forControlEvents:UIControlEventValueChanged];
  pageControl_.autoresizingMask = UIViewAutoresizingFlexibleTopMargin |
                     UIViewAutoresizingFlexibleBottomMargin;
  pageControl_.numberOfPages = 5;
  pageControl_.currentPage = 0;
  [self.view addSubview:pageControl_];
UIBarButtonItem* addButton =
    [[[UIBarButtonItem alloc] initWithTitle:@"添加新页"
```

```
                                      style:UIBarButtonItemStyleBordered
                                     target:self
                                     action:@selector(addDidPush)] autorelease];
    UIBarButtonItem* delButton =
      [[[UIBarButtonItem alloc] initWithTitle:@"删除不喜欢的页"
                                      style:UIBarButtonItemStyleBordered
                                     target:self
                                     action:@selector(delDidPush)] autorelease];
    NSArray* items = [NSArray arrayWithObjects:addButton, delButton, nil];
    [self setToolbarItems:items animated:NO];
}
- (void)pageControlDidChange:(id)sender {
}
- (void)addDidPush {
    if ( 10 > pageControl_.numberOfPages ) {
        ++pageControl_.numberOfPages;
    }
}
- (void)delDidPush {
    if ( 1 < pageControl_.numberOfPages ) {
        --pageControl_.numberOfPages;
        [self pageControlDidChange:pageControl_];
    }
}
@end
```

执行后的效果如图 3-89 所示，子视图界面如图 3-90 所示。单击子视图中的"添加新页"和"删除不喜欢的页"按钮后，可以添加或删除 UIPageControl 控件中的翻页。

图 3-89 根视图效果

图 3-90 根视图效果

实例 104 使用滚动的方式查看屏幕中的内容

实例 104	使用滚动的方式查看屏幕中的内容
源码路径	\daima\104\

实例说明

在 iOS 应用中，因为 iPhone 设备的界面空间有限，所以经常会出现不能完全显示信息的情形。在这个时候，滚动控件 UIScrollView 就可以发挥它的作用，使用后可在添加控件和界面元素时不受设备屏幕边界的限制。

在本实例中包含了一个可滚动视图（UIScrollView），并在 Interface Builder 编辑器中添加了超越屏幕限制的内容，这样便可以使用滚动的方式浏览超过屏幕高度的内容。

具体实现

（1）将可滚动视图（UIScrollView）作为子视图加入到 MainStoryboard.storyboard 中现有的视

图（UIView）中，如图 3-91 所示。

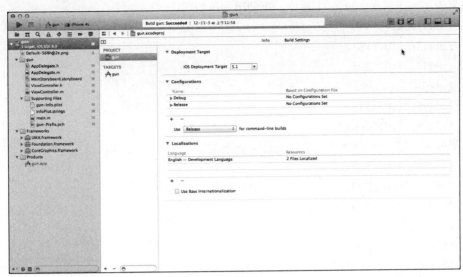

图 3-91　创建的工程

在这个项目中，只需设置可滚动视图对象的一个属性即可。为了访问该对象，需要创建一个与之关联的输出口，把这个输出口命名为"theScroller"。

（2）设计界面。

本实例涉及的内容不多，主要是可滚动的视图及其内容。开发者知道如何在对象库中寻找对象并将其加入到视图中，因此界面设计应很简单。首先，打开该项目的文件 MainStoryboard.storyboard，并确保文档大纲区域可见，方法是依次选择菜单 Editor→Show Document Outline 命令。

选择菜单 View→Utilities→Show Object Library 打开对象库，将一个可滚动视图（UIScrollView）实例拖曳到视图中。将其放在喜欢的位置，并在上方添加一个标题为 Scrolling View 的标签，以免忘记创建的是什么。

将可滚动视图加入到视图后，需要使用一些东西填充，通常编写计算对象位置的代码来将其加入到可滚动视图中。要想将按钮和其他控件加入到屏幕中，首先将添加的每个控件拖曳到可滚动视图对象中，在本实例中添加了 6 个标签。将对象加入可滚动视图中后还有如下两种方案可供选择。此处可以选择对象，然后使用箭头键将对象移到视图可视区域外面的大概位置。依次选择每个对象，并使用 Size Inspector（Option+ Command+5）手工设置其 x 和 y 坐标，如图 3-92 所示。

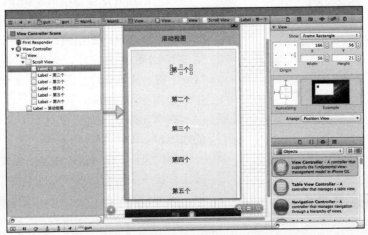

图 3-92　设置每个对象的 x 和 y 坐标

为了帮助我们放置对象，下面是 6 个标签的左边缘中点的 x 和 y 坐标。

如果应用程序将在 iPhone 上运行，可使用如下数字。

- Label1：110，45。
- Label2：110，125。
- Label3：110，205。
- Label4：110，290。
- Label5：110，375。
- Label6：110，460。

如果应用程序将在 iPad 上运行，可使用如下数字。

- Label1：360，130。
- Label2：360，330。
- Label3：360，530。
- Label4：360，730。
- Label5：360，930。
- Label6：360，1130。

从下面的图 3-93 所示的最终视图可知，第 6 个标签不可见，要看到它，需要进行一定的滚动。

图 3-93　最终的界面效果

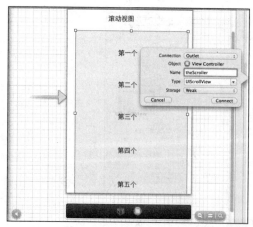

图 3-94　创建到输出口 theScroller 的连接

（3）创建并连接输出口和操作。

这个项目只需要一个输出口，并且不需要任何操作。为了创建这个输出口，切换到助手编辑器。如果需要腾出更多的控件，请隐藏项目导航器。按住"Ctrl"键，从可滚动视图拖曳到文件 ViewController.h 中编译指令 @interface 下方。

在 Xcode 提示时，新建一个名为|"theScroller"的输出口，如图 3-94 所示。

到此为止，Interface Builder 编辑器中的工作就完成了，接下来需要切换到标准编辑器，显示项目导航器，再对文件 ViewController.m 进行具体编码。

（4）实现应用程序逻辑。

为了给可滚动视图添加滚动功能，需要将属性

contentSize 设置为一个 CGSize 值。CGSize 是一个简单的 C 语言数据结构,它包含高度和宽度,可使用函数 CGSize(<width>,<height>)轻松地创建一个这样的对象。例如,要告诉该可滚动视图(theScroller)可水平和垂直分别滚动到 280 点和 600 点,可编写如下代码。

```
self .theScroller.contentSize=CGSizeMake (280.0,600.0);
```

我们并非只能这样做,但我们愿意这样做。如果进行的是 iPhone 开发,需要实现文件 ViewController.m 中的方法 viewDidLoad,其实现代码如下所示。

```
- (void)viewDidLoad
{
    self.theScroller.contentSize=CGSizeMake(280.0,600.0);
    [super viewDidLoad];
    // Do any additional setup after loading the view, typically from a nib.
}
```

如果开发的是一个 iPad 项目,则需要增大 contentSize 的设置,因为 iPad 屏幕更大。所以需要在调用函数 CGSizeMake 时传递参数 900.0 和 1500.0,而不是 280.0 和 600.0。在本实例中,我们使用的宽度正是可滚动视图本身的宽度。为什么这样做呢?因为我们没有理由进行水平滚动。选择的高度旨在演示视图能够滚动。换句话说,这些值可随意选择,开发者根据应用程序包含的内容选择最佳的值即可。

(5)单击 Xcode 工具栏中的按钮 Run,执行后的效果如图 3-95 所示。

图 3-95 执行效果

实例 105 使用滚动的方式查看图片

实例 105	使用滚动的方式查看图片
源码路径	\daima\105\

实例说明

在 iOS 应用中,可以使用 UIScrollView 控件以滚动的方式来查看屏幕中的内容。UIScrollView 控件的常用属性如表 3-2 所示。

表 3-2　　　　　　　　　　　UIScrollView 控件的属性

属　　性	功能描述
CGPoint contentOffSet	监控目前滚动的位置
CGSize contentSize	滚动范围的大小
UIEdgeInsets contentInset	视图在 ScrollView 中的位置
id<UIScrollerViewDelegate> delegate	设置协议
BOOL directionalLockEnabled	指定控件是否只能在一个方向上滚动

属 性	功能描述
BOOL bounces	控制控件遇到边框是否反弹
BOOL alwaysBounceVertical	控制垂直方向遇到边框是否反弹
BOOL alwaysBounceHorizontal	控制水平方向遇到边框是否反弹
BOOL pagingEnabled	控制控件是否整页翻动
BOOL scrollEnabled	控制控件是否能滚动
BOOL showsHorizontalScrollIndicator	控制是否显示水平方向的滚动条
BOOL showsVerticalScrollIndicator	控制是否显示垂直方向的滚动条
UIEdgeInsets scrollIndicatorInsets	指定滚动条在 scrollView 中的位置
UIScrollViewIndicatorStyle indicatorStyle	设定滚动条的样式
float decelerationRate	改变 scrollView 的减速点位置
BOOL tracking	监控当前目标是否正在被跟踪
BOOL dragging	监控当前目标是否正在被拖曳
BOOL decelerating	监控当前目标是否正在减速
BOOL delaysContentTouches	控制视图是否延时调用开始滚动的方法
BOOL canCancelContentTouches	控制控件是否接触取消 touch 的事件
float minimumZoomScale	缩小的最小比例
float maximumZoomScale	放大的最大比例
float zoomScale	设置变化比例
BOOL bouncesZoom	控制缩放的时候是否会反弹
BOOL zooming	判断控件的大小是否正在改变
BOOL zoomBouncing	判断是否正在进行缩放反弹
BOOL scrollsToTop	控制控件滚动到顶部

在本实例中使用了 UIScrollView 滚动控件，使用其 image 属性设置了要查看的图片是 town.jpg。

具体实现

实例文件 UIKitPrjScrollView.m 的具体代码如下所示。

```objc
#import "UIKitPrjScrollView.h"
@implementation UIKitPrjScrollView
- (void)viewDidLoad {
  [super viewDidLoad];
  // ScrollView 的初始化
  UIScrollView* scrollView = [[[UIScrollView alloc] init] autorelease];
  scrollView.frame = self.view.bounds;
  scrollView.autoresizingMask = UIViewAutoresizingFlexibleWidth | UIViewAutoresizingFlexibleHeight;
  // ScrollView 中设置图片
  UIImage* image = [UIImage imageNamed:@"town.jpg"];
  UIImageView* imageView = [[[UIImageView alloc] initWithImage:image] autorelease];
  [scrollView addSubview:imageView];
  scrollView.contentSize = imageView.bounds.size;
  // ScrollView 追加到主画面
  [self.view addSubview:scrollView];
  // 扩大/缩小功能
  scrollView.delegate = self;
  scrollView.minimumZoomScale = 0.1;
  scrollView.maximumZoomScale = 3.0;
}
// 扩大/缩小功能
- (UIView*)viewForZoomingInScrollView:(UIScrollView*)scrollView {
  for ( id subview in scrollView.subviews ) {
    if ( [subview isKindOfClass:[UIImageView class]] ) {
      return subview;
```

```
    }
}
    return nil;
}
@end
```

执行后的效果如图 3-96 所示。

图 3-96 执行效果

实例 106 设置滚动条的颜色

实例 106	设置滚动条的颜色
源码路径	\daima\105\

实例说明

在 iOS 应用中，当使用 UIScrollView 控件实现滚动条效果后，可以使用属性 indicatorStyle 设置希望使用的滚动条指示器的类型。其中默认的效果是在白边界上绘制黑色的滚动条，这在大多数背景下都适用。indicatorStyle 可以设置的风格如下所示。

- UIScrollViewIndicatorStyleDefault：默认样式，灰色线条包围黑色线条。
- UIScrollViewIndicatorStyleBlack：黑色条。
- UIScrollViewIndicatorStyleWhite：白色条。

在本实例中设置了一个 UIScrollView 滚动控件，在屏幕下方设置了一个"Style 切换"按钮，当单击此按钮时会显示不同的滚动条颜色。

具体实现

实例文件 UIKitPrjIndicatorStyle.m 的具体代码如下所示。

```
#import "UIKitPrjIndicatorStyle.h"
@implementation UIKitPrjIndicatorStyle
- (void)dealloc {
    [scrollView_ release];
    [super dealloc];
}
- (void)viewDidLoad {
    [super viewDidLoad];
    // ScrollView 初始化
    scrollView_ = [[UIScrollView alloc] init];
    scrollView_.frame = self.view.bounds;
    scrollView_.autoresizingMask = UIViewAutoresizingFlexibleWidth | UIViewAutoresizingFlexibleHeight;
    // 设置能翻滚的区域
    UIView* view = [[[UIView alloc] init] autorelease];
    view.frame = CGRectMake( 0, 0, 800, 600 );
    view.backgroundColor = [UIColor grayColor];
    scrollView_.contentSize = view.bounds.size;
    [scrollView_ addSubview:view];
    // 将 ScrollView 追加到画面上
```

```
  [self.view addSubview:scrollView_];
  UIBarButtonItem* barButton =
    [[[UIBarButtonItem alloc] initWithTitle:@"Style 切换"
                                      style:UIBarButtonItemStyleBordered
                                     target:self
                                     action:@selector(changeButtonDidPush)] autorelease];
  [self setToolbarItems:[NSArray arrayWithObject:barButton]];
}
- (void)changeButtonDidPush {
  if ( UIScrollViewIndicatorStyleWhite < ++scrollView_.indicatorStyle ) {
    scrollView_.indicatorStyle = UIScrollViewIndicatorStyleDefault;
  }
}
@end
```

执行后的效果如图 3-97 所示。

图 3-97　执行效果

实例 107　将滚动条设置为分页的形式

实例 107	将滚动条设置为分页的形式
源码路径	\daima\105\

实例说明

在 iOS 应用中，当使用 UIScrollView 控件实现滚动条效果后，可以使用属性 pagingEnabled 设置为 YES，这样滚动视图被分割成多个独立区段，此时用户的滚动体验则变成了页面翻转。在编程过程中我们可以用这个属性来进行页面翻转。

在本实例中，使用属性"pagingEnabled=YES"将滚动视图分为了 3 个独立的界面。

具体实现

实例文件 UIKitPrjPaging.m 的具体代码如下所示。

```
#import "UIKitPrjPaging.h"
#pragma mark ----- MyViewController -----
@implementation MyViewController
@synthesize number = number_;
//传入数值初始化 MyViewController 实例的类方法
+ (MyViewController*)myViewControllerWithNumber:(NSInteger)number {
  MyViewController* myViewController = [[[MyViewController alloc] init] autorelease];
  myViewController.number = number;
  return myViewController;
}
- (void)viewDidLoad {
  [super viewDidLoad];
  UILabel* label = [[[UILabel alloc] init] autorelease];
  label.frame = self.view.bounds;
  label.autoresizingMask = UIViewAutoresizingFlexibleWidth | UIViewAutoresizingFlexibleHeight;
  label.backgroundColor = ( self.number % 2 ) ? [UIColor blackColor] : [UIColor whiteColor];
  label.textColor = !( self.number % 2 ) ? [UIColor blackColor] : [UIColor whiteColor];
```

```
    label.textAlignment = UITextAlignmentCenter;
    label.font = [UIFont boldSystemFontOfSize:128];
    label.text = [NSString stringWithFormat:@"%d", self.number];
    [self.view addSubview:label];
}
@end
#pragma mark ----- UIKitPrjPaging -----
@implementation UIKitPrjPaging
static const NSInteger kNumberOfPages = 3;
static const NSInteger kViewHeight = 360;
- (void)viewDidLoad {
    [super viewDidLoad];
    // ScrollView 的初始化
    UIScrollView* scrollView = [[[UIScrollView alloc] init] autorelease];
    scrollView.frame = self.view.bounds;
    scrollView.autoresizingMask = UIViewAutoresizingFlexibleWidth | UIViewAutoresizingFlexibleHeight;
    // 为了能进行画面横向翻滚,设置内容的横向长度
    scrollView.contentSize = CGSizeMake( 320 * kNumberOfPages, kViewHeight );
    // 以页为单位翻滚
    scrollView.pagingEnabled = YES;
    // 隐藏滚动条
    scrollView.showsHorizontalScrollIndicator = NO;
    scrollView.showsVerticalScrollIndicator = NO;
    // 关闭翻滚到顶部
    scrollView.scrollsToTop = NO;
    // 将各画面追加到 ScrollView
    for ( int i = 0; i < kNumberOfPages; ++i ){
        MyViewController* myViewController = [MyViewController myViewControllerWithNumber:i];
        myViewController.view.frame = CGRectMake( 320 * i, 0, 320, kViewHeight );
        [scrollView addSubview:myViewController.view];
    }
    // 将 ScrollView 追加到主画面
    [self.view addSubview:scrollView];
}
@end
```

执行后的效果如图 3-98 所示。

图 3-98 执行效果

第 4 章 文本和表格处理实战

iOS 系统提供了强大的文本和表格功能。通过 iOS 独有的键盘输入系统，可以在智能设备界面中提供各种各样的文本。另外在显示数据时，iOS 通常使用表示视图的方式来罗列数据。本章将通过具体的实例来讲解 iOS 系统中处理文本和表格的基本知识。

实例 108 在屏幕中换行显示文本

实例 108	在屏幕中换行显示文本
源码路径	\daima\108\

实例说明

在 iOS 应用中，可以使用 UITextView 在屏幕中显示文本，并且能够同时显示多行文本。另外还可以使用其 textColor 属性设置文本的颜色，通过 font 属性设置文本的字体和大小。

在本实例中，使用控件 UITextView 在屏幕中同时显示了 12 行文本，并且设置了文本的颜色是白色，设置了字体大小是 32。

具体实现

实例文件 UIKitPrjT11extView.m 的具体代码如下所示。

```
#import "UIKitPrjTextView.h"
@implementation UIKitPrjTextView
- (void)viewDidLoad {
    [super viewDidLoad];
    UITextView* textView = [[[UITextView alloc] init] autorelease];
    textView.frame = self.view.bounds;
    textView.autoresizingMask =
      UIViewAutoresizingFlexibleWidth | UIViewAutoresizingFlexibleHeight;
    //textView.editable = NO; //不可编辑

    textView.backgroundColor = [UIColor blackColor]; //背景为黑色
    textView.textColor = [UIColor whiteColor]; //字符为白色
    textView.font = [UIFont systemFontOfSize:32]; //字体的设置
    textView.text = @"学习 UITextView!\n"
                "第 2 行\n"
                "第 3 行\n"
                "第 4 行\n"
                "第 5 行\n"
                "第 6 行\n"
                "第 7 行\n"
                "第 8 行\n"
                "第 9 行\n"
                "第 10 行\n"
                "第 11 行\n"
                "第 12 行\n";
    [self.view addSubview:textView];
}
@end
```

执行后的效果如图 4-1 所示。

图 4-1 执行效果

实例 109 在屏幕中显示可编辑的文本

实例 109	在屏幕中显示可编辑的文本
源码路径	\daima\108\

实例说明

在 iOS 应用中，当使用 UITextView 控件在屏幕中设置一段文本后，可以将其 editable 属性设置为 YES，这样将这段文本设置为是可以编辑的。在本实例中，使用控件 UITextView 在屏幕中显示了一段文本"亲们，可以编辑这一段文本。"，然后将其 editable 属性设置为 YES。当单击"Edit"按钮后可以编辑这段文本，单击"Done"按钮后可以完成对这段文本的编辑操作。

具体实现

实例文件 UIKitPrjEditableTextView.m 的具体代码如下所示。

```
#import "UIKitPrjEditableTextView.h"
@implementation UIKitPrjEditableTextView
- (void)dealloc {
 [textView_ release];
 [super dealloc];
}
- (void)viewDidLoad {
 [super viewDidLoad];
 textView_ = [[UITextView alloc] init];
 textView_.frame = self.view.bounds;
 textView_.autoresizingMask = UIViewAutoresizingFlexibleWidth |
                              UIViewAutoresizingFlexibleHeight;
 textView_.delegate = self;
 textView_.text = @"亲们，可以编辑这一段文本。";
 [self.view addSubview:textView_];
}
- (void)viewWillAppear:(BOOL)animated {
 [super viewWillAppear:animated];
 [self.navigationController setNavigationBarHidden:NO animated:YES];
 [self.navigationController setToolbarHidden:NO animated:YES];
}
- (void)viewDidAppear:(BOOL)animated {
 [super viewDidAppear:animated];
 [self textViewDidEndEditing:textView_]; //画面显示时设置为非编辑模式
}
- (void)viewWillDisappear:(BOOL)animated {
 [super viewWillDisappear:animated];
 [textView_ resignFirstResponder]; //画面跳转时设置为非编辑模式
}
- (void)textViewDidBeginEditing:(UITextView*)textView {
 static const CGFloat kKeyboardHeight = 216.0;
 // 按钮设置为 [完成]
 self.navigationItem.rightBarButtonItem =
   [[[UIBarButtonItem alloc] initWithBarButtonSystemItem:UIBarButtonSystemItemDone
                              target:self
                              action:@selector(doneDidPush)]
```

```objc
    autorelease];
    [UIView beginAnimations:nil context:nil];
    [UIView setAnimationDuration:0.3];
    // 缩小 UITextView 以免被键盘挡住
    CGRect textViewFrame = textView.frame;
    textViewFrame.size.height = self.view.bounds.size.height - kKeyboardHeight;
    textView.frame = textViewFrame;
    //工具条位置上移
    CGRect toolbarFrame = self.navigationController.toolbar.frame;
    toolbarFrame.origin.y =
      self.view.window.bounds.size.height - toolbarFrame.size.height - kKeyboardHeight;
    self.navigationController.toolbar.frame = toolbarFrame;
    [UIView commitAnimations];
}
- (void)textViewDidEndEditing:(UITextView*)textView {
    // 按钮设置为[编辑]
    self.navigationItem.rightBarButtonItem =
      [[[UIBarButtonItem alloc] initWithBarButtonSystemItem:UIBarButtonSystemItemEdit
                                                    target:self
                                                    action:@selector(editDidPush)] autorelease];
    [UIView beginAnimations:nil context:nil];
    [UIView setAnimationDuration:0.3];
    // 恢复 UITextView 的尺寸
    textView.frame = self.view.bounds;
    // 恢复工具条的位置
    CGRect toolbarFrame = self.navigationController.toolbar.frame;
    toolbarFrame.origin.y =
      self.view.window.bounds.size.height - toolbarFrame.size.height;
    self.navigationController.toolbar.frame = toolbarFrame;
    [UIView commitAnimations];
}
- (void)editDidPush {
    [textView_ becomeFirstResponder];
}
- (void)doneDidPush {
    [textView_ resignFirstResponder];
}
@end
```

执行后的效果如图 4-2 所示。单击"Edit"按钮后可以编辑这段文字，如图 4-3 所示。

图 4-2 执行效果

图 4-3 编辑界面

实例 110　将屏幕中的文本实现编辑状态和非编辑状态之间的切换

实例 110	将屏幕中的文本实现编辑状态和非编辑状态之间的切换
源码路径	\daima\108\

实例 110 将屏幕中的文本实现编辑状态和非编辑状态之间的切换

实例说明

在 iOS 应用中，当使用 UITextView 控件在屏幕中设置一段文本后，使用"编辑/非编辑"模式相互结合的方式，自动调节 UITextView 尺寸。在本实例中，使用控件 UITextView 在屏幕中显示了一段文本"亲们，可以编辑这一段文本。"，然后将其 editable 属性设置为 YES。当单击"Edit"按钮后可以编辑这段文本，单击"Done"按钮后可以完成对这段文本的编辑操作。使用 viewDidAppear:(BOOL)animated 方法设置画面显示时变成非编辑模式，使用 viewWillDisappear:(BOOL)animated 方法设置画面跳转时为非编辑模式。

具体实现

实例文件 UIKitPrjTextViewObserving.m 的具体实现流程如下所示。

（1）设置默认的可编辑文本，定义 viewDidAppear:(BOOL)animated 方法设置画面显示时为非编辑模式，定义 viewWillDisappear:(BOOL)animated 方法设置画面跳转时为非编辑模式。具体代码如下所示。

```objc
#import "UIKitPrjTextViewObserving.h"
@implementation UIKitPrjTextViewObserving
- (void)dealloc {
  [textView_ release];
  [super dealloc];
}
- (void)viewDidLoad {
  [super viewDidLoad];
  // UITextView 的追加
  textView_ = [[UITextView alloc] init];
  textView_.frame = self.view.bounds;
  textView_.delegate = self;
  textView_.autoresizingMask = UIViewAutoresizingFlexibleWidth | UIViewAutoresizingFlexibleHeight;
  textView_.text = @"亲们，可以编辑这一段文本。";
  [self.view addSubview:textView_];
}
- (void)viewDidAppear:(BOOL)animated {
  [super viewDidAppear:animated];
  [self textViewDidEndEditing:textView_]; //画面显示时设置成非编辑模式
}
- (void)viewWillDisappear:(BOOL)animated {
  [super viewWillDisappear:animated];
  [textView_ resignFirstResponder]; //画面跳转时为非编辑模式
}
- (BOOL)textView:(UITextView*)textView shouldChangeTextInRange:(NSRange)range replacementText:(NSString*)text {
  NSLog( @"shouldChangeTextInRange %@", text );
  if ( [text isEqualToString:@"a"] ) {
    return NO; //不可只输入 a 字符
  }
  return YES;
}
```

（2）定义方法 textViewDidBeginEditing，用于在开始编辑文本时调用，通过此方法缩小 UITextView 不让键盘挡住，并且上移工具条的位置。具体代码如下所示。

```objc
- (void)textViewDidBeginEditing:(UITextView*)textView {
  NSLog( @"textViewDidBeginEditing" );
  static const CGFloat kKeyboardHeight = 216.0;
// 将按钮设置成 [编辑]
  self.navigationItem.rightBarButtonItem =
    [[[UIBarButtonItem alloc] initWithBarButtonSystemItem:UIBarButtonSystemItemDone
                                  target:self
                                  action:@selector(doneDidPush)]
autorelease];
```

```objectivec
[UIView beginAnimations:nil context:nil];
[UIView setAnimationDuration:0.3];
// 缩小 UITextView 不让键盘挡住
CGRect textViewFrame = textView.frame;
textViewFrame.size.height = self.view.bounds.size.height - kKeyboardHeight;
textView.frame = textViewFrame;
// 工具条的位置也上移
CGRect toolbarFrame = self.navigationController.toolbar.frame;
toolbarFrame.origin.y =
  self.view.window.bounds.size.height - toolbarFrame.size.height - kKeyboardHeight;
self.navigationController.toolbar.frame = toolbarFrame;
[UIView commitAnimations];
}
```

（3）定义方法 textViewDidEndEditing，用于在开始编辑文本完成时调用，通过此方法恢复 UITextView 的尺寸，并且恢复工具条的位置。具体代码如下所示。

```objectivec
- (void)textViewDidEndEditing:(UITextView*)textView {
 NSLog( @"textViewDidEndEditing" );
 // 按钮设置成［完成］
 self.navigationItem.rightBarButtonItem =
   [[[UIBarButtonItem alloc] initWithBarButtonSystemItem:UIBarButtonSystemItemEdit
                                  target:self
                                  action:@selector(editDidPush)]
autorelease];
 [UIView beginAnimations:nil context:nil];
 [UIView setAnimationDuration:0.3];
 // 恢复 UITextView 的尺寸
 textView.frame = self.view.bounds;
 // 恢复工具条的位置
 CGRect toolbarFrame = self.navigationController.toolbar.frame;
 toolbarFrame.origin.y =
   self.view.window.bounds.size.height - toolbarFrame.size.height;
 self.navigationController.toolbar.frame = toolbarFrame;
 [UIView commitAnimations];
}
- (void)editDidPush {
 [textView_ becomeFirstResponder];
}
- (void)doneDidPush {
 [textView_ resignFirstResponder];
}
@end
```

执行后的效果如图 4-4 所示。

图 4-4 执行效果

实例 111 设置屏幕中文本的对齐方式，确定文本的选择范围

实例 111	设置屏幕中文本的对齐方式，确定文本的选择范围
源码路径	\daima\108\

实例 111　设置屏幕中文本的对齐方式，确定文本的选择范围

实例说明

在 iOS 应用中，当使用 UITextView 控件在屏幕中设置一段文本后，可以使用属性 textAlignment 设置文本的对齐方式。此属性有如下 3 个值。

- UITextAlignmentRight：右对齐。
- UITextAlignmentCenter：居中对齐。
- UITextAlignmentLeft：左对齐。

在本实例中，使用控件 UITextView 在屏幕中显示了一段文本"此文本可编辑"，然后在工具条中添加了 4 个按钮。其中通过按钮"alignment"可以控制文本的对齐方式，通过按钮"Selection"可以获得文本的范围。

具体实现

实例文件 UIKitPrjWorkingWithTheSelection.m 的具体实现流程如下所示。

```
#import "UIKitPrjWorkingWithTheSelection.h"
static const CGFloat kKeyboardHeight = 216.0;
@implementation UIKitPrjWorkingWithTheSelection
- (void)dealloc {
 [textView_ release];
 [super dealloc];
}

- (void)viewDidLoad {
 [super viewDidLoad];
 // UITextView 的追加
 textView_ = [[UITextView alloc] init];
 textView_.frame = self.view.bounds;
 textView_.autoresizingMask = UIViewAutoresizingFlexibleWidth | UIViewAutoresizingFlexibleHeight;
 textView_.text = @"此文本可编辑。";
 [self.view addSubview:textView_];
 // 在工具条中追加按钮
 UIBarButtonItem* hasTextButton =
   [[[UIBarButtonItem alloc] initWithTitle:@"hasText"
                                     style:UIBarButtonItemStyleBordered
                                    target:self
                                    action:@selector(hasTextDidPush)] autorelease];
 UIBarButtonItem* selectionButton =
   [[[UIBarButtonItem alloc] initWithTitle:@"selection"
                                     style:UIBarButtonItemStyleBordered
                                    target:self
                                    action:@selector(selectionDidPush)] autorelease];
 UIBarButtonItem* alignmentButton =
   [[[UIBarButtonItem alloc] initWithTitle:@"alignment"
                                     style:UIBarButtonItemStyleBordered
                                    target:self
                                    action:@selector(alignmentDidPush)] autorelease];
 UIBarButtonItem* scrollButton =
   [[[UIBarButtonItem alloc] initWithTitle:@"top"
                                     style:UIBarButtonItemStyleBordered
                                    target:self
                                    action:@selector(scrollDidPush)] autorelease];
 NSArray* buttons = [NSArray arrayWithObjects:hasTextButton, selectionButton, alignmentButton, scrollButton, nil];
 [self setToolbarItems:buttons animated:YES];
}
- (void)viewDidAppear:(BOOL)animated {
 [super viewDidAppear:animated];
 // 调整工具条位置
 [UIView beginAnimations:nil context:nil];
 [UIView setAnimationDuration:0.3];
 textView_.frame =
   CGRectMake( 0, 0, self.view.bounds.size.width, self.view.bounds.size.height -
```

```
    kKeyboardHeight );
  CGRect toolbarFrame = self.navigationController.toolbar.frame;
  toolbarFrame.origin.y =
    self.view.window.bounds.size.height - toolbarFrame.size.height - kKeyboardHeight;
  self.navigationController.toolbar.frame = toolbarFrame;
  [UIView commitAnimations];
  [textView_ becomeFirstResponder]; //画面显示时显示键盘
}
- (void)viewWillDisappear:(BOOL)animated {
  [super viewWillDisappear:animated];
  // 恢复工具条
  [UIView beginAnimations:nil context:nil];
  [UIView setAnimationDuration:0.3];
  textView_.frame = self.view.bounds;
  CGRect toolbarFrame = self.navigationController.toolbar.frame;
  toolbarFrame.origin.y = self.view.window.bounds.size.height - toolbarFrame.size.height;
  self.navigationController.toolbar.frame = toolbarFrame;
  [UIView commitAnimations];
  [textView_ resignFirstResponder]; //画面隐藏时隐藏键盘
}
- (void)hasTextDidPush {
  UIAlertView* alert = [[[UIAlertView alloc] init] autorelease];
  if ( textView_.hasText ) {
    alert.message = @"textView_.hasText = YES";
  } else {
    alert.message = @"textView_.hasText = NO";
  }
  [alert addButtonWithTitle:@"OK"];
  [alert show];
}
- (void)selectionDidPush {
  UIAlertView* alert = [[[UIAlertView alloc] init] autorelease];
  alert.message = [NSString stringWithFormat:@"location = %d, length = %d",
                   textView_.selectedRange.location, textView_.selectedRange.length];
  [alert addButtonWithTitle:@"OK"];
  [alert show];
}
- (void)alignmentDidPush {
  textView_.editable = NO;
  if ( UITextAlignmentRight < ++textView_.textAlignment ) {
    textView_.textAlignment = UITextAlignmentLeft;
  }
  textView_.editable = YES;
}
- (void)scrollDidPush {
  // NSRange scrollRange = NSMakeRange( 0, 1 );
  [textView_ scrollRangeToVisible:NSMakeRange( 0, 1 )];
}
@end
```

执行后的效果如图 4-5 所示。

图 4-5 执行效果

实例 112　自动处理屏幕中文本的 URL 地址和电话号码

实例 112	自动处理屏幕中文本的 URL 地址和电话号码
源码路径	\daima\108\

实例说明

在 iOS 应用中，当在 UITextView 控件中显示电话号码时，点击这个号码会自动来到拨号界面。当在 UITextView 控件中显示 URL 地址时，点击这个地址会自动来到这个页面。在本实例中，将 textView 的 dataDetectorTypes 属性设置为了 UIDataDetectorTypeAll，这样便实现了自动处理文本中 URL 地址和电话号码的功能。

具体实现

实例文件 UIKitPrjDataDetectorTypes.m 的具体实现流程如下所示。

```
#import "UIKitPrjDataDetectorTypes.h"
@implementation UIKitPrjDataDetectorTypes
- (void)viewDidLoad {
 [super viewDidLoad];
 // UITextView 的追加
 UITextView* textView = [[[UITextView alloc] init] autorelease];
 textView.frame = self.view.bounds;
 textView.editable = NO; //这个必需
 textView.autoresizingMask = UIViewAutoresizingFlexibleWidth | UIViewAutoresizingFlexibleHeight;
 textView.font = [UIFont systemFontOfSize:24];
 textView.text = @"详细如下↓\n"
                 "http://www.apple.com/\n"
                 "联系方式：158-0000-0000\n";
 textView.dataDetectorTypes = UIDataDetectorTypeAll;
 [self.view addSubview:textView];
}
@end
```

执行后的效果如图 4-6 所示。触摸里面的 URL 地址后会进入对应的页面，如图 4-7 所示。

图 4-6　执行效果　　　　　　　　　　　图 4-7　进入对应的页面

实例 113　在屏幕文本中显示密码黑点"."

实例 113	在屏幕文本中显示密码黑点"."
源码路径	\daima\108\

实例说明

在 iOS 应用中，UITextFiled 控件提供的默认样式文本在显示密码时以不可见的黑点形式显示。在本实例中，将属性 secureTextEntry 设置为了 YES，这样在屏幕中的密码便以不可见的黑点形式进行显示。

具体实现

实例文件 UIKitPrjSecureTextEntry.m 的具体实现流程如下所示。

```
#import "UIKitPrjSecureTextEntry.h"
@implementation UIKitPrjSecureTextEntry
- (void)dealloc {
 [textField_ release];
 [super dealloc];
}
- (void)viewDidLoad {
 [super viewDidLoad];
 // UITextField 的追加
 textField_ = [[UITextField alloc] init];
 textField_.frame = self.view.bounds;
 textField_.autoresizingMask = UIViewAutoresizingFlexibleWidth | UIViewAutoresizingFlexibleHeight;
 textField_.backgroundColor = [UIColor whiteColor];
 textField_.text = @"password";
 textField_.secureTextEntry = YES;
 [self.view addSubview:textField_];
}
- (void)viewDidAppear:(BOOL)animated {
 [super viewDidAppear:animated];
 [textField_ becomeFirstResponder]; //< 画面显示时的键盘显示
}
@end
```

执行后的效果如图 4-8 所示。

图 4-8　执行效果

实例 114　自定义 UITableViewCell

实例 114	自定义 UITableViewCell
源码路径	\daima\114\

实例说明

在 iOS 应用中，可以自定义一个 UITableViewCell 表格样式。其实原理就是向行中添加子视图。添加子视图的方法主要有两种：使用代码以及从 .xib 文件加载。当然，后一种方法比较直观。在本实例中自定义了一个 Cell，使得它像 QQ 好友列表的一行一样：左边显示一张图片，在图片的右边

实例 114　自定义 UITableViewCell

显示 3 行标签。

具体实现

（1）运行 Xcode，新建一个 Single View Application，命名为"Custom Cell"。

（2）将图片资源导入到工程。本实例使用了 14 张 50×50 的.png 图片，名称依次是 1、2、…、14，放在一个名为 Images 的文件夹中。将此文件夹拖到工程中，在弹出的窗口中选中 Copy items into…如图 4-9 所示。

（3）创建一个 UITableViewCell 的子类：选中 Custom Cell 目录，依次选择 File→New→New File，在弹出的窗口左边选择 Cocoa Touch，右边选择 Objective-C class。然后单击 Next 按钮，输入类名 CustomCell，Subclass of 选择

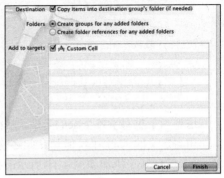

图 4-9　选中 Copy items into…

UITableViewCell。之后选择 Next 和 Create，就建立了两个文件——CustomCell.h 和 CustomCell.m。

（4）创建 CustomCell.xib：依次选择 File→New→New File，在弹出的窗口左边选择 User Interface，右边选择 Empty。单击 Next 按钮，选择 iPhone，再单击 Next 按钮，输入名称"CustomCell"，并选择保存位置。单击 Create 按钮，这样就创建了 CustomCell.xib。

（5）打开 CustomCell.xib，拖一个 Table View Cell 控件到面板上。选中新加的控件，打开 Identity Inspector，选择 Class 为 CustomCell；然后打开 Size Inspector，调整高度为 60。

（6）向新加的 Table View Cell 添加控件，拖放一个 ImageView 控件到左边，并设置大小为 50×50。然后在 ImageView 右边添加 3 个 Label，设置标签字号，最上边的是 14，其余两个是 12。

接下来向文件 CustomCell.h 添加 Outlet 映射，将 ImageView 与 3 个 Label 建立映射，名称分别为 imageView、nameLabel、decLabel 以及 locLable，分别表示头像、昵称、个性签名、地点。然后选中 Table View

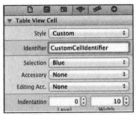

图 4-10　建立映射

Cell，打开 Attribute Inspector，将 Identifier 设置为 CustomCellIdentifier，如图 4-10 所示。

为了充分使用这些标签，开发者还要自己创建一些数据，存在 plist 文件中。

（7）打开文件 CustomCell.h，添加如下 4 个属性。

```
@property (copy, nonatomic) UIImage *image;
@property (copy, nonatomic) NSString *name;
@property (copy, nonatomic) NSString *dec;
@property (copy, nonatomic) NSString *loc;
```

（8）打开文件 CustomCell.m，其中在@implementation 下面添加如下所示的代码：

```
@synthesize image;
@synthesize name;
@synthesize dec;
@synthesize loc;
```

然后在@end 之前添加如下所示的代码：

```
- (void)setImage:(UIImage *)img {
    if (![img isEqual:image]) {
        image = [img copy];
        self.imageView.image = image;
    }
}
-(void)setName:(NSString *)n {
    if (![n isEqualToString:name]) {
        name = [n copy];
        self.nameLabel.text = name;
```

```
    }
-(void)setDec:(NSString *)d {
    if (![d isEqualToString:dec]) {
        dec = [d copy];
        self.decLabel.text = dec;
    }
}
-(void)setLoc:(NSString *)l {
    if (![l isEqualToString:loc]) {
        loc = [l copy];
        self.locLabel.text = loc;
    }
}
```

这相当于重写了各个 set()函数，从而当执行赋值操作时，会执行我们自己写的函数。现在就可以使用自定义的 Cell 了，但是在此之前需先新建一个 plist，用于存储想要显示的数据。在建好的 friendsInfo.plist 中添加如图 4-11 所示的数据。

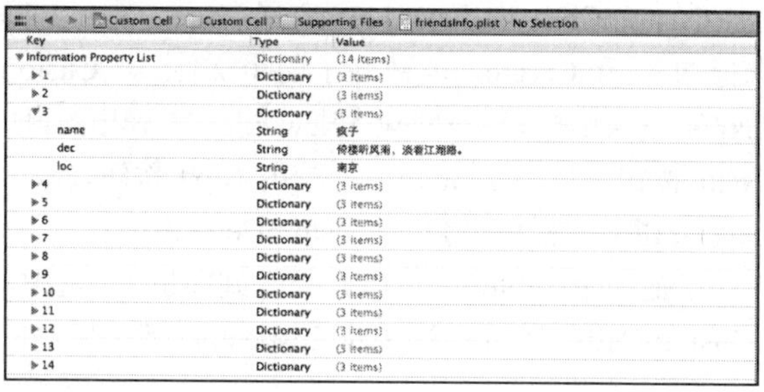

图 4-11　添加数据

在此需要注意每个节点类型的选择。

（9）打开 ViewController.xib，拖一个 Table View 到视图上，并将 Delegate 和 DataSource 都指向 File' Owner。

（10）打开文件 ViewController.h，向其中添加如下所示的代码：

```
#import <UIKit/UIKit.h>
@interface ViewController : UIViewController<UITableViewDelegate, UITableViewDataSource>
@property (strong, nonatomic) NSArray *dataList;
@property (strong, nonatomic) NSArray *imageList;
@end
```

（11）打开文件 ViewController.m，在首部添加如下代码：

```
#import "CustomCell.h"
```

然后在@implementation 后面添加如下代码：

```
@synthesize dataList;
@synthesize imageList;
```

在方法 viewDidLoad 中添加如下所示的代码：

```
- (void)viewDidLoad
{
    [super viewDidLoad];
    // Do any additional setup after loading the view, typically from a nib.
    //加载 plist 文件的数据和图片
    NSBundle *bundle = [NSBundle mainBundle];
    NSURL *plistURL = [bundle URLForResource:@"friendsInfo" withExtension:@"plist"];
    NSDictionary *dictionary = [NSDictionary dictionaryWithContentsOfURL:plistURL];
    NSMutableArray *tmpDataArray = [[NSMutableArray alloc] init];
    NSMutableArray *tmpImageArray = [[NSMutableArray alloc] init];
    for (int i=0; i<[dictionary count]; i++) {
        NSString *key = [[NSString alloc] initWithFormat:@"%i", i+1];
        NSDictionary *tmpDic = [dictionary objectForKey:key];
        [tmpDataArray addObject:tmpDic];
```

实例 114 自定义 UITableViewCell

```
NSString *imageUrl = [[NSString alloc] initWithFormat:@"%i.png", i+1];
    UIImage *image = [UIImage imageNamed:imageUrl];
    [tmpImageArray addObject:image];
}
self.dataList = [tmpDataArray copy];
self.imageList = [tmpImageArray copy];
}
```

在方法 ViewDidUnload 中添加如下所示的代码:

```
self.dataList = nil;
self.imageList = nil;
```

在@end 之前添加如下所示的代码:

```
#pragma mark -
#pragma mark Table Data Source Methods
- (NSInteger)tableView:(UITableView *)tableView numberOfRowsInSection:(NSInteger)section {
    return [self.dataList count];
}
- (UITableViewCell *)tableView:(UITableView *)tableView cellForRowAtIndexPath:(NSIndexPath *)indexPath {
    static NSString *CustomCellIdentifier = @"CustomCellIdentifier";
    static BOOL nibsRegistered = NO;
    if (!nibsRegistered) {
        UINib *nib = [UINib nibWithNibName:@"CustomCell" bundle:nil];
        [tableView registerNib:nib forCellReuseIdentifier:CustomCellIdentifier];
        nibsRegistered = YES;
    }
    CustomCell *cell = [tableView dequeueReusableCellWithIdentifier:CustomCellIdentifier];
    NSUInteger row = [indexPath row];
    NSDictionary *rowData = [self.dataList objectAtIndex:row];
    cell.name = [rowData objectForKey:@"name"];
    cell.dec = [rowData objectForKey:@"dec"];
    cell.loc = [rowData objectForKey:@"loc"];
    cell.image = [imageList objectAtIndex:row];
    return cell;
}
#pragma mark Table Delegate Methods
- (CGFloat)tableView:(UITableView *)tableView heightForRowAtIndexPath:(NSIndexPath *)indexPath {
    return 60.0;
}
- (NSIndexPath *)tableView:(UITableView *)tableView willSelectRowAtIndexPath:(NSIndexPath *)indexPath {
    return nil;
}
```

到此为止，整个实例介绍完毕，执行后的效果如图 4-12 所示。

图 4-12 执行效果

实例 115 拆分表视图

实例 115	拆分表视图
源码路径	\daima\115\

实例说明

在本实例中创建了一个表视图,它包含两个分区,这两个分区的标题分别为 Red 和 Blue,且分别包含常见的红色和绿色花朵的名称。除标题外,每个单元格还包含一幅花朵图像和一个展开箭头。用户触摸单元格时,将出现一个提醒视图,指出选定花朵的名称和颜色。

具体实现

实例文件 ViewController.m 的具体实现代码如下所示。

```objc
#import "ViewController.h"
#define kSectionCount 2
#define kRedSection 0
#define kBlueSection 1
@implementation ViewController
@synthesize redFlowers;
@synthesize blueFlowers;
- (void)didReceiveMemoryWarning
{
    [super didReceiveMemoryWarning];
}
#pragma mark - View lifecycle
- (void)viewDidLoad
{
    self.redFlowers = [[NSArray alloc]
                       initWithObjects:@"aa",@"bb",@"cc",
                       @"dd",nil];
  self.blueFlowers = [[NSArray alloc]
                       initWithObjects:@"ee",@"ff",
                       @"gg",@"hh",@"ii",nil];

    [super viewDidLoad];
}
- (void)viewDidUnload
{
    [self setRedFlowers:nil];
    [self setBlueFlowers:nil];
    [super viewDidUnload];
}
- (void)viewWillAppear:(BOOL)animated
{
    [super viewWillAppear:animated];
}
- (void)viewDidAppear:(BOOL)animated
{
    [super viewDidAppear:animated];
}
- (void)viewWillDisappear:(BOOL)animated
{
    [super viewWillDisappear:animated];
}
- (void)viewDidDisappear:(BOOL)animated
{
    [super viewDidDisappear:animated];
}
-
(BOOL)shouldAutorotateToInterfaceOrientation:(UIInterfaceOrientation)interfaceOrientation
{
    // Return YES for supported orientations
    return (interfaceOrientation != UIInterfaceOrientationPortraitUpsideDown);
```

```objc
}
#pragma mark - Table view data source
- (NSInteger)numberOfSectionsInTableView:(UITableView *)tableView
{
    return kSectionCount;
}
- (NSInteger)tableView:(UITableView *)tableView
    numberOfRowsInSection:(NSInteger)section
{
    switch (section) {
        case kRedSection:
            return [self.redFlowers count];
        case kBlueSection:
            return [self.blueFlowers count];
        default:
            return 0;
    }
}
- (NSString *)tableView:(UITableView *)tableView
titleForHeaderInSection:(NSInteger)section {
    switch (section) {
        case kRedSection:
            return @"红";
        case kBlueSection:
            return @"蓝";
        default:
            return @"Unknown";
    }
}
- (UITableViewCell *)tableView:(UITableView *)tableView
         cellForRowAtIndexPath:(NSIndexPath *)indexPath
{
    UITableViewCell *cell = [tableView
                     dequeueReusableCellWithIdentifier:@"flowerCell"];


    switch (indexPath.section) {
        case kRedSection:
            cell.textLabel.text=[self.redFlowers
                        objectAtIndex:indexPath.row];
            break;
        case kBlueSection:
            cell.textLabel.text=[self.blueFlowers
                        objectAtIndex:indexPath.row];
            break;
        default:
            cell.textLabel.text=@"Unknown";
    }

    UIImage *flowerImage;
    flowerImage=[UIImage imageNamed:
            [NSString stringWithFormat:@"%@%@",
             cell.textLabel.text,@".png"]];
    cell.imageView.image=flowerImage;

    return cell;
}
#pragma mark - Table view delegate
- (void)tableView:(UITableView *)tableView
        didSelectRowAtIndexPath:(NSIndexPath *)indexPath {
    UIAlertView *showSelection;
    NSString    *flowerMessage;

    switch (indexPath.section) {
        case kRedSection:
            flowerMessage=[[NSString alloc]
                            initWithFormat:
                            @"你选择了红色 - %@",
                            [self.redFlowers objectAtIndex: indexPath.row]];
            break;
        case kBlueSection:
```

```
                    flowerMessage=[[NSString alloc]
                                initWithFormat:
                                @"你选择了蓝色 - %@",
                                 [self.blueFlowers objectAtIndex: indexPath.row]];
                break;
            default:
                flowerMessage=[[NSString alloc]
                                initWithFormat:
                                @"我不知道选什么!?"];
                break;
        }
    showSelection = [[UIAlertView alloc]
                        initWithTitle: @"已经选择了"
                        message:flowerMessage
                        delegate: nil
                        cancelButtonTitle: @"Ok"
                        otherButtonTitles: nil];
    [showSelection show];
}
@end
```

执行后的效果如图 4-13 所示。

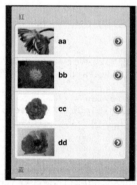

图 4-13　执行效果

实例 116　列表显示 18 条数据

实例 116	列表显示 18 条数据
源码路径	\daima\116\

实例说明

在本实例中，创建了各单元显示内容对象 tableView，并将表格中列表显示的数据存储在数组 dataSource 中。通过 "cell.textLabel.text" 代码设置在单元格中显示 dataSource 中的内容，即列表显示 18 个数据。最后定义了方法 didSelectRowAtIndexPath，通过此方法实现选择某一行数据时的处理动作。

具体实现

实例文件 UIKitPrjSimpleTable.m 的具体实现代码如下所示。

```
#import "UIKitPrjSimpleTable.h"
@implementation UIKitPrjSimpleTable
- (void)dealloc {
    [dataSource_ release];//画面释放时也需释放保存元素的数组
    [super dealloc];
}
- (void)viewDidLoad {
    [super viewDidLoad];
    //初始化表格元素数值
```

```
        dataSource_ = [[NSArray alloc] initWithObjects:
                            @"AAA1", @"AAA2", @"AAA3",
                            @"AAA",  @"AAA5", @"AAA6",
                            @"AAA7", @"AAA8", @"AAA9",
                            @"AAA10", @"AAA11", @"AAA12",
                            @"AAA13", @"AAA14", @"AAA15",
                            @"AAA16", @"AAA17", @"AAA18",
                            nil ];
}
//返回表格行数（本例只有单元数）
- (NSInteger)tableView:(UITableView*)tableView numberOfRowsInSection:(NSInteger)section {
    return [dataSource_ count];
}
//创建各单元显示内容（创建参数 indexPath 指定的单元）
- (UITableViewCell*)tableView:(UITableView*)tableView
  cellForRowAtIndexPath:(NSIndexPath*)indexPath
{
    //为了提供表格显示性能，已创建完成的单元需重复使用
    static NSString* identifier = @"basis-cell";
    //同一形式的单元格重复使用（基本上各形式相同而内容是不同的）
    UITableViewCell* cell = [tableView dequeueReusableCellWithIdentifier:identifier];
    if ( nil == cell ) {
        //初始为空时必须创建
      cell = [[UITableViewCell alloc] initWithStyle:UITableViewCellStyleDefault
                            reuseIdentifier:identifier];
      [cell autorelease];
    }
    //设置单元格中的显示内容
    cell.textLabel.text = [dataSource_ objectAtIndex:indexPath.row];
    return cell;
}
- (void)tableView:(UITableView*)tableView didSelectRowAtIndexPath:(NSIndexPath*)indexPath {
    NSString* message = [dataSource_ objectAtIndex:indexPath.row];
    UIAlertView* alert = [[[UIAlertView alloc] init] autorelease];
    alert.message = message;
    [alert addButtonWithTitle:@"OK"];
    [alert show];
}
@end
```

执行后的效果如图4-14所示。

图4-14 执行效果

实例117 分段显示列表中的数据

实例117	分段显示列表中的数据
源码路径	\daima\116\

实例说明

在iOS项目中，当使用UITableView控件创建一个表格视图后，可以对列表中的数据实现分段

显示。在本实例中，通过方法 numberOfSectionsInTableView 和 tableView:titleForHeaderInSection 实现了分段显示功能。

具体实现

实例文件 UIKitPrjSectionTable.m 的具体实现代码如下所示。

```objc
#import "UIKitPrjSectionTable.h"
@implementation UIKitPrjSectionTable
- (void)dealloc {
  [keys_ release];
  [dataSource_ release];
  [super dealloc];
}
- (id)init {
  if ( (self = [super init]) ) {
    self.title = @"SectionTable"; //追加标题
  }
  return self;
}
- (void)viewDidLoad {
  [super viewDidLoad];
  //创建显示用数据，首先创建段名
  keys_ = [[NSArray alloc] initWithObjects:@"英超", @"西甲", @"意甲", @"德甲", nil];
  //创建各段数据
  NSArray* object1 = [NSArray arrayWithObjects:@"AAA", @"BBB", @"CCC", @"DDD", nil];
  NSArray* object2 = [NSArray arrayWithObjects:@"EEE", @"FFF", nil];
  NSArray* object3 = [NSArray arrayWithObjects:@"GGG", @"HHH", nil];
  NSArray* object4 = [NSArray arrayWithObjects:@"III", @"JJJ", nil];
  NSArray* objects = [NSArray arrayWithObjects:object1, object2, object3, object4, nil];
  //以段名数组、段数据为参数创建数据资源用的字典实例
  dataSource_ = [[NSDictionary alloc] initWithObjects:objects forKeys:keys_];
}
//返回各段的项目数
- (NSInteger)tableView:(UITableView*)tableView numberOfRowsInSection:(NSInteger)section {
  id key = [keys_ objectAtIndex:section];
  return [[dataSource_ objectForKey:key] count];
}
//创建indexPath中指定单元实例
- (UITableViewCell*)tableView:(UITableView*)tableView
  cellForRowAtIndexPath:(NSIndexPath*)indexPath
{
  static NSString* identifier = @"basis-cell";
  UITableViewCell* cell = [tableView dequeueReusableCellWithIdentifier:identifier];
  if ( nil == cell ) {
    cell = [[UITableViewCell alloc] initWithStyle:UITableViewCellStyleDefault
                           reuseIdentifier:identifier];
    [cell autorelease];
  }
  //首先取得单元格的段名
  id key = [keys_ objectAtIndex:indexPath.section];
  //返回对应段及对应位置的数据，并设置到单元中
  NSString* text = [[dataSource_ objectForKey:key] objectAtIndex:indexPath.row];
  cell.textLabel.text = text;
  return cell;
}
//返回段的数目
- (NSInteger)numberOfSectionsInTableView:(UITableView*)tableView {
  return [keys_ count];
}
//返回对应段的段名
- (NSString*)tableView:(UITableView*)tableView titleForHeaderInSection:(NSInteger)section {
  return [keys_ objectAtIndex:section];
}
- (NSArray*)sectionIndexTitlesForTableView:(UITableView*)tableView {
  return keys_;
}
@end
```

执行后的效果如图 4-15 所示。

在上述代码中，如果将 UITableView 的 Style 属性设置为 UITableViewStyleGrouped，则将以分组的样式演示列表中的数据，如图 4-16 所示。

图 4-15　执行效果

图 4-16　分组样式显示

实例 118　删除单元格

实例 118	删除单元格
源码路径	\daima\116\

实例说明

在 iOS 项目中，当使用 UITableView 控件创建一个表格视图后，可以删除表格中的某一个单元格。在具体实现时，只需要调用 UITableView 的 setEditing:animated 方法，将此方法设置为 YES 后，就会在单元格的左侧显示删除按钮 "-"，触摸此删除按钮即可删除这个单元格。另外，当在 iOS 项目中添加或删除单元格时，可以使用属性 UITableViewRowAnimation 设置删除单元格时呈现的动画效果。表 4-1 中列出了 UITableViewRowAnimation 可以支持的动画效果。

表 4-1　　　　　　　　UITableViewRowAnimation 支持的动画效果

动画效果	描述
UITableViewRowAnimationFade	单元格淡出
UITableViewRowAnimationRight	单元格从右侧滑出
UITableViewRowAnimationLeft	单元格从左侧滑出
UITableViewRowAnimationTop	单元格滑动到相邻单元格之上
UITableViewRowAnimationBottom	单元格滑动到相邻单元格之下

在本实例中，演示了使用 setEditing:animated 方法删除单元的过程，并在删除过程中使用 UITableViewRowAnimation 属性设置了删除时呈现的动画效果。

具体实现

实例文件 UIKitPrjDeleteableRow.m 的具体实现代码如下所示。

```
#import "UIKitPrjDeleteableRow.h"
@implementation UIKitPrjDeleteableRow
- (void)dealloc {
  [dataSource_ release];
  [super dealloc];
}
- (void)viewDidLoad {
  [super viewDidLoad];
  dataSource_ = [[NSMutableArray alloc] initWithObjects:
                  @"AAA1", @"AAA2", @"AAA3",
                  @"AAA",  @"AAA5", @"AAA6",
```

```
                          @"AAA7",  @"AAA8",  @"AAA9",
                          @"AAA10", @"AAA11", @"AAA12",
                          @"AAA13", @"AAA14", @"AAA15",
                          @"AAA16", @"AAA17", @"AAA18",
                          nil ];
}
- (void)viewDidAppear:(BOOL)animated {
  [super viewDidAppear:animated];
  [self.tableView setEditing:YES animated:YES];
}
- (NSInteger)tableView:(UITableView*)tableView numberOfRowsInSection:(NSInteger)section {
  return [dataSource_ count];
}
- (UITableViewCell*)tableView:(UITableView*)tableView
  cellForRowAtIndexPath:(NSIndexPath*)indexPath
{
  static NSString* identifier = @"basis-cell";
  UITableViewCell* cell = [tableView dequeueReusableCellWithIdentifier:identifier];
  if ( nil == cell ) {
    cell = [[UITableViewCell alloc] initWithStyle:UITableViewCellStyleDefault
                            reuseIdentifier:identifier];
    [cell autorelease];
  }
  cell.textLabel.text = [dataSource_ objectAtIndex:indexPath.row];
  return cell;
}
// 单元的追加/删除
- (void)tableView:(UITableView*)tableView
  commitEditingStyle:(UITableViewCellEditingStyle)editingStyle
  forRowAtIndexPath:(NSIndexPath*)indexPath
{
  if ( UITableViewCellEditingStyleDelete == editingStyle ) {
    // 从 datasource 删除实际数据
    [dataSource_ removeObjectAtIndex:indexPath.row];
    // 删除表格中的单元
    [tableView deleteRowsAtIndexPaths:[NSArray arrayWithObject:indexPath]
              withRowAnimation:UITableViewRowAnimationLeft];
  }
}
@end
```

执行后的效果如图 4-17 所示。

图 4-17 执行效果

实例 119　添加新的单元格

实例 119	添加新的单元格
源码路径	\daima\116\

实例 119 添加新的单元格

实例说明

在 iOS 项目中，当使用 UITableView 控件创建一个表格视图后，可以继续添加新的单元格表格。在具体实现时，只需要调用 editingStyleforRowAtIndexPath:indexPath 方法即可。当针对最后一个单元格时，此方法的返回值是 UITableViewCellEditingStyleInsert，将其他的单元格情况返回 UITableViewCellEditingStyleDelete。

在本实例中，演示了使用 setEditing:animated 方法添加单元格的过程，设置了操作处理的单元格数据 dataSource，在编辑模式的情况下将最后的 Row 变成添加模式。

具体实现

实例文件 UIKitPrjInsertableRow.m 的具体实现代码如下所示。

```objc
#import "UIKitPrjInsertableRow.h"
@implementation UIKitPrjInsertableRow

- (void)dealloc {
  [dataSource_ release];
  [super dealloc];
}
- (void)viewDidLoad {
  [super viewDidLoad];
  dataSource_ = [[NSMutableArray alloc] initWithObjects:
                        @"aaa", @"bbb", @"ccc",
                        @"添加新单元", nil ];
}
- (void)viewDidAppear:(BOOL)animated {
  [super viewDidAppear:animated];
  [self.tableView setEditing:YES animated:YES];
}
- (NSInteger)tableView:(UITableView*)tableView numberOfRowsInSection:(NSInteger)section {
  return dataSource_.count;
}
- (UITableViewCell*)tableView:(UITableView*)tableView
   cellForRowAtIndexPath:(NSIndexPath*)indexPath
{
  static NSString* identifier = @"basis-cell";
  UITableViewCell* cell = [tableView dequeueReusableCellWithIdentifier:identifier];
  if ( nil == cell ) {
    cell = [[UITableViewCell alloc] initWithStyle:UITableViewCellStyleDefault
                      reuseIdentifier:identifier];
    [cell autorelease];
  }
  cell.textLabel.text = [dataSource_ objectAtIndex:indexPath.row];
  return cell;
}
- (UITableViewCellEditingStyle)tableView:(UITableView*)tableView
   editingStyleForRowAtIndexPath:(NSIndexPath*)indexPath
{
  // 编辑模式的情况下，将最后的 Row 变成插入模式
  if ( tableView.editing && dataSource_.count <= indexPath.row + 1 ) {
    return UITableViewCellEditingStyleInsert;
  } else {
    return UITableViewCellEditingStyleDelete;
  }
}
// 单元的追加/删除
- (void)tableView:(UITableView*)tableView
  commitEditingStyle:(UITableViewCellEditingStyle)editingStyle
   forRowAtIndexPath:(NSIndexPath*)indexPath
{
  if ( UITableViewCellEditingStyleDelete == editingStyle ) {
    // 从 datasource 删除实际数据
    [dataSource_ removeObjectAtIndex:indexPath.row];
    // 删除表格中的单元
    [tableView deleteRowsAtIndexPaths:[NSArray arrayWithObject:indexPath]
```

```
                withRowAnimation:UITableViewRowAnimationLeft];
    } else if ( UITableViewCellEditingStyleInsert == editingStyle ) {
        // 在 dataSource 中追加 1 个数据
        [dataSource_ insertObject:@"新的" atIndex:( dataSource_.count - 1 )];
        // 在表格中追加单元
        [tableView insertRowsAtIndexPaths:[NSArray arrayWithObject:indexPath]
                withRowAnimation:UITableViewRowAnimationBottom];
    }
}
@end
```

执行后的效果如图 4-18 所示。

图 4-18 执行效果

实例 120　移动单元格的位置

实例 120	移动单元格的位置
源码路径	\daima\116\

实例说明

在 iOS 项目中，当使用 UITableView 控件创建一个表格视图后，可以移动表格中的单元格。在具体实现时，只需要调用 tableView:moveRowAtIndexPath:toIndexPath:方法即可。

在本实例中，将移动前的单元索引和移动后的单元索引传入到了 tableView:moveRowAtIndexPath:toIndexPath:方法中，并且使用了方法 exchangeObjectAtIndex:withObjectAtIndex 更改了数据资源的顺序。

具体实现

实例文件 UIKitPrjMoveableRow.m 的具体实现代码如下所示。

```
#import "UIKitPrjMoveableRow.h"
@implementation UIKitPrjMoveableRow
//设置限制单元
- (BOOL)tableView:(UITableView*)tableView canMoveRowAtIndexPath:(NSIndexPath*)indexPath {
    // 最后单元之外的情况下 YES
    return ( dataSource_.count > indexPath.row + 1 );
}
//限制单元移动到最后一个单元格的下方
-(NSIndexPath *)tableView:(UITableView *)tableView
      targetIndexPathForMoveFromRowAtIndexPath:(NSIndexPath *)sourceIndexPath
         toProposedIndexPath:(NSIndexPath *)proposedDestinationIndexPath
{
        if (dataSource_.count > proposedDestinationIndexPath.row + 1) {
            return proposedDestinationIndexPath;
        }else {
            return sourceIndexPath;
        }
}
- (void)tableView:(UITableView*)tableView
    moveRowAtIndexPath:(NSIndexPath*)fromIndexPath toIndexPath:(NSIndexPath*)toIndexPath
{
```

```
    NSUInteger fromRow = fromIndexPath.row;
    NSUInteger toRow = toIndexPath.row;
    while ( fromRow < toRow ) {
      [dataSource_ exchangeObjectAtIndex:fromRow withObjectAtIndex:fromRow+1];
      fromRow++;
    }
    while ( fromRow > toRow ) {
      [dataSource_ exchangeObjectAtIndex:fromRow withObjectAtIndex:fromRow-1];
      fromRow--;
    }
}
@end
```

执行后的效果如图 4-19 所示。

图 4-19 执行效果

实例 121 实现单元格的编辑模式和非编辑模式的切换

实例 121	实现单元格的编辑模式和非编辑模式的切换
源码路径	\daima\116\

实例说明

在 iOS 项目中，当使用 UITableView 控件创建一个表格视图后，可以继续实现单元格的编辑模式和非编辑模式的切换。在 UIViewController 中提供了实现编辑模式和非编辑模式按钮的方法，使用 editButtonItem 方法可以获取"编辑/追加"按钮用的 UIBarButtonItem 实例。

在本实例的导航条中添加了"编辑/完成"按钮，触摸编辑按钮后会变成编辑模式，按钮标题也变成了"完成"。

具体实现

实例文件 UIKitPrjEditingButton.m 的具体实现代码如下所示。

```
#import "UIKitPrjEditingButton.h"
@implementation UIKitPrjEditingButton
- (void)viewDidLoad {
  [super viewDidLoad];
  dataSource_ = [[NSMutableArray alloc] initWithObjects:
                        @"aaa1", @"aaa2", @"aaa3",
                        nil ];
  self.navigationItem.rightBarButtonItem = [self editButtonItem];
}
- (void)viewDidAppear:(BOOL)animated {
  [super viewDidAppear:animated];
  self.tableView.editing = NO;
}
- (void)tableView:(UITableView*)tableView willBeginEditingRowAtIndexPath:(NSIndexPath*)indexPath {
}
- (void)tableView:(UITableView*)tableView didEndEditingRowAtIndexPath:(NSIndexPath*)indexPath {
}
- (void)setEditing:(BOOL)editing animated:(BOOL)animated {
  if ( editing ) {
    // 编辑模式时追加「追加新单元」单元
    NSIndexPath* indexPath = [NSIndexPath indexPathForRow:dataSource_.count inSection:0];
    [dataSource_ addObject:@"添加新单元格"];
    [self.tableView insertRowsAtIndexPaths:[NSArray arrayWithObject:indexPath]
                    withRowAnimation:UITableViewRowAnimationTop];
```

```
        } else {
           // 结束编辑模式时，删除「追加新单元」单元
           NSIndexPath* indexPath = [NSIndexPath indexPathForRow:dataSource_.count-1 inSection:0];
           [dataSource_ removeLastObject];
           [self.tableView deleteRowsAtIndexPaths:[NSArray arrayWithObject:indexPath]
                           withRowAnimation:UITableViewRowAnimationTop];
        }
        [super setEditing:editing animated:YES];
}
@end
```

执行后的效果如图 4-20 所示。

图 4-20　执行效果

实例 122　编辑分组单元格（1）

实例 122	编辑分组单元格（1）
源码路径	\daima\116\

实例说明

在 iOS 项目中，当使用 UITableView 控件创建一个表格视图后，除了可以将单元格分组之外，还可以编辑这些分组单元格。在具体实现时，需要实现方法 tableView: canEditRowAtIndexPath。

在本实例中设置了两组数据，一组是"只读的"，另一组是"可编辑的"。使用 tableView:commitEditingStyle forRowAtIndexPath: 设置单元格为可编辑，并通过 tableView: canEditRowAtIndexPath 方法来处理"可编辑的"分组中的单元格。

具体实现

实例文件 UIKitPrjEditableGroupTable.m 的具体实现代码如下所示。

```
#import "UIKitPrjEditableGroupTable.h"
@implementation UIKitPrjEditableGroupTable

- (void)dealloc {
  [keys_ release];
  [dataSource_ release];
  [super dealloc];
}
- (id)init {
  if ( (self = [super initWithStyle:UITableViewStyleGrouped]) ) {
  }
  return self;
}
- (void)viewDidLoad {
  [super viewDidLoad];
  // 创建显示数据
  keys_ = [[NSArray alloc] initWithObjects:@"只读的", @"可编辑的", nil];
  NSMutableArray* object1 =
    [NSMutableArray arrayWithObjects:@"aaa1", @"aaa2", @"aaa3", nil];
  NSMutableArray* object2 =
    [NSMutableArray arrayWithObjects:@"aaa1", @"aaa2", nil];
  dataSource_ = [[NSMutableArray alloc] initWithObjects:object1, object2, nil];
  self.navigationItem.rightBarButtonItem = [self editButtonItem];
}
- (NSInteger)tableView:(UITableView*)tableView numberOfRowsInSection:(NSInteger)section {
  return [[dataSource_ objectAtIndex:section] count];
```

```objc
}
- (UITableViewCell*)tableView:(UITableView*)tableView
  cellForRowAtIndexPath:(NSIndexPath*)indexPath
{
  static NSString* identifier = @"basis-cell";
  UITableViewCell* cell = [tableView dequeueReusableCellWithIdentifier:identifier];
  if ( nil == cell ) {
    cell = [[UITableViewCell alloc] initWithStyle:UITableViewCellStyleDefault
                             reuseIdentifier:identifier];
    [cell autorelease];
  }
  NSString* text = [[dataSource_ objectAtIndex:indexPath.section] objectAtIndex:indexPath.row];
  cell.textLabel.text = text;
  return cell;
}
- (NSInteger)numberOfSectionsInTableView:(UITableView*)tableView {
  return [keys_ count];
}

- (NSString*)tableView:(UITableView*)tableView titleForHeaderInSection:(NSInteger)section {
  return [keys_ objectAtIndex:section];
}
- (void)tableView:(UITableView*)tableView
  commitEditingStyle:(UITableViewCellEditingStyle)editingStyle
  forRowAtIndexPath:(NSIndexPath*)indexPath
{
  if ( UITableViewCellEditingStyleDelete == editingStyle ) {
    // 删除 dataSource 中的数据
    NSMutableArray* datas = [dataSource_ objectAtIndex:indexPath.section];
    [datas removeObjectAtIndex:indexPath.row];
    [dataSource_ replaceObjectAtIndex:indexPath.section withObject:datas];
    // 删除表格中的单元
    [tableView deleteRowsAtIndexPaths:[NSArray arrayWithObject:indexPath]
              withRowAnimation:UITableViewRowAnimationLeft];
  }
}
- (BOOL)tableView:(UITableView*)tableView canEditRowAtIndexPath:(NSIndexPath*)indexPath {
  return ( 1 == indexPath.section );
}
- (void)tableView:(UITableView*)tableView
  moveRowAtIndexPath:(NSIndexPath*)fromIndexPath toIndexPath:(NSIndexPath*)toIndexPath
{
  // 同一 Section 区域中的处理
  NSMutableArray* datas = [dataSource_ objectAtIndex:fromIndexPath.section];
  NSUInteger fromRow = fromIndexPath.row;
  NSUInteger toRow = toIndexPath.row;
  while ( fromRow < toRow ) {
    [datas exchangeObjectAtIndex:fromRow withObjectAtIndex:fromRow+1];
    fromRow++;
  }
  while ( fromRow > toRow ) {
    [datas exchangeObjectAtIndex:fromRow withObjectAtIndex:fromRow-1];
    fromRow--;
  }
  [dataSource_ replaceObjectAtIndex:fromIndexPath.section with Object:datas];
}
- (void)tableView:(UITableView*)tableView willBeginEditing
RowAtIndexPath:(NSIndexPath*)indexPath {
}

- (void)tableView:(UITableView*)tableView didEndEditing
RowAtIndexPath:(NSIndexPath*)indexPath {
}
@end
```

执行后的效果如图 4-21 所示。

第4章 文本和表格处理实战

图 4-21 执行效果

实例 123 编辑分组单元格（2）

实例 123	编辑分组单元格（2）
源码路径	\daima\116\

实例说明

在 iOS 项目中，当使用 UITableView 控件创建一个表格视图后，除了可以编辑或删除单个单元格外，还可以同时编辑或删除多个单元格。在本实例中，预先设置了 aaaX1、aaa1、aaaX2 和 aaa3 共 4 个单元格，当触摸屏幕上方的"批处理开始"按钮后，会同时删除 aaaX1 和 aaaX2 两个单元格，并增加一个单元格 aaa2。

具体实现

实例文件 UIKitPrjBeginUpdates.m 的具体实现代码如下所示。

```
#import "UIKitPrjBeginUpdates.h"
@implementation UIKitPrjBeginUpdates
- (void)dealloc {
  [dataSource_ release];
  [super dealloc];
}
- (void)viewDidLoad {
  [super viewDidLoad];
  dataSource_ =
    [[NSMutableArray alloc] initWithObjects:@"aaaX1", @"aaa1", @"aaaX2", @"aaa3", nil];
  UIBarButtonItem* barButton =
    [[[UIBarButtonItem alloc] initWithTitle:@"批处理开始"
                                style:UIBarButtonItemStyleBordered
                                target:self
                                action:@selector(batchButtonDidPush:)] autorelease];
  self.navigationItem.rightBarButtonItem = barButton;
}
- (void)batchButtonDidPush:(id)sender {
  // 整理数据
  [dataSource_ setArray:[NSArray arrayWithObjects:@"aaa1", @"aaa2", @"aaa3", nil]];
  //确定删除单元
  NSArray* forDeleting = [NSArray arrayWithObjects:[NSIndexPath indexPathForRow:0 inSection:0],
                                                    [NSIndexPath indexPathForRow:2 inSection:0],
                                                    nil];
  //确定追加位置
  NSArray* forInserting = [NSArray arrayWithObject:[NSIndexPath indexPathForRow:1 inSection:0]];
  [self.tableView beginUpdates];
  // 删除 ITEM X 的内容
  [self.tableView deleteRowsAtIndexPaths:forDeleting withRowAnimation:UITableViewRowAnimationFade];
  // 插入 ITEM 2
  [self.tableView insertRowsAtIndexPaths:forInserting withRowAnimation:UITableView
```

```
RowAnimationFade];
  [self.tableView endUpdates];

  if ( [sender isKindOfClass:[UIBarButtonItem class]] ) {
    UIBarButtonItem* button = sender;
    button.enabled = NO;
  }
}
- (NSInteger)tableView:(UITableView*)tableView numberOfRowsInSection:(NSInteger)section {
  return [dataSource_ count];
}
- (UITableViewCell*)tableView:(UITableView*)tableView
  cellForRowAtIndexPath:(NSIndexPath*)indexPath
{
  static NSString* identifier = @"basis-cell";
  UITableViewCell* cell = [tableView dequeueReusableCellWithIdentifier:identifier];
  if ( nil == cell ) {
    cell = [[UITableViewCell alloc] initWithStyle:UITableViewCellStyleDefault
                          reuseIdentifier:identifier];
    [cell autorelease];
  }
  cell.textLabel.text = [dataSource_ objectAtIndex:indexPath.row];
  return cell;
}
@end
```

执行后的效果如图 4-22 所示，触摸"批处理开始"按钮后的效果如图 4-23 所示。

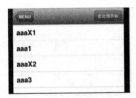

图 4-22 执行效果

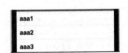

图 4-23 批处理后的效果

实例 124　设置单元格的尺寸和颜色

实例 124	设置单元格的尺寸和颜色
源码路径	\daima\116\

实例说明

在 iOS 项目中，当使用 UITableView 控件创建一个表格视图后，可以自行设置单元格的背景颜色和尺寸大小。在本实例中，通过单元格的 backgroundColor 属性设置了背景颜色，通过其 rowHeight 属性设置单元格的高度尺寸，并且使用 separatorColor 属性设置了单元格中分割线的颜色。

具体实现

实例文件 UIKitPrjBlackCellTable.m 的具体实现代码如下所示。

```
#import "UIKitPrjBlackCellTable.h"
@implementation UIKitPrjBlackCellTable
- (void)viewDidLoad {
  [super viewDidLoad];
  self.tableView.backgroundColor = [UIColor blackColor]; //背景颜色
  self.tableView.rowHeight = 128.0; //单元的尺寸
  self.tableView.separatorColor = [UIColor redColor]; //红色分隔线
  //self.tableView.separatorStyle = UITableViewCellSeparatorStyleNone;
}
- (UITableViewCell*)tableView:(UITableView*)tableView
  cellForRowAtIndexPath:(NSIndexPath*)indexPath
{
  static NSString* identifier = @"basis-cell";
```

```
    UITableViewCell* cell = [tableView dequeueReusableCellWithIdentifier:identifier];
    if ( nil == cell ) {
      cell = [[UITableViewCell alloc] initWithStyle:UITableViewCellStyleDefault
                         reuseIdentifier:identifier];
      [cell autorelease];
      cell.textLabel.textColor = [UIColor whiteColor]; //文本颜色变成白色
      cell.textLabel.textAlignment = UITextAlignmentCenter; //中间对齐
      cell.textLabel.font = [UIFont systemFontOfSize:64]; //字体为 64
    }
    NSString* text = [dataSource_ objectAtIndex:indexPath.row];
    cell.textLabel.text = text;
    return cell;
}
@end
```

执行后的效果如图 4-24 所示。

图 4-24 执行效果

实例 125 在单元格中添加图片

实例 125	在单元格中添加图片
源码路径	\daima\116\

实例说明

在 iOS 项目中,当使用 UITableView 控件创建一个表格视图后,可以在单元格中添加指定的图片。在本实例中,使用 UITableViewCell 的属性 imageView 在单元格中添加了指定的图片。在 iOS 中,属性 imageView 本身是只读的,是通过 imageView 的 image 属性来设置任意的 UIImage 实现添加图片功能的。

具体实现

实例文件 UIKitPrjCellWithImage.m 的具体实现代码如下所示。

```
#import "UIKitPrjCellWithImage.h"
@implementation UIKitPrjCellWithImage
- (void)dealloc {
  [images_ release];
  [dataSource_ release];
  [super dealloc];
}
- (void)viewDidLoad {
  [super viewDidLoad];
  dataSource_ = [[NSArray alloc] initWithObjects:@"aaa", @"bbb", @"ccc", @"ddd", nil];
  images_ = [[NSMutableArray alloc] initWithCapacity:8];
  //将与数据项同名的 png 图片导入并保存在数组中
  for ( NSString* name in dataSource_ ) {
    NSString* imageName = [NSString stringWithFormat:@"%@.png", name];
    UIImage* image = [UIImage imageNamed:imageName];
    [images_ addObject:image];
  }
}
- (NSInteger)tableView:(UITableView*)tableView numberOfRowsInSection:(NSInteger)section {
  return [dataSource_ count];
}
```

```
- (UITableViewCell*)tableView:(UITableView*)tableView
  cellForRowAtIndexPath:(NSIndexPath*)indexPath
{
  static NSString* identifier = @"basis-cell";
  UITableViewCell* cell = [tableView dequeueReusableCellWithIdentifier:identifier];
  if ( nil == cell ) {
    cell = [[UITableViewCell alloc] initWithStyle:UITableViewCellStyleDefault
                            reuseIdentifier:identifier];
    [cell autorelease];
  }
  cell.textLabel.text = [dataSource_ objectAtIndex:indexPath.row];
  //在单元的 imageView 属性中设置图片
  cell.imageView.image = [images_ objectAtIndex:indexPath.row];
  return cell;
}
@end
```

执行后的效果如图 4-25 所示。

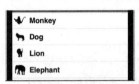

图 4-25　执行效果

实例 126　为单元格中的图片添加注释

实例 126	为单元格中的图片添加注释
源码路径	\daima\116\

实例说明

在 iOS 项目中，当使用 UITableView 控件创建一个表格视图后，除了可以在单元格中添加指定的图片外，还可以添加一些注释来说明这些图片。在本实例以上一个实例为基础，使用 UITableViewCell 的属性 detailTextLabel 设置了注释用的字符串。然后使用 initWithStyle:reuseIdentifier:方法指定了一个设置样式的常量。

具体实现

实例文件 UIKitPrjCellWithDetail.m 的具体实现代码如下所示。

```
#import "UIKitPrjCellWithDetail.h"
@implementation UIKitPrjCellWithDetail
static NSString* kCellStyleDefault = @"style-default";
static NSString* kCellStyleValue1 = @"style-value1";
static NSString* kCellStyleValue2 = @"style-value2";
static NSString* kCellStyleSubtitle = @"style-subtitle";
- (void)dealloc {
  [details_ release];
  [super dealloc];
}
- (void)viewDidLoad {
  [super viewDidLoad];
  // 详细信息文本
  details_ = [[NSArray alloc] initWithObjects:@"猴哥",
                                              @"狗狗",
                                              @"狮王",
                                              @"大鼻子",
                                              nil];

  // 单元样式的默认设置
  cellStyle_ = UITableViewCellStyleSubtitle;
  // 导航条中追加按钮
  UIBarButtonItem* button =
```

```
                [[[UIBarButtonItem alloc] initWithTitle:@"CellStyle"
                                       style:UIBarButtonItemStyleBordered
                                       target:self
                                       action:@selector(buttonDidPush)] autorelease];
    self.navigationItem.rightBarButtonItem = button;
}
- (UITableViewCell*)tableView:(UITableView*)tableView
    cellForRowAtIndexPath:(NSIndexPath*)indexPath
{
    NSString* identifier;
    switch ( cellStyle_ ) {
      case UITableViewCellStyleValue1: identifier = kCellStyleValue1; break;
      case UITableViewCellStyleValue2: identifier = kCellStyleValue2; break;
      case UITableViewCellStyleSubtitle: identifier = kCellStyleSubtitle; break;
      default: identifier = kCellStyleDefault; break;
    }
    UITableViewCell* cell = [tableView dequeueReusableCellWithIdentifier:identifier];
    if ( nil == cell ) {
        cell = [[UITableViewCell alloc] initWithStyle:cellStyle_
                                      reuseIdentifier:identifier];
        [cell autorelease];
    }
    //dataSource_中保存有各单元显示用字符串
    cell.textLabel.text = [dataSource_ objectAtIndex:indexPath.row];
    //images_中保存了各单元中显示用图片
    cell.imageView.image = [images_ objectAtIndex:indexPath.row];
    //details_中保存了各单元详细信息用的字符串
    cell.detailTextLabel.text = [details_ objectAtIndex:indexPath.row];
    return cell;
}
- (void)buttonDidPush {
    if ( UITableViewCellStyleSubtitle < ++cellStyle_ ) {
        cellStyle_ = UITableViewCellStyleDefault;
    }
    [self.tableView reloadData];
}
@end
```

执行后的效果如图4-26所示。

图4-26 执行效果

实例127 在单元格中添加附件

实例127	在单元格中添加附件
源码路径	\daima\116\

实例说明

在iOS项目中，当使用UITableView控件创建一个表格视图后，除了可以在单元格中添加指定的图片外，还可以添加一些复选标志或详细信息按钮等附件。本实例以上一个实例为基础，重写了tableView:cellForRowAtIndexPath方法，并通过属性accessoryType设置了追加内容。

具体实现

实例文件UIKitPrjCellWithAccessory.m的具体实现代码如下所示。

```
#import "UIKitPrjCellWithAccessory.h"
@implementation UIKitPrjCellWithAccessory
- (void)viewDidLoad {
    [super viewDidLoad];
```

```objc
    // 向导航条中追加按钮
    UIBarButtonItem* button =
        [[[UIBarButtonItem alloc] initWithTitle:@"AccesoryType"
                                          style:UIBarButtonItemStyleBordered
                                         target:self
                                         action:@selector(buttonDidPush)] autorelease];
    self.navigationItem.rightBarButtonItem = button;
}
- (UITableViewCell*)tableView:(UITableView*)tableView
    cellForRowAtIndexPath:(NSIndexPath*)indexPath
{
    //如果父类中没有追加创建 UITableViewCell 实例代码时，需要在此追加相关代码
    //父类中已经追加了相关代码，因此此处将省略
    UITableViewCell* cell = [super tableView:tableView cellForRowAtIndexPath:indexPath];
    cell.accessoryType = accessoryType_;
    return cell;
}
- (void)buttonDidPush {
    if ( UITableViewCellAccessoryCheckmark < ++accessoryType_ ) {
        accessoryType_ = UITableViewCellAccessoryNone;
    }
    [self.tableView reloadData];
}
- (void)tableView:(UITableView*)tableView
    accessoryButtonTappedForRowWithIndexPath:(NSIndexPath*)indexPath
{
    //UIKitPrjCellWithDetail 是另外创建的详细画面
    UIViewController* viewController = [[UIKitPrjCellWithDetail alloc] init];
    [self.navigationController pushViewController:viewController animated:YES];
    [viewController release];
}
@end
```

执行后的效果如图 4-27 所示。

图 4-27　执行效果

实例 128　在单元格中添加自定义附件

实例 128	在单元格中添加自定义附件
源码路径	\daima\116\

实例说明

在 iOS 项目中，当使用 UITableView 控件创建一个表格视图后，除了可以在单元格中添加一些复选标志或详细信息按钮等附件外，还可以添加自定义的附件。此自定义功能是通过属性 accessoryView 实现的，本实例演示了使用此属性添加自定义附件的过程。

具体实现

实例文件 UIKitPrjAccessoryView.m 的具体实现代码如下所示。

```objc
#import "UIKitPrjAccessoryView.h"
@implementation UIKitPrjAccessoryView
- (void)viewDidLoad {
    [super viewDidLoad];
    self.navigationItem.rightBarButtonItem = nil;
}
```

```
- (UITableViewCell*)tableView:(UITableView*)tableView
 cellForRowAtIndexPath:(NSIndexPath*)indexPath
{
 //如果父类中没有追加创建 UITableViewCell 实例代码时,需要在此追加相关代码
 UITableViewCell* cell = [super tableView:tableView cellForRowAtIndexPath:indexPath];
 UIButton* button = [UIButton buttonWithType:UIButtonTypeInfoDark];
 [button addTarget:self
           action:@selector(infoDidPush)
  forControlEvents:UIControlEventTouchUpInside];
 cell.accessoryView = button;
 return cell;
}
- (void)infoDidPush {
 //ModalViewController 为另创建的画面
 //UIViewController* viewController = [[ModalViewController alloc] init];
 //[self presentModalViewController:viewController animated:YES];
 //[viewController release];
}
@end
```

执行后的效果如图 4-28 所示。

图 4-28 执行效果

实例 129 设置只在编辑模式下显示附件

实例 129	设置只在编辑模式下显示附件
源码路径	\daima\116\

实例说明

在 iOS 项目中,当使用 UITableView 控件创建一个表格视图后,我们可以设置只有在编辑模式下才显示附件。在具体实现时,是通过 editingAccessoryType 属性单独设置的。在本实例中,演示了只有在编辑模式下才显示附件的具体实现过程,在本实例可编辑时可以让用户编辑计数数字。

具体实现

实例文件 UIKitPrjEditingAccessoryType.m 的具体实现代码如下所示。

```
#import "UIKitPrjEditingAccessoryType.h"
@implementation UIKitPrjEditingAccessoryType
- (void)viewDidLoad {
 [super viewDidLoad];
 // 设置详细信息的文本
 details_ = [[NSArray alloc] initWithObjects:@"8",
                                              @"18",
                                              @"28",
                                              @"38",
                                              nil];
 // 单元样式的默认设置
 cellStyle_ = UITableViewCellStyleValue1;
 // 在导航条中追加编辑按钮
 self.navigationItem.rightBarButtonItem = [self editButtonItem];

 self.tableView.allowsSelection = NO;
 self.tableView.allowsSelectionDuringEditing = YES;
}
- (UITableViewCell*)tableView:(UITableView*)tableView
 cellForRowAtIndexPath:(NSIndexPath*)indexPath
{
 UITableViewCell* cell = [super tableView:tableView cellForRowAtIndexPath:indexPath];
 cell.editingAccessoryType = UITableViewCellAccessoryDetailDisclosureButton;
 return cell;
```

```
}
- (UITableViewCellEditingStyle)tableView:(UITableView*)tableView
  editingStyleForRowAtIndexPath:(NSIndexPath*)indexPath
{
  return UITableViewCellEditingStyleNone;
}
- (BOOL)tableView:(UITableView*)tableView
  shouldIndentWhileEditingRowAtIndexPath:(NSIndexPath*)indexPath
{
  return NO;
}
- (void)tableView:(UITableView*)tableView
  accessoryButtonTappedForRowWithIndexPath:(NSIndexPath*)indexPath
{
//  UIViewController* viewController = [[DetailViewController alloc] init];
//  [self.navigationController pushViewController:viewController animated:YES];
//  [viewController release];
}
@end
```

执行效果如图 4-29 所示。

图 4-29　执行效果

实例 130　向单元格中添加其他控件

实例 130	向单元格中添加其他控件
源码路径	\daima\116\

实例说明

在 iOS 项目中，当使用 UITableView 控件创建一个表格视图后，可以在单元格中继续添加新的控件。在本实例中，在初始化载入方法 viewDidLoad 中设置了单元格中初始显示的数据，然后使用 contentView 属性设置了单元格的内容，并使用 addSubview 方法向单元格中添加了各种 UIView 子类。

具体实现

实例文件 UIKitPrjCustomizedCell.m 的具体实现代码如下所示。

```
#import "UIKitPrjCustomizedCell.h"
#pragma mark ----- Private Methods Definition -----
//定义私有方法
@interface UIKitPrjCustomizedCell ()
- (UIImageView*)imageViewForCell:(const UIView*)cell withFileName:(NSString*)fileName;
- (UISwitch*)switchForCell:(const UIView*)cell;
- (UISlider*)sliderForCell:(const UIView*)cell;
@end
#pragma mark ----- Start Implementation For Methods -----
@implementation UIKitPrjCustomizedCell

- (void)dealloc {
  [sections_ release];
  [dataSource_ release];
  [super dealloc];
}
- (id)init {
  if ( (self = [super initWithStyle:UITableViewStyleGrouped]) ) {
  }
  return self;
}
- (void)viewDidLoad {
  [super viewDidLoad];
  // 创建显示用数据数组
```

第 4 章 文本和表格处理实战

```objc
    sections_ = [[NSArray alloc] initWithObjects: @"姓名", @"必杀技", @"强弱", nil ];
    NSArray* rows1 = [NSArray arrayWithObjects: @"C罗", nil ];
    NSArray* rows2 = [NSArray arrayWithObjects: @"全能", nil ];
    NSArray* rows3 = [NSArray arrayWithObjects: @"速度", @"计数", nil ];
    dataSource_ = [[NSArray alloc] initWithObjects: rows1, rows2, rows3, nil ];
}
- (NSInteger)numberOfSectionsInTableView:(UITableView *)tableView {
    return [sections_ count];
}
- (NSInteger)tableView:(UITableView*)tableView numberOfRowsInSection:(NSInteger)section {
    return [[dataSource_ objectAtIndex:section] count];
}
- (UITableViewCell*)tableView:(UITableView*)tableView
  cellForRowAtIndexPath:(NSIndexPath*)indexPath
{
    static NSString* identifier = @"basis-cell";
    UITableViewCell* cell = [tableView dequeueReusableCellWithIdentifier:identifier];
    if ( nil == cell ) {
        cell = [[UITableViewCell alloc] initWithStyle:UITableViewCellStyleDefault
                            reuseIdentifier:identifier];
        [cell autorelease];
    }
    cell.textLabel.text =
        [[dataSource_ objectAtIndex:indexPath.section] objectAtIndex:indexPath.row];
    switch ( indexPath.section ) {
    case 0: //向单元中追加 UIImageView
        [cell.contentView addSubview:[self imageViewForCell:cell withFileName:@"brows.png"]];
        break;
    case 1: //向单元中追加 UISwitch
        [cell.contentView addSubview:[self switchForCell:cell]];
        break;
    case 2: //向单元中追加 UISlider
        [cell.contentView addSubview:[self sliderForCell:cell]];
        break;
    default:
        break;
    }
    return cell;
}
// 不同单元设置不同的高度
- (CGFloat)tableView:(UITableView*)tableView   heightForRowAtIndexPath:(NSIndexPath*)indexPath {
    if ( 0 == indexPath.section ) {
        return 100.0;
    } else {
        return 44.0;
    }
}
#pragma mark ----- Private Methods -----
//创建包含 UIImageView 实例的单元
- (UIImageView*)imageViewForCell:(const UITableViewCell*)cell withFileName:(NSString*)fileName {
    UIImage* image = [UIImage imageNamed:fileName];
    UIImageView* theImageView = [[[UIImageView alloc] initWithImage:image] autorelease];
    CGPoint newCenter = cell.contentView.center;
    newCenter.x += 80;
    theImageView.center = newCenter;
    theImageView.autoresizingMask = UIViewAutoresizingFlexibleLeftMargin |
                        UIViewAutoresizingFlexibleRightMargin |
                        UIViewAutoresizingFlexibleTopMargin |
                        UIViewAutoresizingFlexibleBottomMargin;
    return theImageView;
}
//创建包含 UISwitch 实例的单元
- (UISwitch*)switchForCell:(const UITableViewCell*)cell {
    UISwitch* theSwitch = [[[UISwitch alloc] init] autorelease];
    theSwitch.on = YES;
    CGPoint newCenter = cell.contentView.center;
    newCenter.x += 80;
    theSwitch.center = newCenter;
    theSwitch.autoresizingMask = UIViewAutoresizingFlexibleLeftMargin |
                        UIViewAutoresizingFlexibleRightMargin |
                        UIViewAutoresizingFlexibleTopMargin |
                        UIViewAutoresizingFlexibleBottomMargin;
    return theSwitch;
}
```

```
//创建包含 UISlider 实例的单元
- (UISlider*)sliderForCell:(const UITableViewCell*)cell {
  UISlider* theSlider = [[[UISlider alloc] init] autorelease];
  theSlider.value = theSlider.maximumValue / 2;
  theSlider.frame = CGRectMake( 0, 0, cell.bounds.size.width / 2, cell.bounds.size.height );
  CGPoint newCenter = cell.contentView.center;
  newCenter.x += 50;
  theSlider.center = newCenter;
  theSlider.autoresizingMask = UIViewAutoresizingFlexibleLeftMargin |
                               UIViewAutoresizingFlexibleRightMargin |
                               UIViewAutoresizingFlexibleTopMargin |
                               UIViewAutoresizingFlexibleBottomMargin;
  return theSlider;
}
@end
```

执行后的效果如图 4-30 所示。

图 4-30　执行效果

实例 131　自定义单元格的背景

实例 131	自定义单元格的背景
源码路径	\daima\116\

实例说明

在 iOS 项目中，当使用 UITableView 控件创建一个表格视图后，我们可以自定义设置单元格的背景。在具体实现时，是通过 UITableView 的属性 backgroundColor 实现的。在具体实现时，需要先将其背景色设置为透明。在本实例中，使用属性 backgroundColor 设置了单元格的背景。

具体实现

实例文件 UIKitPrjCellWithBackgroundView.m 的具体实现代码如下所示。

```
#import "UIKitPrjCellWithBackgroundView.h"
@implementation UIKitPrjCellWithBackgroundView
- (void)dealloc {
  [dataSource_ release];
  [super dealloc];
}
- (void)viewDidLoad {
  [super viewDidLoad];
  //将背景设置为透明
  self.tableView.backgroundColor = [UIColor clearColor];
  //隐藏单元分隔线
  self.tableView.separatorStyle = UITableViewCellSeparatorStyleNone;
  dataSource_ = [[NSArray alloc] initWithObjects:@"aaa1", @"aaa2", @"aaa3", nil];
}
- (NSInteger)tableView:(UITableView*)tableView numberOfRowsInSection:(NSInteger)section {
  return [dataSource_ count];
}
- (UITableViewCell*)tableView:(UITableView*)tableView
  cellForRowAtIndexPath:(NSIndexPath*)indexPath
{
```

```
  static NSString* identifier = @"basis-cell";
  UITableViewCell* cell = [tableView dequeueReusableCellWithIdentifier:identifier];
  if ( nil == cell ) {
    cell = [[UITableViewCell alloc] initWithStyle:UITableViewCellStyleDefault
                          reuseIdentifier:identifier];
    UIImage* image = [UIImage imageNamed:@"frame.png"];
    UIImage* stretchableImage = [image stretchableImageWithLeftCapWidth:30 topCapHeight:30];
    UIImageView* imageView = [[[UIImageView alloc] initWithImage:stretchableImage]
autorelease];
    cell.backgroundView = imageView;
    [cell autorelease];
  }
  cell.textLabel.text = [dataSource_ objectAtIndex:indexPath.row];
  return cell;
}
@end
```

执行后的效果如图 4-31 所示。

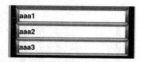

图 4-31 执行效果

实例 132 设置被选中单元格的背景颜色

实例 132	设置被选中单元格的背景颜色
源码路径	\daima\116\

实例说明

在 iOS 项目中，当使用 UITableView 控件创建一个表格视图后，我们可以设置被选中单元格的背景颜色。在具体实现时，是通过 UITableViewCell 的属性 selectionStyle 实现的。在本实例中，首先设置了单元格中显示的数据是 dataSource，然后分别设置了如下 3 个单元格的 3 种被选择时的颜色样式：

- UITableViewCellSelectionStyleBlue；
- UITableViewCellSelectionStyleGray；
- UITableViewCellSelectionStyleNone。

具体实现

实例文件 UIKitPrjSelectionStyle.m 的具体实现代码如下所示。

```
#import "UIKitPrjSelectionStyle.h"
@implementation UIKitPrjSelectionStyle
- (void)dealloc {
  [selectionStyles_ release];
  [dataSource_ release];
  [super dealloc];
}
- (void)viewDidLoad {
  [super viewDidLoad];
  dataSource_ = [[NSArray alloc] initWithObjects:
                        @"Blue",
                        @"Gray",
                        @"None",
                        nil];
  selectionStyles_ = [[NSArray alloc] initWithObjects:
                          [NSNumber
numberWithInteger:UITableViewCellSelectionStyleBlue],
                          [NSNumber
numberWithInteger:UITableViewCellSelectionStyleGray],
                          [NSNumber
numberWithInteger:UITableViewCellSelectionStyleNone],
```

```
                                         nil];
- (NSInteger)tableView:(UITableView*)tableView numberOfRowsInSection:(NSInteger)section {
  return [dataSource_ count];
}

- (UITableViewCell*)tableView:(UITableView*)tableView
  cellForRowAtIndexPath:(NSIndexPath*)indexPath
{
  static NSString* identifier = @"basis-cell";
  UITableViewCell* cell = [tableView dequeueReusableCellWithIdentifier:identifier];
  if ( nil == cell ) {
    cell = [[UITableViewCell alloc] initWithStyle:UITableViewCellStyleDefault
                            reuseIdentifier:identifier];
    [cell autorelease];
  }
  cell.textLabel.text = [dataSource_ objectAtIndex:indexPath.row];
  cell.selectionStyle = [[selectionStyles_ objectAtIndex:indexPath.row] integerValue];
  //cell.textLabel.highlightedTextColor = [UIColor blackColor];
  return cell;
}

@end
```

执行后的效果如图 4-32 所示。

图 4-32 执行效果

实例 133 自动滚动到被选中单元格

实例 133	自动滚动到被选中单元格
源码路径	\daima\116\

实例说明

在 iOS 项目中，当使用 UITableView 控件创建一个表格视图后，我们可以自动滚动到被选中的单元格。在具体实现时，是通过 scrollToNearestSelectedRowAtScrollPosition: animated 方法实现的。在本实例中，将 animated 参数设置为 YES，这样便以动画的方式滚动显示单元格中的内容，并且分别设置了"来到顶部"、"来到当前选择"和"来到底部"3 个按钮的处理事件程序。

具体实现

实例文件 UIKitPrjScroll.m 的具体实现代码如下所示。

```
#import "UIKitPrjScroll.h"
@implementation UIKitPrjScroll
- (void)viewDidLoad {
  [super viewDidLoad];
  UIBarButtonItem* topButton =
    [[[UIBarButtonItem alloc] initWithTitle:@"来到顶部"
                        style:UIBarButtonItemStyleBordered
                        target:self
                        action:@selector(topDidPush)] autorelease];
  UIBarButtonItem* currentButton =
    [[[UIBarButtonItem alloc] initWithTitle:@"来到当前选择"
                        style:UIBarButtonItemStyleBordered
                        target:self
                        action:@selector(currentDidPush)] autorelease];
  UIBarButtonItem* bottomButton =
    [[[UIBarButtonItem alloc] initWithTitle:@"来到底部"
                        style:UIBarButtonItemStyleBordered
                        target:self
```

第4章 文本和表格处理实战

```
                          action:@selector(bottomDidPush)] autorelease];
  NSArray* buttons = [NSArray arrayWithObjects:topButton, currentButton, bottomButton,
nil];
  [self setToolbarItems:buttons animated:YES];
}
- (void)topDidPush {
  //滚动到第1个单元格
  NSIndexPath* indexPath = [NSIndexPath indexPathForRow:0 inSection:0];
  [self.tableView scrollToRowAtIndexPath:indexPath
                  atScrollPosition:UITableViewScrollPositionNone
                     animated:YES];
  //[self.tableView
scrollToNearestSelectedRowAtScrollPosition:UITableViewScrollPositionTop
                                                   animated:YES];
}
- (void)currentDidPush {
  [self.tableView
scrollToNearestSelectedRowAtScrollPosition:UITableViewScrollPositionNone
                                       animated:YES];
}
- (void)bottomDidPush {
  //滚动到最后一个单元格
  NSIndexPath* indexPath = [NSIndexPath indexPathForRow:dataSource_.count-1 inSection:0];
  [self.tableView scrollToRowAtIndexPath:indexPath
                  atScrollPosition:UITableViewScrollPositionNone
                     animated:YES];
  //[self.tableView
scrollToNearestSelectedRowAtScrollPosition:UITableViewScrollPositionBottom
  //                                                 animated:YES];
}
-                              (void)tableView:(UITableView*)tableView
didSelectRowAtIndexPath:(NSIndexPath*)indexPath {
}
@end
```

执行后的效果如图4-33所示。

图4-33 执行效果

实例 134 在单元格中自动排列指定的数据

实例 134	在单元格中自动排列指定的数据
源码路径	\daima\116\

实例说明

在iOS项目中,我们可以在单元格中自动排列任意数据,这一功能是通过UILocalizedIndexedCollation类实现的。通过UILocalizedIndexedCollation类可以方便地为部分指数图表进行组织、整理以及数据本地化处理。通过表视图的数据源,使用排序对象提供的输入节的标题和节索引标题的表视图。UILocalizedIndexedCollation是一个辅助类,是iPhone OS 3.0新引进的类,用来帮助data source组织数据,以便于在有序列表(indexed lists)中显示,当用户点击某一个项目的时候,可以正确显示对应的section。UILocalizedIndexedCollation还可以本地化section的标题。

具体实现

实例文件 UIKitPrjLocalizedIndexedCollation.m 的具体实现代码如下所示。

```objc
#import "UIKitPrjLocalizedIndexedCollation.h"
#pragma mark ----- Character -----
@implementation Character
@synthesize name = name_;
@synthesize job = job_;
@end
@implementation UIKitPrjLocalizedIndexedCollation
@synthesize dataSource = dataSource_;
- (void)viewDidLoad {
  [super viewDidLoad];
  NSMutableArray* dataSourceTemp = [NSMutableArray arrayWithCapacity:22];
  for ( int i = 0; i < 20; ++i ){
    Character* dummy = [[Character alloc] init];
    dummy.name = ( i % 2 ) ? @"例子" : @"演示";
    dummy.job = @"写书的";
    [dataSourceTemp addObject:dummy];
    [dummy release];
  }
  Character* character1 = [[Character alloc] init];
  character1.name = @"王五";
  character1.job = @"公务员";
  [dataSourceTemp addObject:character1];
  [character1 release];
  Character* character2 = [[Character alloc] init];
  character2.name = @"李六";
  character2.job = @"服务业";
  [dataSourceTemp addObject:character2];
  [character2 release];
  UILocalizedIndexedCollation* theCollation = [UILocalizedIndexedCollation currentCollation];
  // 创建与 Section 数目相同的空数组，类型为 NSMutableArray
  NSInteger sectionCount = [[theCollation sectionTitles] count];
  NSMutableArray* sectionArrays = [NSMutableArray arrayWithCapacity:sectionCount];
  for ( int i = 0; i <= sectionCount; ++i ) {
    [sectionArrays addObject:[NSMutableArray arrayWithCapacity:1]];
  }
  // 将数据放入上面创建的数组中
  for ( Character* character in dataSourceTemp ) {
    NSInteger sect = [theCollation sectionForObject:character
                     collationStringSelector:@selector(name)];
    [[sectionArrays objectAtIndex:sect] addObject:character];
  }
  // 将 Section 中数据追加到表格用数组中
  self.dataSource = [NSMutableArray arrayWithCapacity:sectionCount];
  for ( NSMutableArray* sectionArray in sectionArrays ) {
    NSArray* sortedSection = [theCollation sortedArrayFromArray:sectionArray
                              collationStringSelector:@selector(name)];
    [self.dataSource addObject:sortedSection];
  }
}
- (NSArray*)sectionIndexTitlesForTableView:(UITableView*)tableView {
  return [[UILocalizedIndexedCollation currentCollation] sectionIndexTitles];
}
- (NSString*)tableView:(UITableView*)tableView titleForHeaderInSection:(NSInteger)section {
  if ( [[self.dataSource objectAtIndex:section] count] < 1 ) {
    return nil;
  }
  return [[[UILocalizedIndexedCollation currentCollation] sectionTitles] objectAtIndex:section];
}
- (NSInteger)tableView:(UITableView*)tableView
  sectionForSectionIndexTitle:(NSString*)title atIndex:(NSInteger)index
{
  return [[UILocalizedIndexedCollation currentCollation] sectionForSectionIndexTitleAtIndex:index];
}
- (NSInteger)numberOfSectionsInTableView:(UITableView*)tableView {
  return [self.dataSource count];
}
- (NSInteger)tableView:(UITableView*)tableView numberOfRowsInSection:(NSInteger)section {
  return [[self.dataSource objectAtIndex:section] count];
}
```

```
- (UITableViewCell*)tableView:(UITableView*)tableView
  cellForRowAtIndexPath:(NSIndexPath*)indexPath
{
    static NSString* identifier = @"basis-cell";
    UITableViewCell* cell = [tableView dequeueReusableCellWithIdentifier:identifier];
    if ( nil == cell ) {
        cell = [[UITableViewCell alloc] initWithStyle:UITableViewCellStyleDefault
                                      reuseIdentifier:identifier];
        [cell autorelease];
    }
    Character* character =
        [[self.dataSource objectAtIndex:indexPath.section] objectAtIndex:indexPath.row];
    cell.textLabel.text = character.name;
    return cell;
}
@end
```

执行后的效果如图 4-34 所示。

图 4-34 执行效果

实例 135 为每行单元格设置展开子项

实例 135	为每行单元格设置展开子项
源码路径	\daima\135\

实例说明

在 iOS 项目中，当使用 UITableViewCell 实现单元格功能后，我们可以实现可展开的 UITableViewCell 的效果。在本实例中，首先定义了 6 行 tableView 单元格，当单击每行单元格时，会以展开的样式显示此单元格下面的子项。

具体实现

实例文件 FirstViewController.m 的主要实现代码如下所示。

```
- (void)viewDidLoad
{
    [super viewDidLoad];
    self.view.backgroundColor = [UIColor whiteColor];
    [self loadModel];
    _tableView = [[UITableView alloc]initWithFrame:CGRectMake(-10, 0,340,460) style:UITable
ViewStyleGrouped];
    _tableView.delegate = self;
    _tableView.dataSource = self;
    _tableView.separatorColor = [UIColor clearColor];
    _tableView.backgroundColor = [UIColor clearColor];
    _tableView.separatorStyle = UITableViewCellSeparatorStyleNone;
    [self.view addSubview:_tableView];
}
- (void)loadModel{
    _currentRow = -1;
    headViewArray = [[NSMutableArray alloc]init ];
    for(int i = 0;i< 5 ;i++)
      {
          HeadView* headview = [[HeadView alloc] init];
          headview.delegate = self;
          headview.section = i;
```

```objc
            [headview.backBtn setTitle:[NSString stringWithFormat:@"第%d组",i] forState:
UIControlStateNormal];
            [self.headViewArray addObject:headview];
            [headview release];
        }
}
- (void)viewDidUnload
{
    [super viewDidUnload];
    _tableView= nil;
}
#pragma mark - TableViewdelegate&&TableViewdataSource
- (UIView *)tableView:(UITableView *)tableView viewForFooterInSection:(NSInteger)section{
    return nil;
}
- (CGFloat)tableView:(UITableView *)tableView heightForRowAtIndexPath:(NSIndexPath
*)indexPath{
    HeadView* headView = [self.headViewArray objectAtIndex:indexPath.section];

    return headView.open?45:0;
}
- (CGFloat)tableView:(UITableView *)tableView heightForHeaderInSection:(NSInteger)section{
    return 45;
}
- (CGFloat)tableView:(UITableView *)tableView heightForFooterInSection:(NSInteger)section{
    return 0.1;
}
- (UIView*)tableView:(UITableView *)tableView viewForHeaderInSection:(NSInteger)section{
    return [self.headViewArray objectAtIndex:section];
}
- (NSInteger)tableView:(UITableView *)tableView numberOfRowsInSection:(NSInteger)section{
    HeadView* headView = [self.headViewArray objectAtIndex:section];
    return headView.open?5:0;
}

- (NSInteger)numberOfSectionsInTableView:(UITableView *)tableView{
    return [self.headViewArray count];
}
- (UITableViewCell*)tableView:(UITableView *)tableView cellForRowAtIndexPath:(NSIndexPath
*)indexPath{
    static NSString *indentifier = @"cell";
    UITableViewCell *cell = [tableView dequeueReusableCellWithIdentifier:indentifier];
    if (!cell) {
        cell = [[[UITableViewCell alloc] initWithStyle:UITableViewCellStyleDefault
reuseIdentifier:indentifier] autorelease];
        UIButton* backBtn= [[UIButton alloc]initWithFrame:CGRectMake(0, 0, 340, 45)];
        backBtn.tag = 20000;
        [backBtn setBackgroundImage:[UIImage imageNamed:@"btn_on"] forState:UIControl
StateHighlighted];
        backBtn.userInteractionEnabled = NO;
        [cell.contentView addSubview:backBtn];
        [backBtn release];
        UIImageView* line = [[UIImageView alloc]initWithFrame:CGRectMake(0, 44, 340, 1)];
        line.backgroundColor = [UIColor grayColor];
        [cell.contentView addSubview:line];
        [line release];
    }
    UIButton* backBtn = (UIButton*)[cell.contentView viewWithTag:20000];
    HeadView* view = [self.headViewArray objectAtIndex:indexPath.section];
    [backBtn setBackgroundImage:[UIImage imageNamed:@"btn_2_nomal"] forState:
UIControlStateNormal];
if (view.open) {
        if (indexPath.row == _currentRow) {
            [backBtn setBackgroundImage:[UIImage imageNamed:@"btn_nomal"] forState:
UIControlStateNormal];
        }
}
cell.textLabel.text = [NSString stringWithFormat:@"%d-%d",indexPath.section,indexPath.row];
    cell.textLabel.backgroundColor = [UIColor clearColor];
    cell.textLabel.textColor = [UIColor whiteColor];
return cell;
}
- (void)tableView:(UITableView *)tableView didSelectRowAtIndexPath:(NSIndexPath *)indexPath{
    _currentRow = indexPath.row;
    [_tableView reloadData];
}
```

```
#pragma mark - HeadViewdelegate
-(void)selectedWith:(HeadView *)view{
    _currentRow = -1;
    if (view.open) {
        for(int i = 0;i<[headViewArray count];i++)
        {
            HeadView *head = [headViewArray objectAtIndex:i];
            head.open = NO;
            [head.backBtn setBackgroundImage:[UIImage imageNamed:@"btn_momal"] forState:UIControlStateNormal];
        }
        [_tableView reloadData];
        return;
    }
    _currentSection = view.section;
    [self reset];
}
//界面重置
- (void)reset
{
    for(int i = 0;i<[headViewArray count];i++)
    {
        HeadView *head = [headViewArray objectAtIndex:i];

        if(head.section == _currentSection)
        {
            head.open = YES;
            [head.backBtn setBackgroundImage:[UIImage imageNamed:@"btn_nomal"] forState:UIControlStateNormal];

        }else {
            [head.backBtn setBackgroundImage:[UIImage imageNamed:@"btn_momal"] forState:UIControlStateNormal];

            head.open = NO;
        }
    }
    [_tableView reloadData];
}
- (void)dealloc{
    [_tableView release];
    [headViewArray release];
    [super dealloc];
}
@end
```

执行后的效果如图4-35所示。

图4-35 执行效果

实例136 实现气泡样式的聊天对话框效果

实例136	实现气泡样式的聊天对话框效果
源码路径	\daima\136\

实例 136 实现气泡样式的聊天对话框效果

实例说明

在本实例的功能是实现气泡样式的聊天对话框效果，气泡大小可以根据文字输入的多少而自动改变高度。本实例借助了第三方开源代码 http://alexbarinov.github.com/UIBubbleTableView/，读者可以登录此网址查看此开源代码的具体实现。

具体实现

实例文件 ViewController.m 的主要实现代码如下所示。

```
#import "ViewController.h"
#import "UIBubbleTableView.h"
#import "UIBubbleTableViewDataSource.h"
#import "NSBubbleData.h"
@implementation ViewController
- (void)viewDidLoad
{
    [super viewDidLoad];

    bubbleTable.bubbleDataSource = self;

    bubbleData = [[NSMutableArray alloc] initWithObjects:
               [NSBubbleData dataWithText:@"你好，出去玩吗？" andDate:[NSDate dateWithTimeIntervalSinceNow:-300] andType:BubbleTypeMine],
               [NSBubbleData dataWithText:@"明天天气好吗？" andDate:[NSDate dateWithTimeIntervalSinceNow:-280] andType:BubbleTypeSomeoneElse],
               [NSBubbleData dataWithText:@"多云，亲" andDate:[NSDate dateWithTimeIntervalSinceNow:0] andType:BubbleTypeMine],
               [NSBubbleData dataWithText:@"温度怎么样啊" andDate:[NSDate dateWithTimeIntervalSinceNow:300] andType:BubbleTypeSomeoneElse],
               [NSBubbleData dataWithText:@"2 到 21 度，温差比较大，容易感冒.." andDate:[NSDate dateWithTimeIntervalSinceNow:395] andType:BubbleTypeMine],
               [NSBubbleData dataWithText:@"奥，那算了，我不去了." andDate:[NSDate dateWithTimeIntervalSinceNow:400] andType:BubbleTypeMine],
               nil];
}
- (void)viewDidUnload
{
    [super viewDidUnload];
}
- (BOOL)shouldAutorotateToInterfaceOrientation:(UIInterfaceOrientation)interfaceOrientation
{
    return (interfaceOrientation != UIInterfaceOrientationPortraitUpsideDown);
}
#pragma mark - UIBubbleTableViewDataSource implementation
- (NSInteger)rowsForBubbleTable:(UIBubbleTableView *)tableView
{
    return [bubbleData count];
}
- (NSBubbleData *)bubbleTableView:(UIBubbleTableView *)tableView dataForRow:(NSInteger)row
{
    return [bubbleData objectAtIndex:row];
}
@end
```

执行效果如图 4-36 所示。

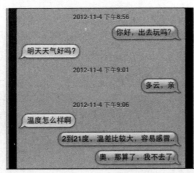

图 4-36 执行效果

实例 137　在搜索框中实现下拉列表效果

实例 137	在搜索框中实现下拉列表效果
源码路径	\daima\137\

实例说明

在本实例中，在搜索框中实现了下拉列表效果。在搜索框上面输入文字时出现一个下拉列表，此功能可以用作搜索自动提示。首先设置在屏幕中显示"Item0"到"Item9"一共 10 行文本，然后通过 searchBarShouldBeginEditing 方法设置搜索框为可编辑状态。输入搜索关键字后，会自动弹出一个提示框，触摸回车键后会在屏幕下方列表显示相关的信息。

具体实现

实例文件 DropDownListViewController.m 的主要实现代码如下所示。

```objc
- (void)viewDidLoad {
    [super viewDidLoad];
    _searchStr = [[NSString alloc] initWithString:@"Item"];
    _ddList = [[DDList alloc] initWithStyle:UITableViewStylePlain];
    _ddList._delegate = self;
    [self.view addSubview:_ddList.view];
    [_ddList.view setFrame:CGRectMake(30, 36, 200, 0)];
}
- (void)didReceiveMemoryWarning {
    // Releases the view if it doesn't have a superview.
    [super didReceiveMemoryWarning];

    // Release any cached data, images, etc that aren't in use.
}
- (void)viewDidUnload {
    // Release any retained subviews of the main view.
    // e.g. self.myOutlet = nil;
}

- (void)dealloc {
    [_ddList release];
    [_searchStr release];
    [super dealloc];
}
- (void)setDDListHidden:(BOOL)hidden {
    NSInteger height = hidden ? 0 : 180;
    [UIView beginAnimations:nil context:nil];
    [UIView setAnimationDuration:.2];
    [_ddList.view setFrame:CGRectMake(30, 36, 200, height)];
    [UIView commitAnimations];
}
#pragma mark -
#pragma mark PassValue protocol
- (void)passValue:(NSString *)value{
    if (value) {
        _searchBar.text = value;
        [self searchBarSearchButtonClicked:_searchBar];
    }
    else {

    }
}
#pragma mark -
#pragma mark SearchBar Delegate Methods
- (void)searchBar:(UISearchBar *)searchBar textDidChange:(NSString *)searchText {
    if ([searchText length] != 0) {
        _ddList._searchText = searchText;
        [_ddList updateData];
        [self setDDListHidden:NO];
    }
    else {
        [self setDDListHidden:YES];
    }
}
```

```
- (BOOL)searchBarShouldBeginEditing:(UISearchBar *)searchBar {
    searchBar.showsCancelButton = YES;
    for(id cc in [searchBar subviews])
    {
      if([cc isKindOfClass:[UIButton class]])
      {
         UIButton *btn = (UIButton *)cc;
         [btn setTitle:@"取消" forState:UIControlStateNormal];
      }
    }
    return YES;
}
- (void)searchBarTextDidBeginEditing:(UISearchBar *)searchBar {
    searchBar.text = @"";
}
- (void)searchBarTextDidEndEditing:(UISearchBar *)searchBar {
    searchBar.showsCancelButton = NO;
    searchBar.text = @"";
}
- (void)searchBarSearchButtonClicked:(UISearchBar *)searchBar {
    [self setDDListHidden:YES];
    self._searchStr = [searchBar text];
    [searchBar resignFirstResponder];
    [_tableView reloadData];
}
- (void)searchBarCancelButtonClicked:(UISearchBar *)searchBar {
    [self setDDListHidden:YES];
    [searchBar resignFirstResponder];
}
```

执行后的效果如图4-37所示。

图4-37 执行效果

实例138 实现一个高度自动适应性的输入框

实例138	实现一个高度自动适应性的输入框
源码路径	\daima\138\

实例说明

本实例的功能是实现输入框随着内容的增多高度可以自动适应的效果。也就是说，TextView支持输入多行文字，并且框的高度随着输入文字而自动变高。本实例的实现过程其实比较简单，借助了开源代码HPGrowingTextView。在iOS应用中，HPGrowingTextView是一个UITextView的扩展，实现了当文本增多和减少时对输入框本身的伸缩和滚动处理。

具体实现

实例文件GrowingTextViewExampleViewController.m的具体实现代码如下所示。

```
#import "GrowingTextViewExampleViewController.h"
@implementation GrowingTextViewExampleViewController
-(id)init
{
    self = [super init];
    if(self){
        [[NSNotificationCenter defaultCenter] addObserver:self
     selector:@selector(keyboardWillShow:)
         name:UIKeyboardWillShowNotification
      object:nil];
```

```objc
        [[NSNotificationCenter defaultCenter] addObserver:self
      selector:@selector(keyboardWillHide:)
         name:UIKeyboardWillHideNotification
       object:nil];
    }
    return self;
}
// 用 loadview 创建一个视图
- (void)loadView {
    self.view = [[[UIView alloc] initWithFrame:[[UIScreen mainScreen] applicationFrame]] autorelease];
    self.view.backgroundColor = [UIColor colorWithRed:219.0f/255.0f green:226.0f/255.0f blue:237.0f/255.0f alpha:1];
    containerView = [[UIView alloc] initWithFrame:CGRectMake(0, self.view.frame.size.height - 40, 320, 40)];
    textView = [[HPGrowingTextView alloc] initWithFrame:CGRectMake(6, 3, 240, 40)];
    textView.contentInset = UIEdgeInsetsMake(0, 5, 0, 5);
    textView.minNumberOfLines = 1;
    textView.maxNumberOfLines = 6;
    textView.returnKeyType = UIReturnKeyGo;
    textView.font = [UIFont systemFontOfSize:15.0f];
    textView.delegate = self;
    textView.internalTextView.scrollIndicatorInsets = UIEdgeInsetsMake(5, 0, 5, 0);
    textView.backgroundColor = [UIColor whiteColor];

    [self.view addSubview:containerView];
    UIImage *rawEntryBackground = [UIImage imageNamed:@"MessageEntryInputField.png"];
    UIImage *entryBackground = [rawEntryBackground stretchableImageWithLeftCapWidth:13 topCapHeight:22];
    UIImageView *entryImageView = [[[UIImageView alloc] initWithImage:entryBackground] autorelease];
    entryImageView.frame = CGRectMake(5, 0, 248, 40);
    entryImageView.autoresizingMask = UIViewAutoresizingFlexibleHeight | UIViewAutoresizingFlexibleWidth;
    UIImage *rawBackground = [UIImage imageNamed:@"MessageEntryBackground.png"];
    UIImage *background = [rawBackground stretchableImageWithLeftCapWidth:13 topCapHeight:22];
    UIImageView *imageView = [[[UIImageView alloc] initWithImage:background] autorelease];
    imageView.frame = CGRectMake(0, 0, containerView.frame.size.width, containerView.frame.size.height);
    imageView.autoresizingMask = UIViewAutoresizingFlexibleHeight | UIViewAutoresizingFlexibleWidth;
    textView.autoresizingMask = UIViewAutoresizingFlexibleWidth;
    // 视图层次
    [containerView addSubview:imageView];
    [containerView addSubview:textView];
    [containerView addSubview:entryImageView];
    UIImage *sendBtnBackground = [[UIImage imageNamed:@"MessageEntrySendButton.png"] stretchableImageWithLeftCapWidth:13 topCapHeight:0];
    UIImage *selectedSendBtnBackground = [[UIImage imageNamed:@"MessageEntrySendButton.png"] stretchableImageWithLeftCapWidth:13 topCapHeight:0];
    UIButton *doneBtn = [UIButton buttonWithType:UIButtonTypeCustom];
    doneBtn.frame = CGRectMake(containerView.frame.size.width - 69, 8, 63, 27);
    doneBtn.autoresizingMask = UIViewAutoresizingFlexibleTopMargin | UIViewAutoresizingFlexibleLeftMargin;
    [doneBtn setTitle:@"Done" forState:UIControlStateNormal];
    [doneBtn setTitleShadowColor:[UIColor colorWithWhite:0 alpha:0.4] forState:UIControlStateNormal];
    doneBtn.titleLabel.shadowOffset = CGSizeMake (0.0, -1.0);
    doneBtn.titleLabel.font = [UIFont boldSystemFontOfSize:18.0f];
    [doneBtn setTitleColor:[UIColor whiteColor] forState:UIControlStateNormal];
    [doneBtn addTarget:self action:@selector(resignTextView) forControlEvents:UIControlEventTouchUpInside];
    [doneBtn setBackgroundImage:sendBtnBackground forState:UIControlStateNormal];
    [doneBtn setBackgroundImage:selectedSendBtnBackground forState:UIControlStateSelected];
    [containerView addSubview:doneBtn];
    containerView.autoresizingMask = UIViewAutoresizingFlexibleWidth | UIViewAutoresizingFlexibleTopMargin;
}
-(void)resignTextView
{
    [textView resignFirstResponder];
}
//下面是代码，来自 http://brettschumann.com/
-(void) keyboardWillShow:(NSNotification *)note{
    // 获得键盘的大小和 loctaion
```

实例 138 实现一个高度自动适应性的输入框

```
    CGRect keyboardBounds;
    [[note.userInfo valueForKey:UIKeyboardFrameEndUserInfoKey] getValue: &keyboardBounds];
    NSNumber *duration = [note.userInfo objectForKey:UIKeyboardAnimationDurationUserInfoKey];
    NSNumber *curve = [note.userInfo objectForKey:UIKeyboardAnimationCurveUserInfoKey];
    // 转换旋转界面.
    keyboardBounds = [self.view convertRect:keyboardBounds toView:nil];
        // 得到一个矩形的 View 框架
        CGRect containerFrame = containerView.frame;
    containerFrame.origin.y = self.view.bounds.size.height - (keyboardBounds.size.height + containerFrame.size.height);
        // 设定动画
        [UIView beginAnimations:nil context:NULL];
        [UIView setAnimationBeginsFromCurrentState:YES];
    [UIView setAnimationDuration:[duration doubleValue]];
    [UIView setAnimationCurve:[curve intValue]];
        // 通过新信息设置视图
        containerView.frame = containerFrame;
        [UIView commitAnimations];
}
-(void) keyboardWillHide:(NSNotification *)note{
    NSNumber *duration = [note.userInfo objectForKey:UIKeyboardAnimationDurationUserInfoKey];
    NSNumber *curve = [note.userInfo objectForKey:UIKeyboardAnimationCurveUserInfoKey];
        // 得到一个矩形的 View 框架
        CGRect containerFrame = containerView.frame;
    containerFrame.origin.y = self.view.bounds.size.height - containerFrame.size.height;
        // 设置动画效果
        [UIView beginAnimations:nil context:NULL];
        [UIView setAnimationBeginsFromCurrentState:YES];
    [UIView setAnimationDuration:[duration doubleValue]];
    [UIView setAnimationCurve:[curve intValue]];
        // 通过新信息设置视图
        containerView.frame = containerFrame;
        [UIView commitAnimations];
}
- (void)growingTextView:(HPGrowingTextView *)growingTextView willChangeHeight:(float)height
{
    float diff = (growingTextView.frame.size.height - height);

    CGRect r = containerView.frame;
    r.size.height -= diff;
    r.origin.y += diff;
    containerView.frame = r;
}
-(BOOL)shouldAutorotateToInterfaceOrientation:(UIInterfaceOrientation)toInterfaceOrientation
{
    return YES;
}
- (void)didReceiveMemoryWarning {
    // 如果没有一个视图则释放
    [super didReceiveMemoryWarning];
}
- (void)viewDidUnload {
}
- (void)dealloc {
    [textView release];
    [containerView release];
    [super dealloc];
}
@end
```

执行后的效果如图 4-38 所示。

图 4-38 执行效果

第 5 章 屏幕显示实战

顾客在购买一款智能设备时,追求的是既界面美观又功能强大的产品。所以对于 iOS 程序员来说,我们的任务是开发既能保证界面美观绚丽、功能又强大的应用程序。在本书第 2 章中已经讲解了界面布局有关的实例,在第 3 章已经讲解了和控件有关的实例,在第 4 章已经讲解了文本和表格处理的基本知识。在本章的内容中,将以前面三章的内容为基础,通过具体范例详细讲解在 iOS 屏幕中显示信息的知识。

实例 139 在屏幕中显示一段文本

实例 139	在屏幕中显示一段文本
源码路径	\daima\139\

实例说明

本实例的功能是,使用本书前面所学的 UILabel 控件在屏幕中显示一段文本。本实例只有一个加载方法 viewDidLoad,设置执行后就在屏幕中显示文本"输出了一段文本"。

具体实现

实例文件 UIKitPrjLabel.m 的具体实现代码如下所示。

```
#import "UIKitPrjLabel.h"
@implementation UIKitPrjLabel
- (void)viewDidLoad {
  [super viewDidLoad];
UILabel* label = [[[UILabel alloc] init] autorelease];
  // 将标签的大小设置成与画面相同的尺寸
  label.frame = self.view.bounds;
  // 设置画面与标签的自动调整属性
  label.autoresizingMask =
    UIViewAutoresizingFlexibleWidth | UIViewAutoresizingFlexibleHeight;
  label.textAlignment = UITextAlignmentCenter;
  label.text = @"输出了一段文本";
  label.backgroundColor = [UIColor whiteColor];
  label.textColor = [UIColor blackColor];
  [self.view addSubview:label];
}
@end
```

执行后的效果如图 5-1 所示。

图 5-1 执行效果

实例 140　绘制字符串

实例 140	绘制字符串
源码路径	\daima\139\

实例说明

本实例的功能是，使用 NSString 的 InstanceMethod 方法绘制指定的字符串"使用 NSString 的 InstanceMethod 绘制字符串，比较不同设置时的显示效果"。在具体实现时，使用了 drawInRect:rect:withFont 方法设置了每行的宽度，这样便实现了换行功能。并且在实例中使用 UILineBreakMode 设置了换行模式，UILineBreakMode 有如下 4 种模式。

- UILineBreakModeWordWrap = 0：以单词为单位换行，以单位为单位截断。
- UILineBreakModeCharacterWrap：以字符为单位换行，以字符为单位截断。
- UILineBreakModeClip：以单词为单位换行，以字符为单位截断。
- UILineBreakModeHeadTruncation：以单词为单位换行。如果是单行，则开始部分有省略号。如果是多行，则中间有省略号，省略号后面有 4 个字符。
- UILineBreakModeTailTruncation：以单词为单位换行。无论是单行还是多行，都是末尾有省略号。
- UILineBreakModeMiddleTruncation：以单词为单位换行。无论是单行还是多行，都是中间有省略号，省略号后面只有 2 个字符。

具体实现

实例文件 UIKitPrjLineBreak.m 的具体实现代码如下所示。

```
#import "UIKitPrjLineBreak.h"
@implementation LineBreakTest
- (id)initWithLineBreakMode:(UILineBreakMode)mode {
  if ( (self = [super init]) ) {
    lineBreakMode_ = mode;
    self.backgroundColor = [UIColor whiteColor];
    self.autoresizingMask = UIViewAutoresizingFlexibleLeftMargin |
                   UIViewAutoresizingFlexibleRightMargin |
                   UIViewAutoresizingFlexibleTopMargin |
                   UIViewAutoresizingFlexibleBottomMargin;
  }
  return self;
}
- (void)drawRect:(CGRect)rect {
  NSString* message =
    @"使用 NSString 的 InstanceMethod 绘制字符串，比较不同设置时的显示效果。";
  UIFont* systemFont = [UIFont systemFontOfSize:18];
  [message drawInRect:rect
          withFont:systemFont
       lineBreakMode:lineBreakMode_];
}
@end
@implementation UIKitPrjLineBreak
- (void)viewDidLoad {
  [super viewDidLoad];
  LineBreakTest* test1 =
    [[[LineBreakTest alloc] initWithLineBreakMode:UILineBreakModeWordWrap] autorelease];
  test1.frame = CGRectMake( 0, 10, 320, 45 );
  [self.view addSubview:test1];
  LineBreakTest* test2 =
    [[[LineBreakTest alloc] initWithLineBreakMode:UILineBreakModeCharacterWrap] autorelease];
  test2.frame = CGRectMake( 0, 80, 320, 45 );
  [self.view addSubview:test2];
  LineBreakTest* test3 =
```

```
            [[[LineBreakTest alloc] initWithLineBreakMode:UILineBreakModeClip] autorelease];
        test3.frame = CGRectMake( 0, 150, 320, 45 );
        [self.view addSubview:test3];
        LineBreakTest* test4 =
            [[[LineBreakTest alloc] initWithLineBreakMode:UILineBreakModeHeadTruncation] autorelease];
        test4.frame = CGRectMake( 0, 220, 320, 45 );
        [self.view addSubview:test4];
        LineBreakTest* test5 =
            [[[LineBreakTest alloc] initWithLineBreakMode:UILineBreak ModeTailTruncation] autorelease];
        test5.frame = CGRectMake( 0, 290, 320, 45 );
        [self.view addSubview:test5];
        LineBreakTest* test6 =
            [[[LineBreakTest alloc] initWithLineBreakMode:UILineBreakModeMiddle Truncation] autorelease];
        test6.frame = CGRectMake( 0, 360, 320, 45 );
        [self.view addSubview:test6];
    }
    @end
```

执行后的效果如图 5-2 所示。

图 5-2　执行效果

实例 141　设置屏幕中文本的横向对齐方式

实例 141	设置屏幕中文本的横向对齐方式
源码路径	\daima\139\

实例说明

在 iOS 应用中，默认的文本对齐方式是居左对齐的。其实我们可以使用 drawInRect: rect:withFont:lineBreakMode:aligment 方法设置水平位置的对齐方式，具体来说有如下 3 种方式。

- UITextAlignmentLeft：左对齐。
- UITextAlignmentRight：右对齐。
- UITextAlignmentCenter：居中对齐。

在本实例中设置了 3 段文本，分别设置了这 3 段文本居左、居中和居右对齐。

具体实现

实例文件 UIKitPrjTextAlignment.m 的具体实现代码如下所示。

```
#import "UIKitPrjTextAlignment.h"
@implementation TextAlignmentTest
- (id)initWithTextAlignment:(UITextAlignment)textAlignment {
    if ( (self = [super init]) ) {
        textAlignment_ = textAlignment;
        self.backgroundColor = [UIColor whiteColor];
        self.autoresizingMask = UIViewAutoresizingFlexibleLeftMargin |
                                UIViewAutoresizingFlexibleRightMargin |
                                UIViewAutoresizingFlexibleTopMargin |
                                UIViewAutoresizingFlexibleBottomMargin;
    }
    return self;
```

```
}
- (void)drawRect:(CGRect)rect {
  NSString* message;
  switch ( textAlignment_ ) {
    case UITextAlignmentLeft:
      message = @"左对齐";
      break;
    case UITextAlignmentCenter:
      message = @"居中对齐";
      break;
    case UITextAlignmentRight:
      message = @"右对齐";
      break;
    default:
      break;
  }
  UIFont* systemFont = [UIFont systemFontOfSize:18];
  [message drawInRect:rect
           withFont:systemFont
      lineBreakMode:UILineBreakModeWordWrap
          alignment:textAlignment_];
}
@end
@implementation UIKitPrjTextAlignment
- (void)viewDidLoad {
  [super viewDidLoad];
  TextAlignmentTest* test1 =
    [[[TextAlignmentTest alloc] initWithTextAlignment:UITextAlignmentLeft] autorelease];
  test1.frame = CGRectMake( 0, 10, 320, 40 );
  [self.view addSubview:test1];
  TextAlignmentTest* test2 =
    [[[TextAlignmentTest alloc] initWithTextAlignment:UITextAlignmentCenter] autorelease];
  test2.frame = CGRectMake( 0, 70, 320, 40 );
  [self.view addSubview:test2];
  TextAlignmentTest* test3 =
    [[[TextAlignmentTest alloc] initWithTextAlignment:UITextAlignmentRight] autorelease];
  test3.frame = CGRectMake( 0, 130, 320, 40 );
  [self.view addSubview:test3];
}
@end
```

执行后的效果如图 5-3 所示。

图 5-3　执行效果

实例 142　缩小文本并设置纵向对齐方式

实例 142	缩小文本并设置纵向对齐方式
源码路径	\daima\139\

实例说明

在 iOS 应用中，可以使用方法 drawAtPoint:forWidth:withFont:fontSize: lineBreakMode:baselineAdjustment 缩小在屏幕中绘制的文本，并且可以使用 baselineAdjustment 设置文本的纵向对齐方式。在本实例中设置了 4 行文本，缩小居底部对齐了第二行文本，缩小居中部对齐了第三行文本，缩小居顶部对齐了第四行文本。

第 5 章 屏幕显示实战

具体实现

实例文件 UIKitPrjBaselineAdjustment.m 的具体实现代码如下所示。

```objc
#import "UIKitPrjBaselineAdjustment.h"
@implementation BaselineAdjustmentTest
- (id)initWithBaselineAdjustment:(UIBaselineAdjustment)baselineAdjustment {
  if ( (self = [super init]) ) {
    baselineAdjustment_ = baselineAdjustment;
    self.backgroundColor = [UIColor whiteColor];
    self.autoresizingMask = UIViewAutoresizingFlexibleLeftMargin |
                            UIViewAutoresizingFlexibleRightMargin |
                            UIViewAutoresizingFlexibleTopMargin |
                            UIViewAutoresizingFlexibleBottomMargin;
  }
  return self;
}
- (void)drawRect:(CGRect)rect {
  NSString* message =
    @"看我一眼吧";
  UIFont* systemFont = [UIFont systemFontOfSize:36]; //< 原来的字体大小
  if ( -1 == baselineAdjustment_ ) {
    [message drawAtPoint:rect.origin
           forWidth:rect.size.width
           withFont:systemFont
           fontSize:36
      lineBreakMode:UILineBreakModeWordWrap
 baselineAdjustment:baselineAdjustment_];
  } else {
    [message drawAtPoint:rect.origin
           forWidth:rect.size.width
           withFont:systemFont
           fontSize:10 //实际绘制的字体大小
      lineBreakMode:UILineBreakModeWordWrap
 baselineAdjustment:baselineAdjustment_];
  }
}
@end
@implementation UIKitPrjBaselineAdjustment
- (void)viewDidLoad {
  [super viewDidLoad];
  BaselineAdjustmentTest* test0 =
    [[[BaselineAdjustmentTest alloc] initWithBaselineAdjustment:-1] autorelease];
  test0.frame = CGRectMake( 0, 10, 320, 40 );
  [self.view addSubview:test0];
  BaselineAdjustmentTest* test1 =
    [[[BaselineAdjustmentTest alloc] initWithBaselineAdjustment:UIBaselineAdjustmentAlignBaselines] autorelease];
  test1.frame = CGRectMake( 0, 70, 320, 40 );
  [self.view addSubview:test1];
  BaselineAdjustmentTest* test2 =
    [[[BaselineAdjustmentTest alloc] initWithBaselineAdjustment:UIBaselineAdjustmentAlignCenters] autorelease];
  test2.frame = CGRectMake( 0, 130, 320, 40 );
  [self.view addSubview:test2];
  BaselineAdjustmentTest* test3 =
    [[[BaselineAdjustmentTest alloc] initWithBaselineAdjustment:UIBaselineAdjustmentNone] autorelease];
  test3.frame = CGRectMake( 0, 190, 320, 40 );
  [self.view addSubview:test3];
}
@end
```

执行后的效果如图 5-4 所示。

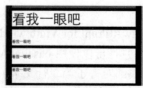

图 5-4　执行效果

实例 143 设置屏幕中的字符串自动缩小

实例 143	设置屏幕中的字符串自动缩小
源码路径	\daima\139\

实例说明

在 iOS 应用中，当屏幕中某行的宽度不能容纳文本时，可以使用 minFontSize 设置字符串的字体大小进行自动缩小处理。在本实例中设置了 1 行文本，然后使用在上一个实例中介绍的方法 drawAtPoint:forWidth:withFont:fontSize: lineBreakMode:baselineAdjustment，并在此方法的基础上添加了 minFontSize 参数和 actualFontSize 参数，这两个参数的具体说明如下所示。

- minFontSize：指定了允许缩小的最小尺寸。
- actualFontSize：是一个输出指针型参数，用于获取实际字符串时采用的字体大小。

具体实现

实例文件 UIKitPrjMinFontSize.m 的具体实现代码如下所示。

```objc
#import "UIKitPrjMinFontSize.h"
@implementation MinFontSizeTest
- (id)init {
  if ( (self = [super init]) ) {
    self.backgroundColor = [UIColor whiteColor];
    self.autoresizingMask = UIViewAutoresizingFlexibleLeftMargin |
                        UIViewAutoresizingFlexibleRightMargin |
                        UIViewAutoresizingFlexibleTopMargin |
                        UIViewAutoresizingFlexibleBottomMargin;
  }
  return self;
}
- (void)drawRect:(CGRect)rect {
  NSString* message =
    @"在宽度不能容纳文本的情况下，可以自动缩小屏幕中文本的字体大小。";
  UIFont* systemFont = [UIFont systemFontOfSize:36];
  CGFloat actualFontSize;
  [message drawAtPoint:rect.origin
           forWidth:rect.size.width
           withFont:systemFont
         minFontSize:6
       actualFontSize:&actualFontSize
        lineBreakMode:UILineBreakModeWordWrap
     baselineAdjustment:UIBaselineAdjustmentAlignCenters];
  NSLog( @"actualFontSize = %f", actualFontSize );
}
@end
@implementation UIKitPrjMinFontSize
- (void)viewDidLoad {
  [super viewDidLoad];
  MinFontSizeTest* test = [[[MinFontSizeTest alloc] init] autorelease];
  test.frame = CGRectMake( 0, 10, 320, 40 );
  [self.view addSubview:test];
}
@end
```

执行后的效果如图 5-5 所示。

图 5-5 执行效果

实例 144 获取绘制文本所需要的空间范围

实例 144	获取绘制文本所需要的空间范围
源码路径	\daima\139\

实例说明

在 iOS 应用中，可以使用 UIKit 中的 NSString 中的类计算绘制字符串时所需要的空间。此功能可以通过如下方法实现。

- sizeWithFont：计算使用指定的字体在第一行中绘制所需要的空间。
- sizeWithFont:forWidth:lineBreakMode：指定宽度最大值计算空间。
- sizeWithFont:constrainedToSize：指定宽度和高度的最大值，计算绘制需要的空间。
- sizeWithFont:constrainedToSize:lineBreakMode：指定宽度和高度的最大值，并指定换行/省略模式后计算绘制需要的空间。
- sizeWithFont:minFontSize:actualFontSize:forWidth:lineBreakMode：计算字符串自动缩小时参数 actualFontSize 中传入 CGFloat 类型指针后，可以获得实际绘制时的字符大小。

在本实例中演示了上述方法的基本用法。

具体实现

实例文件 UIKitPrjSizeWithFont.m 的具体实现代码如下所示。

```
#import "UIKitPrjSizeWithFont.h"
@implementation SizeWithFontTest
- (id)init {
  if ( (self = [super init]) ) {
    self.backgroundColor = [UIColor whiteColor];
    self.autoresizingMask = UIViewAutoresizingFlexibleLeftMargin |
                            UIViewAutoresizingFlexibleRightMargin |
                            UIViewAutoresizingFlexibleTopMargin |
                            UIViewAutoresizingFlexibleBottomMargin;
  }
  return self;
}
- (void)drawRect:(CGRect)rect {
  NSString* message =
    @"sizeWithFont:方法中计算绘制字符串时的字符大小。";
  UIFont* systemFont = [UIFont systemFontOfSize:18];
  CGFloat actualFontSize;
  [message drawAtPoint:rect.origin
           forWidth:rect.size.width
           withFont:systemFont
         minFontSize:6
       actualFontSize:&actualFontSize
        lineBreakMode:UILineBreakModeWordWrap
     baselineAdjustment:UIBaselineAdjustmentAlignCenters];
CGSize size;
  // 计算各种情况下字符的大小
  size = [message sizeWithFont:systemFont];
  NSLog( @"●sizeWithFont: 的运行结果" );
  NSLog( @"size = %f, %f", size.width, size.height );
  // 设置了横向宽度的限制值的情况下（不支持多行）
  size = [message sizeWithFont:systemFont
                forWidth:rect.size.width
             lineBreakMode:UILineBreakModeTailTruncation];
  NSLog( @"●sizeWithFont:forWidth:lineBreakMode: 的运行结果" );
  NSLog( @"size = %f, %f", size.width, size.height );
  // 设置了横向与纵向宽度的限制值的情况下（支持多行）
  size = [message sizeWithFont:systemFont constrainedToSize:rect.size];
  NSLog( @"●sizeWithFont:constrainedToSize: 的运行结果" );
  NSLog( @"size = %f, %f", size.width, size.height );
```

```
   // 设置了横向与纵向宽度且指定了换行/变换方法的情况下
   size = [message sizeWithFont:systemFont
          constrainedToSize:rect.size
             lineBreakMode:UILineBreakModeCharacterWrap];
   NSLog( @"●sizeWithFont:constrainedToSize:lineBreakMode: 的运行结果" );
   NSLog( @"size = %f, %f", size.width, size.height );
   // 使用字体自动缩小设置时
   size = [message sizeWithFont:systemFont
          minFontSize:6
       actualFontSize:&actualFontSize
          forWidth:rect.size.width
          lineBreakMode:UILineBreakModeWordWrap];
   NSLog( @"●sizeWithFont:minFontSize:actualFontSize:forWidth:lineBreakMode: 的运行结果
" );
   NSLog( @"size = %f, %f", size.width, size.height );
   NSLog( @"actualFontSize = %f", actualFontSize );
}
@end
@implementation UIKitPrjSizeWithFont
- (void)viewDidLoad {
  [super viewDidLoad];
  SizeWithFontTest* test1 =
    [[[SizeWithFontTest alloc] init] autorelease];
  test1.frame = CGRectMake( 0, 10, 320, 66 );
  [self.view addSubview:test1];
  SizeWithFontTest* test2 =
    [[[SizeWithFontTest alloc] init] autorelease];
  test2.frame = CGRectMake( 0, 70, 320, 40 );
  [self.view addSubview:test2];
  SizeWithFontTest* test3 =
    [[[SizeWithFontTest alloc] init] autorelease];
  test3.frame = CGRectMake( 0, 130, 320, 40 );
  [self.view addSubview:test3];
}
@end
```

执行后的效果如图 5-6 所示。

图 5-6　执行效果

实例 145　显示系统中的字体

实例 145	显示系统中的字体
源码路径	\daima\139\

实例说明

在 iOS 应用中，可以使用 UIFont 中定义的类方法 systemFontOfSize 来设置字体的大小。在本实例中，除了使用方法 systemFontOfSize 设置字体的大小外，还使用 boldSystemFontOfSize 设置了粗体效果，使用 italicSystemFontOfSize 3 个值设置了斜体效果。

具体实现

实例文件 UIKitPrjSystemFont.m 的具体实现代码如下所示。

```
#import "UIKitPrjSystemFont.h"
@implementation UIKitPrjSystemFont
```

```objc
- (void)dealloc {
  [items_ release];
  [super dealloc];
}
- (void)viewDidLoad {
  [super viewDidLoad];
  items_ = [[NSArray alloc] initWithObjects:
            @"System Font",
            @"Small Font",
            @"Bold Font",
            @"Italic Font",
            @"Button Font Size",
            @"Label Font Size",
            nil];
}
#pragma mark UITableView methods
- (NSInteger)tableView:(UITableView*)tableView
 numberOfRowsInSection:(NSInteger)section
{
  return [items_ count];
}
- (UITableViewCell*)tableView:(UITableView*)tableView
 cellForRowAtIndexPath:(NSIndexPath*)indexPath
{
  static NSString *CellIdentifier = @"Cell";
  UITableViewCell *cell = [tableView dequeueReusableCellWithIdentifier:CellIdentifier];
  if (cell == nil) {
    cell = [[[UITableViewCell alloc] initWithStyle:UITableViewCellStyleDefault reuseIdentifier:CellIdentifier] autorelease];
  }
  NSString* labelCaption = [items_ objectAtIndex:indexPath.row];
  cell.textLabel.text = labelCaption;
  UIFont* font;
  if ( [labelCaption isEqualToString:@"System Font"] ) {
    font = [UIFont systemFontOfSize:[UIFont systemFontSize]];
  } else if ( [labelCaption isEqualToString:@"Small Font"] ) {
    font = [UIFont systemFontOfSize:[UIFont smallSystemFontSize]];
  } else if ( [labelCaption isEqualToString:@"Bold Font"]) {
    font = [UIFont boldSystemFontOfSize:[UIFont systemFontSize]];
  } else if ( [labelCaption isEqualToString:@"Italic Font"]) {
    font = [UIFont italicSystemFontOfSize:[UIFont systemFontSize]];
  } else if ( [labelCaption isEqualToString:@"Button Font Size"]) {
    font = [UIFont systemFontOfSize:[UIFont buttonFontSize]];
  } else if ( [labelCaption isEqualToString:@"Label Font Size"]) {
    font = [UIFont systemFontOfSize:[UIFont labelFontSize]];
  } else {
    font = [UIFont systemFontOfSize:[UIFont systemFontSize]];
  }
  cell.textLabel.font = font;
  return cell;
}
@end
```

执行效果如图 5-7 所示。

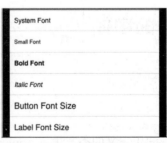

图 5-7 执行效果

实例 146　列表显示系统中所有的字体

实例 146	列表显示系统中所有的字体
源码路径	\daima\139\

实例 146 列表显示系统中所有的字体

实例说明

在 iOS 应用中，可以设置任意的名称来创建 UIFont 实例。在具体实现时是使用 fontWithName:size 方法实现的。在本实例中，使用 fontWithName:size 方法创建了 UIFont 实例，使用列表的方式显示了系统中所有已存在的字体。

具体实现

实例文件 UIKitPrjFontWithName.m 的具体实现代码如下所示。

```objc
#import "UIKitPrjFontWithName.h"
@implementation UIKitPrjFontWithName
- (void)dealloc {
    [familyNames_ release];
    [fontNames_ release];
    [super dealloc];
}
- (id)init {
    if ( (self = [super initWithStyle:UITableViewStyleGrouped]) ) {
        if ( !fontNames_ ) {
            fontNames_ = [[NSMutableDictionary alloc] initWithCapacity:32];
            familyNames_ = [[[UIFont familyNames] sortedArrayUsingSelector:@selector(compare:)] retain];
            for ( id familyName in familyNames_ ) {
                NSArray* fonts = [[UIFont fontNamesForFamilyName:familyName]
                                  sortedArrayUsingSelector:@selector(compare:)];
                [fontNames_ setObject:fonts forKey:familyName];
            }
        }
    }
    return self;
}
#pragma mark UITableView methods
- (NSInteger)numberOfSectionsInTableView:(UITableView*)tableView {
    return [fontNames_ count];
}
- (NSInteger)tableView:(UITableView*)tableView
 numberOfRowsInSection:(NSInteger)section
{
    NSString* familyName = [familyNames_ objectAtIndex:section];
    return [[fontNames_ objectForKey:familyName] count];
}
- (UITableViewCell*)tableView:(UITableView*)tableView
        cellForRowAtIndexPath:(NSIndexPath*)indexPath
{
    static NSString *CellIdentifier = @"Cell";
    UITableViewCell *cell = [tableView dequeueReusableCellWithIdentifier:CellIdentifier];
    if (cell == nil) {
        cell = [[[UITableViewCell alloc] initWithStyle:UITableViewCellStyleDefault reuseIdentifier:CellIdentifier] autorelease];
    }
    NSString* familyName = [familyNames_ objectAtIndex:indexPath.section];
    NSArray* values = [fontNames_ objectForKey:familyName];
    NSString* fontName = [values objectAtIndex:indexPath.row];
    cell.textLabel.text = fontName;
    UIFont* font = [UIFont fontWithName:fontName size:[UIFont labelFontSize]];
    cell.textLabel.font = font;

    return cell;
}
- (NSString*)tableView:(UITableView*)tableView
 titleForHeaderInSection:(NSInteger)section
{
    return [familyNames_ objectAtIndex:section];
}
@end
```

执行后的效果如图 5-8 所示。

图 5-8 执行效果

实例 147 在屏幕中显示不同的颜色

实例 147	在屏幕中显示不同的颜色
源码路径	\daima\147\

实例说明

在 iOS 应用中，可以用一个 UIColor 对象来定义文字的色彩。UIColor 这个类提供了许多不同的方法，可以很轻松地调出任何颜色。可以用静态方法来创建颜色，这样它们会在停止使用后被释放。可以用灰度值、色相或者 RGB 复合值等多种形式来创建颜色。要创建一个简单的 RGB 色彩，可以指定一组 4 个浮点值，分别对应红、绿、蓝和 alpha 值（透明度），取值均 0.0～1.0。这些值表示了 0%（0.0）～100%（1.0）的范围。

另外，类 UIColor 还支持许多静态方法，可以创建系统颜色，这些颜色都经过 iPhone 的校正，以达到尽可能准确的地步。这些方法均来自 UIColor.h，具体定义如下所示：

```
+ (UIColor *)blackColor;         // 0.0 白色
+ (UIColor *)darkGrayColor;      // 0.333 白色
+ (UIColor *)lightGrayColor;     // 0.667 白色
+ (UIColor *)whiteColor;         // 1.0 白色
+ (UIColor *)grayColor;          // 0.5 白色
+ (UIColor *)redColor;           // 1.0, 0.0, 0.0 RGB
+ (UIColor *)greenColor;         // 0.0, 1.0, 0.0 RGB
+ (UIColor *)blueColor;          // 0.0, 0.0, 1.0 RGB
+ (UIColor *)cyanColor;          // 0.0, 1.0, 1.0 RGB
+ (UIColor *)yellowColor;        // 1.0, 1.0, 0.0 RGB
+ (UIColor *)magentaColor;       // 1.0, 0.0, 1.0 RGB
+ (UIColor *)orangeColor;        // 1.0, 0.5, 0.0 RGB
+ (UIColor *)purpleColor;        // 0.5, 0.0, 0.5 RGB
+ (UIColor *)brownColor;         // 0.6, 0.4, 0.2 RGB
+ (UIColor *)clearColor;         // 0.0 白色, 0.0 alpha
```

在本实例中，使用 UIColor 设置了 14 种不同的背景颜色。

具体实现

实例文件 UIKitPrjPresetColor.m 的具体实现代码如下所示。

```
#import "UIKitPrjPresetColor.h"
@implementation UIKitPrjPresetColor
- (void)viewDidLoad {
  [super viewDidLoad];
  const NSUInteger kBarHeight = 30;
  NSUInteger top = 0;
  UILabel* label1 = [[[UILabel alloc] init] autorelease];
  label1.frame = CGRectMake( 0, 0, 320, kBarHeight );
  label1.autoresizingMask = UIViewAutoresizingFlexibleLeftMargin |
                    UIViewAutoresizingFlexibleRightMargin |
                    UIViewAutoresizingFlexibleTopMargin |
                    UIViewAutoresizingFlexibleBottomMargin;
  label1.backgroundColor = [UIColor blackColor];
  [self.view addSubview:label1];
```

```objc
top += kBarHeight;
UILabel* label2 = [[[UILabel alloc] init] autorelease];
label2.frame = CGRectMake( 0, top, 320, kBarHeight );
label2.autoresizingMask = label1.autoresizingMask;
label2.backgroundColor = [UIColor darkGrayColor];
[self.view addSubview:label2];
top += kBarHeight;
UILabel* label3 = [[[UILabel alloc] init] autorelease];
label3.frame = CGRectMake( 0, top, 320, kBarHeight );
label3.autoresizingMask = label1.autoresizingMask;
label3.backgroundColor = [UIColor lightGrayColor];
[self.view addSubview:label3];
top += kBarHeight;
UILabel* label4 = [[[UILabel alloc] init] autorelease];
label4.frame = CGRectMake( 0, top, 320, kBarHeight );
label4.autoresizingMask = label1.autoresizingMask;
label4.backgroundColor = [UIColor whiteColor];
[self.view addSubview:label4];
top += kBarHeight;
UILabel* label5 = [[[UILabel alloc] init] autorelease];
label5.frame = CGRectMake( 0, top, 320, kBarHeight );
label5.autoresizingMask = label1.autoresizingMask;
label5.backgroundColor = [UIColor grayColor];
[self.view addSubview:label5];
top += kBarHeight;
UILabel* label6 = [[[UILabel alloc] init] autorelease];
label6.frame = CGRectMake( 0, top, 320, kBarHeight );
label6.autoresizingMask = label1.autoresizingMask;
label6.backgroundColor = [UIColor redColor];
[self.view addSubview:label6];
top += kBarHeight;
UILabel* label7 = [[[UILabel alloc] init] autorelease];
label7.frame = CGRectMake( 0, top, 320, kBarHeight );
label7.autoresizingMask = label1.autoresizingMask;
label7.backgroundColor = [UIColor greenColor];
[self.view addSubview:label7];
top += kBarHeight;
UILabel* label8 = [[[UILabel alloc] init] autorelease];
label8.frame = CGRectMake( 0, top, 320, kBarHeight );
label8.autoresizingMask = label1.autoresizingMask;
label8.backgroundColor = [UIColor blueColor];
[self.view addSubview:label8];
top += kBarHeight;
UILabel* label9 = [[[UILabel alloc] init] autorelease];
label9.frame = CGRectMake( 0, top, 320, kBarHeight );
label9.autoresizingMask = label1.autoresizingMask;
label9.backgroundColor = [UIColor cyanColor];
[self.view addSubview:label9];
top += kBarHeight;
UILabel* label10 = [[[UILabel alloc] init] autorelease];
label10.frame = CGRectMake( 0, top, 320, kBarHeight );
label10.autoresizingMask = label1.autoresizingMask;
label10.backgroundColor = [UIColor yellowColor];
[self.view addSubview:label10];
top += kBarHeight;
UILabel* label11 = [[[UILabel alloc] init] autorelease];
label11.frame = CGRectMake( 0, top, 320, kBarHeight );
label11.autoresizingMask = label1.autoresizingMask;
label11.backgroundColor = [UIColor magentaColor];
[self.view addSubview:label11];
top += kBarHeight;
UILabel* label12 = [[[UILabel alloc] init] autorelease];
label12.frame = CGRectMake( 0, top, 320, kBarHeight );
label12.autoresizingMask = label1.autoresizingMask;
label12.backgroundColor = [UIColor orangeColor];
[self.view addSubview:label12];
top += kBarHeight;
UILabel* label13 = [[[UILabel alloc] init] autorelease];
label13.frame = CGRectMake( 0, top, 320, kBarHeight );
label13.autoresizingMask = label1.autoresizingMask;
label13.backgroundColor = [UIColor purpleColor];
[self.view addSubview:label13];
top += kBarHeight;
UILabel* label14 = [[[UILabel alloc] init] autorelease];
label14.frame = CGRectMake( 0, top, 320, kBarHeight );
label14.autoresizingMask = label1.autoresizingMask;
```

```
        label14.backgroundColor = [UIColor brownColor];
        [self.view addSubview:label14];
        top += kBarHeight;
        UILabel* label15 = [[[UILabel alloc] init] autorelease];
        label15.frame = CGRectMake( 0, top, 320, kBarHeight );
        label15.autoresizingMask = label1.autoresizingMask;
        label15.backgroundColor = [UIColor clearColor];
        [self.view addSubview:label15];
    }
    @end
```

执行后的效果如图 5-9 所示。

图 5-9　执行效果

实例 148　使用系统颜色

实例 148	使用系统颜色
源码路径	\daima\147\

实例说明

在 iOS 应用中，在屏幕中可以使用系统字体所用到的颜色，并且在使用时可以设置 RGB 的值来实现不同颜色的显示效果。在本实例中，在屏幕中调用了系统颜色。

具体实现

实例文件 UIKitPrjSystemColor.m 的具体实现代码如下所示。

```
#import "UIKitPrjSystemColor.h"
@implementation UIKitPrjSystemColor
- (void)viewDidLoad {
    [super viewDidLoad];
    const NSUInteger kBarHeight = 80;
    NSUInteger top = 0;
    UILabel* label1 = [[[UILabel alloc] init] autorelease];
    label1.frame = CGRectMake( 0, top, 320, kBarHeight );
    label1.autoresizingMask = UIViewAutoresizingFlexibleLeftMargin |
                              UIViewAutoresizingFlexibleRightMargin |
                              UIViewAutoresizingFlexibleTopMargin |
                              UIViewAutoresizingFlexibleBottomMargin;
    label1.backgroundColor = [UIColor lightTextColor];
    [self.view addSubview:label1];
    top += kBarHeight;
    UILabel* label2 = [[[UILabel alloc] init] autorelease];
    label2.frame = CGRectMake( 0, top, 320, kBarHeight );
    label2.autoresizingMask = label1.autoresizingMask;
    label2.backgroundColor = [UIColor darkTextColor];
    [self.view addSubview:label2];

    top += kBarHeight;
    UILabel* label3 = [[[UILabel alloc] init] autorelease];
    label3.frame = CGRectMake( 0, top, 320, kBarHeight );
    label3.autoresizingMask = label1.autoresizingMask;
    label3.backgroundColor = [UIColor groupTableViewBackgroundColor];
    [self.view addSubview:label3];

    top += kBarHeight;
    UILabel* label4 = [[[UILabel alloc] init] autorelease];
    label4.frame = CGRectMake( 0, top, 320, kBarHeight );
    label4.autoresizingMask = label1.autoresizingMask;
```

```
        label4.backgroundColor = [UIColor viewFlipsideBackgroundColor];
        [self.view addSubview:label4];
}
@end
```

执行后的效果如图 5-10 所示。

图 5-10　执行效果

实例 149　在屏幕中自定义颜色

实例 149	在屏幕中自定义颜色
源码路径	\daima\147\

实例说明

在 iOS 应用中，除了可以使用预设颜色和系统颜色外，我们还可以自行创建指定的颜色，在具体实现时，可以使用方法 colorWithRed:green:blue:alpha 来实现，通过此方法分别指定了 RGB 值和 alpha 值，例如下面的代码便是创建了一个 UIColor 实例。

```
UIColor colorWithRed:1.0 green:0.5 blue:0.0 alpha:1.0
```

在本实例中，使用方法 colorWithRed:green:blue:alpha 自行创建了指定的颜色。

具体实现

实例文件 UIKitPrjCreatingColor.m 的具体实现代码如下所示。

```
#import "UIKitPrjCreatingColor.h"
@implementation UIKitPrjCreatingColor
- (void)viewDidLoad {
    [super viewDidLoad];
    UILabel* label1 = [[[UILabel alloc] init] autorelease];
    label1.frame = CGRectMake( 0, 0, 320, 20 );
    label1.autoresizingMask = UIViewAutoresizingFlexibleLeftMargin |
                              UIViewAutoresizingFlexibleRightMargin |
                              UIViewAutoresizingFlexibleTopMargin |
                              UIViewAutoresizingFlexibleBottomMargin;
    label1.backgroundColor = [UIColor colorWithRed:1.0 green:0.5 blue:0.0 alpha:1.0];
    [self.view addSubview:label1];

    UILabel* label2 = [[[UILabel alloc] init] autorelease];
    label2.frame = CGRectMake( 0, 20, 320, 20 );
    label2.autoresizingMask = label1.autoresizingMask;
    label2.backgroundColor = [UIColor colorWithHue:0.5 saturation:1.0 brightness:0.5 alpha:1.0];
    [self.view addSubview:label2];

    UILabel* label3 = [[[UILabel alloc] init] autorelease];
    label3.frame = CGRectMake( 0, 40, 320, 20 );
    label3.autoresizingMask = label1.autoresizingMask;
    label3.backgroundColor = [UIColor colorWithWhite:0.5 alpha:1.0];
```

```
    [self.view addSubview:label3];

    UILabel* label4 = [[[UILabel alloc] init] autorelease];
    label4.frame = CGRectMake( 0, 60, 320, 20 );
    label4.autoresizingMask = label1.autoresizingMask;
    UIColor* red = [UIColor redColor];
    label4.backgroundColor = [red colorWithAlphaComponent:0.5];
    [self.view addSubview:label4];
}
@end
```

执行后的效果如图 5-11 所示。

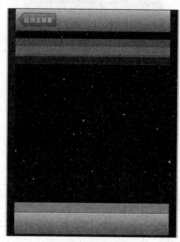

图 5-11 执行效果

实例 150 使用背景图片创建特殊背景

实例 150	使用背景图片创建特殊背景
源码路径	\daima\147\

实例说明

在 iOS 应用中，除了可以使用预设颜色和系统颜色外，还可以使用指定的背景图片创建类似颜色的个性背景。此功能是通过方法 colorWithPatternImage 实现的，此方法可以为标签 UILabel 设置特殊的背景。

在本实例中，使用方法 colorWithPatternImage 设置了图片"dog.jpg"为背景。

具体实现

实例文件 UIKitPrjPatternImage.m 的具体实现代码如下所示。

```
#import "UIKitPrjPatternImage.h"
@implementation UIKitPrjPatternImage
- (void)viewDidLoad {
    [super viewDidLoad];
    UILabel* label = [[[UILabel alloc] init] autorelease];
    label.frame = self.view.bounds;
    label.autoresizingMask = UIViewAutoresizingFlexibleWidth | UIViewAutoresizingFlexibleHeight;
    UIImage* image = [UIImage imageNamed:@"dog.jpg"];
    label.backgroundColor = [UIColor colorWithPatternImage:image];
    [self.view addSubview:label];
}
@end
```

执行效果如图 5-12 所示。

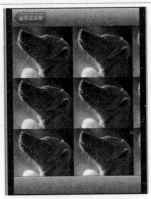

图 5-12 执行效果

实例 151 在屏幕中绘制指定颜色的文字

实例 151	在屏幕中绘制指定颜色的文字
源码路径	\daima\147\

实例说明

在 iOS 应用中，可以使用 UIColor 设置绘图颜色。在本实例中，在 MyView 类中重写了 drawRect 方法，并在其中绘制了字符串和背景。设置的背景图片是 "dog.jpg"，并且绘制了一个框的颜色为灰色的四边形，然后使用 drawRect 方法绘制了文本"我可爱吗，亲？"。

具体实现

实例文件 UIKitPrjDrawing.m 的具体实现代码如下所示。

```
#import "UIKitPrjDrawing.h"
//MyView 类中重写 drawRect: 方法，并在其中进行字符串及背景绘制
@implementation MyView
- (void)drawRect:(CGRect)rect {
UIImage* image = [UIImage imageNamed:@"dog.jpg"];
  UIColor* patternColor = [UIColor colorWithPatternImage:image];
  // 绘制四边形
  CGContextRef context = UIGraphicsGetCurrentContext();
  CGContextBeginPath( context );
  CGContextAddRect( context, rect );
  CGContextClosePath( context );
  [[UIColor grayColor] setStroke]; //设置边框的颜色为灰色
  [patternColor setFill]; //将背景设置为图片背景
  CGContextDrawPath( context, kCGPathEOFillStroke );
  // 绘制字符
  [[UIColor blueColor] set]; //将字符颜色设置为蓝色
  [@"我可爱吗，亲？" drawInRect:rect
                withFont:[UIFont boldSystemFontOfSize:24]
           lineBreakMode:UILineBreakModeClip
               alignment:UITextAlignmentCenter];
}
@end
//创建 MyView 实例并追加到画面（self.view）中
@implementation UIKitPrjDrawing
- (void)viewDidLoad {
  [super viewDidLoad];
  MyView* myView = [[[MyView alloc] init] autorelease];
  myView.frame = CGRectMake( 40, 40, 240, 240 );
  [self.view addSubview:myView];
}
@end
```

执行后的效果如图 5-13 所示。

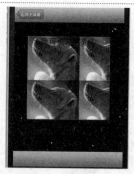

图 5-13 执行效果

实例 152　在屏幕中显示图像

实例 152	在屏幕中显示图像
源码路径	\daima\152\

实例说明

在 iOS 应用中，图像视图（UIImageView）控件用于在屏幕中显示一个图像。可以将图像视图加入到应用程序中，并用于向用户呈现信息。UIImageView 实例还可以创建简单的基于帧的动画，其中包括开始、停止和设置动画播放速度的控件。在使用 Retina 屏幕的设备中，图像视图可利用其高分辨率屏幕。令开发人员兴奋的是，无需编写任何特殊代码，无需检查设备类型，而只需将多幅图像加入到项目中，而图像视图将在正确的时间加载正确的图像。

在本实例中，使用 UIImageView 控件在屏幕中显示一幅指定的图像。

具体实现

实例文件 UIKitPrjUIImageView.m 的具体实现代码如下所示。

```
#import "UIKitPrjUIImageView.h"
@implementation UIKitPrjUIImageView
- (void)viewDidLoad {
    [super viewDidLoad];
    // 读入图片文件
    UIImage* image = [UIImage imageNamed:@"dog.jpg"];
    // UIImageView 的创建
    UIImageView* imageView = [[[UIImageView alloc] initWithImage:image] autorelease];
    // 设置中心位置以及自动调节参数
    imageView.center = self.view.center;
    imageView.autoresizingMask = UIViewAutoresizingFlexibleTopMargin |
                                UIViewAutoresizingFlexibleBottomMargin;
    // 将图片 View 追加到 self.view 中
    [self.view addSubview:imageView];
}
@end
```

执行效果如图 5-14 所示。

图 5-14 执行效果

实例 153 在屏幕中绘制一幅图像

实例 153	在屏幕中绘制一幅图像
源码路径	\daima\152\

实例说明

在 iOS 应用中，如果不使用 UIImageView 显示一个图像，也可以使用 UIImage 直接绘制图片。在具体实现时，可以分别通过 UIImage 的方法 drawAtPoint 或 drawInRect 方法实现。在本实例中，分别演示了使用方法 drawAtPoint 和 drawInRect 绘制图像的过程。

具体实现

实例文件 UIKitPrjUIImage.m 的具体实现代码如下所示。

```
#import "UIKitPrjUIImage.h"
//实现 UIView 子类
@implementation DrawImageTest
- (void)dealloc {
 [image_ release];
 [super dealloc];
}
- (id)initWithImage:(UIImage*)image {
  if ( (self = [super init]) ) {
    image_ = image;
  }
  return self;
}
- (void)drawRect:(CGRect)rect {
//在 drawAtPoint:与 drawInRect:间切换，比较具体效果
 [image_ drawAtPoint:rect.origin];
 //[image_ drawInRect:rect];
}
@end
@implementation UIKitPrjUIImage
- (void)viewDidLoad {
 [super viewDidLoad];
 // 读入图片文件
 UIImage* image = [UIImage imageNamed:@"dog.jpg"];
 // 创建定制的 View
 DrawImageTest* test = [[[DrawImageTest alloc] initWithImage:image] autorelease];
 test.frame = self.view.bounds;
 test.autoresizingMask =
   UIViewAutoresizingFlexibleWidth | UIViewAutoresizingFlexibleHeight;
 [self.view addSubview:test];
}
@end
```

方法 drawAtPoint 绘制的执行效果如图 5-15 所示。方法 drawInRect 绘制的执行效果如图 5-16 所示。

图 5-15 执行效果 1

图 5-16 执行效果 2

实例 154 在屏幕中绘图时设置透明度

实例 154	在屏幕中绘图时设置透明度
源码路径	\daima\152\

实例说明

在 iOS 应用中，当使用 UIImage 在屏幕中绘制图像时，除了可以设置图像的大小尺寸外，还可以指定其透明值，即 alpha 的值。在本实例中，使用 drawInRect:blendMode_alpha:设置了绘制图像时的透明值，并且在 CGBlendMode 中设置了透明类型。

具体实现

实例文件 UIKitPrjBlendMode.m 的具体实现代码如下所示。

```
#import "UIKitPrjBlendMode.h"
@implementation BlendModeTest
@synthesize blendMode = blendMode_;
- (void)dealloc {
  [frontImage_ release];
  [backImage_ release];
  [super dealloc];
}
- (id)init {
  if ( (self = [super init]) ) {
    backImage_ = [UIImage imageNamed:@"back.png"];
    frontImage_ = [UIImage imageNamed:@"dog.jpg"];
    CGRect newFrame = self.frame;
    newFrame.size = frontImage_.size;
    self.frame = newFrame;
  }
  return self;
}
- (void)drawRect:(CGRect)rect {
  [backImage_ drawInRect:rect];
  [frontImage_ drawInRect:rect blendMode:blendMode_ alpha:1.0];
}
- (void)changeMode {
  if ( kCGBlendModeLuminosity < ++blendMode_ ) {
    blendMode_ = kCGBlendModeNormal;
  }
}
@end
#pragma mark ----- Private Methods Definition -----
@interface UIKitPrjBlendMode ()
- (void)changeLabel;
@end
#pragma mark ----- Start Implementation For Methods -----
@implementation UIKitPrjBlendMode
- (void)dealloc {
  [test_ release];
  [label_ release];
  [super dealloc];
}
- (void)viewDidLoad {
  [super viewDidLoad];
  test_ = [[BlendModeTest alloc] init];
  test_.center = self.view.center;
  test_.autoresizingMask = UIViewAutoresizingFlexibleTopMargin |
                  UIViewAutoresizingFlexibleBottomMargin;
  [self.view addSubview:test_];
  label_ = [[UILabel alloc] init];
  label_.frame =
    CGRectMake( 0, self.view.bounds.size.height - 100, self.view.bounds.size.width, 20 );
  label_.autoresizingMask = UIViewAutoresizingFlexibleTopMargin |
                   UIViewAutoresizingFlexibleBottomMargin;
```

实例 154　在屏幕中绘图时设置透明度

```objc
    label_.textAlignment = UITextAlignmentCenter;
    [self.view addSubview:label_];
    [self changeLabel];
    UIImage* imageBack = [UIImage imageNamed:@"back.png"];
    UIImageView* imageViewBack = [[[UIImageView alloc] initWithImage:imageBack] autorelease];
    imageViewBack.frame = CGRectMake( 0, 0, 90, 83 );
    imageViewBack.autoresizingMask = UIViewAutoresizingFlexibleLeftMargin |
                                    UIViewAutoresizingFlexibleRightMargin |
                                    UIViewAutoresizingFlexibleTopMargin |
                                    UIViewAutoresizingFlexibleBottomMargin;
    [self.view addSubview:imageViewBack];
    UIImage* imageFront = [UIImage imageNamed:@"dog.jpg"];
    UIImageView* imageViewFront = [[[UIImageView alloc] initWithImage:imageFront] autorelease];
    imageViewFront.frame = CGRectMake( 320 - 90, 0, 90, 83 );
    imageViewFront.autoresizingMask = imageViewFront.autoresizingMask;
    [self.view addSubview:imageViewFront];
}
#pragma mark ----- Private Methods -----
- (void)changeLabel {
    switch ( test_.blendMode ) {
        case kCGBlendModeMultiply: label_.text = @"kCGBlendModeMultiply"; break;
        case kCGBlendModeScreen: label_.text = @"kCGBlendModeScreen"; break;
        case kCGBlendModeOverlay: label_.text = @"kCGBlendModeOverlay"; break;
        case kCGBlendModeDarken: label_.text = @"kCGBlendModeDarken"; break;
        case kCGBlendModeLighten: label_.text = @"kCGBlendModeLighten"; break;
        case kCGBlendModeColorDodge: label_.text = @"kCGBlendModeColorDodge"; break;
        case kCGBlendModeColorBurn: label_.text = @"kCGBlendModeColorBurn"; break;
        case kCGBlendModeSoftLight: label_.text = @"kCGBlendModeSoftLight"; break;
        case kCGBlendModeHardLight: label_.text = @"kCGBlendModeHardLight"; break;
        case kCGBlendModeDifference: label_.text = @"kCGBlendModeDifference"; break;
        case kCGBlendModeExclusion: label_.text = @"kCGBlendModeExclusion"; break;
        case kCGBlendModeHue: label_.text = @"kCGBlendModeHue"; break;
        case kCGBlendModeSaturation: label_.text = @"kCGBlendModeSaturation"; break;
        case kCGBlendModeColor: label_.text = @"kCGBlendModeColor"; break;
        case kCGBlendModeLuminosity: label_.text = @"kCGBlendModeLuminosity"; break;
        case kCGBlendModeClear: label_.text = @"kCGBlendModeClear"; break;
        case kCGBlendModeCopy: label_.text = @"kCGBlendModeCopy"; break;
        case kCGBlendModeSourceIn: label_.text = @"kCGBlendModeSourceIn"; break;
        case kCGBlendModeSourceOut: label_.text = @"kCGBlendModeSourceOut"; break;
        case kCGBlendModeSourceAtop: label_.text = @"kCGBlendModeSourceAtop"; break;
        case kCGBlendModeDestinationOver: label_.text = @"kCGBlendModeDestinationOver"; break;
        case kCGBlendModeDestinationIn: label_.text = @"kCGBlendModeDestinationIn"; break;
        case kCGBlendModeDestinationOut: label_.text = @"kCGBlendModeDestinationOut"; break;
        case kCGBlendModeDestinationAtop: label_.text = @"kCGBlendModeDestinationAtop"; break;
        case kCGBlendModeXOR: label_.text = @"kCGBlendModeXOR"; break;
        case kCGBlendModePlusDarker: label_.text = @"kCGBlendModePlusDarker"; break;
        case kCGBlendModePlusLighter: label_.text = @"kCGBlendModePlusLighter"; break;
        default: label_.text = @"kCGBlendModeNormal"; break;
    }
}
#pragma mark ----- Responder -----
- (void)touchesEnded:(NSSet*)touches withEvent:(UIEvent*)event {
    [test_ changeMode];
    [self changeLabel];
    [test_ setNeedsDisplay];
}
@end
```

执行后的效果如图 5-17 所示。

图 5-17　执行效果

实例 155　限制图像的缩放区域

实例 155	限制图像的缩放区域
源码路径	\daima\152\

实例说明

在 iOS 应用中，UIImage 除了可以在屏幕中绘制图像后，还可以限制图像的扩大或缩小的区域。在具体实现时，是通过 UIImage 的方法 stretchableImageWithLeftCapWidth: topCapHeight:实现的，通过此方法可以设置限制区域的宽度和高度，在设置后可以获取新的 UIImage。在本实例中，演示了使用的方法 stretchableImageWithLeftCapWidth: topCapHeight:的具体流程。

具体实现

实例文件 UIKitPrjImageWithCap.m 的具体实现代码如下所示。

```
#import "UIKitPrjImageWithCap.h"
@implementation UIKitPrjImageWithCap
- (void)viewDidLoad {
    [super viewDidLoad];
    self.view.backgroundColor = [UIColor whiteColor];
    // 读入红色圆圈的图片
    UIImage* image = [UIImage imageNamed:@"circle.png"];
    // 限制伸缩区域创建标题用的图片
    UIImage* imageWithCap = [image stretchableImageWithLeftCapWidth:30 topCapHeight:30];
    // 没有获取的图片作为按钮的背景
    UIButton* button = [UIButton buttonWithType:UIButtonTypeCustom];
    [button setBackgroundImage:image forState:UIControlStateNormal];
    [button setTitle:@"长UIButton 无标题版" forState:UIControlStateNormal];
    [button sizeToFit];
    button.center = CGPointMake( 160, 50 );
    [self.view addSubview:button];
    
    // 将以覆盖图片作为背景按钮贴在 UIView 新片背景的按钮
    UIButton* buttonWithCap = [UIButton buttonWithType:UIButtonTypeCustom];
    [buttonWithCap setBackgroundImage:imageWithCap forState:UIControlStateNormal];
    [buttonWithCap setTitle:@"长UIButton 有标题版" forState:UIControlStateNormal];
    [buttonWithCap sizeToFit];
    buttonWithCap.center = CGPointMake( 160, 150 );
    [self.view addSubview:buttonWithCap];
}
@end
```

执行后的效果如图 5-18 所示。

图 5-18　执行效果

实例 156　使用 UIImageView 实现动画效果

实例 156	使用 UIImageView 实现动画效果
源码路径	\daima\152\

实例 156　使用 UIImageView 实现动画效果

实例说明

在 iOS 应用中，UIImage 除了可以在屏幕中绘制图像后，还可以在其属性 animationImages 中设置 NSArray 类型的图片数组，然后使用方法 startAnimating 开始动画效果，使用方法 stopAnimating 结束动画效果。在本实例中，演示了使用 UIImageView 实现动画效果的流程。

具体实现

实例文件 UIKitPrjUIImageViewAnimation.m 的具体实现代码如下所示。

```objectivec
#import "UIKitPrjUIImageViewAnimation.h"
@implementation UIKitPrjUIImageViewAnimation
- (void)dealloc {
  [imageView_ release];
  [super dealloc];
}
- (void)viewWillAppear:(BOOL)animated {
 [super viewWillAppear:animated];
 // 动画开始
 [imageView_ startAnimating];
}
- (void)viewWillDisappear:(BOOL)animated {
 [super viewWillDisappear:animated];
 // 动画停止
 [imageView_ stopAnimating];
}
- (void)viewDidLoad {
 [super viewDidLoad];
 self.view.backgroundColor = [UIColor blackColor];

 UIImage* chara1 = [UIImage imageNamed:@"chara1.png"];
 UIImage* chara2 = [UIImage imageNamed:@"chara2.png"];

 imageView_ = [[UIImageView alloc] init];
 imageView_.frame = CGRectMake( 0, 0, 64, 64 );
 // 在 NSArray 数组中设置动画素材用的图片
 imageView_.animationImages = [NSArray arrayWithObjects:chara1, chara2, nil];
 // 将动画素材设置为每隔 0.5 秒改变一次
 imageView_.animationDuration = 0.5;
     imageView_.alpha=0.5;
 imageView_.center = self.view.center;
 imageView_.autoresizingMask = UIViewAutoresizingFlexibleLeftMargin |
                    UIViewAutoresizingFlexibleRightMargin |
                    UIViewAutoresizingFlexibleTopMargin |
                    UIViewAutoresizingFlexibleBottomMargin;
 [self.view addSubview:imageView_];
}
@end
```

执行后的效果如图 5-19 所示。

图 5-19　执行效果

实例 157　在屏幕中实现日历效果

实例 157	在屏幕中实现日历效果
源码路径	\daima\157\

实例说明

在 iOS 应用中，可以使用日期选择框控件在屏幕中实现日历效果。但是这些日期选择框控件不够美观，和我们现实中所看到的界面效果有很大的差别。在本实例中，实现了一个十分简单易用、可以自定义界面的日历代码效果。在本实例的实现代码中，提供了 delegate 的回调方法，通过此方法可以方便地自定义日历的动作，包括选择日期等。

具体实现

实例文件 CKCalendarView.m 的主要实现代码如下所示。

```
- (id)initWithFrame:(CGRect)frame {
    self = [super initWithFrame:frame];
    if (self) {
        self.calendar = [[NSCalendar alloc] initWithCalendarIdentifier:NSGregorianCalendar];
        [self.calendar setLocale:[NSLocale currentLocale]];
        [self.calendar setFirstWeekday:self.calendarStartDay];
        self.cellWidth = DEFAULT_CELL_WIDTH;
        self.layer.cornerRadius = 6.0f;
        self.layer.shadowOffset = CGSizeMake(2, 2);
        self.layer.shadowRadius = 2.0f;
        self.layer.shadowOpacity = 0.4f;
        self.layer.borderColor = [UIColor blackColor].CGColor;
        self.layer.borderWidth = 1.0f;
        UIView *highlight = [[UIView alloc] initWithFrame:CGRectZero];
        highlight.backgroundColor = [UIColor colorWithWhite:1.0 alpha:0.2];
        highlight.layer.cornerRadius = 6.0f;
        [self addSubview:highlight];
        self.highlight = highlight;

        // 实现日历的标题
        UILabel *titleLabel = [[UILabel alloc] initWithFrame:CGRectZero];
        titleLabel.textAlignment = UITextAlignmentCenter;
        titleLabel.backgroundColor = [UIColor clearColor];
        titleLabel.autoresizingMask = UIViewAutoresizingFlexibleBottomMargin | UIViewAutoresizingFlexibleWidth;
        [self addSubview:titleLabel];
        self.titleLabel = titleLabel;

        UIButton *prevButton = [UIButton buttonWithType:UIButtonTypeCustom];
        [prevButton setImage:[UIImage imageNamed:@"left_arrow.png"] forState:UIControlStateNormal];
        prevButton.autoresizingMask = UIViewAutoresizingFlexibleBottomMargin | UIViewAutoresizingFlexibleRightMargin;
        [prevButton addTarget:self action:@selector(moveCalendarToPreviousMonth) forControlEvents:UIControlEventTouchUpInside];
        [self addSubview:prevButton];
        self.prevButton = prevButton;

        UIButton *nextButton = [UIButton buttonWithType:UIButtonTypeCustom];
        [nextButton setImage:[UIImage imageNamed:@"right_arrow.png"] forState:UIControlStateNormal];
        nextButton.autoresizingMask = UIViewAutoresizingFlexibleBottomMargin | UIViewAutoresizingFlexibleLeftMargin;
        [nextButton addTarget:self action:@selector(moveCalendarToNextMonth) forControlEvents:UIControlEventTouchUpInside];
        [self addSubview:nextButton];
        self.nextButton = nextButton;

        // 实现日历本身的主体界面
```

```objc
    UIView *calendarContainer = [[UIView alloc] initWithFrame:CGRectZero];
    calendarContainer.layer.borderWidth = 1.0f;
    calendarContainer.layer.borderColor = [UIColor blackColor].CGColor;
    calendarContainer.autoresizingMask    =    UIViewAutoresizingFlexibleTopMargin    |
UIViewAutoresizingFlexibleWidth;
    calendarContainer.layer.cornerRadius = 4.0f;
    calendarContainer.clipsToBounds = YES;
    [self addSubview:calendarContainer];
    self.calendarContainer = calendarContainer;

    GradientView *daysHeader = [[GradientView alloc] initWithFrame:CGRectZero];
    daysHeader.autoresizingMask = UIViewAutoresizingFlexibleBottomMargin | UIView
AutoresizingFlexibleWidth;
    [self.calendarContainer addSubview:daysHeader];
    self.daysHeader = daysHeader;

    NSMutableArray *labels = [NSMutableArray array];
    for (NSString *day in [self getDaysOfTheWeek]) {
        UILabel *dayOfWeekLabel = [[UILabel alloc] initWithFrame:CGRectZero];
        dayOfWeekLabel.text = [day uppercaseString];
        dayOfWeekLabel.textAlignment = UITextAlignmentCenter;
        dayOfWeekLabel.backgroundColor = [UIColor clearColor];
        dayOfWeekLabel.shadowColor = [UIColor whiteColor];
        dayOfWeekLabel.shadowOffset = CGSizeMake(0, 1);
        [labels addObject:dayOfWeekLabel];
        [self.calendarContainer addSubview:dayOfWeekLabel];
    }
    self.dayOfWeekLabels = labels;

    // 实现 42 个按钮
    NSMutableArray *dateButtons = [NSMutableArray array];
    dateButtons = [NSMutableArray array];
    for (int i = 0; i < 43; i++) {
        DateButton *dateButton = [DateButton buttonWithType:UIButtonTypeCustom];
        [dateButton setTitle:[NSString stringWithFormat:@"%d", i] forState:UIControl
        StateNormal];
        [dateButton addTarget:self action:@selector(dateButtonPressed:)  forControl
        Events:UIControlEventTouchUpInside];
        [dateButtons addObject:dateButton];
    }
    self.dateButtons = dateButtons;

    // 初始化
    self.monthShowing = [NSDate date];
    [self setDefaultStyle];
    }

    [self layoutSubviews]; // 第一个月
    return self;
}

- (void)layoutSubviews {
    [super layoutSubviews];

    CGFloat containerWidth = self.bounds.size.width - (CALENDAR_MARGIN * 2);
    self.cellWidth = (containerWidth / 7.0) - CELL_BORDER_WIDTH;

    CGFloat containerHeight = ([self numberOfWeeksInMonthContainingDate:self.monthShowing]
* (self.cellWidth + CELL_BORDER_WIDTH) + DAYS_HEADER_HEIGHT);
    CGRect newFrame = self.frame;
    newFrame.size.height = containerHeight + CALENDAR_MARGIN + TOP_HEIGHT;
    self.frame = newFrame;

    self.highlight.frame = CGRectMake(1, 1, self.bounds.size.width - 2, 1);

    self.titleLabel.frame = CGRectMake(0, 0, self.bounds.size.width, TOP_HEIGHT);
    self.prevButton.frame = CGRectMake(BUTTON_MARGIN, BUTTON_MARGIN, 48, 38);
    self.nextButton.frame  = CGRectMake(self.bounds.size.width - 48 - BUTTON_MARGIN,
BUTTON_MARGIN, 48, 38);
    self.calendarContainer.frame = CGRectMake(CALENDAR_MARGIN, CGRectGetMaxY(self.
titleLabel.frame), containerWidth, containerHeight);
    self.daysHeader.frame = CGRectMake(0,  0,  self.calendarContainer.frame.size.width,
```

```objc
            DAYS_HEADER_HEIGHT);
    for (UILabel *dayLabel in self.dayOfWeekLabels) {
            dayLabel.frame = CGRectMake(CGRectGetMaxX(lastDayFrame) + CELL_BORDER_WIDTH,
lastDayFrame.origin.y, self.cellWidth, self.daysHeader.frame.size.height);
            lastDayFrame = dayLabel.frame;
    }
    for (DateButton *dateButton in self.dateButtons) {
            [dateButton removeFromSuperview];
    }
    NSDate *date = [self firstDayOfMonthContainingDate:self.monthShowing];
    uint dateButtonPosition = 0;
    while ([self dateIsInMonthShowing:date]) {
            DateButton *dateButton = [self.dateButtons objectAtIndex:dateButtonPosition];
            dateButton.date = date;
            if ([dateButton.date isEqualToDate:self.selectedDate]) {
                dateButton.backgroundColor = self.selectedDateBackgroundColor;
                [dateButton setTitleColor:self.selectedDateTextColor forState:UIControlStateNormal];
            } else if ([self dateIsToday:dateButton.date]) {
                [dateButton setTitleColor:self.currentDateTextColor forState:UIControlStateNormal];
                dateButton.backgroundColor = self.currentDateBackgroundColor;
            } else {
                dateButton.backgroundColor = [self dateBackgroundColor];
                [dateButton setTitleColor:[self dateTextColor] forState:UIControlStateNormal];
            }
            dateButton.frame = [self calculateDayCellFrame:date];
            [self.calendarContainer addSubview:dateButton];
            :date];
            dateButtonPosition++;
    }
}
- (void)setMonthShowing:(NSDate *)aMonthShowing {
    _monthShowing = aMonthShowing;

    NSDateFormatter *dateFormatter = [[NSDateFormatter alloc] init];
    dateFormatter.dateFormat = @"MMMM YYYY";
    self.titleLabel.text = [dateFormatter stringFromDate:aMonthShowing];
    [self setNeedsLayout];
}

- (void)setDefaultStyle {
    self.backgroundColor = UIColorFromRGB(0x393B40);

    [self setTitleColor:[UIColor whiteColor]];
    [self setTitleFont:[UIFont boldSystemFontOfSize:17.0]];

    [self setDayOfWeekFont:[UIFont boldSystemFontOfSize:12.0]];
    [self setDayOfWeekTextColor:UIColorFromRGB(0x999999)];
    [self setDayOfWeekBottomColor:UIColorFromRGB(0xCCCFD5) topColor:[UIColor whiteColor]];

    [self setDateFont:[UIFont boldSystemFontOfSize:16.0f]];
    [self setDateTextColor:UIColorFromRGB(0x393B40)];
    [self setDateBackgroundColor:UIColorFromRGB(0xF2F2F2)];
    [self setDateBorderColor:UIColorFromRGB(0xDAE1E6)];

    [self setSelectedDateTextColor:UIColorFromRGB(0xF2F2F2)];
    [self setSelectedDateBackgroundColor:UIColorFromRGB(0x88B6DB)];

    [self setCurrentDateTextColor:UIColorFromRGB(0xF2F2F2)];
    [self setCurrentDateBackgroundColor:[UIColor lightGrayColor]];
}

- (CGRect)calculateDayCellFrame:(NSDate *)date {
    int row = [self weekNumberInMonthForDate:date] - 1;
    int placeInWeek = ((([self dayOfWeekForDate:date] - 1) - self.calendar.firstWeekday + 8) % 7;

    return CGRectMake(placeInWeek * (self.cellWidth + CELL_BORDER_WIDTH), (row * (self.cellWidth + CELL_BORDER_WIDTH)) + CGRectGetMaxY(self.daysHeader.frame) + CELL_BORDER_WIDTH, self.cellWidth, self.cellWidth);
}

- (void)moveCalendarToNextMonth {
```

```
    NSDateComponents* comps = [[NSDateComponents alloc]init];
    [comps setMonth:1];
    self.monthShowing = [self.calendar dateByAddingComponents:comps toDate:self.
monthShowing options:0];
}

- (void)moveCalendarToPreviousMonth {
    self.monthShowing   =   [[self   firstDayOfMonthContainingDate:self.monthShowing]
dateByAddingTimeInterval:-100000];
}

- (void)dateButtonPressed:(id)sender {
    DateButton *dateButton = sender;
    self.selectedDate = dateButton.date;
    [self.delegate calendar:self didSelectDate:self.selectedDate];
    [self setNeedsLayout];
}

#pragma mark - Theming getters/setters

- (void)setTitleFont:(UIFont *)font {
    self.titleLabel.font = font;
}
- (UIFont *)titleFont {
    return self.titleLabel.font;
}

- (void)setTitleColor:(UIColor *)color {
    self.titleLabel.textColor = color;
}
- (UIColor *)titleColor {
    return self.titleLabel.textColor;
}

- (void)setButtonColor:(UIColor *)color {
    [self.prevButton setImage:[CKCalendarView imageNamed:@"left_arrow.png" withColor:
color] forState:UIControlStateNormal];
    [self.nextButton setImage:[CKCalendarView imageNamed:@"right_arrow.png" withColor:
color] forState:UIControlStateNormal];
}

- (void)setInnerBorderColor:(UIColor *)color {
    self.calendarContainer.layer.borderColor = color.CGColor;
}

- (void)setDayOfWeekFont:(UIFont *)font {
    for (UILabel *label in self.dayOfWeekLabels) {
        label.font = font;
    }
}
- (UIFont *)dayOfWeekFont {
    return (self.dayOfWeekLabels.count > 0) ? ((UILabel *)[self.dayOfWeekLabels last
Object]).font : nil;
}

- (void)setDayOfWeekTextColor:(UIColor *)color {
    for (UILabel *label in self.dayOfWeekLabels) {
        label.textColor = color;
    }
}
- (UIColor *)dayOfWeekTextColor {
    return (self.dayOfWeekLabels.count > 0) ? ((UILabel *)[self.dayOfWeekLabels last
Object]).textColor : nil;
}

- (void)setDayOfWeekBottomColor:(UIColor *)bottomColor topColor:(UIColor *)topColor {
    [self.daysHeader setColors:[NSArray arrayWithObjects:topColor, bottomColor, nil]];
}

- (void)setDateFont:(UIFont *)font {
    for (DateButton *dateButton in self.dateButtons) {
        dateButton.titleLabel.font = font;
```

```objc
    }
}
- (UIFont *)dateFont {
    return (self.dateButtons.count > 0) ? ((DateButton *)[self.dateButtons lastObject]).titleLabel.font : nil;
}

- (void)setDateTextColor:(UIColor *)color {
    for (DateButton *dateButton in self.dateButtons) {
        [dateButton setTitleColor:color forState:UIControlStateNormal];
    }
}
- (UIColor *)dateTextColor {
    return (self.dateButtons.count > 0) ? [((DateButton *)[self.dateButtons lastObject]) titleColorForState:UIControlStateNormal] : nil;
}

- (void)setDateBackgroundColor:(UIColor *)color {
    for (DateButton *dateButton in self.dateButtons) {
        dateButton.backgroundColor = color;
    }
}
- (UIColor *)dateBackgroundColor {
    return (self.dateButtons.count > 0) ? ((DateButton *)[self.dateButtons lastObject]).backgroundColor : nil;
}

- (void)setDateBorderColor:(UIColor *)color {
    self.calendarContainer.backgroundColor = color;
}
- (UIColor *)dateBorderColor {
    return self.calendarContainer.backgroundColor;
}

#pragma mark - Calendar helpers

- (NSDate *)firstDayOfMonthContainingDate:(NSDate *)date {
    NSDateComponents *comps = [self.calendar components:(NSYearCalendarUnit | NSMonthCalendarUnit | NSDayCalendarUnit) fromDate:date];
    [comps setDay:1];
    return [self.calendar dateFromComponents:comps];
}

- (NSArray *)getDaysOfTheWeek {
    NSDateFormatter *dateFormatter = [[NSDateFormatter alloc] init];

    // 调整阵列
    NSArray *weekdays = [dateFormatter shortWeekdaySymbols];
    NSUInteger firstWeekdayIndex = [self.calendar firstWeekday] -1;
    if (firstWeekdayIndex > 0)
    {
        weekdays = [[weekdays subarrayWithRange:NSMakeRange(firstWeekdayIndex, 7-firstWeekdayIndex)]
                    arrayByAddingObjectsFromArray:[weekdays subarrayWithRange:NSMakeRange(0,firstWeekdayIndex)]];
    }
    return weekdays;
}

- (int)dayOfWeekForDate:(NSDate *)date {
    NSDateComponents *comps = [self.calendar components:NSWeekdayCalendarUnit fromDate:date];
    return comps.weekday;
}

- (BOOL)dateIsToday:(NSDate *)date {
    NSDateComponents *otherDay = [self.calendar components:NSEraCalendarUnit|NSYearCalendarUnit|NSMonthCalendarUnit|NSDayCalendarUnit fromDate:date];
    NSDateComponents *today = [self.calendar components:NSEraCalendarUnit|NSYearCalendarUnit|NSMonthCalendarUnit|NSDayCalendarUnit fromDate:[NSDate date]];
```

```
    return ([today day] == [otherDay day] &&
            [today month] == [otherDay month] &&
            [today year] == [otherDay year] &&
            [today era] == [otherDay era]);
}

- (int)weekNumberInMonthForDate:(NSDate *)date {
    NSDateComponents *comps = [self.calendar components:(NSWeekOfMonthCalendarUnit) fromDate:date];
    return comps.weekOfMonth;
}

- (int)numberOfWeeksInMonthContainingDate:(NSDate *)date {
    return [self.calendar rangeOfUnit:NSWeekCalendarUnit inUnit:NSMonthCalendarUnit forDate:date].length;
}

- (BOOL)dateIsInMonthShowing:(NSDate *)date {
    NSDateComponents *comps1 = [self.calendar components:(NSMonthCalendarUnit) fromDate:self.monthShowing];
    NSDateComponents *comps2 = [self.calendar components:(NSMonthCalendarUnit) fromDate:date];
    return comps1.month == comps2.month;
}

- (NSDate *)nextDay:(NSDate *)date {
    NSDateComponents *comps = [[NSDateComponents alloc] init];
    [comps setDay:1];
    return [self.calendar dateByAddingComponents:comps toDate:date options:0];
}

+ (UIImage *)imageNamed:(NSString *)name withColor:(UIColor *)color {
    UIImage *img = [UIImage imageNamed:name];

    UIGraphicsBeginImageContext(img.size);
    CGContextRef context = UIGraphicsGetCurrentContext();
    [color setFill];

    CGContextTranslateCTM(context, 0, img.size.height);
    CGContextScaleCTM(context, 1.0, -1.0);

    CGContextSetBlendMode(context, kCGBlendModeColorBurn);
    CGRect rect = CGRectMake(0, 0, img.size.width, img.size.height);
    CGContextDrawImage(context, rect, img.CGImage);

    CGContextClipToMask(context, rect, img.CGImage);
    CGContextAddRect(context, rect);
    CGContextDrawPath(context,kCGPathFill);

    UIImage *coloredImg = UIGraphicsGetImageFromCurrentImageContext();
    UIGraphicsEndImageContext();

    return coloredImg;
}
@end
```

执行后的效果如图 5-20 所示。

图 5-20 执行效果

实例 158　在屏幕中自定义一个导航条

实例 158	在屏幕中自定义一个导航条
源码路径	\daima\158\

实例说明

在 iOS 应用中，我们可以根据个人爱好自定义一个导航条效果。在本实例中，给导航条的下方加上了阴影效果，并且自定义了导航条的背景图片。在导航条中，设置了背景图片"Title.png"。

具体实现

实例文件 CustomNavigationBar.m 的主要实现代码如下所示。

```
@implementation CustomNavigationBar
- (void)awakeFromNib {
    [super awakeFromNib];
    _image = [UIImage imageNamed:@"Title.png"];
    self.tintColor = [UIColor colorWithRed:46.0 / 255.0 green:149.0 / 255.0 blue:206.0 / 255.0 alpha:1.0];
    // draw shadow
    self.layer.masksToBounds = NO;
    self.layer.shadowOffset = CGSizeMake(0, 3);
    self.layer.shadowOpacity = 0.6;
    self.layer.shadowPath = [UIBezierPath bezierPathWithRect:self.bounds].CGPath;
}

- (void)drawRect:(CGRect)rect
{
    [_image drawInRect:rect];
}
@end
```

实例文件 AppDelegate.m 的主要实现代码如下所示。

```
#import "AppDelegate.h"
@implementation AppDelegate
@synthesize window = _window;
- (id)customControllerWithRootViewController:(UIViewController *)root {
    UINavigationController *nav = [[[NSBundle mainBundle] loadNibNamed:@"Navigation Controller" owner:self options:nil] objectAtIndex:0];
    [nav setViewControllers:[NSArray arrayWithObject:root]];
    return nav;
}
- (BOOL)application:(UIApplication *)application didFinishLaunchingWithOptions:(NSDictionary *)launchOptions
{
    self.window = [[UIWindow alloc] initWithFrame:[[UIScreen mainScreen] bounds]];
    // 开始定制
    UIViewController *mainViewController = [[UIViewController alloc] init];
    mainViewController.view.backgroundColor = [UIColor whiteColor];
    mainViewController.title = @"Title";
    UINavigationController *navigationController = [self customControllerWithRootViewController:mainViewController];
    self.window.rootViewController = navigationController;
    [self.window makeKeyAndVisible];
    return YES;
}
@end
```

执行后的效果如图 5-21 所示。

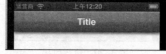

图 5-21　执行效果

实例 159 在屏幕中实现仿 iPhone 锁定界面效果

实例 159	在屏幕中实现仿 iPhone 锁定界面效果
源码路径	\daima\159\

实例说明

我们知道 iPhone 手机的锁定界面非常美观，其实使用前面所学习的 iOS 控件技术也可以实现类似的效果。本实例借助了 http://www.apache.org/licenses/LICENSE-2.0 中的开源框架，在屏幕中实现了仿 iPhone 锁定界面的滑块解锁效果。

具体实现

实例文件 ViewController.m 的主要实现代码如下所示。

```objc
#import "ViewController.h"
#define kMaxTranslation 190.0f
@interface ViewController () {
    CGFloat sliderInitialX;
}
@end
@implementation ViewController
@synthesize animatedLabel;
@synthesize slider;
- (void)viewDidLoad
{
    [super viewDidLoad];
    UIPanGestureRecognizer *pan = [[UIPanGestureRecognizer alloc] initWithTarget:self action:@selector(pan:)];
    [self.slider addGestureRecognizer:pan];
}
- (void)pan:(UIPanGestureRecognizer *)gr
{
    if (gr.state == UIGestureRecognizerStateBegan) {
        [self.animatedLabel stopAnimating];
    }
    if (gr.state == UIGestureRecognizerStateChanged) {
        CGPoint t = [gr translationInView:self.view];  //获取转换信息
        CGRect f = self.slider.frame;
        f.origin.x = MAX(sliderInitialX, MIN(kMaxTranslation, f.origin.x+t.x));
        //滑块界限
        self.slider.frame = f;
        self.animatedLabel.alpha = 1-(self.slider.frame.origin.x/(kMaxTranslation*0.5 - sliderInitialX));  //计算标签透明度
        [gr setTranslation:CGPointZero inView:self.view];  //重新转换

    }
    if (gr.state == UIGestureRecognizerStateEnded) {
    [UIView animateWithDuration:0.1 animations:^{
        CGRect f = self.slider.frame;
        f.origin.x = sliderInitialX;
        self.slider.frame = f;
    } completion:^(BOOL finished) {
        [self.animatedLabel startAnimating];
        self.animatedLabel.alpha = 1.0f;
    }];

    }
}
- (void)viewWillAppear:(BOOL)animated
{
    [super viewWillAppear:animated];
    [self.animatedLabel startAnimating];
    sliderInitialX = self.slider.frame.origin.x;
}
- (void)viewWillDisappear:(BOOL)animated
```

```
{
    [super viewWillDisappear:animated];
    [self.animatedLabel stopAnimating];
}
- (void)viewDidUnload
{
    [self setAnimatedLabel:nil];
    [self setSlider:nil];
    [super viewDidUnload];
    // 释放视图
}
- (BOOL)shouldAutorotateToInterfaceOrientation:(UIInterfaceOrientation)interfaceOrientation
{
    return UIInterfaceOrientationIsPortrait(interfaceOrientation);
}
@end
```

执行效果如图 5-22 所示。

图 5-22　执行效果

第6章 图形、图像和动画实战

图形和图像永远是多媒体的重要组成部分,在智能设备中,图形、图像和动画处理也是极其重要的一个构成部分。在本书前面的章节中,已经讲解了和图像有关的基本知识,在本章将通过具体实例的实现流程,来详细讲解在 iOS 系统中处理图形、图像和动画的基本方法。

实例 160 在屏幕中实现一个简单的动画效果

实例 160	在屏幕中实现一个简单的动画效果
源码路径	\daima\160\

实例说明

在 iOS 应用中,可以使用 UIView 中的动画类实现动画效果。其中最为常用的方法是 beginAnimations 和 commitAnimations。在本实例中,借助了这两个方法实现了简单的动画效果。

具体实现

实例文件 UIKitPrjAnimationMove.m 的具体实现代码如下所示。

```
#import "UIKitPrjAnimationMove.h"
@implementation UIKitPrjAnimationMove
- (void)dealloc {
  [star_ release];
  [super dealloc];
}
- (void)viewDidLoad {
  [super viewDidLoad];
  self.view.backgroundColor = [UIColor blackColor];
  UIImage* image = [UIImage imageNamed:@"star.png"];
  star_ = [[UIImageView alloc] initWithImage:image];
  star_.center = CGPointMake( -100, -100 );
  [self.view addSubview:star_];
}
#pragma mark ----- Responder -----
- (void)touchesEnded:(NSSet*)touches withEvent:(UIEvent*)event {
  star_.center = CGPointMake( -100, -100 ); //图片开始的位置
  [UIView beginAnimations:nil context:NULL]; //动画开始
  [UIView setAnimationDuration:1.0]; //1次动画时间设置为 1.0 秒
  star_.center = CGPointMake( 420, 400 ); //图片移动终点位置
  [UIView commitAnimations]; //动画结束
}
@end
```

执行后的效果如图 6-1 所示。

图 6-1 执行效果

实例 161　设置在屏幕中的动画延迟

实例 161	设置在屏幕中的动画延迟
源码路径	\daima\160\

实例说明

在 iOS 应用中，当使用 UIView 中的动画类实现动画效果后，可以设置动画重复执行或延迟执行。在本实例中，使用方法 setAnimationRepeatCount 设置动画重复执行 10 次，使用方法 setAnimationDelay 设置动画停顿 3 秒后开始执行。

具体实现

实例文件 UIKitPrjAnimationRepeat.m 的具体实现代码如下所示。

```
#import "UIKitPrjAnimationRepeat.h"
@implementation UIKitPrjAnimationRepeat
- (void)dealloc {
  [star_ release];
  [super dealloc];
}
- (void)viewDidLoad {
  [super viewDidLoad];
  self.view.backgroundColor = [UIColor blackColor];
UIImage* image = [UIImage imageNamed:@"star.png"];
  star_ = [[UIImageView alloc] initWithImage:image];
  star_.center = CGPointMake( -100, -100 );
  [self.view addSubview:star_];
}
#pragma mark ----- Responder -----
- (void)touchesEnded:(NSSet*)touches withEvent:(UIEvent*)event {
  star_.center = CGPointMake( -100, -100 ); //图片开始位置

  [UIView beginAnimations:nil context:NULL]; //开始动画
  [UIView setAnimationDuration:1.0]; //1次动画的持续时间设置为1.0秒
  [UIView setAnimationDelay:3.0]; //停顿3秒后动画开始
  [UIView setAnimationRepeatCount:10.0]; //动画重复10回
  star_.center = CGPointMake( 420, 400 ); //设置图片移动的终点位置
  [UIView commitAnimations]; //结束动画
}
@end
```

执行后的效果如图 6-2 所示。

图 6-2　执行效果

实例 162　设置在屏幕中动画的透明度

实例 162	设置在屏幕中动画的透明度
源码路径	\daima\160\

实例说明

在 iOS 应用中,可以通过改变 alpha 值来设置透明度,通过其 setAnimationCurve:方法可以设置动画的弧度。通过改变 alpha 属性值不但可以实现淡入/淡出效果,而且可以实现消融效果画面。在本实例中,设置动画效果的素材图片为 "123.jpg",然后使用方法 setAnimationCurve 设置了动画的弧度,使用方法 setAnimationDuration 设置一次动画持续时间为 2.0 秒,并通过 animationCurve 值在标签中显示当前的动画弧度。

具体实现

实例文件 UIKitPrjAnimationCurve.m 的具体实现代码如下所示。

```objc
#import "UIKitPrjAnimationCurve.h"
@implementation UIKitPrjAnimationCurve
- (void)dealloc {
  [star_ release];
  [label_ release];
  [super dealloc];
}
- (void)viewDidLoad {
  [super viewDidLoad];
  self.view.backgroundColor = [UIColor blackColor];
  // UIImageView 对象的初始化
  UIImage* image = [UIImage imageNamed:@"123.jpg"];
  star_ = [[UIImageView alloc] initWithImage:image];
  star_.center = CGPointMake( self.view.center.x, -100 );
  [self.view addSubview:star_];
  // 标签的创建与初始化
  label_ = [[UILabel alloc] init];
  label_.frame = CGRectMake( 0, self.view.bounds.size.height - 20, 320, 20 );
  label_.autoresizingMask = UIViewAutoresizingFlexibleTopMargin |
                            UIViewAutoresizingFlexibleBottomMargin;
  label_.textAlignment = UITextAlignmentCenter;
  label_.text = @"UIViewAnimationCurveEaseInOut";
  [self.view addSubview:label_];
}
#pragma mark ----- Responder -----
- (void)touchesEnded:(NSSet*)touches withEvent:(UIEvent*)event {
  static UIViewAnimationCurve animationCurve = UIViewAnimationCurveEaseInOut;
  star_.center = CGPointMake( self.view.center.x, -100 ); //图片开始的位置
  star_.alpha = 1.0; //开始时的 alpha 值
  [UIView beginAnimations:nil context:NULL];
  [UIView setAnimationCurve:animationCurve]; //动画弧的设置
  [UIView setAnimationDuration:10.0]; //将一次动画持续时间设置为 2.0 秒
  star_.center = CGPointMake( self.view.center.x, 300 ); //图片移动终点位置
  star_.alpha = 0.0; //结束时的 alpha 值
  [UIView commitAnimations];

  //在标签中显示当前的动画弧
  switch ( animationCurve ) {
    case UIViewAnimationCurveEaseInOut:
      label_.text = @"UIViewAnimationCurveEaseInOut";
      break;
    case UIViewAnimationCurveEaseIn:
      label_.text = @"UIViewAnimationCurveEaseIn";
      break;
    case UIViewAnimationCurveEaseOut:
      label_.text = @"UIViewAnimationCurveEaseOut";
      break;
    case UIViewAnimationCurveLinear:
      label_.text = @"UIViewAnimationCurveLinear";
      break;
    default:
      label_.text = @"-";
      break;
  }
  // 动画弧的修改
  if ( UIViewAnimationCurveLinear < ++animationCurve ) {
    animationCurve = UIViewAnimationCurveEaseInOut;
  }
}
@end
```

执行后的效果如图 6-3 所示。

图 6-3 执行效果

实例 163　设置屏幕中的动画实现放大/缩小/旋转效果

实例 163	设置屏幕中的动画实现放大/缩小/旋转效果
源码路径	\daima\160\

实例说明

在 iOS 应用中，通过 UIView 可以实现几何变化效果。在本实例中，通过 UIView 的属性 transform 分别实现了放大、缩小、旋转效果，并且设置了开始时的 alpha 值，通过 setAnimationRepeatAutoreverses 设置了 reverse（反转）效果，通过 transformScale 实现了放大处理，通过 transformRotate 设置了旋转处理。

具体实现

实例文件 UIKitPrjAnimationTransform.m 的具体实现代码如下所示。

```
#import "UIKitPrjAnimationTransform.h"
@implementation UIKitPrjAnimationTransform
- (void)dealloc {
  [star_ release];
  [super dealloc];
}
- (void)viewDidLoad {
  [super viewDidLoad];
  self.view.backgroundColor = [UIColor blackColor];
  // UIImageView 对象的初始化
  UIImage* image = [UIImage imageNamed:@"123.jpg"];
  star_ = [[UIImageView alloc] initWithImage:image];
  star_.center = CGPointMake( self.view.center.x, -100 );
  [self.view addSubview:star_];
}
#pragma mark ----- Responder -----
- (void)touchesEnded:(NSSet*)touches withEvent:(UIEvent*)event {
  star_.center = CGPointMake( self.view.center.x, -100 ); //图片开始的位置
  star_.alpha = 1.0; //开始时的 alpha 值
  star_.transform = CGAffineTransformIdentity; //初始化 transform

  [UIView beginAnimations:nil context:NULL];
  [UIView setAnimationRepeatAutoreverses:YES]; //设置 reverse（反转）
  [UIView setAnimationDuration:2.0]; //将一次动画持续时间设置为 2.0 秒
  [UIView setAnimationCurve:UIViewAnimationCurveEaseIn];
  star_.center = CGPointMake( self.view.center.x, 300 ); //图片移动终点位置
  star_.alpha = 0.0; //结束时的 alpha 值

  // 下面混合了扩大与旋转
  CGAffineTransform transformScale = CGAffineTransformScale( CGAffineTransformIdentity, 5, 5 );
  CGAffineTransform transformRotate = CGAffineTransformRotate( CGAffineTransformIdentity, M_PI );
  star_.transform = CGAffineTransformConcat( transformScale, transformRotate );
  [UIView commitAnimations];
```

}
@end

执行后的效果如图 6-4 所示。

图 6-4 执行效果

实例 164 检测屏幕中动画的状态

实例 164	检测屏幕中动画的状态
源码路径	\daima\160\

实例说明

在 iOS 应用中，当通过 UIView 实现动画效果后，可以继续使用如下 3 种方法进行完善。

- setAnimationDelegate：将委托设置为自己（当前类）。
- setAnimationWillStartSelector：指定动画开始时调用方法的选择器。
- setAnimationDidStopSelector：指定动画结束时调用方法的选择器。

在本实例中，通过使用上述方法在屏幕中实现了动画效果。

具体实现

实例文件 UIKitPrjAnimationObserving.m 的具体实现代码如下所示。

```
#import "UIKitPrjAnimationObserving.h"
#pragma mark ----- Private Methods Definition -----
@interface UIKitPrjAnimationObserving ()
- (void)startAnimation;
- (void)animationDidStop:(NSString*)animationID finished:(NSNumber*)finished context:
(void*)context;
@end
#pragma mark ----- Start Implementation For Methods -----
@implementation UIKitPrjAnimationObserving

- (void)dealloc {
    [star_ release];
    [super dealloc];
}
- (void)viewDidLoad {
    [super viewDidLoad];
    self.view.backgroundColor = [UIColor blackColor];
    // UIImageView 对象的初始化
    UIImage* image = [UIImage imageNamed:@"123.jpg"];
    star_ = [[UIImageView alloc] initWithImage:image];
    star_.center = CGPointMake( self.view.center.x, -100 );
    [self.view addSubview:star_];
}

- (void)viewWillAppear:(BOOL)animated {
```

```
    [super viewWillAppear:animated];
    [self startAnimation]; //画面显示时开始动画
}

#pragma mark ----- Private Methods -----
- (void)startAnimation {
    star_.center = CGPointMake( self.view.center.x, -100 ); //图片开始的位置
    star_.transform = CGAffineTransformIdentity; //初始化 transform
    [UIView beginAnimations:nil context:NULL];
    [UIView setAnimationDelegate:self]; //将委托设置为自己（当前类）
    [UIView setAnimationDidStopSelector:@selector(animationDidStop:finished:context:)];
    [UIView setAnimationDuration:2.0]; //将一次动画持续时间设置为 2.0 秒
    [UIView setAnimationCurve:UIViewAnimationCurveEaseIn];
    star_.center = CGPointMake( self.view.center.x, 300 ); //图片移动终点位置
    CGAffineTransform transformScale = CGAffineTransformScale( CGAffineTransformIdentity, 5, 5 );
    CGAffineTransform transformRotate = CGAffineTransformRotate( CGAffineTransformIdentity, M_PI );
    star_.transform = CGAffineTransformConcat( transformScale, transformRotate );
    [UIView commitAnimations];
}
- (void)animationDidStop:(NSString*)animationID finished:(NSNumber*)finished context:(void*)context {
    // 如果动画正常结束时（非强制取消）再次开始动画
    if ( [finished boolValue] ) {
        [self startAnimation];
    }
}
@end
```

执行后的效果如图 6-5 所示。

图 6-5　执行效果

实例 165　在屏幕中实现过渡动画效果

实例 165	在屏幕中实现过渡动画效果
源码路径	\daima\160\

实例说明

在 iOS 应用中，当通过 UIView 实现动画效果后，可以继续使用其方法 setAnimationTransition:forView:cache 实现在两个动画之间过渡切换的效果。在本实例中，通过使用 touchesEnded 方法实现了触摸画面时显示过渡动画的效果，在具体实现时首先创建了下一个画面（UIView），然后暂时将动画置为无效状态，并通过 if 语句切换过渡动画效果。

具体实现

实例文件 UIKitPrjTransition.m 的具体实现代码如下所示。

```
#import "UIKitPrjTransition.h"
//UIView 中指定的 tag 的常量
static const NSInteger kTagViewForTransitionTest = 1;
#pragma mark ----- Private Methods Definition -----
//声明私有方法
```

```objc
@interface UIKitPrjTransition ()
- (UIView*)nextView;
- (void)animationDidStop;
@end
#pragma mark ----- Start Implementation For Methods -----
@implementation UIKitPrjTransition
- (void)viewDidLoad {
  [super viewDidLoad];
  [self.view addSubview:[self nextView]];//显示初始图片
}
#pragma mark ----- Responder -----
//触摸画面时显示过渡动画效果
- (void)touchesEnded:(NSSet*)touches withEvent:(UIEvent*)event {
//如果非动画可运行状态则返回
  if ( ![UIView areAnimationsEnabled] ) {
    [self.nextResponder touchesEnded:touches withEvent:event];
    return;
  }
  //过渡动画的初始设置
  static UIViewAnimationTransition transition = UIViewAnimationTransitionFlipFromLeft;
  UIView* nextView = [self nextView];//创建下一个画面(UIView)
  [UIView beginAnimations:nil context:NULL];
  [UIView setAnimationDelegate:self];
  [UIView setAnimationDidStopSelector:@selector(animationDidStop)];
  [UIView setAnimationDuration:10.0];
  [UIView setAnimationTransition:transition forView:self.view cache:YES];
  [[self.view viewWithTag:kTagViewForTransitionTest] removeFromSuperview];
  [self.view addSubview:nextView];
  [UIView commitAnimations];
  //暂时将动画设置为无效状态
  [UIView setAnimationsEnabled:NO];
  //切换过渡动画效果
  if ( UIViewAnimationTransitionCurlDown < ++transition ) {
    transition = UIViewAnimationTransitionFlipFromLeft;
  }
}
#pragma mark ----- Private Methods -----
//创建下一画面的私有方法
- (UIView*)nextView {
  static BOOL isFront = YES;
  UIImage* image;
  if ( isFront ) {
    image = [UIImage imageNamed:@"dog.jpg"]; //表面图片
  } else {
    image = [UIImage imageNamed:@"town.jpg"]; //里层图片
  }
  isFront = ( YES != isFront );
  UIView* view = [[[UIImageView alloc] initWithImage:image] autorelease];
  view.tag = kTagViewForTransitionTest;
  view.frame = self.view.bounds;
  view.autoresizingMask =
    UIViewAutoresizingFlexibleWidth | UIViewAutoresizingFlexibleHeight;
  view.contentMode = UIViewContentModeScaleAspectFill;
  return view;
}
//动画结束时被调用,重新将动画设置为有效
- (void)animationDidStop {
  [UIView setAnimationsEnabled:YES];
}
@end
```

执行后的效果如图6-6所示。

图6-6 执行效果

第 6 章　图形、图像和动画实战

实例 166　联合使用滑块和步进控件实现动画效果

实例 166	联合使用滑块和步进控件实现动画效果
源码路径	\daima\166\

实例说明

在本实例中，将通过一个具体的实现过程，来演示联合使用图像动画、滑块和步进控件的方法。本实例将使用这些新 UI 元素（和一些介绍过的控件）来创建一个用户控制的动画。

经过本书前面内容的学习我们了解到，图像视图可以显示图像文件和简单动画，而滑块让用户能够以可视化方式从指定范围内选择一个值。在这个项目中，我们将使用一系列图像视图（UIImageView）实例创建一个循环动画，还将使用一个滑块（UISlider）让用户能够设置动画的播放速度。动画的内容是一个跳跃的小兔子，我们可以控制每秒跳多少次。跳跃速度通过滑块设置，并显示在一个标签（UILabel）中；步进控件提供了另一种以特定的步长调整速度的途径。用户还可使用按钮（UIButton）开始或停止播放动画。

在具体实现之前，我们需要考虑如下两个问题。

（1）动画是使用一系列图像创建的。在这个项目中提供了一个 20 帧的动画，当然读者也可以使用自己的图像。

（2）虽然滑块和步进控件让用户能够以可视化方式输入指定范围内的值，但对其如何设置该值用户没有太大的控制权。例如最小值必须小于最大值，但是我们无法控制沿哪个方向拖曳滑块将增大或减小设置的值。这些局限性并非障碍，而只是意味着我们可能需要做一些计算（或试验）才能获得所需的行为。

具体实现

（1）启动 Xcode，然后在左侧导航选择第一项"Create a new Xcode project"，如图 6-7 所示。

（2）在弹出的新界面中选择项目类型和模板。在 New Project 窗口的左侧，确保选择了项目类型 iOS 中的 Application；在右边的列表中选择 Single View Application，再单击"Next"按钮，如图 6-8 所示。

图 6-7　创建一个 Xcode 工程

（3）在 Product Name 文本框中输入"lianhe"。对于公司标识符，可以将其设置为我们的域名，但顺序相反（笔者再次使用的是 com.guan）。保留文本框 Class Prefix 为空，并确保从下拉列表 Device Family 中选择了 iPhone 或 iPad，这里选择的是 iPhone，如图 6-9 所示。然后，单击 Next 按钮。

实例 166　联合使用滑块和步进控件实现动画效果

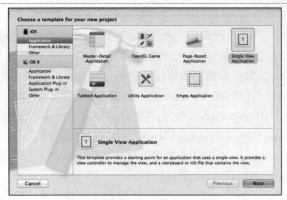

图 6-8　选择 Single View Application

图 6-9　指定应用程序的名称和目标设备

（4）在 Xcode 提示时指定存储位置，再单击 Create 按钮创建项目。这将创建一个简单的应用程序结构，它包含一个应用程序委托、一个窗口、一个视图（在故事板场景中定义的）和一个视图控制器。几秒钟后，项目窗口将打开，如图 6-10 所示。

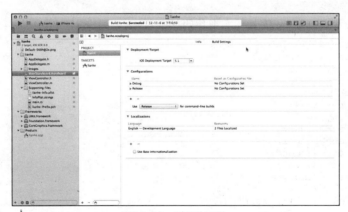

图 6-10　创建的工程

本项目使用了 20 帧存储为 PNG 文件的动画，这些动画帧包含在项目文件夹 "lianhe" 的文件夹 Images 中，如图 6-11 所示。

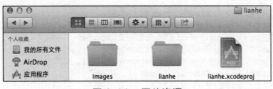

图 6-11　图片资源

由于我们预先知道需要这些图像，因此可立即将其加入到项目中。为此，在 Xcode 的项目导航器中展开项目编组，再展开项目代码编组"lianhe"，然后将文件夹 Images 拖曳到该编组中。在 Xcode 提示时，务必选择必要时复制资源并新建编组。

现在可以在 Interface Builder 编辑器中轻松地访问这些图像文件了，而无需编写代码。在本实例中需要为多个对象提供输出口和操作。在此总共需要 9 个输出口，具体说明如下所示。

- 用 5 个图像视图（UIImageView），它们包含动画的 5 个副本，分别通过 bunnyView1、bunnyView2、bunnyView3、bunnyView4 和 bunnyView5 引用这些图像视图。
- 使用滑块控件（UISlider）设置播放速度，将连接到 speedSlider，而播放速度本身将输出到一个名为"hopsPerSecond"的标签（UILabel）中。
- 使用步进控件（UIStepper），它提供了另一种设置动画播放速度的途径，将通过 speedStepper 来访问它。
- 用于开始和停止播放动画的按钮（UIButton）将连接到输出口 toggleButton。

setSpeed 在用户修改了滑块值时，需要调整动画速度时被调用；setIncrement 的用途与 setSpeed 类似，在用户按下步进控件时被调用；toggleAnimation 用于开始和停止播放动画。

实例文件 ViewController.h 的具体实现代码如下所示。

```
#import <UIKit/UIKit.h>
@interface ViewController : UIViewController
@property (strong, nonatomic) IBOutlet UIImageView *bunnyView1;
@property (strong, nonatomic) IBOutlet UIImageView *bunnyView2;
@property (strong, nonatomic) IBOutlet UIImageView *bunnyView3;
@property (strong, nonatomic) IBOutlet UIImageView *bunnyView4;
@property (strong, nonatomic) IBOutlet UIImageView *bunnyView5;
@property (strong, nonatomic) IBOutlet UISlider *speedSlider;
@property (strong, nonatomic) IBOutlet UIStepper *speedStepper;
@property (strong, nonatomic) IBOutlet UILabel *hopsPerSecond;
@property (strong, nonatomic) IBOutlet UIButton *toggleButton;
- (IBAction)toggleAnimation:(id)sender;
- (IBAction)setSpeed:(id)sender;
- (IBAction)setIncrement:(id)sender;
@end
```

实例文件 ViewController.m 的具体实现代码如下所示。

```
#import "ViewController.h"
@implementation ViewController
@synthesize bunnyView1;
@synthesize bunnyView2;
@synthesize bunnyView3;
@synthesize bunnyView4;
@synthesize bunnyView5;
@synthesize speedSlider;
@synthesize speedStepper;
@synthesize hopsPerSecond;
@synthesize toggleButton;
- (void)didReceiveMemoryWarning
{
    [super didReceiveMemoryWarning];
    // Release any cached data, images, etc that aren't in use.
}

#pragma mark - View lifecycle
- (void)viewDidLoad
{
    NSArray *hopAnimation;
    hopAnimation=[[NSArray alloc] initWithObjects:
            [UIImage imageNamed:@"frame-1.png"],
            [UIImage imageNamed:@"frame-2.png"],
            [UIImage imageNamed:@"frame-3.png"],
            [UIImage imageNamed:@"frame-4.png"],
            [UIImage imageNamed:@"frame-5.png"],
            [UIImage imageNamed:@"frame-6.png"],
            [UIImage imageNamed:@"frame-7.png"],
```

```objc
                    [UIImage imageNamed:@"frame-8.png"],
                    [UIImage imageNamed:@"frame-9.png"],
                    [UIImage imageNamed:@"frame-10.png"],
                    [UIImage imageNamed:@"frame-11.png"],
                    [UIImage imageNamed:@"frame-12.png"],
                    [UIImage imageNamed:@"frame-13.png"],
                    [UIImage imageNamed:@"frame-14.png"],
                    [UIImage imageNamed:@"frame-15.png"],
                    [UIImage imageNamed:@"frame-16.png"],
                    [UIImage imageNamed:@"frame-17.png"],
                    [UIImage imageNamed:@"frame-18.png"],
                    [UIImage imageNamed:@"frame-19.png"],
                    [UIImage imageNamed:@"frame-20.png"],
                    nil
                    ];
    self.bunnyView1.animationImages=hopAnimation;
    self.bunnyView2.animationImages=hopAnimation;
    self.bunnyView3.animationImages=hopAnimation;
    self.bunnyView4.animationImages=hopAnimation;
    self.bunnyView5.animationImages=hopAnimation;
    self.bunnyView1.animationDuration=1;
    self.bunnyView2.animationDuration=1;
    self.bunnyView3.animationDuration=1;
    self.bunnyView4.animationDuration=1;
    self.bunnyView5.animationDuration=1;
    [super viewDidLoad];
}
- (void)viewDidUnload
{
    [self setBunnyView1:nil];
    [self setBunnyView2:nil];
    [self setBunnyView3:nil];
    [self setBunnyView4:nil];
    [self setBunnyView5:nil];
    [self setSpeedSlider:nil];
    [self setSpeedStepper:nil];
    [self setHopsPerSecond:nil];
    [self setToggleButton:nil];
    [super viewDidUnload];
    // Release any retained subviews of the main view.
    // e.g. self.myOutlet = nil;
}
- (void)viewWillAppear:(BOOL)animated
{
    [super viewWillAppear:animated];
}
- (void)viewDidAppear:(BOOL)animated
{
    [super viewDidAppear:animated];
}
- (void)viewWillDisappear:(BOOL)animated
{
    [super viewWillDisappear:animated];
}
- (void)viewDidDisappear:(BOOL)animated
{
    [super viewDidDisappear:animated];
}
- (BOOL)shouldAutorotateToInterfaceOrientation:(UIInterfaceOrientation)interfaceOrientation
{
    // Return YES for supported orientations
    return (interfaceOrientation != UIInterfaceOrientationPortraitUpsideDown);
}
- (IBAction)toggleAnimation:(id)sender {
    if (bunnyView1.isAnimating) {
        [self.bunnyView1 stopAnimating];
        [self.bunnyView2 stopAnimating];
        [self.bunnyView3 stopAnimating];
        [self.bunnyView4 stopAnimating];
        [self.bunnyView5 stopAnimating];
```

```
                    [self.toggleButton setTitle:@"跳跃!"
                                forState:UIControlStateNormal];
        } else {
            [self.bunnyView1 startAnimating];
            [self.bunnyView2 startAnimating];
            [self.bunnyView3 startAnimating];
            [self.bunnyView4 startAnimating];
            [self.bunnyView5 startAnimating];
            [self.toggleButton setTitle:@"停下!"
                                forState:UIControlStateNormal];
        }
}
- (IBAction)setSpeed:(id)sender {
    NSString *hopRateString;
    self.bunnyView1.animationDuration=2-self.speedSlider.value;
    self.bunnyView2.animationDuration=
    self.bunnyView1.animationDuration+((float)(rand()%11+1)/10);
    self.bunnyView3.animationDuration=
    self.bunnyView1.animationDuration+((float)(rand()%11+1)/10);
    self.bunnyView4.animationDuration=
    self.bunnyView1.animationDuration+((float)(rand()%11+1)/10);
    self.bunnyView5.animationDuration=
    self.bunnyView1.animationDuration+((float)(rand()%11+1)/10);
    [self.bunnyView1 startAnimating];
    [self.bunnyView2 startAnimating];
    [self.bunnyView3 startAnimating];
    [self.bunnyView4 startAnimating];
    [self.bunnyView5 startAnimating];
    [self.toggleButton setTitle:@"Sit Still!"
                       forState:UIControlStateNormal];
    hopRateString=[[NSString alloc]
               initWithFormat:@"%1.2f hps",1/(2-self.speedSlider.value)];
    self.hopsPerSecond.text=hopRateString;
}
- (IBAction)setIncrement:(id)sender {
    self.speedSlider.value=self.speedStepper.value;
    [self setSpeed:nil];
}
@end
```

执行效果如图 6-12 所示，跳跃后的效果如图 6-13 所示。

图 6-12　初始效果

图 6-13　跳跃后的效果

实例 167　实现全屏显示效果

实例 167	实现全屏显示效果
源码路径	\daima\167\

实例 167　实现全屏显示效果

实例说明

在 iOS 应用中，硬件设备的屏幕大小是一定的，例如，iPhone 的屏幕是 320 像素×480 像素。要想实现全屏显示效果，需要将屏幕中的导航条隐藏起来。在本实例中，将图片"dog.jpg"作为素材文件，然后使用方法 setStatusBarHidden:animated 隐藏了屏幕中的状态条，使用方法 setNavigationBarHidden:animated 隐藏了屏幕中的导航条，使用方法 setToolbarHidden:animated 隐藏了屏幕中的工具条。

具体实现

实例文件 UIKitPrjFullScreen.m 的具体实现代码如下所示。

```objc
#import "UIKitPrjFullScreen.h"
@implementation UIKitPrjFullScreen
- (void)viewDidLoad {
    [super viewDidLoad];
    // 追加图片
    UIImage* image = [UIImage imageNamed:@"dog.jpg"];
    UIImageView* imageView = [[[UIImageView alloc] initWithImage:image] autorelease];
    imageView.frame = self.view.bounds;
    imageView.autoresizingMask = UIViewAutoresizingFlexibleWidth | UIViewAutoresizingFlexibleHeight;
    imageView.contentMode = UIViewContentModeScaleAspectFill;
    [self.view addSubview:imageView];
}
- (void)viewWillAppear:(BOOL)animated {
    [super viewWillAppear:animated];
    fullScreen_ = NO;
    [self.navigationController setNavigationBarHidden:NO animated:NO];
    [self.navigationController setToolbarHidden:NO animated:NO];
}
#pragma mark ----- Responder -----
- (void)touchesEnded:(NSSet*)touches withEvent:(UIEvent*)event {
    fullScreen_ = !fullScreen_;
    // 状态栏→导航栏的顺序
    [[UIApplication sharedApplication] setStatusBarHidden:fullScreen_ animated:YES];
    [self.navigationController setNavigationBarHidden:fullScreen_ animated:YES];
    [self.navigationController setToolbarHidden:fullScreen_ animated:YES];
    CGRect applicationFrame = [[UIScreen mainScreen] applicationFrame];
    CGRect bounds = [[UIScreen mainScreen] bounds];
    NSLog( @"applicationFrame( %f, %f, %f, %f )", applicationFrame.origin.x, applicationFrame.origin.y, applicationFrame.size.width, applicationFrame.size.height );
    NSLog( @"bounds( %f, %f, %f, %f )", bounds.origin.x, bounds.origin.y, bounds.size.width, bounds.size.height );
}
@end
```

执行后的效果如图 6-14 所示。

图 6-14　执行后的效果

实例 168　实现渐变样式的全屏效果切换

实例 168	实现渐变样式的全屏效果切换
源码路径	\daima\167\

实例说明

在 iOS 应用中，当实现全屏显示效果时，可以实现渐变半透明的切换效果。在本实例中，首先将图片"dog.jpg"作为素材文件，然后将属性 statusBarStyle 设置为 UIStatusBarStyleBlackTranslucent，使状态条半透明化。接着将 navigationController.barStyle 设置为 UIBarStyleBlack，将 navigation-Controller 的 translucent 设置为 YES，这样便分别实现了状态条/导航条/工具条的透明效果；最后，设置 wantsFullScreenLayout 为 YES。

具体实现

实例文件 UIKitPrjFullScreenWithTransparent.m 的具体实现代码如下所示。

```objc
#import "UIKitPrjFullScreenWithTransparent.h"
@implementation UIKitPrjFullScreenWithTransparent
- (void)viewDidLoad {
  [super viewDidLoad];
  // 图像追加
  UIImage* image = [UIImage imageNamed:@"dog.jpg"];
  UIImageView* imageView = [[[UIImageView alloc] initWithImage:image] autorelease];
  imageView.frame = self.view.bounds;
  imageView.autoresizingMask = UIViewAutoresizingFlexibleWidth | UIViewAutoresizingFlexibleHeight;
  imageView.contentMode = UIViewContentModeScaleAspectFill;
  [self.view addSubview:imageView];
}
- (void)viewWillAppear:(BOOL)animated {
  [super viewWillAppear:animated];
  fullScreen_ = NO;
  [self.navigationController setNavigationBarHidden:NO animated:NO];
  [self.navigationController setToolbarHidden:NO animated:NO];
  // 使状态条/导航条/工具条透明
  [UIApplication sharedApplication].statusBarStyle = UIStatusBarStyleBlackTranslucent;
  self.navigationController.navigationBar.barStyle = UIBarStyleBlack;
  self.navigationController.navigationBar.translucent = YES;
  self.navigationController.toolbar.barStyle = UIBarStyleBlack;
  self.navigationController.toolbar.translucent = YES;
  // 如果不指定此属性，状态条下将不能绘制图像
  self.wantsFullScreenLayout = YES;
}
#pragma mark ----- Responder -----
- (void)touchesEnded:(NSSet*)touches withEvent:(UIEvent*)event {
  fullScreen_ = !fullScreen_;

  BOOL needAnimation = YES;
  if ( needAnimation ) {
    [UIView beginAnimations:nil context:NULL];
    [UIView setAnimationDuration:0.3];
  }
  [[UIApplication sharedApplication] setStatusBarHidden:fullScreen_ animated:needAnimation];
  self.navigationController.navigationBar.alpha = fullScreen_ ? 0.0 : 1.0;
  self.navigationController.toolbar.alpha = fullScreen_ ? 0.0 : 1.0;
  if ( needAnimation ) {
    [UIView commitAnimations];
  }
}
@end
```

执行后的效果如图 6-15 所示。

图 6-15 执行效果

实例 169 设置屏幕中的元素随着设备旋转而自动适应

实例 169	设置屏幕中的元素随着设备旋转而自动适应
源码路径	\daima\167\

实例说明

在 iOS 应用中,当设备的方向旋转时,屏幕中的元素会随之自动适应。在具体实现时,只需设置方法 shouldAutorotateToInterfaceOrientation 返回 YES 即可实现。在本实例中,演示了通过此方法实现自适应的过程。

具体实现

实例文件 UIKitPrjRotate.m 的具体实现代码如下所示。

```
#import "UIKitPrjRotate.h"
@implementation UIKitPrjRotate
- (void)viewDidLoad {
  [super viewDidLoad];
  self.title = @"Rotate";
  self.view.backgroundColor = [UIColor blackColor];
  UIImage* image = [UIImage imageNamed:@"dog.jpg"];
  UIImageView* imageView = [[[UIImageView alloc] initWithImage:image] autorelease];
  imageView.frame = CGRectMake( 30, 0, 240, 240 );
  imageView.contentMode = UIViewContentModeScaleAspectFit;

  [self.view addSubview:imageView];
}
- (void)viewWillAppear:(BOOL)animated {
  [super viewWillAppear:animated];
  [self.navigationController setNavigationBarHidden:NO animated:NO];
  [self.navigationController setToolbarHidden:NO animated:NO];
}
- (BOOL)shouldAutorotateToInterfaceOrientation:(UIInterfaceOrientation)interfaceOrientation {
  return YES;
}
@end
```

执行效果如图 6-16 所示。

图 6-16 执行效果

实例 170 设置界面旋转时自动调整图像尺寸

实例 170	设置界面旋转时自动调整图像尺寸
源码路径	\daima\167\

实例说明

在 iOS 应用中,当设备的方向旋转时,屏幕中的元素会随之自动适应。但是在自适应过程中,因为屏幕宽度和高度的不同,可能会造成尺寸比例不协调的情况发生。此时可以使用自动调整尺寸属性来修饰界面。在本实例中,使用属性 contentMode 设置当 imageView_ 的尺寸发生变化时,其中的图像比例不会改变。使用属性 autoresizingMask 设置随着母体 View 的变化改变图像宽度与高度。

具体实现

实例文件 UIKitPrjRotateAndAutoresizing.m 的具体实现代码如下所示。

```
#import "UIKitPrjRotateAndAutoresizing.h"
@implementation UIKitPrjRotateAndAutoresizing
- (void)dealloc {
  [imageView_ release];
  [super dealloc];
}
#pragma mark ----- Override Methods -----
- (void)viewDidLoad {
  [super viewDidLoad];
  self.title = @"Rotate";
  self.view.backgroundColor = [UIColor blackColor];
  //创建图像的 UIImageView 实例
  UIImage* image = [UIImage imageNamed:@"dog.jpg"];
  imageView_ = [[UIImageView alloc] initWithImage:image];
  //设置此项后,当 imageView_ 的尺寸发生变化时,其中的图像比例不会改变
  imageView_.contentMode = UIViewContentModeScaleAspectFill;
  //将 clipsToBounds 设置成 YES 后,超出 frame 以外的图像不再绘制
  imageView_.clipsToBounds = YES;
  //向 autoresizingMask 设置此两常量后,将随着母体 View 的变化改变图像宽度与高度
  imageView_.autoresizingMask = UIViewAutoresizingFlexibleWidth | UIViewAutoresizingFlexibleHeight;
  [self.view addSubview:imageView_];
}
- (void)viewWillAppear:(BOOL)animated {
  [super viewWillAppear:YES];
  [self.navigationController setNavigationBarHidden:NO animated:NO];
  [self.navigationController setToolbarHidden:NO animated:NO];
}
- (void)viewDidAppear:(BOOL)animated {
  [super viewDidAppear:YES];
  imageView_.frame = CGRectInset( self.view.bounds, 20, 20 );
}
- (BOOL)shouldAutorotateToInterfaceOrientation:(UIInterfaceOrientation)
```

```
interfaceOrientation {
  return YES;
}
@end
```

执行后的效果如图 6-17 所示。

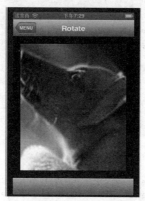

图 6-17 执行效果

实例 171 定制屏幕中的旋转图像

实例 171	定制屏幕中的旋转图像
源码路径	\daima\167\

实例说明

在本实例中，分别初始化了实例变量 imageView1_ 和 imageView2_，在里面放置图像。然后追加了第一个图像"face.jpg"，并通过 autoresizingMask 属性设置所有的自动调整参数，并设置将此图像布置在画面的上半部分。接下来追加了第二个图像"dog.jpg"，并将此图像布置在画面的下半部分。最后通过 willAnimateRotationToInterfaceOrientation 方法定制了画面旋转后的布局，在实例中分别处理了画面朝向为纵向的情况和画面朝向为横向的情况。

具体实现

实例文件 UIKitPrjRotateManual.m 的具体实现代码如下所示。

```
#import "UIKitPrjRotateManual.h"
@implementation UIKitPrjRotateManual
- (void)dealloc {
  [imageView1_ release];
  [imageView2_ release];
  [super dealloc];
}
#pragma mark ----- Override Methods -----
//初始化实例变量 imageView1_以及 imageView2_，其中放置图像
- (void)viewDidLoad {
  [super viewDidLoad];
  self.title = @"Rotate";
  self.view.backgroundColor = [UIColor blackColor];
  //追加第一个图像
  UIImage* image1 = [UIImage imageNamed:@"face.jpg"];
  imageView1_ = [[UIImageView alloc] initWithImage:image1];
  imageView1_.contentMode = UIViewContentModeScaleAspectFill;
  imageView1_.clipsToBounds = YES;
  //在 autoresizingMask 属性中设置所有的自动调整参数
  imageView1_.autoresizingMask = UIViewAutoresizingFlexibleWidth | UIViewAutoresizingFlexibleHeight
               | UIViewAutoresizingFlexibleLeftMargin | UIView
```

```objc
AutoresizingFlexibleRightMargin
                                | UIViewAutoresizingFlexibleTopMargin | UIView
AutoresizingFlexibleBottomMargin;
    CGSize frameSize = self.view.frame.size;
    //将图像布置在画面的上半部分
    imageView1_.frame = CGRectMake( 0, 0, frameSize.width, frameSize.height / 2 );
    [self.view addSubview:imageView1_];
    //追加第二个图像
    UIImage* image2 = [UIImage imageNamed:@"dog.jpg"];
    imageView2_ = [[UIImageView alloc] initWithImage:image2];
    imageView2_.contentMode = UIViewContentModeScaleAspectFill;
    imageView2_.clipsToBounds = YES;
    imageView2_.autoresizingMask = imageView1_.autoresizingMask;
    //将图像布置在画面下半部分
    imageView2_.frame = CGRectMake( 0, frameSize.height / 2, frameSize.width, frameSize.height
/ 2 );
    [self.view addSubview:imageView2_];
}
- (void)viewWillAppear:(BOOL)animated {
    [super viewWillAppear:YES];
    [self.navigationController setNavigationBarHidden:NO animated:NO];
    [self.navigationController setToolbarHidden:NO animated:NO];
}
- (BOOL)shouldAutorotateToInterfaceOrientation:(UIInterfaceOrientation)
interfaceOrientation {
    return YES;
}
- (void)willRotateToInterfaceOrientation:(UIInterfaceOrientation)toInterfaceOrientation
    duration:(NSTimeInterval)duration
{
    NSLog( @"willRotateToInterfaceOrientation" );
}
//定制画面旋转后的布局
- (void)willAnimateRotationToInterfaceOrientation:(UIInterfaceOrientation)
interfaceOrientation
    duration:(NSTimeInterval)duration
{
    CGSize frameSize = self.view.frame.size;
    //画面旋转后根据画面的朝向进行布局定制
    switch ( interfaceOrientation ) {
        case UIInterfaceOrientationPortrait:
        case UIInterfaceOrientationPortraitUpsideDown:
            //画面朝向为纵向的情况下
            //将第一个图像布置在画面上方
            imageView1_.frame = CGRectMake( 0, 0, frameSize.width, frameSize.height / 2 );
            //将第二个图像布置在画面下方
            imageView2_.frame = CGRectMake( 0, frameSize.height / 2, frameSize.width, frameSize.
height / 2 );
            break;
        default:
            //画面朝向为横向的情况下
            //将第一个图像布置在画面左半部分
            imageView1_.frame = CGRectMake( 0, 0, frameSize.width / 2, frameSize.height );
            //将第二个图像布置在画面右半部分
            imageView2_.frame = CGRectMake( frameSize.width / 2, 0, frameSize.width / 2,
frameSize.height );
            break;
    }
}
- (void)didRotateFromInterfaceOrientation:(UIInterfaceOrientation)fromInterfaceOrienta-
tion {
    NSLog( @"didRotateFromInterfaceOrientation" );
}
@end
```

执行效果如图 6-18 所示。

图 6-18 执行效果

实例 172 同时实现屏幕自适应功能和全屏功能

实例 172	同时实现屏幕自适应功能和全屏功能
源码路径	\daima\167\

实例说明

本实例是对前面几个实例功能的融合,同时实现了屏幕自适应功能和全屏功能。在本实例中设置了动画标志,当动画标志为 YES 时开始定义动画,并且设置了动画的持续时间,将动画持续时间设置成与状态条消失时间相同,并且还设置了状态条、导航条、工具条的透明效果。

具体实现

实例文件 UIKitPrjRotateAndFullScreen.m 的具体实现代码如下所示。

```
#import "UIKitPrjRotateAndFullScreen.h"
@implementation UIKitPrjRotateAndFullScreen
- (void)viewDidLoad {
  [super viewDidLoad];
  // 图像追加
  UIImage* image = [UIImage imageNamed:@"dog.jpg"];
  UIImageView* imageView = [[[UIImageView alloc] initWithImage:image] autorelease];
  imageView.frame = self.view.bounds;
  imageView.autoresizingMask = UIViewAutoresizingFlexibleWidth | UIViewAutoresizingFlexibleHeight;
  imageView.contentMode = UIViewContentModeScaleAspectFill;
  [self.view addSubview:imageView];
}
- (void)viewWillAppear:(BOOL)animated {
  [super viewWillAppear:animated];
  //实例变量,记录显示是否为全屏显示
  //此实例变量在@interface 中定义
  fullScreen_ = NO;
  //显示导航条、工具条
  [self.navigationController setNavigationBarHidden:NO animated:NO];
  [self.navigationController setToolbarHidden:NO animated:NO];
  // 使状态条/导航条/工具条透明
  [UIApplication sharedApplication].statusBarStyle = UIStatusBarStyleBlackTranslucent;
  self.navigationController.navigationBar.barStyle = UIBarStyleBlack;
  self.navigationController.navigationBar.translucent = YES;
  self.navigationController.toolbar.barStyle = UIBarStyleBlack;
  self.navigationController.toolbar.translucent = YES;
  // 指定此属性,使状态条下也可绘制图像
  self.wantsFullScreenLayout = YES;
}
//重写 UIViewController 中此方法,让其返回 YES 即可实现画面旋转
- (BOOL)shouldAutorotateToInterfaceOrientation:(UIInterfaceOrientation)interfaceOrientation {
```

第 6 章 图形、图像和动画实战

```
    return YES;
}
#pragma mark ----- Responder -----
- (void)touchesEnded:(NSSet*)touches withEvent:(UIEvent*)event {
    //全屏显示与非全屏显示间的切换
    fullScreen_ = !fullScreen_;
    //设置动画标志
    BOOL needAnimation = YES;
    //动画标志为 YES 时，动画定义开始
    if ( needAnimation ) {
        //动画定义开始
        [UIView beginAnimations:nil context:NULL];
        //设置动画持续时间，将动画持续时间设置成与状态条消失时间相同
        [UIView setAnimationDuration:0.3];
    }
    //隐藏状态条（状态条逐渐消失）
    [[UIApplication sharedApplication] setStatusBarHidden:fullScreen_ animated:needAnimation];
    // API 参考中不推荐直接改变 navigationBar 的 alpha 值
    self.navigationController.navigationBar.alpha = fullScreen_ ? 0.0 : 1.0;
    self.navigationController.toolbar.alpha = fullScreen_ ? 0.0 : 1.0;
    //动画结束
    if ( needAnimation ) {
        [UIView commitAnimations];
    }
    if ( !fullScreen_ ) {
        // 若不进行如下设置，当全屏状态下旋转后再解除全屏状态时导航条的位置将偏移
        [self.navigationController setNavigationBarHidden:YES animated:NO];
        [self.navigationController setToolbarHidden:YES animated:NO];
        [self.navigationController setNavigationBarHidden:NO animated:NO];
        [self.navigationController setToolbarHidden:NO animated:NO];
    }
}
@end
```

执行效果如图 6-19 所示。

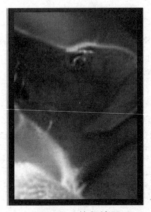

图 6-19　执行效果

实例 173　创建可旋转和调整大小的界面

实例 173	创建可旋转和调整大小的界面
源码路径	\daima\173\

实例说明

在本实例中，将使用 Interface Builder 内置的工具来指定视图如何适应旋转。因为本实例完全依赖于 Interface Builder 工具来支持界面旋转和大小调整，所以几乎所有的功能都是在 Size Inspector 中使用自动调整大小和锚工具完成的。本实例将使用一个标签（UILabel）和几个按钮（UIButton），可以将它们换成其他界面元素，读者将发现旋转和大小调整处理适用于整个 iOS 对象库。

实例 173 创建可旋转和调整大小的界面

具体实现

（1）首先启动 Xcode，并使用 Apple 模板 Single View Application 新建一个名为"xuanzhuan"的项目，如图 6-20 所示。

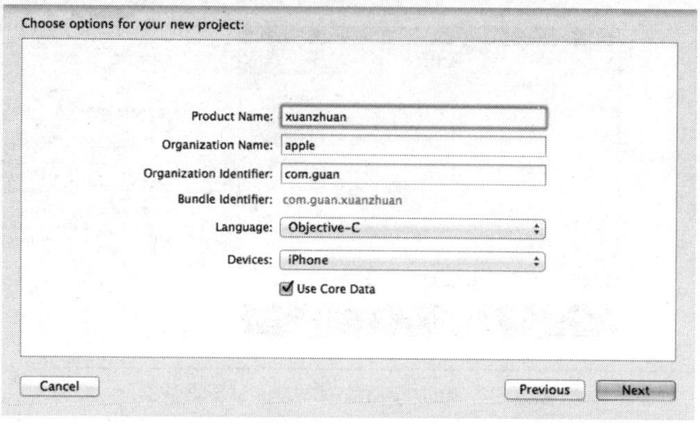

图 6-20 创建工程

虽然所有与 UI 相关的工作都将在 Interface Builder 中完成，但还需确保方法 shouldAutorotateToInterfaceOrientation 对要支持的所有朝向都返回 true。

（2）启用旋转。

打开视图控制器的实现文件 ViewController.m，并找到方法 shouldAutorotateToInterface-Orientation。在该方法中返回 YES，以支持所有的 iOS 屏幕朝向，具体代码如下所示。

```
-(BOOL) shouldAutorotateToInterfaceOrientation:
    (UlInterfaceOrientation) interfaceOrientation
    {
    return YES;
}
```

（3）设计灵活的界面。

在创建可旋转和调整大小的界面时，开头与创建其他 iOS 界面一样，只需拖放即可。依次选择菜单 View→Utilities→Show Object Library，打开对象库，拖曳一个标签（UILabel）和 4 个按钮（UIButton）到视图 SimpleSpin 中。将标签放在视图顶端居中，并将其标题改为 SimpleSpin。按如下方式给按钮命名以便能够区分它们——"摸我 1"、"摸我 2"、"摸我 3"和"摸我 4"，并将它们放在标签下方，如图 6-21 所示。创建可旋转的应用程序界面与创建其他应用程序界面的方法相同。

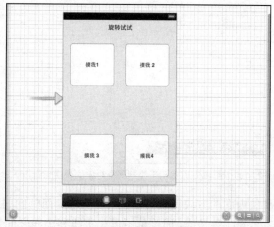

图 6-21 创建可旋转的应用程序界面

为了查看旋转后该界面是什么样的，可模拟横向效果。为此在文档大纲中选择视图控制器，再打开 Attributes Inspector（Option+ Command+4）；在 Simulated Metrics 部分，将 Orientation 的设置改为 Landscape，Interface Builder 编辑器将相应地调整，如图 6-22 所示。查看完毕后，务必将朝向改回到 Portrait 或 Inferred。

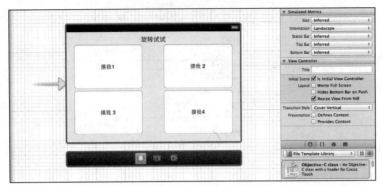

图 6-22　修改模拟的朝向以测试界面旋转

此时旋转后的视图不太正确，其原因是加入到视图中的对象默认锚定其左上角。这说明无论屏幕的朝向如何，对象左上角相对于视图左上角的距离都保持不变。另外在默认情况下，对象不能在视图中调整大小。因此，无论是在纵向还是横向模式下，所有元素的大小都保持不变，哪怕它们不适合视图。

为了修复这种问题并创建出与 iOS 设备相称的界面，需要使用 Size Inspector（大小检查器）。

（4）Size Inspector 中的 Autosizing。

在配置应用程序的外观和功能方面，Attributes Inspector 和 Connections Inspector 很有用；在另一方面，Size Inspector（Option+ Command+5）在很大程度上说还是"旁观者"。到目前为止，我们只是偶尔使用它来设置控件的坐标，而从未使用它来启用功能。

自动旋转和自动调整大小完全是通过 Size Inspector 中的 Autosizing 设置控制的，如图 6-23 所示。这个颇具欺骗性的"方块中的方块"界面让我们能够告诉 Interface Builder 要如何锚定控件以及控件可在哪些方向（水平或垂直）上调整大小。

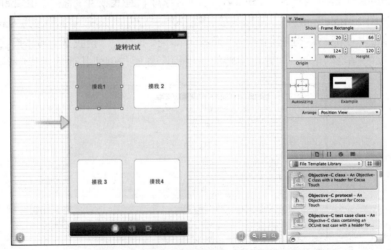

图 6-23　Autosizing 设置用于控制屏幕对象的属性 anchor 和 size

为理解其工作原理，想象内面的方块代表当前的界面元素，而外面的方块代表包含该元素的视图。内面方块和外面方块之间的线条是锚，通过单击可在实线和虚线之间切换。实线表示设置了锚，意味着当界面旋转时这些距离将保持不变。

在内面的方块中有两个双向箭头，它们代表水平和垂直大小调整，单击这些箭头将在虚线和实线之间切换，实线箭头意味着控件可在相应的方向上调整大小。在默认情况下，对象的左上角被锚定且不能调整大小，图6-23显示了这种配置。

（5）指定界面的Autosizing设置。

为了使用合适的Autosizmg属性来修改simplespin界面，需要选择每个界面元素，按快捷键"option+Command+5"打开size Inspector，再按下面的描述配置其锚定和大小调整属性。

- 标签SimpleSpin：这个标签应显示在视图顶端并居中，因此其上边缘与视图上边缘的距离应保持不变，大小也应保持不变（Anchor设置为Top，Resizing设置为None）。
- 点我1：该按钮的左边缘与视图左边缘的距离应保持不变，但应让它在需要时上下浮动。它应能够水平调整大小以填满更大的水平空间（Anchor设置为Left，Resizing设置为Horizontal）。
- 点我2：该按钮右边缘与视图右边缘之间的距离应保持不变，但应允许它在需要时上下浮动。它应能够水平调整大小以填满更大的水平空间（Anchor设置为Right，Resizing设置为Horizontal）。
- 点我3：该按钮左边缘与视图左边缘之间的距离应保持不变，其下边缘与视图下边缘之间的距离也应如此。它应能够水平调整大小以填满更大的水平空间，Anchor设置为Left和Bottom，Resizing设置为Horizontal。
- 点我4：该按钮右边缘与视图右边缘之间的距离应保持不变，其下边缘与视图下边缘之间的距离也应如此。它应能够水平调整大小以填满更大的水平空间（Anchor设置为Right和Bottom，Resizing设置为Horizontal）。

此时运行该应用程序（或模拟横向模式）并预览结果，随着设备的移动，界面元素将自动调整大小，如图6-24所示。

图6-24 执行效果

实例174　屏幕旋转时调整控件的框架

实例174	屏幕旋转时调整控件的框架
源码路径	\daima\174\

实例说明

在iOS应用中，有时使用Interface Builder难以满足现实项目的需求，如果界面包含间距不规则的控件且布局紧密，将难以按我们预期的方式显示。另外，我们还可能想在不同朝向下调整界面，使其看起来截然不同。例如将原本位于视图顶端的对象放到视图底部。在这两种情况下，我们可能想调整控件的框架以适合旋转后的iOS设备屏幕。

本实例演示了旋转时调整控件的框架的方法，整个实现逻辑很简单：当设备旋转时判断它将旋转到哪个朝向，然后设置每个要调整其位置或大小的 UI 元素的 frame 属性。本实例将创建两次界面，在 Interface Builder 编辑器中创建该界面的第一个版本后，我们将使用 Size Inspector 获取其中每个元素的位置和大小，这些值将用于设置纵向模式下界面元素的框架。然后旋转该界面，调整所有控件的大小和位置，使其适合新朝向，并再次收集所有的框架值。最后，我们实现一个方法，它在设备朝向发生变化时自动设置每个控件的框架值。

具体实现

（1）规划变量和连接。

在本演示实例中，将手工调整 3 个 UI 元素的大小和位置：两个按钮（UIButton）和一个标签（UILabel）。将以编程方式访问它们，因此首先需要编辑头文件和实现文件，在其中包含对应于每个 UI 元素的输出口：buttonOne、buttonTwo 和 viewLabel。我们需要实现一个方法，但它不是由 UI 触发的操作。我们将编写 willRotateToInterfaceOrientation: toInterfaceOrientation:duration:的实现，每当界面需要旋转时都将自动调用它。

（2）启用旋转。

即使不利用 Interface Builder 的自动调整"大小/自动"旋转功能，也必须在方法 shouldAutorotateToInterfaceOrientation:中启用旋转。所以需要修改文件 ViewController.m，使其包含在本章上一个示例中添加的实现，具体代码如下所示。

```
- (BOOL)shouldAutorotateToInterfaceOrientation:(UIInterfaceOrientation)
interfaceOrientation
{
    // Return YES for supported orientations
    return YES;
}
```

（3）设计界面。

接下来开始调整本项目的框架，跟踪界面元素的坐标和大小。虽然在 Interface Builder 编辑器中设计界面，但是必须记录每个界面元素的位置，这是因为每当屏幕旋转时，都需要重新指定所有界面元素的位置。没有恢复到默认位置的方法，因此即使对于我们创建的初始布局，也必须使用 x 和 y 坐标以及大小来指定，以便需要时能够恢复到默认布局。

单击文件 MainStoryboard.storyboard 开始设计视图，具体流程如下所示。

第 1 步：禁用自动调整大小。

首先，单击以选择视图，并按 Option+ Command+4 打开 Attributes Inspector。在 View 部分取消选中复选框 Autoresize Subviews，如图 6-25 所示。

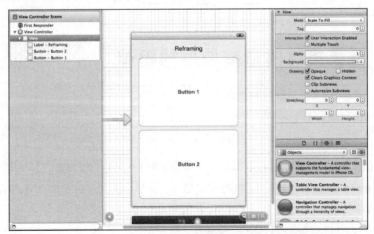

图 6-25　手工调整控件的大小和位置时应禁用自动调整大小

如果没有禁用视图的自动调整大小功能，则应用程序代码调整 UI 元素的大小和位置的同时，iOS 也将尝试这样做，但是结果可能极其混乱。

第 2 步：第一次设计视图。

接下来需要像创建其他应用程序一样设计视图，在对象库中单击并拖曳这些元素到视图中。将标签的文本设置为"改变框架"，并将其放在视图顶端；将按钮的标题分别设置为"点我 1"和"点我 2"，并将它们放在标签下方。最终的布局应该如图 6-26 所示。

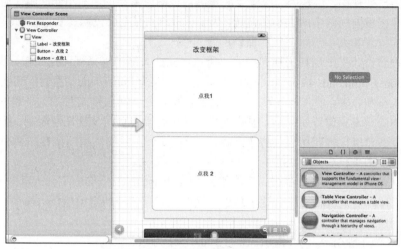

图 6-26　像创建其他应用程序一样设计视图

在获得所需的布局后，通过 Size Inspector 获悉每个 UI 元素的 frame 属性值。首先选择标签，并按 Option+ Command+5 打开 Size Inspector。单击 Origin 方块左上角，将其设置为度量坐标的原点。然后确保在下拉列表 Show 中选择了 Frame Rectangle，如图 6-27 所示。

然后将该标签的 X、Y、W（宽度）和 H（高度）属性值记录下来，它们表示视图中对象的 frame 属性。对两个按钮重复上述过程。对于每个 u 元素，都将获得 4 个值。在此列出我们使用的框架值（包括 iPhone 项目和 iPad 项目）供读者参考。

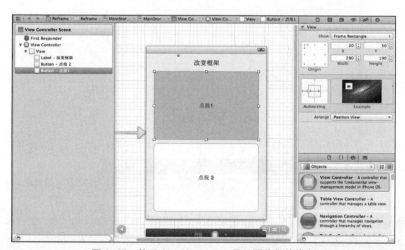

图 6-27　使用 Size Inspector 显示要收集的信息

iPhone 项目中的框架值如下所示。

- 标签：X 为 95.0，Y 为 15.0，W 为 130.0，H 为 20.0。
- 点我 1：X 为 20.0，Y 为 50.0，W 为 280.0，H 为 190.0。

- 点我 2：X 为 20.0，Y 为 250.0，W 为 280.0，H 为 190.0。

iPad 项目中的框架值如下所示。

- 标签：X 为 275.0，Y 为 20.0，W 为 225.0，H 为 60.0。
- 点我 1：X 为 20.0，Y 为 168.0，W 为 728.0，H 为 400.0。
- 点我 2：X 为 20.0，Y 为 584.0，W 为 728.0，H 为 400.0。

第 3 步：重新排列视图。

接下来重新排列视图，这是因为收集了配置纵向视图所需要的所有 frame 属性值，但是还没有定义标签和按钮在横向视图中的大小和位置。为了获取这些信息，我们需要以横向模式重新排列视图，收集所有的位置和大小信息，然后撤销所做的修改。

接下来必须将设计视图切换为横向模式。所以在文档大纲中选择视图控制器，再在 Attributes Inspector（Option+Command+4）中将 Orientation 的设置改为 Landscape。当切换到横向模式后，调整所有元素的大小和位置，使其与开发者希望它们在设备处于横向模式时的大小和位置相同。由于我们将以编程方式来设置位置和大小，因此对开发者如何排列它们没有任何限制。我们将"点我 1"放在顶端，并使其宽度比视图稍小；将"点我 2"放在底部，并使其宽度比视图稍小；将标签"改变框架"放在视图中央，如图 6-28 所示。

与前面一样，获得所需的视图布局后，使用 Size Inspector (Option+Command+5)收集每个 UI 元素的 x 和 y 坐标以及宽度和高度。这里列出在横向模式下使用的框架值供大家参考。

iPhone 项目中的框架值如下所示。

- 标签：X 为 175.0，Y 为 140.0，W 为 130.0，H 为 20.0。
- 点我 1：X 为 20.0，Y 为 20.0，W 为 440.0，H 为 100.0。
- 点我 2：x 为 20.0，Y 为 180.0，w 为 440.0，H 为 100.0。

iPad 项目中的框架值如下所示。

- 标签：X 为 400.0，Y 为 340.0，W 为 225.0，H 为 60.0。
- 点我 1：X 为 20.0，Y 为 20.0，W 为 983.0，H 为 185.0。
- 点我 2：X 为 20.0，Y 为 543.0，W 为 983.0，H 为 185.0。

图 6-28 排列

收集横向模式下的 frame 属性值后，撤销对视图所做的修改。为此，可不断选择菜单 Edit→Undo（Command+Z），直到恢复到为纵向模式设计的界面。保存文件 MainStoryboard.storyboard。

（4）创建并连接输出口。

在编写调整框架的代码前，还需将标签和按钮连接到我们在这个项目开头规划的输出口。为此切换到助手编辑器模式，再按住 Control 键，从每个 UI 元素拖曳到接口文件 ViewController.h，并正确地命名输出口（viewLabel、buttonOne 和 buttonTwo）。下面的图 6-29 显示了从"改变框架"

标签到输出口 viewLabel 的连接。

（5）实现应用程序逻辑。

创建视图并记录纵向视图和横向视图中标签和按钮框架的值后，剩下的唯一工作就是检测 iOS 设备即将旋转并相应地调整框架。

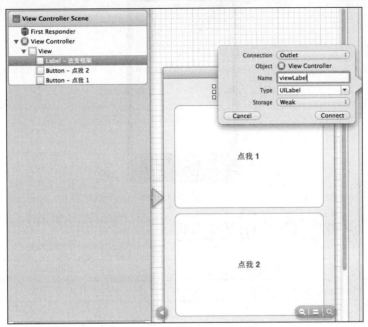

图 6-29　创建与标签和按钮相关联的输出口

每当需要旋转 iOS 界面时，都将自动调用方法 willRotateToInterfaceOrientation:toInterfaceOrientation:duration:。我们将把参数 toInterfaceOrientation 同各种 iOS 朝向常量进行比较，以确定应使用横向还是纵向视图的框架值。

在 Xcode 中打开文件 ViewController.m，并添加如下所示的代码。

```
-(void)willRotateToInterfaceOrientation:
    (UIInterfaceOrientation)toInterfaceOrientation
    duration:(NSTimeInterval)duration {

    [super willRotateToInterfaceOrientation:toInterfaceOrientation
                                   duration:duration];

    if (toInterfaceOrientation == UIInterfaceOrientationLandscapeRight ||
        toInterfaceOrientation == UIInterfaceOrientationLandscapeLeft) {
        self.viewLabel.frame=CGRectMake(175.0,140.0,130.0,20.0);
        self.buttonOne.frame=CGRectMake(20.0,20.0,440.0,100.0);
        self.buttonTwo.frame=CGRectMake(20.0,180.0,440.0,100.0);
    } else {
        self.viewLabel.frame=CGRectMake(95.0,15.0,130.0,20.0);
        self.buttonOne.frame=CGRectMake(20.0,50.0,280.0,190.0);
        self.buttonTwo.frame=CGRectMake(20.0,250.0,280.0,190.0);
    }
}
```

上述代码的实现逻辑很简单。首先需要通知父对象：视图要旋转了，向父对象 super 发送消息 willRotateToInterfaceOrientation:toInterfaceOrientation:duration:。将传入的参数 toInterfaceOrientation 同横向模式常量进行比较。如果与其中一种模式匹配，则将 frame 属性设置为函数 CGRectMake() 返回的结果，从标签和 Builder 中收集到 X、Y、W 和 H 值。如果创建的是 iPad 应用程序，并决定使用这些值，则需使用这些值替换程序清单上述代码中的 iPhone 值。最后处理了"纵向模式"朝向，如果设备没有旋转到任何一种横向模式，则肯定被旋转到纵向模式。同样，我们使用在 Interface

Builder 的 Size Inspector 中收集到的值来设置 frame 属性。

到此为止，整个实例介绍完毕，运行并旋转 iOS 模拟器，这样在用户旋转设备时会自动重新排列界面了。执行效果如图 6-30 所示。

图 6-30　执行效果

实例 175　屏幕旋转时切换视图

实例 175	屏幕旋转时切换视图
源码路径	\daima\175\

实例说明

在 iOS 项目应用中，有一些应用程序可以根据设备的朝向显示完全不同的用户界面。例如，iPhone 应用程序 Music 在纵向模式下显示一个可滚动的歌曲列表，而在横向模式下显示一个可快速滑动的 CoverFlow 式专辑视图。通过在手机旋转时切换视图，可以创建外观剧烈变化的应用程序。通过本实例，演示了在 Interface Builder 编辑器中灵活地管理横向和纵向视图的知识。

具体实现

（1）规划变量和连接。

这个应用程序不会提供任何真正的用户界面元素，但我们需要以编程方式访问两个 UIView 实例，其中一个视图用于纵向模式（portraitView），另一个用于横向模式（landscapeView）。与本章上一个项目一样，也是实现一个方法，但它不由任何界面元素触发。

（2）添加一个常量用于表示度到弧度的转换系数。

在这个练习后面，我们需要调用一个特殊的 Core Graphics 方法来指定如何旋转视图。在调用这个方法时，需要传入一个以弧度而不是度为单位的参数。也就是说，不需要将视图旋转 90 度，而必须告诉它要旋转 1.57 弧度。为了实现这种转换，需要定义一个表示转换系数的常量，将度数与该常量相乘得到弧度数。为了定义该常量，在文件 ViewController.m 中将下面的代码行添加到 #import 代码行的后面。

#define kDeg2Rad (3.1415926/180.0)

（3）启用旋转。

与本章的前两个实例一样，需要确保视图控制器的 shouldAutorotateToInterfaceOrientation: 的行为与期望的一致。不同于前面两个实现，这次将只允许在两个横向模式和非倒转纵向模式之间旋转。修改文件 ViewController.m，在其中包含如下所示的代码。

```
-
(BOOL)shouldAutorotateToInterfaceOrientation:(UIInterfaceOrientation)
```

```
interfaceOrientation
{
    return (interfaceOrientation != UIInterfaceOrientationPortraitUpsideDown);
}
```

其实可以将参数 interfaceOrientation 同 UIInterfaceOrientationPortrait、UIInterfaceOrientationLandscapeRight 和 UIInterfaceOrientationLandscapeLeft 进行比较，但这与此处的"非倒转纵向模式"的含义相同。

（4）设计界面。

采用切换视图的方式时，对视图的设计没有任何限制，可像在其他应用程序中一样创建视图。唯一的不同是，如果有多个由同一个视图控制器处理的视图，将需要定义针对所有界面元素的输出口。

打开文件 MainStoryboard.storyboard，从对象库中拖曳一个 UIView 实例到文档大纲中，并将它放在与视图控制器同一级的地方，而不要将其放在现有视图中，如图 6-31 所示。

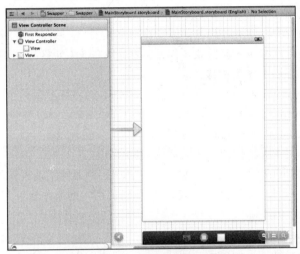

图 6-31　在场景中再添加一个视图

然后打开默认视图并在其中添加一个标签，然后设置背景色，以方便区分视图。这就完成了一个视图的设计，还需要设计另一个视图。但是在 Interface Builder 中，只能编辑被分配给视图控制器的视图。

在文档大纲中，将刚创建的视图拖出视图控制器的层次结构，将其放到与视图控制器同一级的地方。在文档大纲中，将第二个视图拖曳到视图控制器上。这样就可编辑该视图了，并且指定了独特的背景色，并添加了一个标签（如 Landscape View）。设计好第二个视图后，重新调整视图层次结构，将纵向视图嵌套在视图控制器中，并将横向视图放在与视图控制器同一级的地方。

下面的图 6-32 显示了最终的横向视图和纵向视图。

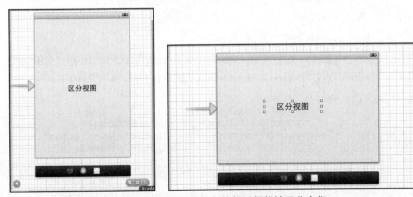

图 6-32　对两个视图进行编辑以便能够区分它们

（5）创建并连接输出口。

为完成界面方面的工作，需要将两个视图连接到两个输出口。默认视图（嵌套在视图控制器中的视图）将连接到 portraitView，而第二个视图将连接到 landscpaeView。切换到助手编辑器模式，并确保文档大纲可见。

由于我们要连接的是视图而不是界面元素，因此建立这些连接的最简单方式是，按住 Control 键，并从文档大纲中的视图拖曳到文件 ViewController.h。

按住 Control 键，并从默认（嵌套）视图拖曳到 ViewController.h 中代码行@interface 下方。为该视图创建一个名为 portraitView 的输出口，对第二个视图重复上述操作，并将输出口命名为 landscapeView。

（6）实现应用程序逻辑。

在很大程度上说，切换视图实际上比前一个项目中实现的框架调整逻辑更容易，但有一点例外：虽然将把其中一个视图用作横向视图，但它并不知道这一点。

要成功地显示横向视图，必须对其进行旋转并指定其大小。原因是视图没有内置的逻辑指出它是横向视图；它只知道自己将在纵向模式下显示，但包含的 UI 元素超出了屏幕边缘。这样当每次改变朝向时，都需要执行以下 3 个步骤。

- 切换视图。
- 通过属性 transform 将视图旋转到合适的朝向。
- 通过属性 bounds 设置视图的原点和大小。

例如，假设要旋转到主屏幕按钮位于右边的横向模式。

第 1 步：首先需要切换视图。为此可以将表示视图控制器的当前视图的属性 self.view 设置为实例变量 landscapeVIew。如果仅这样做，视图将正确切换，但不会旋转到横向模式。以纵向方式显示横向视图很不美观，例如：

```
self.view=self.landscapeView;
```

第 2 步：为了处理旋转，需要设置视图的 transform 属性。该属性决定了在显示视图前应该如何变换它。为了满足这里的需求，必须将视图旋转 90 度（对于主屏幕按钮在右边的横向模式）、旋转-90 度（对于主屏幕按钮位于左边的横向模式）和 0 度（对于纵向模式）。所幸的是为了处理旋转，Core Graphics 的 C 语言函数 CGAffineTransformMakeRotation()接受一个以弧度为单位的角度，并向 transform 属性提供一个合适的结构，例如：

```
self.view.transform=CGAffineTransformMakeRotation (deg2rad *(90));
```

第 3 步：最后一步是设置视图的属性 bounds。bounds 指定了视图变换后的原点和大小。iPhone 纵向视图的原点坐标为（0,0），而宽度和高度分别是 320.0 和 460.0（iPad 为 768.0 和 1004.0）；横向视图的原点坐标也是（0,0），但是宽度和高度分别为 480.0 和 300.0（iPad 为 1024 和 748.0）。与属性 frame 一样，也使用 CGRectMake()的结果来设置 bounds 属性，例如：

```
self.view.bounds=CGRectMake (0.0,0.0,480.0,320.0);
```

这样了解所需的步骤后，接下来开始看具体的实现。

本实例的所有核心功能都是在方法 willRotateToInterfaceOrientation: toInterfaceOrientation: duration:中实现的。打开实现文件 ViewController.m，此方法的具体代码如下所示。

```
-(void)willRotateToInterfaceOrientation:
(UIInterfaceOrientation)toInterfaceOrientation
                    duration:(NSTimeInterval)duration {
    [super willRotateToInterfaceOrientation:toInterfaceOrientation
                          duration:duration];

    if (toInterfaceOrientation == UIInterfaceOrientationLandscapeRight) {
        self.view=self.landscapeView;
        self.view.transform=CGAffineTransformMakeRotation(kDeg2Rad*(90));
        self.view.bounds=CGRectMake(0.0,0.0,480.0,300.0);
```

```
    } else if (toInterfaceOrientation == UIInterfaceOrientationLandscapeLeft) {
        self.view=self.landscapeView;
        self.view.transform=CGAffineTransformMakeRotation(kDeg2Rad*(-90));
        self.view.bounds=CGRectMake(0.0,0.0,480.0,300.0);
    } else {
        self.view=self.portraitView;
        self.view.transform=CGAffineTransformMakeRotation(0);
        self.view.bounds=CGRectMake(0.0,0.0,320.0,460.0);
    }
}
```

在上述代码中，第 5～第 6 行将界面旋转消息发送给父对象，让其做出合适的反应。第 8～第 11 行处理向右旋转（主屏幕按钮位于右边的横向模式），第 12～第 15 行处理向左旋转（主屏幕按钮位于左边的横向模式）。最后，第 16～第 19 行将视图配置为默认朝向——纵向。

到此为止，整个实例介绍完毕，执行后的效果如图 6-33 所示。

图 6-33　执行效果

实例 176　实现一个图片浏览工具

实例 176	实现一个图片浏览工具
源码路径	\daima\176\

实例说明

本实例实现了一个简单的图片浏览工具功能，预先设置了如图 6-34 所示的素材图片。

图 6-34　素材图片

在本实例中，在文件 **DayViewController.h** 中建立了 UIScrollView 对象 scrollView1 和 scrollView2。这样在屏幕中实现了两个浏览界面，其中屏幕上方可以滑动浏览多幅图片，下方只显示一幅图片。

具体实现

实例文件 **DayAppDelegate.h** 的具体实现代码如下所示。

```
#import <UIKit/UIKit.h>
```

```objc
@class DayViewController;
@interface DayAppDelegate : NSObject <UIApplicationDelegate> {
    UIWindow *window;
    DayViewController *viewController;
}
@property (nonatomic, retain) IBOutlet UIWindow *window;
@property (nonatomic, retain) IBOutlet DayViewController *viewController;

@end
```

实例文件 DayViewController.h 的具体实现代码如下所示。

```objc
#import <UIKit/UIKit.h>
#import <Foundation/Foundation.h>
@interface DayViewController : UIViewController {
IBOutlet UIScrollView *scrollView1;
IBOutlet UIScrollView *scrollView2;
}
@property (nonatomic, retain) UIScrollView *scrollView1;
@property (nonatomic, retain) UIScrollView *scrollView2;
@end
```

实例文件 DayViewController.m 的具体实现代码如下所示。

```objc
#import "DayViewController.h"
@implementation DayViewController
@synthesize scrollView1, scrollView2;
const CGFloat kScrollObjHeight  = 175.0;
const CGFloat kScrollObjWidth   = 280.0;
const NSUInteger kNumImages     = 5;
- (void)layoutScrollImages
{
    UIImageView *view = nil;
    NSArray *subviews = [scrollView1 subviews];
    CGFloat curXLoc = 0;
    for (view in subviews)
    {
        if ([view isKindOfClass:[UIImageView class]] && view.tag > 0)
        {
            CGRect frame = view.frame;
            frame.origin = CGPointMake(curXLoc, 0);
            view.frame = frame;

            curXLoc += (kScrollObjWidth);
        }
    }
    [scrollView1 setContentSize:CGSizeMake((kNumImages * kScrollObjWidth), [scrollView1 bounds].size.height)];
}
- (void)viewDidLoad {
  [super viewDidLoad];
    self.view.backgroundColor = [UIColor viewFlipsideBackgroundColor];
    NSUInteger i;
    for (i = 1; i <= kNumImages; i++){
        NSString *imageName = [NSString stringWithFormat:@"image0%d.jpg", i];

        UIImage *image = [UIImage imageNamed:imageName];
        UIImageView *imageView = [[UIImageView alloc] initWithImage:image];
        CGRect rect = imageView.frame;
        rect.size.height = kScrollObjHeight;
        rect.size.width = kScrollObjWidth;
        imageView.frame = rect;
        imageView.tag = i;
        [scrollView1 addSubview:imageView];
        [imageView release];
    }
     [self layoutScrollImages];
    scrollView2.clipsToBounds = YES;
    scrollView2.indicatorStyle = UIScrollViewIndicatorStyleWhite;
    UIImageView *imageView = [[UIImageView alloc] initWithImage:
                              [UIImage imageNamed:@"image00.jpg"]];
    [scrollView2 addSubview:imageView];
     [scrollView2 setContentSize:
```

```
            CGSizeMake(imageView.frame.size.width, imageView.frame.size.height)];
            [scrollView2 setScrollEnabled:YES];
            [imageView release];
}
- (void)didReceiveMemoryWarning {
    [super didReceiveMemoryWarning];

}
- (void)viewDidUnload {
}
- (void)dealloc {
    [scrollView1 release];
    [scrollView2 release];
    [super dealloc];
}
@end
```

执行后的效果如图 6-35 所示。

图 6-35 执行效果

实例 177 实现"烟花烟花满天飞"效果

实例 177	实现"烟花烟花满天飞"效果
源码路径	\daima\177\

实例说明

本实例实现了"烟花烟花满天飞"效果，预先设置了如图 6-36 所示的素材图片。

图 6-36 素材图片

在本实例中设置了两个视图界面，实现了主视图 MainView.xib 和说明视图 FlipsideView.xib 之间的灵活切换。并且为了实现动画效果，引入了关键帧动画框架 QuartzCore.framework。

具体实现

（1）编写文件 FlipsideViewController.h，此文件的功能是实现主页视图的按钮，具体代码如下所示。

```
#import <UIKit/UIKit.h>
@protocol FlipsideViewControllerDelegate;
@interface FlipsideViewController : UIViewController {
    id <FlipsideViewControllerDelegate> delegate;
}
@property (nonatomic, assign) id <FlipsideViewControllerDelegate> delegate;
```

```
- (IBAction)done:(id)sender;
@end

@protocol FlipsideViewControllerDelegate
- (void)flipsideViewControllerDidFinish:(FlipsideViewController *)controller;
@end
```

（2）编写文件 FlipsideViewController.m，此文件的功能是说明视图 FlipsideView.xib 定义功能方法，具体代码如下所示。

```
#import "FlipsideViewController.h"
@implementation FlipsideViewController
@synthesize delegate;
- (void)viewDidLoad {
    [super viewDidLoad];
    self.view.backgroundColor = [UIColor viewFlipsideBackgroundColor];
}
- (IBAction)done:(id)sender {
    [self.delegate flipsideViewControllerDidFinish:self];
}
- (void)didReceiveMemoryWarning {
    [super didReceiveMemoryWarning];
}
- (void)viewDidUnload {
}
- (void)dealloc {
    [super dealloc];
}
@end
```

（3）编写主视图的头文件 MainViewController.h，通过此文件构建了一个"烟花烟花满天飞"的动画界面，具体代码如下所示。

```
#import "FlipsideViewController.h"
#import <QuartzCore/QuartzCore.h>
@interface MainViewController : UIViewController <FlipsideViewControllerDelegate> {
}
- (IBAction)showInfo:(id)sender;
@end
```

（4）文件 MainViewController.m 是文件 MainViewController.h 的实现，具体代码如下所示。

```
#import "MainViewController.h"
@implementation MainViewController
- (void)viewDidLoad {
    [super viewDidLoad];
    UIImageView* FireView = [[UIImageView alloc] initWithFrame:self.view.frame];
    FireView.animationImages = [NSArray arrayWithObjects
    [UIImage imageNamed:@"fire01.png"],
              [UIImage imageNamed:@"fire02.png"],
              [UIImage imageNamed:@"fire03.png"],
              [UIImage imageNamed:@"fire04.png"],
              [UIImage imageNamed:@"fire05.png"],
                                  nil];
    FireView.animationDuration = 1.75;
    FireView.animationRepeatCount = 0;
    [FireView startAnimating];
    [self.view addSubview:FireView];
    [FireView release];
}
- (void)flipsideViewControllerDidFinish:(FlipsideViewController *)controller {
    [self dismissModalViewControllerAnimated:YES];
}
- (IBAction)showInfo:(id)sender {
    FlipsideViewController *controller = [[FlipsideViewController alloc] initWithNibName:@"FlipsideView" bundle:nil];
    controller.delegate = self;
    controller.modalTransitionStyle = UIModalTransitionStyleFlipHorizontal;
    [self presentModalViewController:controller animated:YES];
    [controller release];
}
- (void)didReceiveMemoryWarning {
    [super didReceiveMemoryWarning];
}
- (void)viewDidUnload {
}
```

```
- (void)dealloc {
    [super dealloc];
}
@end
```

执行后的效果如图 6-37 所示。

图 6-37　执行效果

实例 178　实现"漫天飞雪"效果

实例 178	实现"漫天飞雪"效果
源码路径	\daima\178\

实例说明

本实例实现了"漫天飞雪"效果，预先设置了素材图片 flake.png，如图 6-38 所示。

图 6-38　素材图片

在本实例中设置了两个视图界面，实现了主视图 MainView.xib 和说明视图 FlipsideView.xib 之间的灵活切换。并且为了实现动画效果，引入了关键帧动画框架 QuartzCore.framework。

具体实现

（1）编写文件 FlipsideViewController.h，此文件的功能是实现主页视图的按钮，具体代码如下所示。

```
#import <UIKit/UIKit.h>
@protocol FlipsideViewControllerDelegate;
@interface FlipsideViewController : UIViewController {
    id <FlipsideViewControllerDelegate> delegate;
}
@property (nonatomic, assign) id <FlipsideViewControllerDelegate> delegate;
- (IBAction)done:(id)sender;
@end
@protocol FlipsideViewControllerDelegate
- (void)flipsideViewControllerDidFinish:(FlipsideViewController *)controller;
@end
@end
```

(2) 编写文件 FlipsideViewController.m，此文件的功能是说明视图 FlipsideView.xib 定义功能方法，具体代码如下所示。

```objc
#import "FlipsideViewController.h"
@implementation FlipsideViewController
@synthesize delegate;
- (void)viewDidLoad {
    [super viewDidLoad];
    self.view.backgroundColor = [UIColor viewFlipsideBackgroundColor];
}
- (IBAction)done:(id)sender {
    [self.delegate flipsideViewControllerDidFinish:self];
}
- (void)didReceiveMemoryWarning {
    [super didReceiveMemoryWarning];
}
- (void)viewDidUnload {
}
- (void)dealloc {
    [super dealloc];
}
@end
```

(3) 编写主视图的头文件 MainViewController.h，通过此文件构建了一个"漫天飞雪"的动画界面，具体代码如下所示。

```objc
#import "FlipsideViewController.h"
#import <QuartzCore/QuartzCore.h>
@interface MainViewController : UIViewController <FlipsideViewControllerDelegate> {
    IBOutlet UIImage* flakeImage;
}
@property (nonatomic, retain) UIImage* flakeImage;
- (IBAction)showInfo:(id)sender;
@end
```

(4) 文件 MainViewController.m 是文件 MainViewController.h 的实现，具体代码如下所示。

```objc
#import "MainViewController.h"
@implementation MainViewController
@synthesize flakeImage;
- (void)viewDidLoad {
    [super viewDidLoad];
    self.view.backgroundColor = [UIColor colorWithRed:0.5 green:0.5 blue:1.0 alpha:1.0];
    flakeImage = [UIImage imageNamed:@"flake.png"];
    [NSTimer scheduledTimerWithTimeInterval:(0.05) target:self selector:@selector(onTimer) userInfo:nil repeats:YES];
}
- (void)onTimer
{
    UIImageView* flakeView = [[UIImageView alloc] initWithImage:flakeImage];
    int startX = round(random() % 320);
    int endX = round(random() % 320);
    double scale = 1 / round(random() % 100) + 1.0;
    double speed = 1 / round(random() % 100) + 1.0;
    flakeView.frame = CGRectMake(startX, -100.0, 25.0 * scale, 25.0 * scale);
    flakeView.alpha = 0.25;
    [self.view addSubview:flakeView];
    [UIView beginAnimations:nil context:flakeView];
    [UIView setAnimationDuration:5 * speed];
    flakeView.frame = CGRectMake(endX, 500.0, 25.0 * scale, 25.0 * scale);
    [UIView commitAnimations];
}
- (void)flipsideViewControllerDidFinish:(FlipsideViewController *)controller {
    [self dismissModalViewControllerAnimated:YES];
}
- (IBAction)showInfo:(id)sender {
    FlipsideViewController *controller = [[FlipsideViewController alloc] initWithNibName:@"FlipsideView" bundle:nil];
    controller.delegate = self;
    controller.modalTransitionStyle = UIModalTransitionStyleFlipHorizontal;
```

```
        [self presentModalViewController:controller animated:YES];
        [controller release];
}
- (void)didReceiveMemoryWarning {
    [super didReceiveMemoryWarning];
}
- (void)viewDidUnload {
}
- (void)dealloc {
    [flakeImage release];
    [super dealloc];
}
@end
```

执行后的效果如图 6-39 所示。

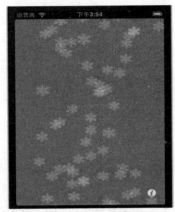

图 6-39　执行效果

实例 179　在屏幕中绘制一个三角形

实例 179	在屏幕中绘制一个三角形
源码路径	\daima\179\

实例说明

在本实例的功能是，在屏幕中绘制一个三角形。当触摸屏幕中的 3 点后，会在这 3 点绘制一个三角形。在具体实现时，定义三角形的 3 个 CGPoint 点对象 firstPoint、secondPoint 和 thirdPoint，然后使用 drawRect 方法将这 3 个点连接起来。

具体实现

（1）编写文件 ViewController.h，此文件的功能是布局视图界面中的元素，本实例比较简单，只用到了 UIViewController。具体代码如下所示。

```
#import <UIKit/UIKit.h>
@interface ViewController : UIViewController
@end
```

（2）文件 ViewController.m 是文件 ViewController.h 的实现，具体代码如下所示。

```
#import "ViewController.h"
#import "TestView.h"
@implementation ViewController
- (void)didReceiveMemoryWarning
{
    [super didReceiveMemoryWarning];
    // 释放任何没有使用的缓存的数据、图像
}
#pragma mark - View lifecycle
```

```objc
- (void)viewDidLoad
{
    [super viewDidLoad];
        // 加载视图
    TestView *view = [[TestView alloc]initWithFrame:self.view.frame];
    self.view = view;
    [view release];
}
- (void)viewDidUnload
{
    [super viewDidUnload];
}
- (void)viewWillAppear:(BOOL)animated
{
    [super viewWillAppear:animated];
}
- (void)viewDidAppear:(BOOL)animated
{
    [super viewDidAppear:animated];
}
- (void)viewWillDisappear:(BOOL)animated
{
     [super viewWillDisappear:animated];
}
- (void)viewDidDisappear:(BOOL)animated
{
     [super viewDidDisappear:animated];
}
- (BOOL)shouldAutorotateToInterfaceOrientation:(UIInterfaceOrientation)interfaceOrientation
{
    // 返回支持的方向
    return (interfaceOrientation != UIInterfaceOrientationPortraitUpsideDown);
}
@end
```

(3) 编写头文件 TestView.h，此文件定义了三角形的 3 个 CGPoint 点对象 firstPoint、secondPoint 和 thirdPoint。具体代码如下所示。

```objc
#import <UIKit/UIKit.h>
@interface TestView : UIView
{
    CGPoint firstPoint;
    CGPoint secondPoint;
    CGPoint thirdPoint;
    NSMutableArray *pointArray;
}
@end
```

(4) 文件 TestView.m 是文件 TestView.h 的实现，具体代码如下所示。

```objc
#import "TestView.h"
@implementation TestView
- (id)initWithFrame:(CGRect)frame
{
    self = [super initWithFrame:frame];
    if (self) {
        // 初始化代码
        self.backgroundColor = [UIColor whiteColor];
        pointArray = [[NSMutableArray alloc]initWithCapacity:3];
        UILabel *label = [[UILabel alloc]initWithFrame:CGRectMake(0, 0, 320, 40)];
        label.text = @"任意点击屏幕内的 3 点以确定一个三角形";
        [self addSubview:label];
        [label release];
    }
    return self;
}
//如果执行了自定义绘制，则只覆盖 drawrect:
//一个空的实现产生不利的影响会表现在动画上
- (void)drawRect:(CGRect)rect
{
    // 绘制代码
    CGContextRef context = UIGraphicsGetCurrentContext();
    CGContextSetRGBStrokeColor(context, 0.5, 0.5, 0.5, 1.0);
```

```
        // 绘制更加明显的线条
        CGContextSetLineWidth(context, 2.0);
        // 画一条连接起来的线条
        CGPoint addLines[] =
        {
            firstPoint,secondPoint,thirdPoint,firstPoint,
        };
        CGContextAddLines(context, addLines, sizeof(addLines)/sizeof(addLines[0]));
        CGContextStrokePath(context);
}
- (void)touchesBegan:(NSSet *)touches withEvent:(UIEvent *)event
{
}
- (void)touchesMoved:(NSSet *)touches withEvent:(UIEvent *)event
{
}
- (void)touchesEnded:(NSSet *)touches withEvent:(UIEvent *)event
{
    UITouch * touch = [touches anyObject];
    CGPoint point = [touch locationInView:self];
    [pointArray addObject:[NSValue valueWithCGPoint:point]];
    if (pointArray.count > 3) {
        [pointArray removeObjectAtIndex:0];
    }
    if (pointArray.count==3) {
        firstPoint = [[pointArray objectAtIndex:0]CGPointValue];
        secondPoint = [[pointArray objectAtIndex:1]CGPointValue];
        thirdPoint = [[pointArray objectAtIndex:2]CGPointValue];
    }
    NSLog(@"%@",[NSString stringWithFormat:@"1:%f/%f\n2:%f/%f\n3:%f/%f",firstPoint.x,firstPoint.y,secondPoint.x,secondPoint.y,thirdPoint.x,thirdPoint.y]);
    [self setNeedsDisplay];
}
-(void)dealloc{
    [pointArray release];
    [super dealloc];
}
@end
```

执行后的效果如图 6-40 所示。

图 6-40　执行效果

实例 180　在屏幕中实现颜色选择器/调色板功能

实例 180	在屏幕中实现颜色选择器/调色板功能
源码路径	\daima\180\

实例说明

在本实例的功能是，在屏幕中实现颜色选择器/调色板功能，让我们可以十分简单地使用颜色选择器。在本实例中没有用到任何图片素材,在颜色选择器上面可以根据饱和度(saturation)和亮度(brightness)来选择某个色系,十分类似于 PhotoShop 上的颜色选择器。

具体实现

(1)编写文件 ILColorPickerDualExampleControllerr.m,此文件的功能是实现一个随机颜色效果,具体代码如下所示。

```
#import "ILColorPickerDualExampleController.h"
@implementation ILColorPickerDualExampleController
#pragma mark - View lifecycle
- (void)viewDidLoad
{
    [super viewDidLoad];
    // 建立一个随机颜色
    UIColor *c=[UIColor colorWithRed:(arc4random()%100)/100.0f
                               green:(arc4random()%100)/100.0f
                                blue:(arc4random()%100)/100.0f
                               alpha:1.0];
    colorChip.backgroundColor=c;
    colorPicker.color=c;
    huePicker.color=c;
}
#pragma mark - ILSaturationBrightnessPickerDelegate implementation

-(void)colorPicked:(UIColor *)newColor forPicker:(ILSaturationBrightnessPickerView *)picker
{
    colorChip.backgroundColor=newColor;
}
@end
```

(2)编写文件 UIColor+GetHSB.m,此文件通过 CGColorSpaceModel 设置了颜色模式值,具体代码如下所示。

```
#import "UIColor+GetHSB.h"
@implementation UIColor(GetHSB)
-(HSBType)HSB
{
    HSBType hsb;
    hsb.hue=0;
    hsb.saturation=0;
    hsb.brightness=0;
    CGColorSpaceModel model=CGColorSpaceGetModel(CGColorGetColorSpace([self CGColor]));
    if ((model==kCGColorSpaceModelMonochrome) || (model==kCGColorSpaceModelRGB))
    {
        const CGFloat *c = CGColorGetComponents([self CGColor]);
        float x = fminf(c[0], c[1]);
        x = fminf(x, c[2]);
        float b = fmaxf(c[0], c[1]);
        b = fmaxf(b, c[2]);
        if (b == x)
        {
            hsb.hue=0;
            hsb.saturation=0;
            hsb.brightness=b;
        }
        else
        {
            float f = (c[0] == x) ? c[1] - c[2] : ((c[1] == x) ? c[2] - c[0] : c[0] - c[1]);
            int i = (c[0] == x) ? 3 : ((c[1] == x) ? 5 : 1);

            hsb.hue=((i - f /(b - x))/6);
            hsb.saturation=(b - x)/b;
            hsb.brightness=b;
        }
    }
    return hsb;
}
```

执行后的效果如图 6-41 所示。

实例 181 在屏幕中实现滑动颜色选择器/调色板功能

图 6-41 执行效果

实例 181 在屏幕中实现滑动颜色选择器/调色板功能

实例 181	在屏幕中实现滑动颜色选择器/调色板功能
源码路径	\daima\181\

实例说明

在本实例的功能是，在屏幕中实现滑动颜色选择器/调色板功能，并且在选择颜色时，还有放大镜查看功能，这样我们可以清楚地看到选择了哪个颜色。除此之外，本实例还可以调整调色板颜色的亮度。

具体实现

实例文件 **RSBrightnessSlider.m** 的具体代码如下所示。

```
#import "RSBrightnessSlider.h"
#import "RSColorPickerView.h"
#import "ANImageBitmapRep.h"
/**
 * 为背景创建默认的绘制位图
 */
CGContextRef RSBitmapContextCreateDefault(CGSize size){
    size_t width = size.width;
    size_t height = size.height;
    size_t bytesPerRow = width * 4;       // 每行的字节 argb
    bytesPerRow += (16 - bytesPerRow%16)%16; //确保是 16 的倍数
    CGColorSpaceRef colorSpace = CGColorSpaceCreateDeviceRGB();
    CGContextRef ctx = CGBitmapContextCreate(NULL,           //自动配置
                                             width,          //宽度
                             height,       //高度
                             8,            //每个的尺寸
                             bytesPerRow,  //每行的字节大小
                    colorSpace,    //CGColorSpaceRef 空间
       kCGImageAlphaPremultipliedFirst );//CGBitmapInfo 对象 bitmapInfo
    CGColorSpaceRelease(colorSpace);
    return ctx;
}
/**
 *返回有滑块的、沙漏状的图像，看上去有点像:
 *
 *  6_____5
 *   \   /
 *  7 \ / 4
 *  ->||<--- cWidth (Center Width)
 *    ||
```

```
 *     8 /  \ 3
 *      /    \
 *   1 ------ 2
 */
UIImage* RSHourGlassThumbImage(CGSize size, CGFloat cWidth){
    //设置大小
    CGFloat width = size.width;
    CGFloat height = size.height;
    //设置背景
    CGContextRef ctx = RSBitmapContextCreateDefault(size);
    //设置颜色
    CGContextSetFillColorWithColor(ctx, [UIColor blackColor].CGColor);
    CGContextSetStrokeColorWithColor(ctx, [UIColor whiteColor].CGColor);
    //绘制滑块，看上面的图的点的个数
    CGFloat yDist83 = sqrtf(3)/2*width;
    CGFloat yDist74 = height - yDist83;
    CGPoint addLines[] = {
        CGPointMake(0, -1),                          //Point 1
        CGPointMake(width, -1),                      //Point 2
        CGPointMake(width/2+cWidth/2, yDist83),      //Point 3
        CGPointMake(width/2+cWidth/2, yDist74),      //Point 4
        CGPointMake(width, height+1),                //Point 5
        CGPointMake(0, height+1),                    //Point 6
        CGPointMake(width/2-cWidth/2, yDist74),      //Point 7
        CGPointMake(width/2-cWidth/2, yDist83)       //Point 8
    };
    //填充路径
    CGContextAddLines(ctx, addLines, sizeof(addLines)/sizeof(addLines[0]));
    CGContextFillPath(ctx);
    //笔画路径
    CGContextAddLines(ctx, addLines, sizeof(addLines)/sizeof(addLines[0]));
    CGContextClosePath(ctx);
    CGContextStrokePath(ctx);
    CGImageRef cgImage = CGBitmapContextCreateImage(ctx);
    CGContextRelease(ctx);
    UIImage* image = [UIImage imageWithCGImage:cgImage];
    CGImageRelease(cgImage);
    return image;
}
/**
 * 返回的图像下图：
 *
 * +-----+
 * | +-+ | ----------------------
 * | | | |                       |
 * ->| |<--- loopSize.width      loopSize.height
 * | | | |                       |
 * | +-+ | ----------------------
 * +-----+
 */
UIImage* RSArrowLoopThumbImage(CGSize size, CGSize loopSize){
    //设置矩形
    CGRect outsideRect = CGRectMake(0, 0, size.width, size.height);
    CGRect insideRect;
    insideRect.size = loopSize;
    insideRect.origin.x = (size.width - loopSize.width)/2;
    insideRect.origin.y = (size.height - loopSize.height)/2;
    //设置背景
    CGContextRef ctx = RSBitmapContextCreateDefault(size);
    //设置颜色
    CGContextSetFillColorWithColor(ctx, [UIColor blackColor].CGColor);
    CGContextSetStrokeColorWithColor(ctx, [UIColor whiteColor].CGColor);
    CGMutablePathRef loopPath = CGPathCreateMutable();
    CGPathAddRect(loopPath, nil, outsideRect);
    CGPathAddRect(loopPath, nil, insideRect);
    //填充路径
    CGContextAddPath(ctx, loopPath);
    CGContextEOFillPath(ctx);
    //笔画路径
    CGContextAddRect(ctx, insideRect);
    CGContextStrokePath(ctx);
    CGImageRef cgImage = CGBitmapContextCreateImage(ctx);
```

实例 181　在屏幕中实现滑动颜色选择器/调色板功能

```
    CGPathRelease(loopPath);
    CGContextRelease(ctx);

    UIImage* image = [UIImage imageWithCGImage:cgImage];
    CGImageRelease(cgImage);
    return image;
}
@implementation RSBrightnessSlider
-(id)initWithFrame:(CGRect)frame {
    self = [super initWithFrame:frame];
    if (self) {
        self.minimumValue = 0.0;
        self.maximumValue = 1.0;
        self.continuous = YES;
        self.enabled = YES;
        self.userInteractionEnabled = YES;
        [self addTarget:self action:@selector(myValueChanged:) forControlEvents:UIControlEventValueChanged];
    }
    return self;
}
-(void)setUseCustomSlider:(BOOL)use {
    if (use) {
        [self setupImages];
    }
}
-(void)myValueChanged:(id)notif {
    [colorPicker setBrightness:self.value];
}
-(void)setupImages {
    ANImageBitmapRep *myRep = [[ANImageBitmapRep alloc] initWithSize:BMPointMake(self.frame.size.width, self.frame.size.height)];
    for (int x = 0; x < myRep.bitmapSize.x; x++) {
        CGFloat percGray = (CGFloat)x / (CGFloat)myRep.bitmapSize.x;
        for (int y = 0; y < myRep.bitmapSize.y; y++) {
            [myRep setPixel:BMPixelMake(percGray, percGray, percGray, 1.0) atPoint:BMPointMake(x, y)];
        }
    }
    [self setMinimumTrackImage:[myRep image] forState:UIControlStateNormal];
    [self setMaximumTrackImage:[myRep image] forState:UIControlStateNormal];

    [myRep release];
}
-(void)setColorPicker:(RSColorPickerView*)cp {
    colorPicker = cp;
    if (!colorPicker) { return; }
    self.value = [colorPicker brightness];
}
@end
```

执行后的效果如图 6-42 所示。

图 6-42　执行效果

实例 182 在屏幕中实现网格化视图效果

实例 182	在屏幕中实现网格化视图效果
源码路径	\daima\182\

实例说明

在本实例的功能是，在屏幕中实现网格化视图（Grid View）效果，并且每个网格视图支持点击动作。本实例使用了 UIGridViewCell 控件和 UIViewController 控件。

具体实现

（1）编写文件 RootViewController.h，此文件是本实例的根视图界面的实现，具体代码如下所示。

```
#import <UIKit/UIKit.h>
#import "UIGridView.h"
#import "UIGridViewDelegate.h"
@interface RootViewController : UIViewController<UIGridViewDelegate> {
}
@property (nonatomic, retain) IBOutlet UIGridView *table;
@end
```

（2）文件 RootViewController.m 是文件 RootViewController.h 的实现，具体代码如下所示。

```
#import "RootViewController.h"
#import "Cell.h"
@implementation RootViewController
@synthesize table;
- (void)didReceiveMemoryWarning {
    [super didReceiveMemoryWarning];
}
- (void)viewDidUnload {
    [super viewDidUnload];
}
- (void)dealloc {
    [super dealloc];
}
- (CGFloat) gridView:(UIGridView *)grid widthForColumnAt:(int)columnIndex
{
    return 80;
}
- (CGFloat) gridView:(UIGridView *)grid heightForRowAt:(int)rowIndex
{
    return 80;
}
- (NSInteger) numberOfColumnsOfGridView:(UIGridView *) grid
{
    return 4;
}
- (NSInteger) numberOfCellsOfGridView:(UIGridView *) grid
{
    return 33;
}
- (UIGridViewCell *) gridView:(UIGridView *)grid cellForRowAt:(int)rowIndex AndColumnAt:(int)columnIndex
{
    Cell *cell = (Cell *)[grid dequeueReusableCell];
    
    if (cell == nil) {
        cell = [[Cell alloc] init];
    }
    
    cell.label.text = [NSString stringWithFormat:@"(%d,%d)", rowIndex, columnIndex];
    
    return cell;
}
- (void) gridView:(UIGridView *)grid didSelectRowAt:(int)rowIndex AndColumnAt:(int)colIndex
```

实例 182 在屏幕中实现网格化视图效果

```objc
{
    NSLog(@"%d, %d clicked", rowIndex, colIndex);
}
@end
```

（3）编写文件 UIGridView.m，此文件在屏幕中实现了网格效果，具体代码如下所示。

```objc
#import "UIGridView.h"
#import "UIGridViewDelegate.h"
#import "UIGridViewCell.h"
#import "UIGridViewRow.h"
@implementation UIGridView
@synthesize uiGridViewDelegate;
- (id)initWithFrame:(CGRect)frame {
    self = [super initWithFrame:frame];
    if (self) {
        [self setUp];
    }
    return self;
}
- (id) initWithCoder:(NSCoder *)aDecoder
{
    self = [super initWithCoder:aDecoder];
    if (self) {
        [self setUp];
        self.separatorStyle = UITableViewCellSeparatorStyleNone;
    }
    return self;
}
- (void) setUp
{
    self.delegate = self;
    self.dataSource = self;
}
- (void)dealloc {
    self.delegate = nil;
    self.dataSource = nil;
    self.uiGridViewDelegate = nil;
    [super dealloc];
}

- (UIGridViewCell *) dequeueReusableCell
{
    UIGridViewCell* temp = tempCell;
    tempCell = nil;
    return temp;
}
// UITableViewController 细节
- (CGFloat)tableView:(UITableView *)tableView heightForHeaderInSection:(NSInteger)section
{
    return 0.0;
}
- (NSInteger)numberOfSectionsInTableView:(UITableView *)tableView
{
    return 1;
}
- (NSInteger)tableView:(UITableView *)tableView numberOfRowsInSection:(NSInteger)section
{
    int residue =  ([uiGridViewDelegate numberOfCellsOfGridView:self] % [uiGridViewDelegate numberOfColumnsOfGridView:self]);

    if (residue > 0) residue = 1;

    return ([uiGridViewDelegate numberOfCellsOfGridView:self] / [uiGridViewDelegate numberOfColumnsOfGridView:self]) + residue;
}
- (CGFloat)tableView:(UITableView *)tableView  heightForRowAtIndexPath:(NSIndexPath *)indexPath
{
    return [uiGridViewDelegate gridView:self heightForRowAt:indexPath.row];
}
- (UITableViewCell *)tableView:(UITableView *)tableView cellForRowAtIndexPath:(NSIndexPath *)indexPath
{
```

```objc
        static NSString *CellIdentifier = @"UIGridViewRow";
        UIGridViewRow *row = (UIGridViewRow *)[tableView dequeueReusableCellWithIdentifier:CellIdentifier];
        if (row == nil) {
            row = [[[UIGridViewRow alloc] initWithStyle:UITableViewCellStyleDefault reuseIdentifier:CellIdentifier] autorelease];
        }
        int numCols = [uiGridViewDelegate numberOfColumnsOfGridView:self];
        int count = [uiGridViewDelegate numberOfCellsOfGridView:self];
        CGFloat x = 0.0;
        CGFloat height = [uiGridViewDelegate gridView:self heightForRowAt:indexPath.row];
        for (int i=0;i<numCols;i++) {
            if ((i + indexPath.row * numCols) >= count) {
                if ([row.contentView.subviews count] > i) {
                    ((UIGridViewCell *)[row.contentView.subviews objectAtIndex:i]).hidden = YES;
                }
                continue;
            }
            if ([row.contentView.subviews count] > i) {
                tempCell = [row.contentView.subviews objectAtIndex:i];
            } else {
                tempCell = nil;
            }
            UIGridViewCell *cell = [uiGridViewDelegate gridView:self cellForRowAt:indexPath.row
                    AndColumnAt:i];
            if (cell.superview != row.contentView) {
             [cell removeFromSuperview];
             [row.contentView addSubview:cell];
             [cell addTarget:self action:@selector(cellPressed:) forControlEvents: UIControlEventTouchUpInside];
            }
            cell.hidden = NO;
            cell.rowIndex = indexPath.row;
            cell.colIndex = i;
            CGFloat thisWidth = [uiGridViewDelegate gridView:self widthForColumnAt:i];
            cell.frame = CGRectMake(x, 0, thisWidth, height);
            x += thisWidth;
        }
        row.frame = CGRectMake(row.frame.origin.x,row.frame.origin.y,x,height);
        return row;
}
- (IBAction) cellPressed:(id) sender
{
    UIGridViewCell *cell = (UIGridViewCell *) sender;
    [uiGridViewDelegate gridView:self didSelectRowAt:cell.rowIndex AndColumnAt:cell.colIndex];
}
@end
```

（4）文件 **UIGridViewCell.m** 实现单元格，具体代码如下所示。

```objc
#import "UIGridViewCell.h"
@implementation UIGridViewCel
@synthesize rowIndex;
@synthesize colIndex;
@synthesize view;
- (void) addSubview:(UIView *)v
{
    [super addSubview:v];
    v.exclusiveTouch = NO;
    v.userInteractionEnabled = NO;
}
@end
```

（5）文件 **UIGridViewCell.m** 实现单元格的行，具体代码如下所示。

```objc
#import "UIGridViewRow.h"
@implementation UIGridViewRow
- (id)initWithStyle:(UITableViewCellStyle)style reuseIdentifier:(NSString *)reuseIdentifier {
    if (self = [super initWithStyle:UITableViewCellStyleDefault reuseIdentifier:reuseIdentifier]) {
        self.selectionStyle = UITableViewCellSelectionStyleNone;
        self.userInteractionEnabled = YES;
```

```
        }
    return self;
}
- (void)dealloc {
    [super dealloc];
}
@end
```

执行后的效果如图 6-43 所示。

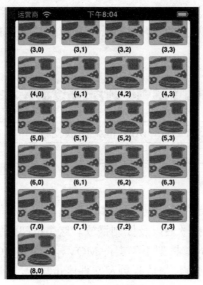

图 6-43　执行效果

第 7 章 多媒体应用实战

在移动智能设备应用中,多媒体是一个重要的应用领域。从严格意义上讲,多媒体包含了屏保、动画、图片、音频、视频和相机等应用。在本书的第 6 章中已经详细介绍了图形图像和动画方面的基本知识。本章将通过几个典型实例的实现过程,详细介绍 Android 中多媒体应用中音频、视频、相册和摄像方面的基本知识。

实例 183　使用 MediaPlayer Framework 框架播放视频

实例 183	使用 MediaPlayer Framework 框架播放视频
源码路径	\daima\183\

实例说明

在 iOS 应用中,Media Player 框架用于播放本地和远程资源中的视频和音频。在应用程序中可使用它打开模态 iPod 界面、选择歌曲以及控制播放。这个框架让您能够与设备提供的所有内置多媒体功能集成。iOS 的 MediaPlayer 框架不仅支持 MOV、MP4 和 3GP 格式,还支持其他视频格式。该框架还提供控件播放、设置回放点、播放视频及文件停止功能,同时对播放各种视频格式的 iPhone 屏幕窗口进行尺寸调整和旋转。

用户可以利用 iOS 中的通知来处理已完成的视频,还可以利用 bada 中 IPlayerEventListener 接口的虚拟函数来处理。在 bada 中,用户可以利用上述 Osp::Media::Player 类来播放视频。Osp::Media 命名空间支持 H264、H.263、MPEG 和 VC-1 视频格式。与音频播放不同,在播放视频时,应显示屏幕。为显示屏幕,借助 Osp::Ui::Controls::OverlayRegion 类来使用 OverlayRegion。OverlayRegion 还可用于照相机预览。

本实例演示了使用 MediaPlayer Framework 框架播放视频的方法。

具体实现

(1) 打开 Xcode,创建一个名为 "BigBuckBunny" 的工程项目。
(2) 然后导入 MediaPlayer Framework 框架,如图 7-1 所示。

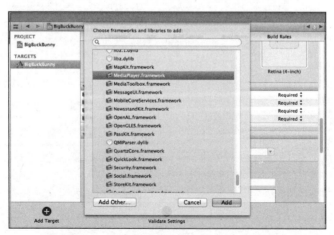

图 7-1　导入 MediaPlayer Framework 框架

实例 183　使用 MediaPlayer Framework 框架播放视频

（3）导入 MediaPlayer 框架后声明 playMovie 方法，代码如下所示。

```
#import 《UIKit/UIKit.h》
#import 《MediaPlayer/MediaPlayer.h》
@interface BigBuckBunnyViewController : UIViewController {
}
-(IBAction)playMovie:(id)sender;
@end
```

（4）实现 playMovie 方法播放视频，具体代码如下所示。

```
-(IBAction)playMovie:(id)sender
{
NSString *filepath = [[NSBundle mainBundle] pathForResource:@"big-buck-bunny-clip" ofType:@"m4v"];
NSURL    *fileURL = [NSURL fileURLWithPath:filepath];
MPMoviePlayerController *moviePlayerController = [[MPMoviePlayerController alloc] initWithContentURL:fileURL];
 [self.view addSubview:moviePlayerController.view];
 [moviePlayerController play];
}
```

如前所述，明确地分配内存给 moviePlayerController 对象，但没有释放该内存，这是编程中的大忌。因为是电影，所以仍然会在此方法执行完毕的时间播放。这样自动释放也不安全，因为我们不知道电影在 autorelease 池释放时电影是否在播放。幸运的是，MPMoviePlayerController 对象是预置来处理这种情况下，在电影播放结束时注册一个名为 MPMoviePlayerPlaybackDidFinishNotification 的通知到 NSNotificationCenter。为了接受这个通知，必须注册一个"观察员"，以应对具体的通知。因此，修改 playMovie 的方法如下。

```
-(IBAction)playMovie:(id)sender
{
NSString *filepath  =  [[NSBundle mainBundle] pathForResource:@"big-buck-bunny-clip" ofType:@"m4v"];
NSURL    *fileURL   =  [NSURL fileURLWithPath:filepath];
MPMoviePlayerController  *moviePlayerController  =  [[MPMoviePlayerController alloc] initWithContentURL:fileURL];
 [[NSNotificationCenter defaultCenter] addObserver:self
selector:@selector(moviePlaybackComplete:)
name:MPMoviePlayerPlaybackDidFinishNotification
object:moviePlayerController];
 [self.view addSubview:moviePlayerController.view];
moviePlayerController.fullscreen = YES;
 [moviePlayerController play];
}
```

现在需要创建 moviePlaybackComplete（刚刚注册的通知），添加以下方法 playMovie:

```
- (void)moviePlaybackComplete:(NSNotification *)notification
{
MPMoviePlayerController *moviePlayerController = [notification object];
 [[NSNotificationCenter defaultCenter] removeObserver:self
name:MPMoviePlayerPlaybackDidFinishNotification
object:moviePlayerController];
 [moviePlayerController.view removeFromSuperview];
 [moviePlayerController release];
}
```

（5）自定义动画显示大小，代码如下所示。

```
[moviePlayerController.view setFrame:CGRectMake(38, 100, 250, 163)];
```

当然 MPMoviePlayerController 还有其他属性的设置，比如：缩放模式，通常包含如下 4 种缩放模式。

- MPMovieScalingModeNone。
- MPMovieScalingModeAspectFit。
- MPMovieScalingModeAspectFill。
- MPMovieScalingModeFill。

然后再次设置为：

```
moviePlayerController.scalingMode = MPMovieScalingModeFill;
```
这样整个实例介绍完毕，执行后可以播放视频，如图 7-2 所示。

图 7-2　执行效果

实例 184　使用 Core Image 框架处理照片

实例 184	使用 Core Image 框架处理照片
源码路径	\daima\184\

实例说明

Core Image 框架是从 iOS 5.0 新增的，它提供了一些非破坏性方法，让我们能够将滤镜应用于图像以及执行其他类型的图像分析（包括人脸识别）。如果在应用程序中添加神奇的图像效果，而又不想了解图像操纵背后复杂的数学知识，那么 Core Image 将是最好的选择。

要在应用程序中使用 Core Image，首先需要添加 Core Image 框架，再导入其接口文件。

```
#import<CoreImage/CoreImage.h>
```

本实例演示了使用 Core Image 框架处理照片的方法。

具体实现

（1）首先需要导入 CoreImage.framework 框架，进行 Mac（不是 iOS）开发的同学请导入 QuartzCore.framework 框架。

（2）然后看如下 3 个主要的类。

- CIContext：与 Core Graphics 和 OpenGL context 类似，所有 Core Image 的处理流程都通过它来进行。
- CIImage：用来存放图片数据，可以通过 UIImage、图片文件或像素数据创建。
- CIFilter：通过它来定义过滤器的详细属性。

CIContext 有两种初始化方法，分别对应 GPU 和 CPU，具体代码如下所示。

```
// 创建基于 GPU 的 CIContext 对象
context = [CIContext contextWithOptions: nil];
// 创建基于 CPU 的 CIContext 对象
context = [CIContext contextWithOptions: [NSDictionary dictionaryWithObject:[NSNumber numberWithBool:YES]
forKey:kCIContextUseSoftwareRenderer]];
```

一般采用第一种基于 GPU 的，因为效率要比 CPU 高很多，但是要注意的是基于 GPU 的 CIContext 对象无法跨应用访问。比如打开 UIImagePickerController 要选张照片进行美化，如果直接在 UIImagePickerControllerDelegate 的委托方法里调用 CIContext 对象进行处理，那么系统会自动将其降为基于 CPU 的,速度会变慢,所以正确的方法应该是在委托方法里先把照片保存下来,然后回到主类里再来处理(代码里将会看到)。

初始化 CIImage 的方法有很多,常用的方法有如下两种。

```
// 通过图片路径创建
CIImage NSString *filePath = [[NSBundle mainBundle] pathForResource:@"image" ofType:@"png"];
NSURL *fileNameAndPath = [NSURL fileURLWithPath:filePath];
beginImage = [CIImage imageWithContentsOfURL:fileNameAndPath];
// 通过 UIImage 对象创建
CIImage UIImage *gotImage = ...; beginImage = [CIImage imageWithCGImage:gotImage.CGImage];
```

下面是 CIFilter 初始化的代码:

```
// 创建过滤器
filter = [CIFilter filterWithName:@"CISepiaTone"];
[filter setValue:beginImage forKey:kCIInputImageKey];
[filter setValue:[NSNumber numberWithFloat:slideValue] forKey:@"inputIntensity"];
```

在上述代码中:

第一行指定使用哪一个过滤器,通过[CIFilter filterNamesInCategory: kCICategoryBuiltIn]能得到所有过滤器的列表;

第二行指定需要处理的图片;

第三行指定过滤参数,每个过滤器的参数都不一样,可以在官方文档里搜索"Core Image Filter Reference"查看。

得到过滤后的图片并输出:

```
CIImage *outputImage = [filter outputImage];
CGImageRef cgimg = [context createCGImage:outputImage fromRect:[outputImage extent]];
UIImage *newImg = [UIImage imageWithCGImage:cgimg];
imgV setImage:newImg];
CGImageRelease(cgimg);
```

在上述代码中:

第一行通过[filter outputImage]可以得到过滤器输出的图片;

第二行通过 CIContext 的方法 createCGImage: fromRect:得到 CGImage;

第三行转化为 UIImage,这样就可以根据需要显示在界面上了。

至此一个过滤周期就完成了,简单来说分以下几个步骤。

第 1 步:初始化 CIContext、CIImage。

第 2 步:初始化 CIFilter 并设置参数。

第 3 步:得到输出的图片。

第 4 步:将图片转化成能显示的 UIImage 类型。

如果想一张图片有多种过滤效果就需要重复第 2 步和第 3 步,并且要将上一个过滤器输出的图片作为下一个过滤器的参数。整个过程非常简单,只需几行代码就可以得到丰富的效果。在本实例中实现了 3 种效果,实现这 3 种效果的核心代码如下所示。

```
- (IBAction)changeValue:(id)sender
{
    self.slider2.value = 0.0;
    self.slider3.value = 0.0;
    float slideValue = self.slider.value;
    // 设置过滤器参数
    [filter setValue:beginImage forKey:kCIInputImageKey];
    [filter setValue:[NSNumber numberWithFloat:slideValue] forKey:@"inputIntensity"];
    // 得到过滤后的图片
    CIImage *outputImage = [filter outputImage];

    // 转换图片
    CGImageRef cgimg = [context createCGImage:outputImage fromRect:[outputImage extent]];
    UIImage *newImg = [UIImage imageWithCGImage:cgimg];
    // 显示图片
    [imgV setImage:newImg];
```

```objc
    // 释放C对象
    CGImageRelease(cgimg);

}
- (IBAction)changeValue2:(id)sender
{
    self.slider.value = 0.0;
    self.slider3.value = 0.0;

    float slideValue = self.slider2.value;
    // 设置过滤器参数
    [filter2 setValue:beginImage forKey:kCIInputImageKey];
    [filter2 setValue:[NSNumber numberWithFloat:slideValue] forKey:@"inputAngle"];
    // 得到过滤后的图片
    CIImage *outputImage = [filter2 outputImage];
    // 转换图片
    CGImageRef cgimg = [context createCGImage:outputImage fromRect:[outputImage extent]];
    UIImage *newImg = [UIImage imageWithCGImage:cgimg];
    // 显示图片
    [imgV setImage:newImg];
    // 释放C对象
    CGImageRelease(cgimg);
}
- (IBAction)changeValue3:(id)sender
{
    self.slider.value = 0.0;
    self.slider2.value = 0.0;
    float slideValue = self.slider3.value;
    // 设置过滤器参数
    [filter3 setValue:beginImage forKey:kCIInputImageKey];
    [filter3 setValue:[NSNumber numberWithFloat:slideValue] forKey:@"inputAngle"];

    // 得到过滤后的图片
    CIImage *outputImage = [filter3 outputImage];
    // 转换图片
    CGImageRef cgimg = [context createCGImage:outputImage fromRect:[outputImage extent]];
    UIImage *newImg = [UIImage imageWithCGImage:cgimg];
    // 显示图片
    [imgV setImage:newImg];
    // 释放C对象
    CGImageRelease(cgimg);
}
```

执行后的效果如图 7-3 所示。

图 7-3　执行效果

实例 185　创建一个多功能播放器

实例 185	创建一个多功能播放器
源码路径	\daima\185\

实例说明

　　本实例的重点是创建一个用于测试多媒体类的应用程序，而不是创建一个真实的应用程序。最

实例 185 创建一个多功能播放器

终的应用程序将能够播放嵌入式或全屏视频,录制并播放音频,浏览并显示照片库(相机)中的图像,将滤镜应用于图像,选择并播放音乐库中的音乐。

因为本实例要实现的功能很多,请务必不要遗漏将定义的任何连接和属性。首先创建一个应用程序骨架,然后通过实现前面讨论的功能使其变得丰满起来。本应用程序包含 5 个主要部分,具体说明如下所示。

(1)设置一个视频播放器,它在用户按下一个按钮时播放一个 MPEG-4 视频文件,还有一个开关可用于切换到全屏模式。

(2)创建一个有播放功能的录音机。

(3)添加一个按钮、一个开关和一个 UIImageView,按钮用于显示照片库或相机,UIImageView 用于显示选定的照片,而开关用于指定图像源。

(4)选择图像后,用户可对其应用滤镜(CIFilter)。

(5)用户可以从音乐库中选择歌曲以及开始和暂停播放。并且还可以使用一个标签在屏幕上显示当前播放的歌曲名。

具体实现

(1)创建项目。

首先,在 Xcode 中使用模板 Single Vew Application 新建一个项目,并将其命名为 MediaPlayground。本应用程序中总共需要添加 3 个额外的框架,以支持多媒体播放(MediaPlayer.Framework)、声音播放/录制(AVFoundation.framework)以及对图像应用滤镜(CoreImage.framework)。

选择项目 MediaPlayground 的顶级编组,并确保选择了目标 MediaPlayground。接下来,单击编辑器中的 Summary 标签,在该选项卡中向下滚动,以找到 Linked Frameworks and Libraries 部分。单击列表下方的+按钮,并在出现的列表中选择 MediaPlayer.framework,再单击 Add 按钮。

对 AVFoundation.framework 和 CoreImage.framework 重复上述操作。添加框架后,将它们拖放到编组 Frameworks 中。接下来需要在项目中添加两个多媒体文件:movie.m4v 和 norecording.wav。其中第一个文件用于演示电影播放器,而第二个是在没有录音时将在录音机中播放的声音。

在本实例的项目文件夹中将文件夹 Media 拖曳到 Xcode 中的项目代码编组中,以便能够在应用程序中直接访问它。在 Xcode 询问时,请务必选择复制文件并新建编组。最后的项目代码编组应类似于图 7-4 所示的结果。

为了让本应用程序正确运行,需要设置很多输出口和操作。首先来看输出口/变量,然后再看操作。对于多媒体播放器,需要一个连接到开关的输出口:toggleFullScreen,该开关切换到全屏模式。还需要一个引用 MPMoviePlayerController 实例的属性/实例变量:moviePlayer,这不是输出口,需要使用代码(而不是通过 Interface Builder 编辑器)创建它。

图 7-4 项目代码编组

为了使用 AV Foundation 录制和播放音频,需要一个连接到 Record 按钮的输出口,以便能够将该按钮的名称在 Record 和 Stop 之间切换。把这个输出口命名为 recordButton。然后需要声明指向录音机(AVAudioRecorder)和音频播放器(AVAudioPlayer)的属性/实例变量:audioRecorder 和 audioPlayer。同样,这两个属性无需暴露为输出口,因为没有 UI 元素连接到它们。

为了实现播放音乐功能,需要连接到 Play Music 按钮和按钮的输出口(分别是 musicPlayButton 和 displayNowPlaying),其中按钮的名称将在 Play 和 Pause 之间切换,而标签将显示当前播放的歌曲的名称。与其他播放器/录音机一样,还需要一个指向音乐播放器本身的属性:musicPlayer。

为了显示图像,启用相机的开关连接到输出口 toggleCamera,而显示选定图像的图像视图将连接到 displayImageView。

接下来开始看具体操作，在此总共需要定义 7 个操作：playMovie、recordAudio、playAudio、chooseImage、applyFilter、chooseMusic 和 playMusic。每个操作都将由一个名称与之类似的按钮触发。

（2）设计界面。本应用程序包括 7 个按钮（UIButton）、2 个开关（UISwitch）、3 个标签（UILabel）和 1 个 UIImageView。另外，需要给嵌入式视频播放器预留控件，该播放器将以编程方式加入。

iPad 开发人员将发现，这些工作完成起来比在 iPhone 中容易得多。在视图可包含的控件数方面，这里已达到了极限。

下面的图 7-5 显示了应用程序界面的一种设计方案。开发人员可以采用这种设计，也可根据自己的喜好对其进行修改，但务必将录音按钮的标题设置为"录音"，而对于播放音乐库中音乐的按钮，务必将其标题设置为"播放音乐"。以编程方式修改这些标题，因此这些按钮的标题必须与指定的一致。最后，务必在视图底部添加一个标签，并将其默认文本设置为"没有播放的音乐"。该标签将显示用户选择的歌曲的名称。

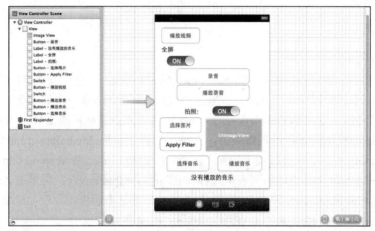

图 7-5　设计的 UI 界面

在此需要注意，可能需要使用 Attributes Inspector（Option+ Command+4）将 UIImageView 的模式设置为 Aspect Fill 或 Aspect Scale，以确保在视图中正确显示照片。

（3）创建并连接输出口和操作。

创建好视图后，切换到助手编辑器模式为建立连接做好准备。下面按界面中控件排列的顺序（从上到下）列出所需的输出口和操作，本项目需要如下所示的输出口。

- 全屏播放电影开关（UISwitch）：toggleFullScreen。
- Record Audio 按钮（UIButton）：recordButton。
- 相机/照片库切换开关（UISwitch）：toggleCamera。
- 图像视图（UIImageView）：displayImageView。
- Play Music 按钮（UIButton）：musicPlayButton。
- 显示当前歌曲名称的标签（UILabel）：displayNowPlaying。

本项目需要如下所示的操作。

- 播放视频按钮（UIButton）：playMovie。
- 录音按钮（UIButton）：recordAudio。
- 播放录音按钮（UIButton）：playAudio。
- 选择图片按钮（UIButton）：chooseImage。
- Apple Filter 按钮（UIButton）：applyFilter。
- 选择音乐按钮（UIButton）：chooseMusic。

- 播放音乐按钮（InButton）：playMusic。

选择文件 MainStoryboard.storyboard，然后切换到助手编辑器界面，按住 Control 键，从切换全屏模式的开关拖曳到文件 ViewController.h 中代码行@interface 下方。在 Xcode 提示时，将输出口命名为 toggleFullscreen。

然后不断重复上述操作，在文件 ViewController.h 中依次创建并连接前面列出的输出口。创建并连接全部 6 个输出口后，开始创建并连接操作。首先，按住 Control 键，并从"播放视频"按钮拖曳到添加的最后一个编译指令@property 下方。在 Xcode 提示时，新建一个名为 playMovie 的操作。

对其他每个按钮重复上述操作，直到在文件 ViewController.h 中新建了 7 个操作。

（4）实现电影播放器。在本实例中，将使用本章前面介绍的 MPMoviePlayerController 类。只需实现如下 3 个方法即可播放电影。

- initWithContentURL：使用提供的 NSURL 对象初始化电影播放器，为播放做好准备。
- play：开始播放选定的电影文件。
- setFullscreen:animated：以全屏模式播放电影。

由于电影播放控制器本身实现了用于控制播放的 GUI，不需要实现额外的功能。然而，如果愿意，还可调用众多其他的方法来控制播放，例如 stop。要想使用电影播放器，必须导入 Media Player 框架的接口文件。为此，修改文件 ViewController.h，在现有#import 代码行后面添加如下代码行。

```
#import <MediaPlayer/MediaPlayer.h>
```

现在可以创建 MPMoviePlayerController 并使用它来播放视频文件了。要想播放电影文件，第一步是声明并初始化一个电影播放器（MPMoviePlayerController）对象。在方法 viewDidLoad 中设置表示电影播放器的实例方法/属性。

首先在文件 ViewController.h 中添加属性 moviePlayer，用于表示 MPMoviePlayerController 实例。为此，在其他属性声明后面添加如下代码行。

```
@property (strong, nonatomic) MPMoviePlayerController*moviePlayer;
```

接下来，在文件 ViewController.m 中的编译指令@implementation 后面添加配套的@synthesize 编译指令：

```
@synthesize moviePlayer;
```

然后在方法 viewDidUnload 中将该属性设置为 nil，从而将电影播放器删除：

```
[self setMoviePlayer:nil];
```

有了可在整个类中使用的属性 moviePlayer 后，下一步是初始化它。为此将方法 viewDidLoad 修改成如下所示的代码。

```
-(void)viewDidLoad
{
//Set up the movie player
NSString kmovieFile=[[NSBundle mainBundle]
pathForResource:@"movie"ofType:@"m4v"];
   self.moviePlayer=[[MPMoviePlayerController alloc]
   initWithContentURL:[NSURL
      fileURLWithPath:
       movieFile]];
    self.moviePlayer.allowsAirPlay=YES;
    [self.moviePlayer.view setFrame:
   CGRectMake(145.0, 20.0, 155.0,100.0)];
[super viewDidLoad];
}
```

在上述代码中，第 4～第 5 行声明了一个名为 movieFile 的字符串变量，并将其设置为前面添加到项目中的电影文件（movie.m4v）的路径。

第 6～第 9 行分配 moviePlayer，并使用一个 NSURL 实例初始化它，该 NSURL 包含 movieFile 提供的路径。使用一行代码完成该任务后，如果愿意就可立即调用 moviePlayer 对象的 play 方法，并看到电影播放。第 10 行为视频播放启用了 AirPlay。第 11～第 12 行设置电影播放器的尺寸，再将视图 moviePlayer 加入到应用程序主视图中。如果编写的是 iPad 应用程序，需要稍微调整尺寸，将这些值替换为 415.0、50.0、300.0 和 250.0。

这样就准备好了电影播放器，可在应用程序的任何地方使用它来播放视频文件 movie.m4v，但我们知道应在哪里使用它——方法 playMovie 中。

要在应用程序 MediaPlayground 中添加电影播放功能，需要实现方法 playMovie，它将被前面添加到界面中的按钮"播放视频"调用。在文件 ViewController.m 中，按照如下代码实现方法 playMovie。

```
- (IBAction)playMovie:(id)sender {
    [self.view addSubview:self.moviePlayer.view];
    [[NSNotificationCenter defaultCenter] addObserver:self
                        selector:@selector(playMovieFinished:)
                        name:MPMoviePlayerPlaybackDidFinishNotification
                        object:self.moviePlayer];
    if ([self.toggleFullscreen isOn]) {
        [self.moviePlayer setFullscreen:YES animated:YES];
    }

    [self.moviePlayer play];
}
```

在上述代码中，第 2 行将 moviePlayer 的视图加入到当前视图中，其坐标是在方法 viewDidLoad 中指定的。当播放完多媒体后，MPMoviePlayerController 将发送 MPMoviePlayerPlaybackDidFinish Notification。第 3～第 6 行为对象 moviePlayer 注册该通知，并请求通知中心接到这种通知后调用方法 playMovieFinished。总之，电影播放器播放完电影（或用户停止播放）时调用 playMovieFinished 方法。第 8～第 10 行使用 UISwitch 的实例方法 isOn 检查开关 toggleFullscreen 是否开启。如果是开的，则使用方法 setFullscreen:animated 将电影放大到覆盖整个屏幕；否则什么也不做，而电影将在前面指定的框架内播放。最后，第 12 行开始播放。

为了在电影播放完毕后进行清理，将对象 moviePlayer 从视图中删除。如果开发人员不介意电影播放器遮住界面，可以不这样做；但是删除电影播放器可避免它分散用户的注意力。为了执行清理工作，在文件 ViewController.m 中，通过如下代码实现方法 playMediaFinished，此方法由通知中心触发。

```
-(void)playMovieFinished:(NSNotification*)theNotification
{
    [[NSNotificationCenter defaultCenter]
     removeObserver:self
     name:MPMoviePlayerPlaybackDidFinishNotification
     object:self.moviePlayer];

    [self.moviePlayer.view removeFromSuperview];
}
```

在此方法中需要完成如下任务。
- 在第 3～第 6 行告诉通知中心可以停止监控通知 MPMoviePlayerPlaybackDidFinish Notification。由于已使用电影播放器播放完视频，将其保留到用户再次播放没有意义。
- 在第 8 行从应用程序主视图中删除电影播放器视图。
- 在第 10 行释放电影播放器。

现在可以在该应用程序中播放电影了，单击 Xcode 工具栏中的 Run 按钮，按"播放视频"按钮即可播放，如图 7-6 所示。

（5）实现音频录制和播放。

本项目的第二部分将在应用程序中添加录制和播放音频的功能，不同于电影播放器，此功能将使用框架 AV Foundation 中的类来实现。为了实现录音机，将使用 AVAudioRecorder 类及其如下方法来实现。

图 7-6 播放视频

- initWithURL:settings:error：该方法接收一个指向本地文件的 NSURL 实例和一个包含一些设置的 NSDictionary 作为参数，并返回一个可供使用的录音机。
- record：开始录音。
- stop：结束录音过程。

播放功能是由 AVAudioPlayer 实现的，使用的方法与前面类似（这并非巧合）。

- initWithContentsOfURL:error：创建一个音频播放器对象，该对象可用于播放 NSURL 对象指向的文件的内容。
- play：播放音频。

首先要为使用 AV Foundation 框架做好准备。要使用 AV Foundation 框架，必须导入两个接口文件：AVFoundation.h 和 CoreAudioTypes.h。在文件 ViewController.h 中，在现有#import 代码行后面添加如下代码行。

```
#import <AVFoundation/AVFoundation.h>
#import<CoreAudio/CoreAudioTypes.h>
```

接下来需要实现录音功能。为了添加录音功能，需要创建方法 recordAudio:。在这个应用程序中，录音过程将一直持续下去，直到用户再次按下相应的按钮。为了实现这种功能，必须在两次调用方法 recordAudio:之间将录音机对象持久化。为确保这一点，将在类 ViewController 中添加实例变量/属性 audioRecorder，用于存储 AVAudioRecorder 对象。为此，在文件 ViewController.h 中添加一个新属性：

```
@property (strong, nonatomic) AVAudioRecorder *audioRecorder;
```

然后在文件 ViewController.m 中，在现有编译指令@synthesize 后面添加一个配套的@synthesize 编译指令：

```
@synthesize audioRecorder;
```

然后，在方法 viewDidUnload 中将该属性设置为 nil，从而将录音机删除：

```
[self setAudioRecorder:nil];
```

接下来，在方法 viewDidLoad 中分配并初始化录音机，让我们能够随时随地地使用它。为此，在文件 ViewController.m 的方法 viewDidLoad 中，添加如下所示的代码。

```
//Set up the audio recorder
        NSURL *soundFileURL=[NSURL fileURLWithPath:
    [NSTemporaryDirectory()
        stringByAppendingString:@" sound.caf"]];
NSDictionary 'soundSetting;
soundSetting=  [NSDictionary dictionaryWithObjectsAndKeys:
[NSNumber numberWithFloat: 44100.O],AVSampleRateKey,
[N$Number numberWithlnt: kAudioFormatMPEG4AACl,AVFormatIDKey,
[NSNumber numberWithlnt: 2],AVNumberOfChannelsKey,
[NSNumber numberWithlnt: AVAudioOualityHigh],
```

```
AVEncoderAudioOualityKey,nil];
self.audjoRecorder= [[AVAudioRecorder alloc]
initWithURL: soundFileURL
settings: soundSetting
error: nil];
 [super viewDidLoad];
}
```

在上述代码中，首先声明了一个 URL（soundFileURL），并将其初始化成指向要存储录音的声音文件。使用函数 NSTemporaryDirectory0 获取临时目录的路径（应用程序将把录音存储到这里），再在它后面加上声音文件名：sound.caf。

然后创建了一个 NSDictionary 对象，它包含用于配置录音格式的键和值。接下来使用 soundFileURL 和存储在字典 soundSettings 中的设置初始化录音机 audioRecorder。此处将参数 error 设置成了 nil，因为在这个例子中我们不关心是否发生了错误。如果发生错误，将返回传递给这个参数的值。

分配并初始化 audioRecorder 后，需要做的只是实现 recordAudio，以便根据需要调用 record 和 stop。为了让程序更有趣，在用户按下 recordButton 按钮时，将其标题在"录音"和"停止录音"之间切换。

在文件 ViewController.m 中，按如下代码修改方法 recordAudio。

```
- (IBAction) recordAudio: (id) sender{
  if ([self. recordButton. titleLabel.text
isEqualToString:@"Record Audio"]){
    [self.audioRecorder record];
    [self.recordButton setTitle:@"停止录音"
fo rState:UICont rolStateNormal];
    } else{
    [self,audioRecorder stop];
    [self.recordButton setTitle:@" 录音"
forState:UIControlStateNormal];
     }
   }
```

上述代码只是初步实现，在后面实现音频播放功能时将修改这个方法，因为它非常适合用于加载录制的音频，为播放做好准备。在上述代码中，第 2 行这个方法首先检查按钮 recordButton 的标题。如果是"录音"，则使用[self audioRecorder record]开始录音（第 4 行），并将 recordButton 的标题设置为"停止录音"（第 5～第 6 行）；否则，说明正在录音，使用[self.audioRecorder stop]结束录音（第 8 行），并将按钮的标题恢复到"录音"（第 9～第 10 行）。

为了实现音频播放器，创建一个可在整个应用程序中使用的实例变量/属性（audioPlayer），然后在 viewDidLoad 中使用默认声音初始化它，这样即使用户没有录音，也有可以播放的声音。

首先，在文件 ViewController.h 添加这个新属性：

`@property (strong, nonatomic) AVAudioPlayer *audioPlayer;`

接下来，在文件 ViewController.m 中，在现有编译指令@synthesize 后面添加配套的@synthesize 编译指令：

`@synthesize audioPlayer;`

在方法 viewDidUnload 中将该属性设置为 nil，从而将音频播放器删除：

`[self setAudioPlayer:nil];`

现在，在方法 viewDidLoad 中分配并初始化音频播放器。为此，在方法 viewDidLoad 中添加如下所示的代码，这样使用默认声音初始化了音频播放器。

```
1: - (void)viewDidLoad
2:   {
3://Set up the movie player
4:NSString  kmovieFile=[[NSBundle mainBundle]
5:pathForResource:@"movie" ofType:@"m4v"];
6:self.moviePlayer=[[MPMoviePlayerController alloc]
7:initWithContentURL: [NSURL
```

```
 8:     fileURLWithPath:
 9:     movieFile]];
10:    self.moviePlayer.allowsAirPlay=YES;
11:    [self .moviePlayer.view setFrame:
12:    CGRectMake(145.0, 20.0, 155.0,100.0)];
13:
14:
15:    //Set up the audio recorder
16:    NSURL *soundFileURL=[NSURL fileURLWithPath:
17:    [NSTemporaryDirectory()
18:    stringByAppendingString:@" sound.caf"]];
19:
20:    NSDictionary *soundSetting;
22:  soundsetting[NSNumber numberWithFloat:y 44100.O],AVSampleRateKey,
22:    [NSNumber numberWithFloat:44100.0],AVSampleRateKey,
23:    [NSNumber numberWithInt: kAudioFormatMPEG4AAC] ,AVFormatIDKey,
24:    [NSNumber numberWithInt:2],AVNumberOfChannelsKey,
25:    [NSNumber numberWithInt: AVAudioQualityHigh],
26:    AVEncoderAudioQualityKey,nil];
27:
28:    self.audioRecorder=[[AVAudioRecorder alloc]
29:    initWithURL: soundFileURL
30:    settings: soundSetting
31:    error: nil];
32:
33:    //Set up the audio player
34:    NSURL *noSoundFileURL=[NSURL fileURLWithPath:
35:    [[NSBundle mainBundle]
36:    pathForResource:@"norecording" ofType:@"wav'
37:    self.audioPlayer= [[AVAudioPlayer alloc]
38:    lnitWithContentsOfURL:noSoundFileURL error:nil]
39:
40:    [super viewDidLoad];
41:  }
```

在上述代码中，音频播放器设置代码始于第34行。此处创建了一个NSURL(noSoundFileURL)，它指向文件norecording.wav，这个文件包含在前面创建项目时添加的文件夹Media中。第37行分配一个音频播放器实例（audioPlayer），并使用noSoundFileURL的内容初始化它。现在可以使用对象audioPlayer来播放默认声音了。

要播放audioPlayer指向的声音，只需向它发送消息play。为此，在方法playAudio中添加这样做的代码。下面的代码显示了这个方法的完整实现。

```
- (IBAction)playAudio:(id)sender {
//    self.audioPlayer.delegate=self;
    [self.audioPlayer play];
}
```

如果现在运行这个应用程序，将能够录制声音，但每次按"录音"按钮时，播放的都是norecording.wav。这是因为我们没有加载录制的声音。

为了加载录音，最佳的选择是在用户单击"停止录音"按钮时，在方法recordAudio中加载。为此，按照如下代码修改方法recordAudio。

```
- (IBAction)recordAudio:(id)sender {
   if ([self.recordButton.titleLabel.text
            isEqualToString:@"录音"]) {
        [self.audioRecorder record];
        [self.recordButton setTitle:@"停止录音"
                    forState:UIControlStateNormal];
    } else {
        [self.audioRecorder stop];
        [self.recordButton setTitle:@"Record Audio"
                    forState:UIControlStateNormal];
    // Load the new sound in the audioPlayer for playback
    NSURL *soundFileURL=[NSURL fileURLWithPath:
            [NSTemporaryDirectory()
            stringByAppendingString:@"sound.caf"]];
    self.audioPlayer = [[AVAudioPlayer alloc]
            initWithContentsOfURL:soundFileURL error:nil];
```

第7章 多媒体应用实战

}

在上述代码中，第12～第14行用于获取并存储临时目录的路径，再使用它来初始化一个NSURL对象：soundFileURL，使其指向录制的声音文件 sound.caf。第15～第16行用于分配音频播放器 audio Player，并使用 soundFileURL 的内容来初始化它。

此时再次运行该应用程序，看看结果如何。现在，当按下 Play Audio 按钮时，如果还未录音，将听到默认声音，如果已经录制过声音，将听到录制的声音。

为了使用 UIImagePickerController，无需导入任何新的接口文件，但必须将类声明为遵守多个协议，具体地说是协议 UIImagePickerControllerDelegate 和 UINavigationControllerDelegate。

在文件 ViewController.h 中，修改代码行@interface，使其包含这些协议：

```
@interface ViewController :UIViewController
<UIImagePickerControllerDelegate,UINavigationControllerDelegate>
```

现在可以使用本章开头介绍的方法实现 UIImagePickerController 了。实际上，我们将使用的代码与我们见过的代码很像。

当用户触摸按钮"选择图片"时，应用程序将调用方法 chooseImage。在该方法中，需要分配 UIImagePickerController，配置它将用于浏览的媒体类型（相机或图片库），设置其委托并显示它。方法 chooseImage 的实现代码如下所示。

```
- (IBAction)chooseImage:(id)sender {
    UIImagePickerController *imagePicker;
    imagePicker = [[UIImagePickerController alloc] init];
    if ([self.toggleCamera isOn]) {
        imagePicker.sourceType=UIImagePickerControllerSourceTypeCamera;
    } else {
        imagePicker.sourceType=UIImagePickerControllerSourceTypePhotoLibrary;
    }
     imagePicker.delegate=self;
    [[UIApplication sharedApplication] setStatusBarHidden:YES];
    [self presentModalViewController:imagePicker animated:YES];
}
```

在上述代码中，第2～第3行分配并初始化了一个 UIntagePickerController 实例，并将其赋给变量 imagePicker。第5～第9行判断开关 toggleCamera 的状态，如果为开，则将图像选择器的 sourceType 属性设置为 UIImagePickerControllerSourceTypeCamera，否则将其设置为 UIImagePickerController SourceTypePhotoLibrary。换句话说，用户可使用这个开关指定从图片库还是相机获取图像。第10行将图像选择器委托设置为 ViewController，这意味着需要实现一些支持方法，以便在用户选择照片后做相应的处理。第12行隐藏应用程序的状态栏，这是必要的，因为照片库和相机界面都将以全屏模式显示。第13行将 imagePicker 视图显示在现有视图上面。

如果仅编写上述代码，则用户触摸按钮 Choose Image 并选择图像时，什么也不会发生。为对用户选择图像做出响应，需要实现委托方法 imagePickerControUer:didFinishPickingMediaWithInfo。

在文件 ViewController.m 中，添加委托方法 imagePickerController:didFinishPickingMediaWithInfo，具体代码如下所示。

```
- (void)imagePickerController:(UIImagePickerController *)picker
        didFinishPickingMediaWithInfo:(NSDictionary *)info {
    [[UIApplication sharedApplication] setStatusBarHidden:NO];
    [self dismissModalViewControllerAnimated:YES];
    self.displayImageView.image=[info objectForKey:
                        UIImagePickerControllerOriginalImage];
}
```

当用户选择图像后，就可重新显示状态栏（第3行），再使用 dismissModaMewController Animated 关闭图像选择器（第4行）。第5～第6行完成了其他所有的工作！为访问用户选择的 UIImage，使用 UIImagePickerControllerOriginalImage 键从字典 info 从提取它，再将其赋给 displayImageView 的属性 image，这将在应用程序视图中显示该图像。

要完成应用程序 MediaPlayground 的图像选择部分,还必须考虑一种情形:用户单击图像选择器中的"取消"按钮,这将不会选择任何图像。委托方法 imagePickerControllerDidCancel 正是针对这种情形的。请实现这个方法,使其重新显示状态栏,并调用 dismissModalViewControllerAnimated 将图像选择器关闭。下面的代码列出了这个简单方法的完整实现。

```
- (void)imagePickerControllerDidCancel:(UIImagePickerController *)picker {
    [[UIApplication sharedApplication] setStatusBarHidden:NO];
    [self dismissModalViewControllerAnimated:YES];
}
```

现在,可以运行该应用程序,并使用按钮 Choose Image 来显示照片库和相机中的照片了。

(6) 实现 Core Image 滤镜。

通过使用 Core Image,可以轻松地在其应用程序中添加高级图像功能。事实上,在本章所做的工作中,实现滤镜是最简单的。首先,在文件 ViewController.h 中,导入框架 Core Image 的接口文件。为此,在其他#import 语句后面添加如下代码行:

```
#import<CoreImage/CoreImage.h>
```

现在可以使用 Core Image 创建并配置滤镜,再将其应用于应用程序的 UIImageView 显示的图像了。

要应用滤镜,首先需要一个 CIImage 实例,但现在只有一个 UIImageView。我们必须做些转换工作,以便应用滤镜并显示结果。这在前面介绍过,不应对涉及的代码感到陌生。方法 applyFilter 的实现代码如下所示。

```
- (IBAction)applyFilter:(id)sender {
    CIImage *imageToFilter;
    imageToFilter=[[CIImage alloc]
              initWithImage:self.displayImageView.image];

    CIFilter *activeFilter = [CIFilter filterWithName:@"CISepiaTone"];
    [activeFilter setDefaults];
    [activeFilter setValue: [NSNumber numberWithFloat: 0.75]
              forKey: @"inputIntensity"];
    [activeFilter setValue: imageToFilter forKey: @"inputImage"];
    CIImage *filteredImage=[activeFilter valueForKey: @"outputImage"];
    CIContext *context = [CIContext contextWithOptions:[NSDictionary dictionary]];
    CGImageRef cgImage = [context createCGImage:filteredImage fromRect:[imageToFilter extent]];
    UIImage *myNewImage = [UIImage imageWithCGImage:cgImage];
    self.displayImageView.image = myNewImage;
    CGImageRelease(cgImage);
}
```

在上述代码中,第 2~第 3 行声明了一个名为"imageToFilter"的 CIImage,然后分配它,并使用对象 displayImageView(UIImageView)包含的 UIImage 初始化它。第 6 行声明并初始化一个 Core Image 滤镜:CISepiaTone。第 7 行设置该滤镜的默认值;对于要使用的任何滤镜,都应这样做。第 8~第 9 行配置滤镜的 InputIntensity 键,将其值设置为 0.75。本章前面说过,Xcode 文档 Core Image Filter Reference 列出了各种键。第 10 行使用滤镜的 inputImage 键设置滤镜将应用到的图像(imageToFilter),而第 11 行获取应用滤镜后的结果,并将其存储到一个新的 CIImage(filteredImage)中。第 13 行使用 UIImage 类的方法 imageWithCIImage,将应用滤镜后的图像转换为一个 UIImage(myNewImage)。第 14 行将 displayImageView 的属性 image 设置为 myNewImage,从而显示应用滤镜后的图像。

此时可以运行该应用程序,选择一张照片,再单击 Apple Filter 按钮。棕色滤镜将导致照片的颜色饱和度接近零,使其看起来像张老照片。

(7) 访问并播放音乐库。

为了完成这个项目,我们将实现对 iOS 设备音乐库的访问:选择声音文件,再播放它们。

首先,将使用 MPMediaPickerController 类来选择要播放的音乐。这里只调用这个类的一个方法——initWithMediaTypes，用于初始化多媒体选择器并限制选择器显示的文件，可以使用如下属性来配置这种对象的行为。

- prompt：用户选择歌曲时向其显示的一个字符串。
- allowsPickingMultipleItems：指定用户只能选择一个声音文件还是可选择多个。

需要遵守 MPMediaPickerControllerDelegate 协议，以便能够在用户选择播放列表后采取相应的措施。还将添加该协议的方法 mediaPicker:didPickMediaItems。

为了播放音频，将使用 MPMusicPlayerController 类，它可使用多媒体选择器返回的播放列表。开始和暂停播放将使用 4 个方法。

- iPodMusicPlayer：这个类方法将音乐播放器初始化为 iPod 音乐播放器，这种播放器能够访问音乐库。
- setQueueWithItemCollection：使用多媒体选择器返回的播放列表对象(MPMediaItemCollection)设置播放队列。
- play：开始播放音乐。
- pause：暂停播放音乐。

由于多媒体选择器与电影播放器一样，也使用框架 Media Player，因此准备工作已完成了一半，无需再导入其他接口文件。然而，必须将类声明为遵守协议 MPMediaPickerControllerDelegate，这样才能响应用户选择。为此，在文件 ViewController.h 中，在@interface 代码行中包含这个协议：

```
@interface ViewController:UIViewController
<MPMediaPickerControllerDelegate,UIImagePickerControllerDelegate,
UINavigationControllerDelegate>
```

在多媒体选择器中做出选择时，要采取有意义的措施，需要播放选择的音乐文件。与电影播放器、录音机和音频播放器一样，要新建一个可在应用程序任何地方访问的音乐播放器对象。

为此，添加一个属性/实例变量（musicPlayer），它是一个 MPMusicPlayerController 实例：

```
@property (strong, nonatomic) MPMusicPlayerController *musicPlayer;
```

接下来，在文件 ViewController.m 中，在现有编译指令@synthesize 后面添加配套的编译指令@synthesize:

```
@synthesize musicPlayer;
```

在方法 viewDidUnload 中，将该属性设置为 nil，以删除音乐播放器：

```
[self setMusicPlayer:nil];
```

这就设置好了指向音乐播放器的属性，但还需创建一个音乐播放器实例。与电影播放器、音频播放器和录音机一样，我们也在方法 viewDidLoad 中完成这项工作。修改方法 viewDidLoad，使用 MPMusicPlayerController 类的方法 iPodMusicPlayer 新建一个音乐，此方法的最终代码如下所示。

```
- (void)viewDidLoad
{
    //Setup the movie player
    NSString *movieFile = [[NSBundle mainBundle]
                    pathForResource:@"movie" ofType:@"m4v"];
    self.moviePlayer = [[MPMoviePlayerController alloc]
                    initWithContentURL: [NSURL
                                    fileURLWithPath:
                                    movieFile]];
    self.moviePlayer.allowsAirPlay=YES;
    [self.moviePlayer.view setFrame:
                CGRectMake(145.0, 20.0, 155.0 , 100.0)];

    //Setup the audio recorder
    NSURL *soundFileURL=[NSURL fileURLWithPath:
                [NSTemporaryDirectory()
                    stringByAppendingString:@"sound.caf"]];
    NSDictionary *soundSetting;
    soundSetting = [NSDictionary dictionaryWithObjectsAndKeys:
            [NSNumber numberWithFloat: 44100.0],AVSampleRateKey,
```

实例 185 创建一个多功能播放器

```
                [NSNumber numberWithInt: kAudioFormatMPEG4AAC],AVFormatIDKey,
                [NSNumber numberWithInt: 2],AVNumberOfChannelsKey,
                [NSNumber numberWithInt: AVAudioQualityHigh],
                    AVEncoderAudioQualityKey,nil];
    self.audioRecorder = [[AVAudioRecorder alloc]
                          initWithURL: soundFileURL
                          settings: soundSetting
                          error: nil];
    //Setup the audio player
    NSURL *noSoundFileURL=[NSURL fileURLWithPath:
                           [[NSBundle mainBundle]
                            pathForResource:@"norecording" ofType:@"wav"]];
    self.audioPlayer = [[AVAudioPlayer alloc]
                        initWithContentsOfURL:noSoundFileURL error:nil];
    //Setup the music player
      self.musicPlayer=[MPMusicPlayerController iPodMusicPlayer];
    [super viewDidLoad];
}
```

在上述代码中，只有第 42 行是新增的，它创建一个 MPMusicPlayerController 实例，并将其赋给属性 musicPlayer。

在这个应用程序中，用户触摸按钮"选择音乐"时，将触发操作 chooseMusic，而该操作将显示多媒体选择器。要使用多媒体选择器，需要采取的步骤与使用图像选择器时类似：实例化选择器并配置其行为，然后将其作为模态视图加入应用程序视图中。用户使用完多媒体选择器后，将把它返回的播放列表加入音乐播放器，并关闭选择器视图；如果用户没有选择任何多媒体，则只需关闭选择器视图即可。

在实现文件 ViewController.m 中，方法 chooseMusic 的实现代码如下所示。

```
- (IBAction)chooseMusic:(id)sender {
    MPMediaPickerController *musicPicker;
      [self.musicPlayer stop];
      self.displayNowPlaying.text=@"No Song Playing";
      [self.musicPlayButton setTitle:@"Play Music"
                            forState:UIControlStateNormal];
      musicPicker = [[MPMediaPickerController alloc]
                     initWithMediaTypes: MPMediaTypeMusic];
      musicPicker.prompt = @"Choose Songs to Play" ;
      musicPicker.allowsPickingMultipleItems = YES;
      musicPicker.delegate = self;
      [self presentModalViewController:musicPicker animated:YES];
}
```

在上述代码中，第 2 行声明了 MPMediaPickerController 实例 musicPicker。接下来，第 4～第 7 行确保调用选择器时，音乐播放器将停止播放当前歌曲，界面中 nowPlaying 标签的文本被设置为默认字符串 No Song Playing，且播放按钮的标题为 PlayMusic。这些代码行并非必不可少，但可确定界面与应用程序中实际发生的情况同步。第 8～第 9 行分配并初始化多媒体选择器控制器实例。初始化时使用的是常量 MPMediaTypeMusic，该常量指定了用户使用选择器可选择的文件类型（音乐）。第 9 行指定一条将显示在音乐选择器顶部的消息。第 10 行将属性 allowsPickingMultipleItems 设置为一个布尔值（YES 或 NO），它决定了用户能否选择多个多媒体文件。第 11 行设置音乐选择器的委托。换句话说，它告诉 musicPicker 对象到 ViewController 中去查找 MPMediaPickerControllerDelegate 协议方法。第 12 行使用视图控制器 musicPicker 将音乐库显示在应用程序视图的上面。

为了获取多媒体选择器返回的播放列表（一个 MPMediaItemCollection 对象）并执行清理工作，需要在实现文件中添加委托协议方法 mediaPicker:didPickMediaItems，具体代码如下所示。

```
- (void)mediaPicker: (MPMediaPickerController *)mediaPicker
  didPickMediaItems:(MPMediaItemCollection *)mediaItemCollection {
     [musicPlayer setQueueWithItemCollection: mediaItemCollection];
     [self dismissModalViewControllerAnimated:YES];
}
```

用户在多媒体选择器中选择歌曲后，将调用该方法，并通过一个 MPMediaItemCollection。对象

（mediaItemCollection）将选择的歌曲传递给它。实际上，可以将 mediaItemCollection 对象视为一个多媒体文件播放列表。在上述代码中，第 1 行使用该播放列表对音乐播放器实例 musicPlayer 进行了配置，这是通过 setQueueWithItemCollection 完成的。为了执行清理工作，第 2 行关闭了模态视图。

要完成多媒体选择器的实现，还需考虑一种情形：用户在没有选择任何多媒体文件的情况下退出多媒体选择器（在没有选择任何文件的情况下轻按 Done 按钮）。为了处理这种情形，需要添加委托协议方法 mediaPickerDidCancel。与图像选择器一样，只需在该方法中关闭模态视图控制器即可。为此，在文件 ViewController.m 中添加这个方法，此方法的代码如下所示。

```
- (void)mediaPickerDidCancel:(MPMediaPickerController *)mediaPicker {
    [self dismissModalViewControllerAnimated:YES];
}
```

这样实现多媒体选择器后，剩下的唯一任务是添加音乐播放器并确保显示了相应的歌曲名。由于已经在视图控制器的 viewDidLoad 方法中创建了 musicPlayer 对象，且在方法 mediaPicdidPickMediaItems 中设置了音乐播放器的播放列表，因此余下的唯一工作是在方法 playMusic 中开始播放和暂停播放。并且在需要时，将 musicPlayButton 按钮的标题在 Play Music 和 Pause Music 之间进行切换。作为最后的点睛之笔，将访问 MPMusicPlayerController 对象 musicPlayer 的属性 no、Ⅳ PlayingItem。该属性是一个 MPMediaItem 对象，而这种对象包含一个字符串属性 MPMediaItemPropertyTitle，其值为当前播放的多媒体文件的名称（如果有的话）。

将上述内容组合在一起后，playMusic 的实现代码如下所示。

```
- (IBAction)playMusic:(id)sender {
    if ([self.musicPlayButton.titleLabel.text
                isEqualToString:@"Play Music"]) {
        [self.musicPlayer play];
        [self.musicPlayButton setTitle:@"Pause Music"
                              forState:UIControlStateNormal];
        self.displayNowPlaying.text=[self.musicPlayer.nowPlayingItem
valueForProperty:MPMediaItemPropertyTitle];
    } else {
        [self.musicPlayer pause];
        [self.musicPlayButton setTitle:@"Play Music"
                              forState:UIControlStateNormal];
        self.displayNowPlaying.text=@"No Song Playing";
    }
}
```

在上述代码中，第 2 行检查 musicPlayButton 的标题是否为 Play Music。如果是，第 4 行开始播放，第 5～第 6 行将该按钮的标题重置为 Pause Music，而第 7～第 8 行将标签 displayNowPlaying 的文本设置为当前歌曲的名称。如果按钮 musicPlayButton 的标题不是 Play Music（第 10 行），将暂停播放音乐，将该按钮的标题重置为 Play Music，并将标签的文本改为 No Soon Playing。实现该方法后，保存文件 ViewController.m，并在 iOS 设备上运行该应用程序，以便对其进行测试。按"选择音乐"按钮将打开多媒体选择器。创建播放列表后，按多媒体选择器中的"完成"按钮，再按"播放音乐"按钮开始播放选择的歌曲。当前歌曲的名称将显示在界面底部。

> **注意** 如果在模拟器上测试音乐播放功能，则不会成功。要测试这些功能，必须使用实际设备。

实例 186 使用系统内的相册

实例 186	使用系统内的相册
源码路径	\daima\186\

实例说明

在 iOS 应用中,可以使用 UIImagePickerController 实现从系统照片中进行选择的功能。在本实例中创建并初始化了 UIImagePickerController,并设置了 UI 的显示格式。UIImagePickerController 可以设置如下 3 种用户界面的显示样式。

- UIImagePickerControllerSourceTypePhotoLibrary:从相册中打开选择照片界面。
- UIImagePickerControllerSourceTypeCamera:启动摄像头打开摄影界面。
- UIImagePickerControllerSourceTypeSavedPhotosAlbum:直接打开保存的照片列表,如果是拥有摄像头的设备则打开相册。

除此之外,本实例中还将 UIImagePickerController 的属性 allowsEditing 设置为了 YES,这样可以对选取的照片实现放大或缩小操作。

具体实现

实例文件 UIKitPrjImagePicker.m 的具体代码如下所示。

```
#import "UIKitPrjImagePicker.h"
@implementation UIKitPrjImagePicker
- (void)dealloc {
  [imageView_ release];
  [super dealloc];
}
- (void)viewDidLoad {
  [super viewDidLoad];
  UIBarButtonItem* barButton =
    [[[UIBarButtonItem alloc] initWithBarButtonSystemItem:UIBarButtonSystemItemCamera
                                          target:self
                                          action:@selector(barButtonDidPush)]
autorelease];
  [self setToolbarItems:[NSArray arrayWithObject:barButton] animated:NO];
  imageView_ = [[UIImageView alloc] init];
  imageView_.frame = self.view.bounds;
  imageView_.contentMode = UIViewContentModeScaleAspectFit;
  imageView_.autoresizingMask =
    UIViewAutoresizingFlexibleWidth | UIViewAutoresizingFlexibleHeight;
  [self.view addSubview:imageView_];
}
- (void)barButtonDidPush {
  UIActionSheet* sheet = [[[UIActionSheet alloc] init] autorelease];
  sheet.delegate = self;
  [sheet addButtonWithTitle:@"PhotoLibrary"];
  [sheet addButtonWithTitle:@"Camera"];
  [sheet addButtonWithTitle:@"SavedPhotosAlbum"];
  [sheet addButtonWithTitle:@"取消"];
  sheet.cancelButtonIndex = 3;
  [sheet showFromToolbar:self.navigationController.toolbar];
}
- (void)actionSheet:(UIActionSheet*)actionSheet clickedButtonAtIndex:(NSInteger)buttonIndex {
  if ( buttonIndex == actionSheet.cancelButtonIndex ) {
  } else {
    UIImagePickerControllerSourceType sourceType = buttonIndex;
    if ( [UIImagePickerController isSourceTypeAvailable:sourceType] ) {
      UIImagePickerController* picker = [[[UIImagePickerController alloc] init]
autorelease];
      picker.delegate = self;
      picker.sourceType = sourceType;
      picker.allowsEditing = YES;
      //picker.cameraViewTransform = CGAffineTransformMakeScale(3.0, 1.0);
      [self presentModalViewController:picker animated:YES];
    } else {
      NSLog( @"%d is not available.", sourceType );
    }
  }
}
- (void)imagePickerController:(UIImagePickerController*)picker
```

```
  didFinishPickingMediaWithInfo:(NSDictionary*)info
{
  // 取得选择的照片
  UIImage* image = [info objectForKey:UIImagePickerControllerEditedImage];
  if ( !image ) {
    image = [info objectForKey:UIImagePickerControllerOriginalImage];
  }
  imageView_.image = image;
  // 保存于相册时的处理如下
  UIImageWriteToSavedPhotosAlbum( image,
                    self,
                    @selector(image:didFinishSavingWithError:contextInfo:),
                    NULL );
  //关闭相册
  [self dismissModalViewControllerAnimated:YES];
}

- (void)imagePickerControllerDidCancel:(UIImagePickerController*)picker {
  //取消时的处理动作，关闭相册
  [self dismissModalViewControllerAnimated:YES];
}
- (void)image:(UIImage*)image
    didFinishSavingWithError:(NSError*)error contextInfo:(void*)contextInfo
{
  if ( error ) {
    // error 非 nil 保存失败
    NSLog( [error localizedDescription] );
  } else {
    // nil 时保存成功
  }
}
@end
```

执行后的效果如图 7-7 所示。选取照片后可以放大或缩小，如图 7-8 所示。

图 7-7　执行效果

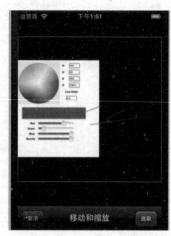

图 7-8　放大或缩小照片

实例 187　实现录制视频功能

实例 187	实现录制视频功能
源码路径	\daima\186\

实例说明

在 iOS 应用中，我们可以使用设备自带的摄像头进行拍照或录制视频功能。在录制视频时，除了要将属性 sourceType 设置成与照片设置时一样的 UIImagePickerControllerSourceTypeCamera 外，还需要在 UIImagePickerController 的 mediaType 属性中包含 kUTTypeMovie 常量。因为这个常量不属于 UIKit，所以在使用前需要先将 MobileCoreServices.framework 框架包含进项目中。在本实例中，演示了录制视频的过程，并且使用属性 "videoMaximumDuration = 10" 设置了允许录制的最长时间为 10 秒。

具体实现

实例文件 UIKitPrjWithVideo.m 的具体代码如下所示。

```objc
#import "UIKitPrjWithVideo.h"
#import <MobileCoreServices/MobileCoreServices.h>
@implementation UIKitPrjWithVideo
- (void)viewDidLoad {
 [super viewDidLoad];
 UIBarButtonItem* barButton =
   [[[UIBarButtonItem alloc] initWithBarButtonSystemItem:UIBarButtonSystemItemCamera
                                                  target:self
                                                  action:@selector(barButtonDidPush)]
autorelease];
 [self setToolbarItems:[NSArray arrayWithObject:barButton] animated:NO];
}
- (void)barButtonDidPush {
 UIActionSheet* sheet = [[[UIActionSheet alloc] init] autorelease];
 sheet.delegate = self;
 [sheet addButtonWithTitle:@"PhotoLibrary"];
 [sheet addButtonWithTitle:@"Video"];
 [sheet addButtonWithTitle:@"取消"];
 sheet.cancelButtonIndex = 2;
 [sheet showFromToolbar:self.navigationController.toolbar];
}
- (void)actionSheet:(UIActionSheet*)actionSheet clickedButtonAtIndex:(NSInteger)buttonIndex {
 if ( buttonIndex == actionSheet.cancelButtonIndex ) {
 } else {
   UIImagePickerControllerSourceType sourceType = buttonIndex;
   if ( [UIImagePickerController isSourceTypeAvailable:sourceType] ) {

     UIImagePickerController* picker = [[[UIImagePickerController alloc] init] autorelease];
     picker.delegate = self;
     picker.sourceType = sourceType;
     picker.videoQuality = UIImagePickerControllerQualityTypeLow;
     picker.videoMaximumDuration = 10;
       //调查指定 sourceType 是否支持视频
     NSArray* mediaTypes = [UIImagePickerController availableMediaTypesForSourceType:sourceType];
     if ( [mediaTypes containsObject:(NSString*)kUTTypeMovie] ) {
         //支持视频的情况下，将 kUTTypeMovie 追加到 mediaTypes 属性中
       picker.mediaTypes = [NSArray arrayWithObject:(NSString*)kUTTypeMovie];
     } else {
       NSLog( @"%@ is not available.", kUTTypeMovie );
     }
     [self presentModalViewController:picker animated:YES];
   } else {
     NSLog( @"%d is not available.", sourceType );
   }
 }
}
- (void)imagePickerController:(UIImagePickerController*)picker
 didFinishPickingMediaWithInfo:(NSDictionary*)info
{
 // 判定选择的是否为视频
 NSString* mediaType = [info objectForKey:UIImagePickerControllerMediaType];
 if ( [mediaType isEqualToString:(NSString*)kUTTypeMovie] ) {
   NSURL* mediaURL = [info objectForKey:UIImagePickerControllerMediaURL];
   NSString* mediaPath = [mediaURL path];
   if ( UIVideoAtPathIsCompatibleWithSavedPhotosAlbum( mediaPath ) ) {
       //将视频保存于相册中
       //方法 UISaveVideoAtPathToSavedPhotosAlbum 的第 2 个参数以后与保存照片时完全相同，此处省略说明
     UISaveVideoAtPathToSavedPhotosAlbum( mediaPath,
                                          self,
```

```
    @selector(video:didFinishSavingWithError:contextInfo:),
                                      NULL );
      } else {
        NSLog( @"不能保存到相册中的处理" );
      }
    } else {
      NSLog( @"非视频时的处理" );
    }
    [self dismissModalViewControllerAnimated:YES];
}
- (void)video:(NSString*)videoPath
    didFinishSavingWithError:(NSError*)error contextInfo:(void*)contextInfo
{
  if ( error ) {
    // error 非空（nil）时保存失败
    NSLog( [error localizedDescription] );
  } else {
    // 否则保存成功
  }
}

@end
```

执行后的效果如图 7-9 所示。

图 7-9 执行效果

实例 188　设置屏幕中视频的画面

实例 188	设置屏幕中视频的画面
源码路径	\daima\186\

实例说明

在 iOS 应用中，当在屏幕中展示视频时，我们可以设置显示视频的画面效果。例如可以使用 UIImagePickerController 的属性 CameraOverlayView 来指定任意的 UIView 子类。在本实例中，隐藏了摄像工具条，实现了重叠的摄影效果。

具体实现

实例文件 UIKitPrjCameraOverlay.m 的具体代码如下所示。

```
#import "UIKitPrjCameraOverlay.h"
#pragma mark ----- CameraOverlayView -----
@implementation CameraOverlayView
@synthesize pickerController = pickerController_;
- (void)touchesEnded:(NSSet*)touches withEvent:(UIEvent*)event {
  [self.pickerController takePicture];
}
@end
#pragma mark ----- UIKitPrjCameraOverlay -----
```

实例 188 设置屏幕中视频的画面

```objc
@implementation UIKitPrjCameraOverlay
- (void)viewDidLoad {
  [super viewDidLoad];

  UIBarButtonItem* barButton =
    [[[UIBarButtonItem alloc] initWithBarButtonSystemItem:UIBarButtonSystemItemCamera
                                       target:self
                                       action:@selector(barButtonDidPush)]
autorelease];
  [self setToolbarItems:[NSArray arrayWithObject:barButton] animated:NO];
  imageView_ = [[UIImageView alloc] init];
  imageView_.frame = self.view.bounds;
  imageView_.contentMode = UIViewContentModeScaleAspectFit;
  imageView_.autoresizingMask =
    UIViewAutoresizingFlexibleWidth | UIViewAutoresizingFlexibleHeight;
  [self.view addSubview:imageView_];
}
- (void)barButtonDidPush {
  UIActionSheet* sheet = [[[UIActionSheet alloc] init] autorelease];
  sheet.delegate = self;
  [sheet addButtonWithTitle:@"Camera"];
  [sheet addButtonWithTitle:@"取消"];
  sheet.cancelButtonIndex = 1;
  [sheet showFromToolbar:self.navigationController.toolbar];
}

- (void)actionSheet:(UIActionSheet*)actionSheet clickedButtonAtIndex:(NSInteger)buttonIndex {
  if ( buttonIndex == actionSheet.cancelButtonIndex ) {
  } else {
    if ( [UIImagePickerController isSourceTypeAvailable:UIImagePickerControllerSourceTypeCamera] ) {
      UIImagePickerController* picker = [[[UIImagePickerController alloc] init] autorelease];
      picker.delegate = self;
      picker.sourceType = UIImagePickerControllerSourceTypeCamera;
        //隐藏默认的摄影控制
      picker.showsCameraControls = NO;
        //创建重叠用的 UIView 实例
      UIImage* image = [UIImage imageNamed:@"sniper.png"];
        //创建并初始化 CameraOverlayView 实例（关于 CameraOverlayView 后面有介绍）
      CameraOverlayView* overlayView = [[[CameraOverlayView alloc] initWithImage:image]
autorelease];
      overlayView.pickerController = picker;
      overlayView.frame = picker.view.bounds;
      overlayView.contentMode = UIViewContentModeScaleAspectFill;
      overlayView.autoresizingMask =
        UIViewAutoresizingFlexibleWidth | UIViewAutoresizingFlexibleHeight;
      overlayView.alpha = 0.5;
      overlayView.userInteractionEnabled = YES;
      picker.cameraOverlayView = overlayView;
      [self presentModalViewController:picker animated:YES];
    } else {
      NSLog( @"UIImagePickerControllerSourceTypeCamera is not available." );
    }
  }
}

- (void)imagePickerController:(UIImagePickerController*)picker
  didFinishPickingMediaWithInfo:(NSDictionary*)info
{
  // 获取选择或者拍摄的照片
  UIImage* image = [info objectForKey:UIImagePickerControllerEditedImage];
  if ( !image ) {
    image = [info objectForKey:UIImagePickerControllerOriginalImage];
  }
  imageView_.image = image;

  [self dismissModalViewControllerAnimated:YES];
}
@end
```

执行后的效果如图 7-10 所示。

图 7-10 执行效果

实例 189 剪辑系统内的视频

实例 189	剪辑系统内的视频
源码路径	\daima\189\

实例说明

在 iOS 应用中，我们可以剪辑处理系统内的视频文件，这一功能是通过类 UIVideoEditorController 实现的。此类的常用方法和属性如下所示。

- isSourceTypeAvailable 方法：检查制定源是否可用。
- availableMediaTypesForSourceType 方法：检查可用媒体（视频还是只能是图片）。
- mediaTypes property：设置界面媒体属性。
- presentViewController:animated:completion 方法：显示界面使用。
- availableMediaTypesForSourceType：指定源可用的媒体种类。
- isSourceTypeAvailable：指定源是否在设备上可用。
- sourceType：运行相关接口前需要指明源类型，必须有效，否则抛出异常。picker 显示的时候改变这个值，picker 会通过相应改变来适应。默认 UIImagePickerControllerSourceTypePhotoLibrary。
- allowsEditing：设置是否可编辑。
- mediaType：指示 picker 中显示的媒体类型。设置每种类型之前应用 availableMediaTypesForSourceType:检查一下。如果为空或者 array 中类型都不可用，则会发生异常。默认 kUTTypeImage，只能显示图片。
- videoQuality：视频拍摄选取时的编码质量，只有 mediaTypes 包含 kUTTypeMovie 时有效。
- videoMaximumDuration：video 最大记录时间，默认 10 分钟，只有当 mediaTypes 包含 kUTTypeMovie 时有效。
- showsCameraControls：设置 picker 是否显示默认的 camera controls，默认是 YES，设置成 NO 可以隐藏默认的 controls 来使用自定义的 overlay view，从而可以实现多选而不是选一张 picker 就 dismiss 了。如果 UIImagePickerControllerSourceTypeCamera 源无效，则 NSInvalidArgumentException 异常。
- cameraOverlayView：自定义的用于显示在 picker 之上的 view。只有当源是 UIImagePickerControllerSourceTypeCamera 时有效。其他时候使用抛出 NSInvalidArgumentException 异常。
- cameraViewTransform：预先动画，只影响预先图像，对自定义的 overlay view 和默认的 picker 无效。只有当 picker 的源是 UIImagePickerControllerSourceTypeCamera 时有效，否则 NSInvalidArgumentException 异常。
- takePicture：使用摄像头选取一个图片。

本实例演示了剪辑视频的基本方法。

具体实现

实例文件 UIKitPrjVideoEditor.m 的具体代码如下所示。

```objc
#import "UIKitPrjVideoEditor.h"
@implementation UIKitPrjVideoEditor
- (void)viewDidLoad {
  [super viewDidLoad];
  UIBarButtonItem* barButton =
    [[[UIBarButtonItem alloc] initWithBarButtonSystemItem:UIBarButtonSystemItemAction
                                        target:self
                                        action:@selector(barButtonDidPush)]
autorelease];
  [self setToolbarItems:[NSArray arrayWithObject:barButton] animated:NO];
}
- (void)barButtonDidPush {
  UIActionSheet* sheet = [[[UIActionSheet alloc] init] autorelease];
  sheet.delegate = self;
  [sheet addButtonWithTitle:@"VideoEditor"];
  [sheet addButtonWithTitle:@"取消"];
  sheet.cancelButtonIndex = 1;
  [sheet showFromToolbar:self.navigationController.toolbar];
}
- (void)actionSheet:(UIActionSheet*)actionSheet clickedButtonAtIndex:(NSInteger)buttonIndex {
  if ( buttonIndex == actionSheet.cancelButtonIndex ) {
  } else {
    UIVideoEditorController* videoEditor = [[[UIVideoEditorController alloc] init]
autorelease];
    videoEditor.delegate = self;
      //准备使用 Resources 目录中的 test.MOV 视频文件
    NSString* videoPath = [[NSBundle mainBundle] pathForResource:@"test" ofType:@"MOV"];
    if ( [UIVideoEditorController canEditVideoAtPath:videoPath] ) {
      videoEditor.videoPath = videoPath;
      [self presentModalViewController:videoEditor animated:YES];
    } else {
      NSLog( @"can't edit video at %@", videoPath );
    }
  }
}

- (void)videoEditorControllerDidCancel:(UIVideoEditorController*)editor {
  [editor dismissModalViewControllerAnimated:YES];
}

- (void)videoEditorController:(UIVideoEditorController*)editor
  didSaveEditedVideoToPath:(NSString*)editedVideoPath
{
  // 保存到相册中
  if ( UIVideoAtPathIsCompatibleWithSavedPhotosAlbum( editedVideoPath ) ) {
    UISaveVideoAtPathToSavedPhotosAlbum( editedVideoPath,
                                        self,
@selector(video:didFinishSavingWithError:contextInfo:),
                                        NULL );
  } else {
    NSLog( @"不能保存到相册时的处理" );
  }
  [editor dismissModalViewControllerAnimated:YES];
}

- (                    void)videoEditorController:(UIVideoEditorController*)editor did
FailWithError:(NSError*)error {
  NSLog( @"%X", error );
  [editor dismissModalViewControllerAnimated:YES];
}

- (void)video:(NSString*)videoPath
  didFinishSavingWithError:(NSError*)error contextInfo:(void*)contextInfo
{
    if ( error ) {
        // error 非空（nil）时保存失败
        NSLog( [error localizedDescription] );
```

```
    } else {
        // 否则保存成功
    }
}
@end
```

执行后的效果如图 7-11 所示。

图 7-11 执行效果

实例 190　开发一个音频播放器

实例 190	开发一个音频播放器
源码路径	\daima\190\

实例说明

本实例的功能是在设备中播放指定的音频文件 "music.mp3"。在本实例中设置了 "播放" 和 "停止" 两个按钮，通过这两个按钮可以对音频文件 "music.mp3" 实现播放控制处理。在实现播放功能之前，需要先导入多媒体框架 AVFoundation.framework。

具体实现

（1）编写文件 DayViewController.h 实现视图界面，具体代码如下所示。

```
#import <UIKit/UIKit.h>
#import <AVFoundation/AVFoundation.h>
@interface DayViewController : UIViewController {
    AVAudioPlayer *audioPlayer;
}
@property (nonatomic, retain) AVAudioPlayer *audioPlayer;
-(IBAction)play;
-(IBAction)stop;
@end
```

（2）文件 DayViewController.m 是文件 DayViewController.h 的实现，具体代码如下所示。

```
#import "DayViewController.h"
@implementation DayViewController
@synthesize audioPlayer;
// 加载视图.
- (void)viewDidLoad {
    [super viewDidLoad];
    NSString *filePath = [[NSBundle mainBundle] pathForResource:@"music"
                                                ofType:@"mp3"];

    NSURL *fileURL = [[NSURL alloc] initFileURLWithPath:filePath];
    self.audioPlayer = [[AVAudioPlayer alloc]
                        initWithContentsOfURL:fileURL error:nil];
    [self.audioPlayer prepareToPlay];
    [filePath release];
    [fileURL release];
```

```
}
-(IBAction)play {
    self.audioPlayer.currentTime = 0;
    [self.audioPlayer play];
}
-(IBAction)stop {
    [self.audioPlayer stop];
}
- (void)didReceiveMemoryWarning {
    // 释放视图
    [super didReceiveMemoryWarning];

    // 释放数据
}
- (void)viewDidUnload {
}
- (void)dealloc {
    [audioPlayer release];
    [super dealloc];
}
@end
```

执行后的效果如图 7-12 所示。

图 7-12 执行效果

实例 191 在屏幕中实现一个电子琴效果

实例 191	在屏幕中实现一个电子琴效果
源码路径	\daima\191\

实例说明

本实例和上一个实例类似，功能是在屏幕中实现一个电子琴效果。在文件 Day-Info.plist 中设置屏幕横向显示，并且在实现播放功能之前，也需要先导入多媒体框架 AVFoundation.framework。

具体实现

编写文件 DayViewController.h 实现视图界面，具体代码如下所示。

```
#import <UIKit/UIKit.h>
#import <AudioToolbox/AudioToolbox.h>
#import <Foundation/Foundation.h>
@interface DayViewController : UIViewController {
    NSString *soundFile;
}
@property(nonatomic,retain) NSString *soundFile;
- (void)playSound:(NSString*)soundKey;
- (IBAction)DO:(id)sender;
- (IBAction)RE:(id)sender;
- (IBAction)MI:(id)sender;
- (IBAction)FA:(id)sender;
- (IBAction)SO:(id)sender;
```

```objc
- (IBAction)LA:(id)sender;
- (IBAction)SI:(id)sender;
- (IBAction)C:(id)sender;
- (IBAction)D:(id)sender;
- (IBAction)E:(id)sender;
- (IBAction)F:(id)sender;
- (IBAction)G:(id)sender;
@end
```

文件 DayViewController.m 是文件 DayViewController.h 的实现,具体代码如下所示。

```objc
#import "DayViewController.h"
@implementation Day19ViewController
@synthesize soundFile;
-(void)playSound:(NSString*)soundKey{
    NSString *path = [NSString stringWithFormat:@"%@%@",[[NSBundle mainBundle] resourcePath],soundKey];
    NSLog(@"%@\n", path);
    SystemSoundID soundID;
    NSURL *filePath = [NSURL fileURLWithPath:path isDirectory:NO];
    AudioServicesCreateSystemSoundID((CFURLRef)filePath, &soundID);
    AudioServicesPlaySystemSound(soundID);
}
- (IBAction)DO:(id)sender{
    soundFile = [NSString stringWithFormat:@"/001.mp3"];
    [self playSound: soundFile];
}
- (IBAction)RE:(id)sender{
    soundFile = [NSString stringWithFormat:@"/002.mp3"];
    [self playSound: soundFile];
}
- (IBAction)MI:(id)sender{
    soundFile = [NSString stringWithFormat:@"/003.mp3"];
    [self playSound: soundFile];
}
- (IBAction)FA:(id)sender{
    soundFile = [NSString stringWithFormat:@"/004.mp3"];
    [self playSound: soundFile];
}
- (IBAction)SO:(id)sender{
    soundFile = [NSString stringWithFormat:@"/005.mp3"];
    [self playSound: soundFile];
}
- (IBAction)LA:(id)sender{
    soundFile = [NSString stringWithFormat:@"/006.mp3"];

    [self playSound: soundFile];
}
- (IBAction)SI:(id)sender{
    soundFile = [NSString stringWithFormat:@"/007.mp3"];
    [self playSound: soundFile];

}
- (IBAction)C:(id)sender{
    soundFile = [NSString stringWithFormat:@"/C.mp3"];
    [self playSound: soundFile];
}
- (IBAction)D:(id)sender{
    soundFile = [NSString stringWithFormat:@"/D.mp3"];
    [self playSound: soundFile];
}
- (IBAction)E:(id)sender{
    soundFile = [NSString stringWithFormat:@"/E.mp3"];
    [self playSound: soundFile];
}
- (IBAction)F:(id)sender{
    soundFile = [NSString stringWithFormat:@"/F.mp3"];
    [self playSound: soundFile];
}
- (IBAction)G:(id)sender{
    soundFile = [NSString stringWithFormat:@"/G.mp3"];
    [self playSound: soundFile];
}
// Override to allow orientations other than the default portrait orientation.
-
```

```
(BOOL)shouldAutorotateToInterfaceOrientation:(UIInterfaceOrientation)interfaceOrienta
tion {
    // Return YES for supported orientations
    return (interfaceOrientation == UIInterfaceOrientationLandscapeLeft);
}
- (void)didReceiveMemoryWarning {
    // Releases the view if it doesn't have a superview.
    [super didReceiveMemoryWarning];
    // Release any cached data, images, etc that aren't in use.
}
- (void)viewDidUnload {
    // Release any retained subviews of the main view.
    // e.g. self.myOutlet = nil;
}
- (void)dealloc {
     [soundFile release];
    [super dealloc];
}
@end
```

执行效果如图 7-13 所示。

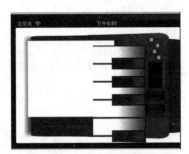

图 7-13 执行效果

实例 192 在屏幕中实现一个 DJ 混音器

实例 192	在屏幕中实现一个 DJ 混音器
源码路径	\daima\192\

实例说明

本实例和前面的两个实例类似，功能是在屏幕中实现一个 DJ 混音器。本实例为混音器添加了指定的背景图片和声音文件，并且在实现播放功能之前，也需要先导入多媒体框架 AVFoundation.framework。

具体实现

编写文件 DayViewController.h 实现视图界面，具体代码如下所示。

```
#import <UIKit/UIKit.h>
#import <Foundation/Foundation.h>
#import <AVFoundation/AVFoundation.h>
@interface DayViewController : UIViewController {
    IBOutlet UISlider *guitarVolumeControl;
    IBOutlet UISlider *beatsVolumeControl;
    AVAudioPlayer *guitarPlayer;
    AVAudioPlayer *beatsPlayer;
    IBOutlet UISwitch *includeGuitar;
    IBOutlet UISwitch *includeBeats;
}
@property (nonatomic, retain) UISlider *guitarVolumeControl;
@property (nonatomic, retain) UISlider *beatsVolumeControl;
@property (nonatomic, retain) AVAudioPlayer *guitarPlayer;
@property (nonatomic, retain) AVAudioPlayer *beatsPlayer;
@property (nonatomic, retain) UISwitch *includeGuitar;
@property (nonatomic, retain) UISwitch *includeBeats;
```

```objc
-(IBAction)guitarVolumeChange;
-(IBAction)beatsVolumeChange;
-(IBAction)guitarSwitch;
-(IBAction)beatsSwitch;
@end
```

文件 DayViewController.m 是文件 DayViewController.h 的实现,具体代码如下所示。

```objc
#import "DayViewController.h"
@implementation DayViewController
@synthesize guitarVolumeControl,beatsVolumeControl;
@synthesize guitarPlayer, beatsPlayer;
@synthesize includeGuitar,includeBeats;
// Implement viewDidLoad to do additional setup after loading the view, typically from a nib.
- (void)viewDidLoad {
    [super viewDidLoad];
    NSString *guitarfilePath = [[NSBundle mainBundle] pathForResource:@"guitar"
                        ofType:@"caf"];
    NSURL *guitarfileURL = [[NSURL alloc] initFileURLWithPath:guitarfilePath];
    self.guitarPlayer = [[AVAudioPlayer alloc]
                        initWithContentsOfURL:guitarfileURL error:nil];
    [self.guitarPlayer prepareToPlay];
    [guitarfilePath release];
    [guitarfileURL release];
    NSString *beatsfilePath = [[NSBundle mainBundle] pathForResource:@"beats"
ofType:@"caf"];
    NSURL *beatsfileURL = [[NSURL alloc] initFileURLWithPath:beatsfilePath];
    self.beatsPlayer = [[AVAudioPlayer alloc]
                        initWithContentsOfURL:beatsfileURL error:nil];
    [self.beatsPlayer prepareToPlay];
    [beatsfilePath release];
    [beatsfileURL release];
}
-(IBAction)guitarVolumeChange{
    guitarPlayer.volume = guitarVolumeControl.value;

}
-(IBAction)beatsVolumeChange{
    beatsPlayer.volume = beatsVolumeControl.value;

}
-(IBAction)guitarSwitch{
    if (includeGuitar.on) {
    self.guitarPlayer.currentTime = 0;
        self.guitarPlayer.numberOfLoops=5;
        [self.guitarPlayer play];
    }else {
        [self.guitarPlayer stop];
    }
}
-(IBAction)beatsSwitch{
if (includeBeats.on) {
    self.beatsPlayer.currentTime = 0;
    self.beatsPlayer.numberOfLoops=0;
    [self.beatsPlayer play];
}else {
    [self.beatsPlayer stop];
}
}
- (void)didReceiveMemoryWarning {
    [super didReceiveMemoryWarning];
}
- (void)viewDidUnload {
}
- (void)dealloc {
    [guitarVolumeControl release];
    [beatsVolumeControl release];
    [guitarPlayer release];
    [beatsPlayer release];
    [includeBeats release];
    [includeGuitar release];
    [super dealloc];
}
@end
```

执行后的效果如图 7-14 所示。

实例 193　在屏幕中实现一个音乐选择器

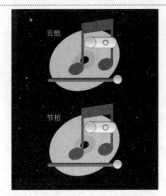

图 7-14　执行效果

实例 193　在屏幕中实现一个音乐选择器

实例 193	在屏幕中实现一个音乐选择器
源码路径	\daima\193\

实例说明

本实例和前面的两个实例类似，功能是在屏幕中实现一个音乐选择器功能。首先准备了音频素材文件 aaa.mp3、bbb.mp3、ccc.mp3 和 ddd.mp3，然后在界面中添加了一个 UIPickerView 控件，在此控件中列表显示了前面的 4 个 MP3 文件供用户选择。选择某个文件，单击下方的"播放"按钮后会播放这首音乐。同样和前面的两个实例一样，在实现播放功能之前，也需要先导入多媒体框架 AVFoundation.framework。

具体实现

编写文件 DayViewController.h 实现视图界面，具体代码如下所示。

```
#import <UIKit/UIKit.h>
#import <Foundation/Foundation.h>
#import <AVFoundation/AVFoundation.h>
@interface DayViewController : UIViewController{
    AVAudioPlayer *player;
    IBOutlet UIPickerView *musicPicker;
    IBOutlet UIProgressView *currentPositionProgress;
    IBOutlet UILabel *durationLabel;
    IBOutlet UILabel *currentPositionLabel;
    IBOutlet UISlider *volumeSlider;
    NSArray *musicColumn;
    NSString *selectedMusic;
}
@property(nonatomic, retain) AVAudioPlayer *player;

@property (nonatomic,retain) UIPickerView *musicPicker;
@property(nonatomic, retain) UIProgressView *currentPositionProgress;
@property(nonatomic, retain) UILabel *durationLabel;
@property(nonatomic, retain) UILabel *currentPositionLabel;
@property(nonatomic, retain) UISlider *volumeSlider;
@property (nonatomic,retain) NSArray *musicColumn;
@property (nonatomic,retain) NSString *selectedMusic;
- (IBAction)play;
- (IBAction)pause;
- (IBAction)setVolume;
- (void)updateDisplay;
@end
```

文件 DayViewController.m 是文件 DayViewController.h 的实现，具体代码如下所示。

```
#import "DayViewController.h"
```

第 7 章 多媒体应用实战

```objc
#import <AudioToolbox/AudioServices.h>
@implementation DayViewController
@synthesize player;
@synthesize musicPicker;
@synthesize currentPositionProgress;
@synthesize durationLabel, currentPositionLabel;
@synthesize volumeSlider;
@synthesize musicColumn,selectedMusic;
- (IBAction)play{
    NSInteger row = [musicPicker selectedRowInComponent:0];
    if(selectedMusic != [musicColumn objectAtIndex:row]){
        selectedMusic = [musicColumn objectAtIndex:row];
        NSString *soundFilePath = [[NSBundle mainBundle] pathForResource: selectedMusic ofType: @"mp3"];
        NSURL *fileURL = [[[NSURL alloc] initFileURLWithPath: soundFilePath] autorelease];
        self.player = [[[AVAudioPlayer alloc] initWithContentsOfURL: fileURL error: nil] autorelease];
        [self.player prepareToPlay];
        self.currentPositionLabel.text = @"0:00";
        //iOS4 环境内使用,播放器时间计算,简易方法
            //durationLabel.text = [self timeToString:self.player.duration];
        //通用播放器时间计算,开始
        NSTimeInterval musiclength = player.duration - player.currentTime;
        musiclength = musiclength/60;
        NSInteger musiclengthint = (int)(musiclength);
        NSTimeInterval secondspositionlength = (musiclength - musiclengthint);
        secondspositionlength = secondspositionlength *60;
        NSInteger secondsintlength = (int)secondspositionlength;
        NSLog(@"time %i and %i and %f", musiclengthint, secondspositionlength, secondsintlength);
        NSString *totaltime;
        if(secondsintlength < 10){
            totaltime = [[NSString alloc] initWithFormat:@"0%i:0%i", musiclengthint, secondsintlength];
        }
        else{
            totaltime = [[NSString alloc] initWithFormat:@"0%i:%i", musiclengthint, secondsintlength];
        }
        self.durationLabel.text = totaltime;
        //通用播放器时间计算,结束
        [self updateDisplay];
        [NSTimer scheduledTimerWithTimeInterval:.1
            selector:@selector(updateDisplay)
            userInfo:nil repeats:YES];
    }
    [self.player play];
}
- (void)updateDisplay {
    self.currentPositionProgress.progress = self.player.currentTime / self.player.duration;
    //iOS4 环境内使用,播放器时间计算,简易方法
    //self.currentPositionLabel.text = [self timeToString:self.player.currentTime];
    //通用播放器时间计算,开始
    NSTimeInterval position = player.currentTime / 60;
    NSInteger positionint = (int)(position);
    NSTimeInterval secondsposition = (position - positionint);
    secondsposition = secondsposition *60;
    NSInteger secondsint = (int)secondsposition;
    NSLog(@"time %i and %i and %f", positionint, secondsint, secondsposition);
    NSString *currenttime;
    if(secondsposition < 10){
        currenttime = [[NSString alloc] initWithFormat:@"0%i:0%i", positionint, secondsint];
    }
    else{
        currenttime = [[NSString alloc] initWithFormat:@"%i:%i", positionint, secondsint];
    }
    self.currentPositionLabel.text = currenttime;
    //通用播放器时间计算,结束
}

- (IBAction)pause{
    [self.player stop];
}
```

实例 193　在屏幕中实现一个音乐选择器

```objc
- (IBAction)setVolume {
    self.player.volume = self.volumeSlider.value;
}
-(NSInteger)numberOfComponentsInPickerView:(UIPickerView *)pickerView
{
    return 1;
}
-(NSInteger)pickerView:(UIPickerView *)pickerView numberOfRowsInComponent:(NSInteger)component
{
    return [musicColumn count];
}
-(UIView *)pickerView:(UIPickerView *)pickerView
          titleForRow:(NSInteger)row
         forComponent:(NSInteger)component
{
    return [musicColumn objectAtIndex:row];
}
// Implement viewDidLoad to do additional setup after loading the view, typically from a nib.
- (void)viewDidLoad {
    [super viewDidLoad];

    NSArray *array=[[NSArray alloc] initWithObjects:@"aaa",
                        @"bbb",
                        @"ccc",
                        @"eee",
                        @"ddd",
                        @"fff",
                        nil];
    self.musicColumn = array;
     [array release];
}
- (void)didReceiveMemoryWarning {
    [super didReceiveMemoryWarning];
}
- (void)viewDidUnload {
}
- (void)dealloc {
    [player release];
    [musicPicker release];

    [currentPositionProgress release];
    [durationLabel release];
    [currentPositionLabel release];
    [volumeSlider release];
    [musicColumn release];
    [selectedMusic release];
    [super dealloc];
}
@end
```

执行后的效果如图 7-15 所示。

图 7-15　执行效果

第7章 多媒体应用实战

实例 194　在屏幕中听声音

实例 194	在屏幕中听声音
源码路径	\daima\194\

实例说明

在本实例的工具栏中添加了多个按钮：后退、听故事、暂停、前进和短音测试，并且添加了两个多媒体框架：AudioToolbox.framework 和 AVFoundation.framework。当单击"听故事"按钮后，会使用 AudioToolbox 来播放音频文件"hongoumeng.mp3"；当单击"短音测试"按钮，会播放音频文件"damn.caff"。

具体实现

编写文件 ViewController.h 实现视图界面，具体代码如下所示。

```
#import <UIKit/UIKit.h>
#import <AudioToolbox/AudioToolbox.h>
#import <AVFoundation/AVFoundation.h>
@interface ViewController : UIViewController <NSXMLParserDelegate,AVAudioPlayerDelegate> {
    SystemSoundID soundID;
    AVAudioPlayer *player;
}
@property (weak) UITextField *tf;
@property (nonatomic, strong) AVAudioPlayer *player;
- (IBAction)playShortSound;
- (IBAction)playLongSound;
- (IBAction) pause;
- (IBAction)skipForward;
- (IBAction)skipBack;
- (IBAction) scrub:(id)sender;
- (IBAction)speak;
-(void)parser:(NSXMLParser*)parser foundCharacters:(NSString*)string;
@end
```

文件 ViewController.m 是文件 ViewController.h 的实现，具体代码如下所示。

```
#import "ViewController.h"
#import <AVFoundation/AVFoundation.h>
@implementation ViewController
@synthesize tf;
@synthesize player;
- (IBAction)playShortSound {
    if (soundID == 0) {
        NSString *path = [[NSBundle mainBundle] pathForResource:@"damn" ofType:@"caff"];
        NSURL *url = [NSURL fileURLWithPath:path];

        AudioServicesCreateSystemSoundID ((__bridge CFURLRef)url, &soundID);
    }
    AudioServicesPlaySystemSound(soundID);
}
- (void)pause {
    [player pause];
}
- (void)play {
    [player play];
}
- (IBAction)playLongSound {
    if (!player) {
        NSError *error = nil;
        NSString  *path  =  [[NSBundle  mainBundle]  pathForResource:@"hongloumeng" ofType:@"mp3"];
        NSURL *url = [NSURL fileURLWithPath:path];

        player = [[AVAudioPlayer alloc] initWithContentsOfURL:url error:&error];
        player.delegate = self;
    }
    if (player.playing) {
        //[self pause];
```

```
        } else if (player) {
            [self play];
        }
    }
    - (IBAction)skipForward {
        player.currentTime = player.currentTime + 30.0;
    }
    - (IBAction)skipBack {
        player.currentTime = player.currentTime - 30.0;
    }
    - (IBAction)scrub:(id)sender
    {
        UISlider *slider = (UISlider *)sender;
        player.currentTime = player.duration * slider.value;
    }
    - (void)audioPlayerBeginInterruption:(AVAudioPlayer *)thePlayer{
        if (thePlayer == player) {
            [self pause];
        }
    }
    - (void)audioPlayerEndInterruption:(AVAudioPlayer *)thePlayer{
        if (thePlayer == player) {
            [self play];
        }
    }
    - (void)disposeSound {
        if (soundID) {
            AudioServicesDisposeSystemSoundID(soundID);
            soundID = 0;
        }
        player = nil;
    }
    - (void)didReceiveMemoryWarning {
        [self disposeSound];
      [super didReceiveMemoryWarning];
    }
    #pragma mark - View lifecycle
    - (void)viewDidUnload
    {

    }
    @end
```

执行后的效果如图7-16所示。

图7-16 执行效果

实例195 播放本地的视频文件

实例195	播放本地的视频文件
源码路径	\daima\195\

实例说明

在本实例中，使用类 MPMoviePlayerController 播放了一个指定的本地视频文件"sample_iTunes.mov"。为了可以让播放的视频全屏显示，特意隐藏了屏幕中的状态栏，并且为了横向显示播放界面，特意在文件 Video-Info.plist 中添加了如下键值。

```
Status bar is initially hidden=YES
```

具体实现

编写文件 ViewController.h 实现视图界面，具体代码如下所示。

```objc
#import <UIKit/UIKit.h>
#import <MediaPlayer/MediaPlayer.h>
@interface ViewController : UIViewController{
    MPMoviePlayerController *mpcontroller;
}
@property(nonatomic ,strong)MPMoviePlayerController *mpcontroller;
@end
```

文件 ViewController.m 是文件 ViewController.h 的实现，具体代码如下所示。

```objc
#import "ViewController.h"
@implementation ViewController
@synthesize mpcontroller;
- (void)didReceiveMemoryWarning
{
    [super didReceiveMemoryWarning];
}
#pragma mark - View lifecycle
- (void)viewDidLoad
{
    NSString *loc = [[NSBundle mainBundle] pathForResource:@"sample_iTunes"
                                                    ofType:@"mov"];//电影文件的位置
    NSURL *url=[NSURL fileURLWithPath:loc];
    //初始化播放器
    mpcontroller = [[MPMoviePlayerController alloc] initWithContentURL:url];
    //把播放器的视图添加到当前视图下（作为子视图）
    [self.view addSubview:mpcontroller.view];
    //设置 frame,让它显示在屏幕上，分别是 X、Y、宽度和高度。你可以再调整
    //mpcontroller.view.frame = CGRectMake(0, 0, 480, 300);
    mpcontroller.view.frame = CGRectMake(-80, 80, 480, 300);
    [mpcontroller.view
setTransform:CGAffineTransformMakeRotation(90.0f*(M_PI/180.0f))];
    //设置电影结束后的回调方法（方法名为 callbackFunction）。注册自己为 observer
    //当 MPMoviePlayerPlaybackDidFinishNotification 事件发生时，就调用指定的方法
    [[NSNotificationCenter defaultCenter] addObserver:self
                                    selector:@selector(callbackFunction:)
name:MPMoviePlayerPlaybackDidFinishNotification object:mpcontroller];
    //设置播放器的一些属性
    mpcontroller.fullscreen = YES;//全屏
    mpcontroller.scalingMode = MPMovieScalingModeFill;
    //mpcontroller.controlStyle = MPMovieControlStyleNone;
    //播放电影
    [mpcontroller play];
    [super viewDidLoad];
    // Do any additional setup after loading the view, typically from a nib.
}
//电影结束后的回调方法
-(void)callbackFunction:(NSNotification*)notification{
    MPMoviePlayerController* video = [notification object];//也可以直接使用 mpcontrol
    //从通知中心注销自己
    [[NSNotificationCenter defaultCenter] removeObserver:self
name:MPMoviePlayerPlaybackDidFinishNotification object:video];
    video = nil;
}
- (void)viewDidUnload
{
    [super viewDidUnload];
}
- (void)viewWillAppear:(BOOL)animated
{
    [super viewWillAppear:animated];
}

- (void)viewDidAppear:(BOOL)animated
{
    [super viewDidAppear:animated];
```

```
}
- (void)viewWillDisappear:(BOOL)animated
{
    [super viewWillDisappear:animated];
}
- (void)viewDidDisappear:(BOOL)animated
{
    [super viewDidDisappear:animated];
}
-
(BOOL)shouldAutorotateToInterfaceOrientation:(UIInterfaceOrientation)interfaceOrienta
tion
{
    return (interfaceOrientation == UIInterfaceOrientationLandscapeRight);
}
@end
```

执行后的效果如图 7-17 所示。

图 7-17　执行效果

实例 196　在播放界面中叠加视频

实例 196	在播放界面中叠加视频
源码路径	\daima\196\

实例说明

本实例是一个 iPad 项目，在项目中特意定义了一个名为 CommentView 的类，通过此类实现了一个气泡效果，这样在屏幕中以气泡的样式注释视频播放界面。在气泡中显示了我们预先设置的文字，显示的文字用 shoutOutTexts 表示。

具体实现

编写文件 **ViewController.h** 实现视图界面，具体代码如下所示。

```
#import <UIKit/UIKit.h>
#import <MediaPlayer/MPMoviePlayerController.h>
#import "CommentView.h"

@interface ViewController : UIViewController {
    UIView *viewForMovie;
    MPMoviePlayerController *player;
    NSArray *shoutOutTexts;
    NSArray *shoutOutTimes;
    int position;
}
@property (nonatomic, retain) IBOutlet UIView *viewForMovie;
@property (nonatomic, retain) MPMoviePlayerController *player;

- (NSURL *)movieURL;
- (void)checkShoutouts:(NSTimer*)theTimer;
- (void)removeView:(NSTimer*)theTimer;
- (BOOL)isTimeForNextShoutout;
```

@end

文件 ViewController.m 是文件 ViewController.h 的实现，主要代码如下所示。

```objc
#import "ViewController.h"
@implementation ViewController
@synthesize player;
@synthesize viewForMovie;
- (void)didReceiveMemoryWarning
{
    [super didReceiveMemoryWarning];
}
#pragma mark - View lifecycle
- (void)viewDidLoad
{
    [super viewDidLoad];
    shoutOutTexts = [NSArray
                     arrayWithObjects:
                     @"这是视频\n 播放器的高级\n 功能展示 ",
                     @"设置 iPod\n 连接与电脑\n 同步!",
                     nil];
    
    shoutOutTimes = [NSArray
                     arrayWithObjects:
                     [[NSNumber alloc] initWithInt: 2],
                     [[NSNumber alloc] initWithInt: 60],
                     nil];
    position = 0;
    [NSTimer
       scheduledTimerWithTimeInterval:1.0f
       target:self
       selector:@selector(checkShoutouts:)
       userInfo:nil
       repeats:YES];
    self.player = [[MPMoviePlayerController alloc] init];
    self.player.contentURL = [self movieURL];
    self.player.view.frame = self.viewForMovie.bounds;
    self.player.view.autoresizingMask =
    UIViewAutoresizingFlexibleWidth |
    UIViewAutoresizingFlexibleHeight;
    [self.viewForMovie addSubview:player.view];
    [self.player play];
}
-(NSURL *)movieURL
{
    NSBundle *bundle = [NSBundle mainBundle];
    NSString *moviePath =
    [bundle
     pathForResource:@"sample_iTunes"
     ofType:@"mov"];
    if (moviePath) {
        return [NSURL fileURLWithPath:moviePath];
    } else {
        return nil;
    }
}
- (void)checkShoutouts:(NSTimer*)theTimer {
    if ([self isTimeForNextShoutout]) {
        CommentView *commentView = [[CommentView alloc]
          initWithText:[shoutOutTexts objectAtIndex:position++]];
        [self.player.view addSubview:commentView];
        [NSTimer scheduledTimerWithTimeInterval:4.0f
          target:self
          selector:@selector(removeView:)
          userInfo:commentView
          repeats:NO];
    }
}
-(BOOL)isTimeForNextShoutout {
    int count = [shoutOutTimes count];
    if (position < count) {
        int timecode = [[shoutOutTimes
```

```
                        objectAtIndex:position] intValue];
            if (self.player.currentPlaybackTime >= timecode) {
                return YES;
            }
        }
        return NO;
}
```

创建一个名为 CommentView 的类，通过此类将气泡叠加在视频播放界面中。实现文件 CommentView.h 的代码如下所示。

```
#import <UIKit/UIKit.h>
@interface CommentView : UIView {
}
- (id)initWithFrame:(CGRect)frame andText:(NSString *) text;
- (id)initWithText:(NSString *) text;
@end
```

文件 CommentView.m 是文件 CommentView.h 的实现，具体代码如下所示。

```
#import "CommentView.h"
@implementation CommentView
- (id)initWithFrame:(CGRect)frame andText:(NSString *) text {
    if ((self = [super initWithFrame:frame])) {
            UIImage *image = [UIImage imageNamed:@"comment.png"];
            UIImageView *imageView = [[UIImageView alloc] initWithImage:image];
            [self addSubview:imageView];
            CGRect rect = CGRectMake(20, 20, 200.0f, 90.0f);
            UILabel *label = [[UILabel alloc] initWithFrame:rect];
            label.text = text;
            label.numberOfLines = 3;
            label.adjustsFontSizeToFitWidth = YES;
            label.textAlignment = UITextAlignmentCenter;
            label.backgroundColor = [UIColor clearColor];
             [self addSubview:label];
        }
    return self;
}
- (id)initWithText:(NSString *) text {
    if ((self = [super init])) {
            UIImage *image = [UIImage imageNamed:@"comment.png"];
            UIImageView *imageView = [[UIImageView alloc] initWithImage:image];
            [self addSubview:imageView];

            CGRect rect = CGRectMake(20, 20, 200.0f, 90.0f);
            UILabel *label = [[UILabel alloc] initWithFrame:rect];
            label.text = text;
            label.numberOfLines = 3;
            label.adjustsFontSizeToFitWidth = YES;
            label.textAlignment = UITextAlignmentCenter;
            label.backgroundColor = [UIColor clearColor];
            [self addSubview:label];
        }
    return self;
}
@end
```

执行后的效果如图 7-18 所示。

图 7-18 执行效果

第 8 章　互联网应用实战

21 世纪的前 10 年被称为信息时代，互联网是信息时代铸造的产物。互联网的推出直接改变了人们的日常生活。现在人们已经越来越离不开网上冲浪和发送邮件等互联网应用了。在本节的内容中，将通过几个典型实例的实现过程，来详细介绍在 Android 系统中和互联网应用相关的基本知识。

实例 197　使用 Web 视图获取网络信息

实例 197	使用 Web 视图获取网络信息
源码路径	\daima\197\

实例说明

在 iOS 应用中，可以将 Web 视图视为没有边框的 Safari 窗口，将其加入到应用程序中并以编程方式进行控制。通过使用这个类，可以用免费方式显示 HTML，加载网页以及支持两个手指张合与缩放手势。Web 视图还可用于实现各种类型的文件，而无需有关这些文件格式的知识。

- HTML、图像和 CSS。
- Word 文档（.doc/.docx）。
- Excel 电子表格（.xls/.xlsx）。
- Keynote 演示文稿（.key.zip）。
- Numbers 电子表格（.numbers.zip）。
- Pages 文档（.pages.zip）。
- PDF 文件（.pdf）。
- PowerPoint 演示文稿（.ppt/.pptx）。

我们可以将这些文件作为资源加入到项目中，并在 Web 视图中显示它们，访问远程服务器中的这些文件或读取 iOS 设备存储空间中的这些文件。

本实例的功能是获取 FloraPhotographs.com 的花朵照片和花朵信息。该应用程序让用户轻按分段控件（ljLSegmentedControl）中的一种花朵颜色，然后从网站 FloraPhotographs.com 取回一朵这样颜色的花朵，并在 Web 视图中显示它。随后用户可以使用开关 UISwitch 来显示和隐藏另一个视图，该视图包含有关该花朵的详细信息。最后，一个标准按钮（UIButton）让用户能够从网站取回另一张当前选定颜色的花朵照片。

具体实现

（1）启动 Xcode，然后在左侧导航选择第一项 "Create a new Xcode project"，如图 8-1 所示。

（2）在弹出的新界面中选择项目类型和模板。在 New Project 窗口的左侧，确保选择了项目类型 iOS 中的 Application，在右边的列表中选择 Single View Application，再单击 "Next" 按钮，如图 8-2 所示。

（3）在 Product Name 文本框中输入 "lianhe"。对于公司标识符，可以将其设置为域名，但顺序相反（本书再次使用的是 com.guan）。保留文本框 Class Prefix 为空，并确保从下拉列表 Device Family 中选择了 iPhone 或 iPad，本书再次选择的是 iPhone，如图 8-3 所示。然后，单击 Next 按钮。

实例 197　使用 Web 视图获取网络信息

图 8-1　创建一个 Xcode 工程

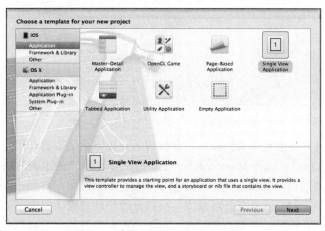

图 8-2　选择 Single View Application

图 8-3　指定应用程序的名称和目标设备

（4）在 Xcode 提示时指定存储位置，再单击 Create 按钮创建项目。这将创建一个简单的应用程序结构，它包含一个应用程序委托、一个窗口、一个视图（在故事板场景中定义的）和一个视图控制器。几秒钟后，项目窗口将打开。同以前一样，这里的重点也是视图（已包含在 MainStoryboard.storyboard 中）和视图控制器类 ViewController，如图 8-4 所示。

（5）规划变量和连接。要创建这个基于 Web 的图像查看器，需要 3 个输出口和两个操作。分段控件将被连接到一个名为 colorChoice 的输出口，因为我们将使用它来确定用户选择的颜色。包含花朵图像的 Web 视图将连接到输出口 flowerView，而包含详细信息的 Web 视图将连接到输出口 flowerDetailView。

应用程序必须使用操作来完成两项工作：获取并显示一幅花朵图像以及显示/隐藏有关花朵的

343

详细信息，其中前者将通过操作 getFlower 来完成，而后者将使用操作 toggleFlowerDetail 来处理。

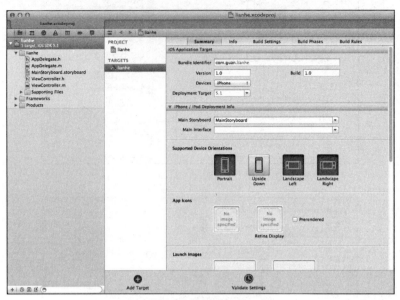

图 8-4　新建的工程

（6）添加分段控件。当确定需要的输出口和操作后，接下来需要设计用户界面。首先需要为设计 UI 配置好 Xcode 工作区：选择 MainStoryboard.storyboard，在 Interface Builder 编辑器中打开它。如果有必要，需要关闭项目导航器以腾出更多的显示空间。要在用户界面中添加分段控件，选择菜单 View→Utilities→Object Library 打开对象库，找到分段控件对象（UISegmentedControl），并将其拖曳到视图中，将它放在视图顶部附近并居中。由于该控件最终用于选择颜色，单击并拖曳一个标签（UILabel）到视图中，将其放在分段控件的上方，并将其文本改为"选择一种颜色"。

在默认情况下，分段控件有两段，其标题分别为 First 和 Second。可双击这些标题并在视图中直接编辑它们，但这不太能够满足我们的要求。在这个项目中，我们需要一个有 4 段的分段控件，每段的文本分别为红、绿、黄和蓝，这些是用户可请求从网站 FloraPhotographs 获取的花朵颜色。显然，要提供所有这些选项，还需添加几段。

分段控件包含的分段数可在 Attributes Inspector 中配置。为此，选择添加到视图中的分段控件，并按 Option+ Command+4 打开 Attributes Inspector。然后在文本框 Segments 中，将数字从 2 增加到 4，用户将立刻能够看到新增的段。在该检查器中，文本框 Segments 下方有一个下拉列表，从中可选择每个段。用户可通过该下拉列表选择一段，再在 Title 文本框中指定其标题。还可以添加图像资源，并指定每段显示的图像。

在 Attributes Inspector 中，除颜色和其他属性外，还有 4 个指定分段控件样式的选项，在属性下拉列表 Style 中选择 Plain、Bordered、Bar 或 Bezeled。分段控件的外观在视图中很可能不合适。为使其大小更合适，可使用控件周围的手柄放大或缩小它。另外，还可使用 Size Inspector（Option+ Command+5）中的 Width 选项调整每段的宽度，如图 8-5 所示。

（7）添加开关。接下来要添加的 UI 元素是开关（UISwitch）。这个应用程序使用的开关的功能是：显示和隐藏包含花朵详细信息的 Web 视图（flowerDetailView）。为添加这个开关，从 Library 将开关（UISwitch）拖曳到视图中，并将它放在屏幕的右边缘，并位于分段控件下方。

与分段控件一样，通过一个屏幕标签提供基本的使用指南很有帮助。为此，将一个标签（UILabel）拖曳到视图中，并将其放在开关左边，再将其文本改为 Show Photo Details。现在，视图应该如图 8-6 所示，但开关很可能显示 ON。

实例 197　使用 Web 视图获取网络信息

图 8-5　使用 Size Inspector 调整每个分段的宽度

图 8-6　添加一个用于显示/隐藏花朵详细信息的开关

　　这个开关只有两个选项：默认状态是开还是关；在开关处于开状态时，应用哪种自定义色调。加入到视图中的开关的默认状态为 ON，但如果想将其默认状态设置为 OFF，则要修改默认状态，选择开关并按 Option+Command+4 打开 Attributes Inspector，再使用下拉列表 State 将默认状态改为 OFF。这就完成了开关的设置。

　　（8）添加 Web 视图。这个应用程序依赖于两个 Web 视图，其中一个显示花朵图像；而另一个显示有关花朵的详细信息（可显示/隐藏它），包含详细信息的 Web 视图将显示在图像上面，因此首先添加主 Web 视图 flowerView。

　　要在应用程序中添加 Web 视图（UIWebView），在对象库中找到它并拖曳到视图中。Web 视图将显示一个可调整大小的矩形，可通过拖曳将其放到任何地方。由于这是将在其中显示花朵图像的 Web 视图，因此将其上边缘放在屏幕中央附近，再调整大小，使其宽度与设备屏幕相同，且完全覆盖视图的下半部分。

　　然后重复上述操作，添加另一个用于显示花朵详细信息的 Web 视图（flowerDetailView），但将其高度调整为大约 0.5 英寸，将其放在屏幕底部并位于 flowerView 的上面，如图 8-7 所示。

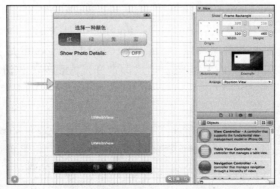

图 8-7　在视图中添加两个 Web 视图（UIWebView）

345

此时可以在文档大纲区域拖曳对象，以调整堆叠顺序。元素离层次结构顶部越近，就越排在后面。要访问 Web 视图的属性，需要选择添加的 Web 视图之一，再按 Option+ Command+4 打开 Attributes Inspector，如图 8-8 所示。

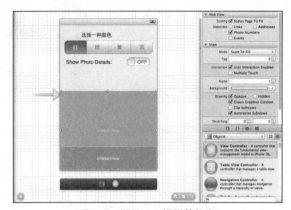

图 8-8　配置 Web 视图的行为

有两类复选框可供选择：Scaling（缩放）和 Detection（检测），其中检测类复选框包括 Phone Number（电话号码）、Address（地址）、Events（事件）和 Link（链接）。如果选中了复选框 Scales Page to Fit，大网页将缩小到与定义的区域匹配；如果选中了检测类复选框，iOS 数据检测器将发挥作用，给它认为是电话号码、地址、日期或 Web 链接的内容添加下画线。

对于第二个 Web 视图，我们不希望使用这种设置，因此选择应用程序中显示花朵详细信息的 Web 视图，并使用 Attributes Inspector 确保不会进行缩放。另外，如果修改该 Web 视图的属性，使其 Alpha 值大约为 0.65 则在照片上面显示详细信息时，将生成漂亮的透明效果。

（9）完成界面设计。现在，该界面只缺少一个按钮（UIButton），它让用户能够随时手工触发 getFlower 方法。如果没有该按钮，则在需要看到新花朵图像时，用户必须使用分段控件切换颜色。该按钮只是触发一个操作（getFlower）。然后拖放一个按钮到视图中，并将它放在屏幕中央 Web 视图的上方。将该按钮的标题改为"获取图片"。至此，界面便设计好了。

（10）创建并连接输出口和操作。在这个项目中，需要连接的界面元素有很多：分段控件、开关、按钮和 Web 视图都需要有到视图控制器的合适连接。下面列举需要使用的输出口和操作。需要的输出口包括如下 3 项。

- 用于指定颜色的分段控件（UISegmentedControl）：colorChoice。
- 显示花朵本身的 Web 视图（UIWebView）：flowerView。
- 显示花朵详细信息的 Web 视图（UIWebView）：flowerDetailView。

需要的操作包括如下 2 项。

- 在用户单击 Get New Flower 按钮时获取新花朵：getFlower。
- 根据开关的设置显示/隐藏花朵详细信息：toggleFlowerDetail。

接下来开始准备工作区，要确保选择了 MainStoryboard.storyboard 后再打开助手编辑器。如果空间不够，隐藏项目导航器和文档大纲区域。这里假定开发人员熟悉该流程，因此从现在开始，将快速介绍连接的创建。

第 1 步：添加输出口。

首先按住 Control 键，并从用于选择颜色的分段控件拖曳到文件 ViewController.h 中编译指令 @interface 的下方。在 Xcode 提示时，将连接类型设置为输出口；将名称设置为 colorChoice，保留其他设置为默认值。这让我们能够在代码中轻松地获悉当前选择的颜色。

继续生成其他的输出口。将主（较大的）Web 视图连接到输出口 flowerView，方法是按住 Control 键，并将它拖曳到 ViewController.h 中编译指令 @property 下方。最后，以同样的方式将第二个 Web

视图连接到输出口 flowerDetailView，如图 8-9 所示。

第 2 步：添加操作。

此应用程序 UI 触发的操作有两个：toggleFlowerDetailgetFlower，用于"隐藏/显示"花朵的详细信息；标准按钮触发 getFlower，以加载新图像。这两种操作很简单。但有时除用户可能执行的显而易见的操作外，还需考虑他们在使用界面时期望发生的情况。

在这个应用程序中，向用户显示的界面很简单。用户应该能够立即意识到他们可选一种颜色，再单击按钮以显示这种颜色的花朵。但应用程序是否应比较聪明，在用户选择颜色后立即显示新花朵呢？为何用户改变颜色后还需单击按钮？通过将 UISegmentedControl 的 Value Changed 事件连接到按钮触发的方法 getFlower 可实现这种功能，而无需多编写一行代码。

首先将开关（UISwitch）连接到操作 toggleFlowerDetail，方法是按住 Control 键，并从开关拖曳到 ViewController.h 中编译指令 @property 下方。确保操作由事件 Value Changed 触发，如图 8-10 所示。

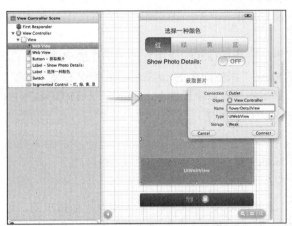

图 8-9　将 Web 视图连接到合适的输出口　　　　图 8-10　并将事件指定为 ValueChanged

接下来按住 Control 键，从按钮拖曳到刚创建的 IBAction 代码行下方。在 Xcode 提示时配置一个新操作：getFlower，并将触发事件指定为 Touch Up Inside。最后还需要将分段控件（UISegmentedControl）连接到新添加的操作 getFlower，并将触发事件指定为 Value Changed，这样用户只需选择颜色就可加载新的花朵图像。

为此，切换到标准编辑器，并确保文档大纲区域可见（选择菜单 Editor→Show Document Outline）。选择分段控件，并按 Option+Command+6（或选择菜单 View→Utilities→Connections Inspector）打开 Connections Inspector。再从 Value Changed 旁边的圆圈拖曳到文档大纲区域中的 View Control 图标，如图 8-11 所示。然后松开鼠标，并在 Xcode 提示时选择 getFlower。

图 8-11　将分段控件的 Value Changed 事件连接到方法 getFlower

设计好界面并建立连接后，接口文件 ViewController.h 的代码如下所示。

```
#import <UIKit/UIKit.h>
@interface ViewController : UIViewController
@property (strong, nonatomic) IBOutlet UISegmentedControl *colorChoice;
@property (strong, nonatomic) IBOutlet UIWebView *flowerView;
@property (strong, nonatomic) IBOutlet UIWebView *flowerDetailView;
- (IBAction)getFlower:(id)sender;
- (IBAction)toggleFlowerDetail:(id)sender;
@end
```

（11）实现应用程序逻辑。

视图控制器需要通过两个操作方法实现如下两个功能。

- toggleflowerDetailView：判断开关的状态是开还是关，并显示或隐藏 Web 视图 flowerDetail View。
- getFlower：将一幅花朵图像加载到 Web 视图 flowerView 中，并加载有关该照片的详细信息，同时将这个照片的详细信息加载到 Web 视图 flowerDetailView 中。

下面首先编写其中较容易的方法：toggleFlowerDetail。

第 1 步：隐藏和显示详细信息 Web 视图。

对从 UIView 派生而来的对象来说，一个很有用的特征是在 iOS 应用程序界面中隐藏或显示它。由于用户在屏幕上看到的几乎任何东西都是从 UIView 类派生而来的，这意味着可隐藏和显示标签、按钮、文本框、图像以及其他视图。要隐藏对象，只需将其布尔值属性 hidden 设置为 TRUE 或 YES（它们的含义相同）。因此，要隐藏 flowerDetailView，代码如下。

```
self.flowerDetailView.hidden=YES;
```

要重新显示它，只需执行相反的操作——将 hidden 属性设置为 FALSE 或 NO。

```
self.flowerDetailView.hidden=NO;
```

要实现方法 toggleFlowerDetail:的逻辑，需要确定开关的当前状态。正如本章前面指出的，可通过方法 isOn 来检查开关的状态，如果开关的状态为开，该方法将返回 TRUE/YES，否则将返回 FALSE/NO。

由于没有创建与开关对应的输出口，因此将在方法中使用变量 sender 来访问它。当操作方法 toggleFlowerDetail 被调用时，该变量被设置为一个这样的引用，即指向触发操作的对象，也就是开关。要检查开关的状态是否为开，可编写如下代码。

```
if([sender isOn]){<switch is on>}else{<switch is off>}
```

根据一个布尔值决定隐藏还是显示 flowerDetailView，而这个布尔值是从开关的 isOn 方法返回的。这可转换为如下两个条件。

- 如果[sender isOn]为 YES，则应显示该 Web 视图（flowerDetailView.hidden=NO）。
- 如果[sender isOn]为 NO，则应隐藏该 Web 视图（flowerDetailView.hidden=YES）。

换句话说，开关的状态与要给 Web 视图的 hidden 属性设置的值正好相反。在 C 语言（和 Objective-C）中，要对布尔值取反，只需在它前面加上一个惊叹号（!）。因此，要决定显示还是隐藏 flowerDetailView，只需将 hidden 属性设置为! [send isOn]。

在 ViewController.m 中，实现 Xcode 添加的方法存根 toggleFlowerDetail。完整的代码如下所示。

```
- (IBAction)toggleFlowerDetail:(id)sender {
  self.flowerDetailView.hidden=![sender isOn];
    /*
    if ([sender isOn]) {
    flowerDetailView.hidden=NO;
    } else {
    flowerDetailView.hidden=YES;
    }
    */
}
```

（12）加载并显示花朵图像和详细信息。

为取回花朵图像，需要利用 FloraPhotographs 专门提供的一项功能。为与该网站交互，需要采取如下 4 个步骤来完成。

第 1 步：从分段控件获取选定的颜色。

第 2 步：生成一个被称为会话 ID 的随机数，让 FloraPhotographs.com 能够跟踪请求。

第 3 步：请求 URL http://www.floraphotographs.com/showrandomios.php?color<color>&session=<session ID>，其中<color>和<session ID>分别是选定 B<J 颜色和生成的随机数。这个 URL 将返回一张花朵照片。

第 4 步：请求 URL http://www.floraphotographs.com/detailios.php?session=<session ID>，其中<session ID>是第 3 步使用的随机数。该 URL 将返回前一步请求的花朵照片的详细信息。

先了解实现这些功能的代码，再讨论实现背后的细节。添加方法 getFlower 的实现，具体代码如下所示。

```
- (IBAction)getFlower:(id)sender {
   NSURL *imageURL;
    NSURL *detailURL;
    NSString *imageURLString;
    NSString *detailURLString;
    NSString *color;
    int sessionID;
    color=[self.colorChoice titleForSegmentAtIndex:
           self.colorChoice.selectedSegmentIndex];
    sessionID=random()%50000;
    imageURLString=[[NSString alloc] initWithFormat:
@"http://www.floraphotographs.com/showrandomios.php?color=%@&session=%d"
                   ,color,sessionID];
    detailURLString=[[NSString alloc] initWithFormat:@"http://www.floraphotographs.com/detailios.php?session=%d"
                   ,sessionID];
    imageURL=[[NSURL alloc] initWithString:imageURLString];
    detailURL=[[NSURL alloc] initWithString:detailURLString];
    [self.flowerView loadRequest:[NSURLRequest requestWithURL:imageURL]];
    [self.flowerDetailView loadRequest:[NSURLRequest requestWithURL:detailURL]];
    self.flowerDetailView.backgroundColor=[UIColor clearColor];
}
```

（13）修复应用程序加载时的界面问题。

实现方法 getFlower 后，便可运行应用程序，且应用程序的一切都将正常工作，只是应用程序启动时，两个 Web 视图是空的，且显示了详细信息视图，尽管开关被设置为 OFF。

为修复这种问题，可在应用程序启动后立刻加载一幅图像，并将 flowerDetailView.hidden 设置为 YES。所以将视图控制器的 viewDidLoad 改为如下所示的代码。

```
- (void)viewDidLoad
{
   self.flowerDetailView.hidden=YES;
    [self getFlower:nil];
    [super viewDidLoad];
}
```

经过上述操作，self.flowerDetailView.hidden=YES 将隐藏详细信息视图。通过使用[self getFlower: nil]，可在视图控制器（被称为 self）中调用 getFlower:，并将一幅花朵图像加载到 Web 视图中。方法 getFlower:接受一个参数，因此向它传递 nil，就像前一章所做的那样（在方法 getFlower: 中没有使用这个值，因此提供参数 nil 不会导致任何问题）。

图 8-12　执行效果

要测试 FlowerWeb 应用程序的最终版本，在 Xcode 中单击按钮 Run。运行后会发现可以缩放 Web 视图并使用手指进行滚动。执行效果如图 8-12 所示。

实例 198 在屏幕中显示指定的网页

实例 198	在屏幕中显示指定的网页
源码路径	\daima\198\

实例说明

在 iOS 应用中，可以使用 UIWebView 控件在屏幕中显示指定的网页。在本实例中，首先在工具条中追加活动指示器，然后使用 requestWithURL 设置了要显示的网页是 http://www.apple.com。并且为了实现良好的体验，特意在载入页面时使用了状态监视功能。在具体实现时，使用 UIActivityIndicatorView 向用户展示"处理中"的图标。

具体实现

实例文件 UIKitPrjWebViewSimple.m 的具体代码如下所示。

```
#import "UIKitPrjWebViewSimple.h"
@implementation UIKitPrjWebViewSimple
- (void)dealloc {
 [activityIndicator_ release];
 if ( webView_.loading ) [webView_ stopLoading];
 webView_.delegate = nil; //Apple 文档中推荐, release 前需要如此编写
 [webView_ release];
 [super dealloc];
}
- (void)viewDidLoad {
 [super viewDidLoad];
 self.title = @"明确显示通信状态";
 // UIWebView 的设置
 webView_ = [[UIWebView alloc] init];
 webView_.delegate = self;
 webView_.frame = self.view.bounds;
 webView_.autoresizingMask =
   UIViewAutoresizingFlexibleWidth | UIViewAutoresizingFlexibleHeight;
 webView_.scalesPageToFit = YES;
 [self.view addSubview:webView_];
 // 在工具条中追加活动指示器
 activityIndicator_ = [[UIActivityIndicatorView alloc] init];
 activityIndicator_.frame = CGRectMake( 0, 0, 20, 20 );
 UIBarButtonItem* indicator =
   [[[UIBarButtonItem alloc] initWithCustomView:activityIndicator_] autorelease];
 UIBarButtonItem* adjustment =
   [[[UIBarButtonItem alloc] initWithBarButtonSystemItem:UIBarButtonSystemItemFlexibleSpace
                                                  target:nil
                                                  action:nil] autorelease];
 NSArray* buttons = [NSArray arrayWithObjects:adjustment, indicator, adjustment, nil];
 [self setToolbarItems:buttons animated:YES];
}

- (void)viewDidAppear:(BOOL)animated {
 [super viewDidAppear:animated];
 //Web 页面显示
    NSURLRequest* request =
   [NSURLRequest requestWithURL:[NSURL URLWithString:@"http://www.apple.com"]];
 [webView_ loadRequest:request];
}
- (void)webViewDidStartLoad:(UIWebView*)webView {
 [activityIndicator_ startAnimating];
}
- (void)webViewDidFinishLoad:(UIWebView*)webView {
 [activityIndicator_ stopAnimating];
}
- (void)webView:(UIWebView*)webView didFailLoadWithError:(NSError*)error {
```

```
    [activityIndicator_ stopAnimating];
}
@end
```

执行后的效果如图 8-13 所示。

图 8-13 执行效果

实例 199 控制屏幕中的网页

实例 199	控制屏幕中的网页
源码路径	\daima\199\

实例说明

在 iOS 应用中，当使用 UIWebView 控件在屏幕中显示指定的网页后，我们可以设置一些链接来控制访问页，例如 "返回上一页"、"进入下一页" 等。此类功能是通过如下方法实现的。

- reload：重新读入页面。
- stopLoading：读入停止。
- goBack：返回前一画面。
- goForward：进入下一画面。

在本实例的屏幕中，添加了 "返回" 和 "向前" 两个按钮，并且设置了 "重载" 和 "停止" 图标，共同实现了网页控制功能。

具体实现

实例文件 UIKitPrjWebView.m 的具体代码如下所示。

```
#import "UIKitPrjWebView.h"
@implementation UIKitPrjWebView
- (void)dealloc {
  if ( webView_.loading ) [webView_ stopLoading];
  webView_.delegate = nil;
  [webView_ release];
  [reloadButton_ release];
  [stopButton_ release];
  [backButton_ release];
  [forwardButton_ release];
  [super dealloc];
}
- (void)viewDidLoad {
  [super viewDidLoad];
  self.title = @"UIWebView 演示";
  // UIWebView 的设置
  webView_ = [[UIWebView alloc] init];
  webView_.delegate = self;
  webView_.frame = self.view.bounds;
```

```objc
  webView_.autoresizingMask =
    UIViewAutoresizingFlexibleWidth | UIViewAutoresizingFlexibleHeight;
  webView_.scalesPageToFit = YES;
  [self.view addSubview:webView_];
  // 工具条中追加按钮
  reloadButton_ =
    [[UIBarButtonItem alloc] initWithBarButtonSystemItem:UIBarButtonSystemItemRefresh
                                  target:self
                                  action:@selector(reloadDidPush)];
  stopButton_ =
    [[UIBarButtonItem alloc] initWithBarButtonSystemItem:UIBarButtonSystemItemStop
                                  target:self
                                  action:@selector(stopDidPush)];
  backButton_ =
    [[UIBarButtonItem alloc] initWithTitle:@"返回"
                                  style:UIBarButtonItemStyleBordered
                                  target:self
                                  action:@selector(backDidPush)];
  forwardButton_ =
    [[UIBarButtonItem alloc] initWithTitle:@"向前"
                                  style:UIBarButtonItemStyleBordered
                                  target:self
                                  action:@selector(forwardDidPush)];
  NSArray* buttons =
    [NSArray arrayWithObjects:backButton_, forwardButton_, reloadButton_, stopButton_,
nil];
  [self setToolbarItems:buttons animated:YES];
}
- (void)reloadDidPush {
  [webView_ reload]; //< 重新读入页面
}
- (void)stopDidPush {
  if ( webView_.loading ) {
    [webView_ stopLoading]; //< 读入停止
  }
}
- (void)backDidPush {
  if ( webView_.canGoBack ) {
    [webView_ goBack]; //< 返回前一画面
  }
}
- (void)forwardDidPush {
  if ( webView_.canGoForward ) {
    [webView_ goForward]; //< 进入下一画面
  }
}
- (void)updateControlEnabled {
  // 统一更新活动指示以及按钮状态
  [UIApplication sharedApplication].networkActivityIndicatorVisible = webView_.loading;
  stopButton_.enabled = webView_.loading;
  backButton_.enabled = webView_.canGoBack;
  forwardButton_.enabled = webView_.canGoForward;
}
- (void)viewDidAppear:(BOOL)animated {
  // 画面显示结束后读入 Web 页面画面
  [super viewDidAppear:animated];
  NSURLRequest* request =
    [NSURLRequest requestWithURL:[NSURL URLWithString:@"http://www.apple.com/"]];
  [webView_ loadRequest:request];
  [self updateControlEnabled];
}
- (void)viewWillDisappear:(BOOL)animated {
  // 画面关闭时状态条的活动指示器设置成 OFF
  [super viewWillDisappear:animated];
  [UIApplication sharedApplication].networkActivityIndicatorVisible = NO;
}
- (void)webViewDidStartLoad:(UIWebView*)webView {
  [self updateControlEnabled];
}
```

```objc
- (void)webViewDidFinishLoad:(UIWebView*)webView {
    [self updateControlEnabled];
}
- (void)webView:(UIWebView*)webView didFailLoadWithError:(NSError*)error {
    [self updateControlEnabled];
}
@end
```

执行效果如图 8-14 所示。

图 8-14　执行效果

实例 200　在网页中加载显示 PDF、Word 和 JPEG 图片

实例 200	在网页中加载显示 PDF、Word 和 JPEG 图片
源码路径	\daima\200\

实例说明

在 iOS 应用中，当使用 UIWebView 控件在屏幕中显示指定的网页后，我们可以在网页中加载显示 PDF、Word 和图片等格式的文件。在本实例的屏幕中，通过使用 loadData:MIMEType:textEncodingName:baseURL 方法，分别显示了指定的 JPEG 图片、PDF 文件和 Word 文件。

具体实现

实例文件 UIKitPrjWebViewLoadData.m 的具体代码如下所示。

```objc
#import "UIKitPrjWebViewLoadData.h"
@implementation UIKitPrjWebViewLoadData
- (void)dealloc {
    [activityIndicator_ release];
    if ( webView_.loading ) [webView_ stopLoading];
    webView_.delegate = nil;
    [webView_ release];
    [super dealloc];
}

- (void)viewDidLoad {
    [super viewDidLoad];
    self.title = @"loadData";
    // UIWebView 的设置
    webView_ = [[UIWebView alloc] init];
    webView_.delegate = self;
    webView_.frame = self.view.bounds;
    webView_.autoresizingMask =
        UIViewAutoresizingFlexibleWidth | UIViewAutoresizingFlexibleHeight;
    [self.view addSubview:webView_];
    // 工具条的设置
    activityIndicator_ = [[UIActivityIndicatorView alloc] init];
    activityIndicator_.frame = CGRectMake( 0, 0, 20, 20 );
```

```
    UIBarButtonItem* indicator =
      [[[UIBarButtonItem alloc] initWithCustomView:activityIndicator_] autorelease];
    UIBarButtonItem* adjustment =
      [[[UIBarButtonItem alloc] initWithBarButtonSystemItem:UIBarButtonSystemItemFlexibleSpace
                                                    target:nil
                                                    action:nil] autorelease];
    NSArray* buttons = [NSArray arrayWithObjects:adjustment, indicator, adjustment, nil];
    [self setToolbarItems:buttons animated:YES];
}
- (void)viewDidAppear:(BOOL)animated {
    [super viewDidAppear:animated];
    /*NSString* path;
    if ( path = [[NSBundle mainBundle] pathForResource:@"sample" ofType:@"pdf"] ) {
      NSData* data = [NSData dataWithContentsOfFile:path];
      [webView_ loadData:data MIMEType:@"application/pdf" textEncodingName:nil baseURL:nil];
    } else {
      NSLog( @"file not found." );
    }
    if ( path = [[NSBundle mainBundle] pathForResource:@"dog" ofType:@"jpg"] ) {
        NSData* data = [NSData dataWithContentsOfFile:path];
        [webView_ loadData:data MIMEType:@"image/jpeg" textEncodingName:nil baseURL:nil];
    } else {
        NSLog( @"file not found." );
    }
    */
    NSString* path = [[NSBundle mainBundle] pathForResource:@"sample.doc" ofType:nil];
    NSURL* url = [NSURL fileURLWithPath:path];
    NSURLRequest* request = [NSURLRequest requestWithURL:url];
    [webView_ loadRequest:request];
}
- (void)updateControlEnabled {
    if ( webView_.loading ) {
      [activityIndicator_ startAnimating];
    } else {
      [activityIndicator_ stopAnimating];
    }
}
- (void)webViewDidStartLoad:(UIWebView*)webView {
    NSLog( @"webViewDidStartLoad" );
    [self updateControlEnabled];
}
- (void)webViewDidFinishLoad:(UIWebView*)webView {
    NSLog( @"webViewDidFinishLoad" );
    [self updateControlEnabled];
}
- (void)webView:(UIWebView*)webView didFailLoadWithError:(NSError*)error {
    NSLog( @"didFailLoadWithError:%d", error.code );
    NSLog( @"%@", error.localizedDescription );
    [self updateControlEnabled];
}
@end
```

执行后的效果如图 8-15 所示。

图 8-15 执行效果

实例 201 在网页中加载 HTML 代码

实例 201	在网页中加载 HTML 代码
源码路径	\daima\201\

实例说明

在 iOS 应用中，当使用 UIWebView 控件在屏幕中显示指定的网页后，我们可以在网页中加载显示 HTML 代码。在本实例的屏幕中，通过使用 UIWebView 的 loadHTMLString: baseURL 方法加载显示了如下 HTML 代码。

```
"<b>【手机号码】</b><br />"
"000-0000-0000<hr />"
"<b>【主页】</b><br />"
"http://www.apple.com/"
```

具体实现

实例文件 UIKitPrjLoadHTMLString.m 的具体代码如下所示。

```
#import "UIKitPrjLoadHTMLString.h"
@implementation UIKitPrjLoadHTMLString
- (void)dealloc {
  [webView_ release];
  [super dealloc];
}
- (void)viewDidLoad {
  [super viewDidLoad];
  self.title = @"loadHTMLString";
  // UIWebView 的设置
  webView_ = [[UIWebView alloc] init];
  webView_.frame = self.view.bounds;
  webView_.autoresizingMask =
    UIViewAutoresizingFlexibleWidth | UIViewAutoresizingFlexibleHeight;
  webView_.dataDetectorTypes = UIDataDetectorTypeAll;
  [self.view addSubview:webView_];
}
- (void)viewDidAppear:(BOOL)animated {
  [super viewDidAppear:animated];
  NSString* html = @"<b>【手机号码】</b><br />"
                    "000-0000-0000<hr />"
                    "<b>【主页】</b><br />"
                    "http://www.apple.com/";
  [webView_ loadHTMLString:html baseURL:nil];
}
@end
```

执行后的效果如图 8-16 所示。

图 8-16 执行效果

实例 202 在网页中实现触摸处理

实例 202	在网页中实现触摸处理
源码路径	\daima\202\

实例说明

在 iOS 应用中,当使用 UIWebView 控件在屏幕中显示指定的网页后,我们可以通过触摸的方式浏览指定的网页。在具体实现时,是通过 webView:shouldStartLoadWithRequest:navigationType 方法实现的。NavigationType 包括如下所示的可选参数值。

- UIWebViewNavigationTypeLinkClicked:链接被触摸时请求这个链接。
- UIWebViewNavigationTypeFormSubmitted:form 被提交时请求这个 form 中的内容。
- UIWebViewNavigationTypeBackForward:当通过 goBack 或 goForward 进行页面转移时移动目标 URL。
- UIWebViewNavigationTypeReload:当页面重新导入时导入这个 URL。
- UIWebViewNavigationTypeOther:使用 loadRequest 方法读取内容。

在本实例中,预先准备了 4 个 ".html" 文件,然后在 iPhone 设备中通过触摸的方式浏览这 4 个 ".html" 文件。

具体实现

(1) 文件 top.htm 的具体代码如下所示。

```html
<html>
  <head>
    <title>首页</title>
      <meta charset="utf-8">
      <meta name="viewport" content="width=device-width" />
  </head>
  <body>
    <h1>3 个颜色</h1>
    <hr />
    <h2>准备选择哪一件衣服?</h2>
    <ol>
      <li /><a href="page1.htm">红色衣服</a>
      <li /><a href="page2.htm">银色衣服</a>
      <li /><a href="page3.htm">黑色衣服</a>
    </ol>
  </body>
</html>
```

(2) 文件 page1.html 的具体代码如下所示。

```html
<html>
  <head>
    <title>PAGE 1</title>
      <meta charset="utf-8">
      <meta name="viewport" content="width=device-width" />
  </head>
  <body>
    <h1>红色衣服</h1>
    <hr />
    <h2>没有任何东西。</h2>
    <ol>
      <li /><a href="top.htm">返回</a>
    </ol>
  </body>
</html>
```

（3）文件 page2.html 的具体代码如下所示。

```html
<html>
  <head>
    <title>PAGE 2</title>
  </head>
  <body>
    <h1>银色衣服</h1>
    <hr />
    <h2>两件衣服</h2>
    <ol>
      <li /><a href="page1.htm">红色衣服</a>
      <li /><a href="page3.htm">黑色衣服</a>
      <li /><a href="top.htm">返回</a>
    </ol>
  </body>
</html>
```

（4）文件 page3.html 的具体代码如下所示。

```html
<html>
  <head>
    <title>HTML 的标题</title>
  </head>
  <body>
    <h1>黑色衣服</h1>
    <hr />
    <h2>有一个黑色的围巾！</h2>
    <ol>
      <li /><a href="top.htm">返回</a>
    </ol>
    <form action="document.title">
      <input type="submit" value="执行 JavaScript" />
    </form>
  </body>
</html>
```

（5）实例文件 UIKitPrjHTMLViewer.m 的具体代码如下所示。

```objc
#import "UIKitPrjHTMLViewer.h"
#pragma mark ----- Private Methods Definition -----
@interface UIKitPrjHTMLViewer ()
- (void)loadHTMLFile:(NSString*)path;
@end
#pragma mark ----- Start Implementation For Methods -----
@implementation UIKitPrjHTMLViewer
- (void)dealloc {
  [activityIndicator_ release];
  if ( webView_.loading ) [webView_ stopLoading];
  webView_.delegate = nil;
  [webView_ release];
  [super dealloc];
}
- (void)viewDidLoad {
  [super viewDidLoad];
  self.title = @"HTMLViewer";
  // UIWebView 的设置
  webView_ = [[UIWebView alloc] init];
  webView_.delegate = self;
  webView_.frame = self.view.bounds;
  webView_.autoresizingMask =
    UIViewAutoresizingFlexibleWidth | UIViewAutoresizingFlexibleHeight;
  [self.view addSubview:webView_];
  // 工具条的设置
  activityIndicator_ = [[UIActivityIndicatorView alloc] init];
  activityIndicator_.frame = CGRectMake( 0, 0, 20, 20 );
  UIBarButtonItem* indicator =
    [[[UIBarButtonItem alloc] initWithCustomView:activityIndicator_] autorelease];
```

```objc
        UIBarButtonItem* adjustment =
          [[[UIBarButtonItem alloc] initWithBarButtonSystemItem:UIBarButtonSystemItemFlexibleSpace
                                                        target:nil
                                                        action:nil] autorelease];
        NSArray* buttons = [NSArray arrayWithObjects:adjustment, indicator, adjustment, nil];
        [self setToolbarItems:buttons animated:YES];
}
//读入指定 HTML 文件的私有方法
- (void)loadHTMLFile:(NSString*)path {
    NSArray* components = [path pathComponents];
    NSString* resourceName = [components lastObject];
    NSString* absolutePath;
    if ( absolutePath == [[NSBundle mainBundle] pathForResource:resourceName ofType:nil] )
    {
        NSData* data = [NSData dataWithContentsOfFile:absolutePath];
        [webView_ loadData:data MIMEType:@"text/html" textEncodingName:@"utf-8" baseURL:nil];
    } else {
        NSLog( @"%@ not found.", resourceName );
    }
}
- (void)updateControlEnabled {
    if ( webView_.loading ) {
        [activityIndicator_ startAnimating];
    } else {
        [activityIndicator_ stopAnimating];
    }
}
- (BOOL)webView:(UIWebView*)webView
    shouldStartLoadWithRequest:(NSURLRequest*)request
    navigationType:(UIWebViewNavigationType)navigationType
{
    //触摸链接后,进入 href 属性为 URL 的下一画面
    if ( UIWebViewNavigationTypeLinkClicked == navigationType ) {
        NSString* url = [[request URL] path];
        [self loadHTMLFile:url];
        return FALSE;
    } else if ( UIWebViewNavigationTypeFormSubmitted == navigationType ) {
        NSString* url = [[request URL] path];
        NSArray* components = [url pathComponents];
        NSString* resultString = [webView stringByEvaluatingJavaScriptFromString:[components lastObject]];
        UIAlertView* alert = [[[UIAlertView alloc] init] autorelease];
        alert.message = resultString;
        [alert addButtonWithTitle:@"OK"];
        [alert show];
        return FALSE;
    }
    return TRUE;
}
//画面显示后,首先显示 top.htm
- (void)viewDidAppear:(BOOL)animated {
    [super viewDidAppear:animated];
    [self loadHTMLFile:@"top.htm"];
}

- (void)webViewDidStartLoad:(UIWebView*)webView {
    NSLog( @"webViewDidStartLoad" );
    [self updateControlEnabled];
}

- (void)webViewDidFinishLoad:(UIWebView*)webView {
    NSLog( @"webViewDidFinishLoad" );
    [self updateControlEnabled];
}

- (void)webView:(UIWebView*)webView didFailLoadWithError:(NSError*)error {
    NSLog( @"didFailLoadWithError:%d", error.code );
    NSLog( @"%@", error.localizedDescription );
    [self updateControlEnabled];
}
```

@end

执行效果如图 8-17 所示。

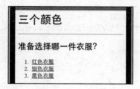

图 8-17　执行效果

实例 203　在屏幕中显示 CSDN 主页

实例 203	在屏幕中显示 CSDN 主页
源码路径	\daima\203\

实例说明

在本实例中，使用 UIWebView 控件在屏幕中显示了 CSDN 主页。

具体实现

（1）使用 UIWebView 的 loadRequest 方法加载一个 URL 地址，它需要一个 NSURLRequest 参数。我们定义一个方法用来加载 URL。在 UIWebViewDemoViewController 中定义下面方法。

```
- (void)loadWebPageWithString:(NSString*)urlString
{
    NSURL *url =[NSURL URLWithString:urlString];
    NSLog(urlString);
    NSURLRequest *request =[NSURLRequest requestWithURL:url];
    [webView loadRequest:request];
}
```

（2）在界面上放置 3 个控件：一个 textfield、一个 button、一个 uiwebview，布局如图 8-18 所示。

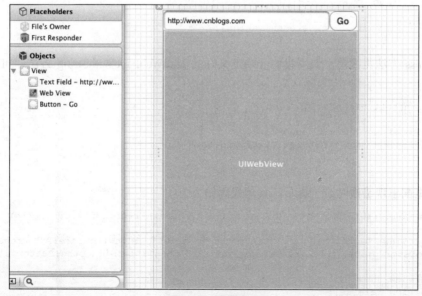

图 8-18　界面布局

（3）在代码中定义相关的控件。其中，webView 用于展示网页，textField 用于实现地址栏，

activityIndicatorView 用于加载的动画，buttonPress 用于实现按钮的点击事件。

```
@interface UIWebViewDemoViewController :UIViewController<UIWebViewDelegate> {
    IBOutlet UIWebView *webView;
    IBOutlet UITextField *textField;
    UIActivityIndicatorView *activityIndicatorView;
}
- (IBAction)buttonPress:(id) sender;
- (void)loadWebPageWithString:(NSString*)urlString;
@end
```

设置 UIWebView，初始化 UIActivityIndicatorView，具体代码如下所示。

```
- (void)viewDidLoad
{
    [super viewDidLoad];
    webView.scalesPageToFit =YES;
    webView.delegate =self;
    activityIndicatorView = [[UIActivityIndicatorView alloc]
                    initWithFrame : CGRectMake(0.0f, 0.0f, 32.0f, 32.0f)] ;
    [activityIndicatorView setCenter: self.view.center] ;
    [activityIndicatorView setActivityIndicatorViewStyle:
UIActivityIndicatorViewStyleWhite] ;
    [self.view addSubview : activityIndicatorView] ;
    [self buttonPress:nil];
    // Do any additional setup after loading the view from its nib.
}
```

（4）在 UIWebView 中主要有下面几个委托方法。

- (void)webViewDidStartLoad:(UIWebView *)webView：开始加载的时候执行该方法。
- (void)webViewDidFinishLoad:(UIWebView *)webView：加载完成的时候执行该方法。
- (void)webView:(UIWebView *)webView didFailLoadWithError:(NSError *)error：加载出错的时候执行该方法。

我们可以将 activityIndicatorView 放置到前面两个委托方法中，具体代码如下所示。

```
- (void)webViewDidStartLoad:(UIWebView *)webView
{
    [activityIndicatorView startAnimating] ;
}
- (void)webViewDidFinishLoad:(UIWebView *)webView
{
    [activityIndicatorView stopAnimating];
}
```

buttonPress 方法很简单，调用我们开始定义好的 loadWebPageWithString 方法即可，代码如下所示。

```
- (IBAction)buttonPress:(id) sender
{
    [textField resignFirstResponder];
    [self loadWebPageWithString:textField.text];
}
```

当请求页面出现错误时给予提示，具体代码如下：

```
- (void)webView:(UIWebView *)webView didFailLoadWithError:(NSError *)error
{
    UIAlertView *alterview = [[UIAlertView alloc] initWithTitle:@"" message:[error
localizedDescription]  delegate:nil cancelButtonTitle:nil otherButtonTitles:@"OK",
nil];
    [alterview show];
    [alterview release];
}
```

到此为止，整个实例介绍完毕，执行后的效果如图 8-19 所示。

图 8-19 执行效果

实例 204　一个简单的网页浏览器

实例 204	一个简单的网页浏览器
源码路径	\daima\204\

实例说明

在本实例中，结合前面的几个实例，使用 UIWebView 控件实现了一个浏览器的功能。本实例比较简单，在视图中设置了一个 WebViewViewController 对象，并设置了一个 UITextField 输入框。当在输入框中输入"www."格式的网址后，单击"go"按钮会来到这个地址。

具体实现

（1）新建一个工程，在界面中拖动 3 个控件——WebView、TextField、Button，把 TextField 和 button 放到 WebView 上面，如图 8-20 所示。

（2）开始声明输出口。右键单击某个控件，然后拖曳到 WebViewViewController.h 文件的 @interface 和@end 之间。然后弹出一个框框，输入控件名字"webView"，如图 8-21 所示。

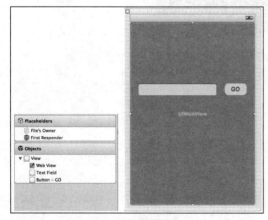

图 8-20　界面视图

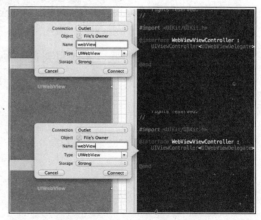

图 8-21　输入控件名字"webView"

然后输入控件名字"textField"，如图 8-22 所示。

继续输入控件名字"Button"，如图 8-23 所示。

（3）给 button 添加一个方法，如图 8-24 所示。

（4）声明一个 UIActivityIndicatorView 对象和一个 loadWebPageWithString 方法，并添加上 UIWebViewDelegate 协议。具体代码如下所示。

```
#import <UIKit/UIKit.h>
@interface WebViewViewController : UIViewController<UIWebViewDelegate>
@property (strong, nonatomic) IBOutlet UIWebView *webView;
@property (strong, nonatomic) IBOutlet UITextField *textField;
@property (strong, nonatomic) IBOutlet UIButton *button;
@property (strong,nonatomic) UIActivityIndicatorView *activityIndicatorView;
- (IBAction)buttonPressed:(id)sender;
-(void)loadWebPageWithString:(NSString *)urlString;
@end
```

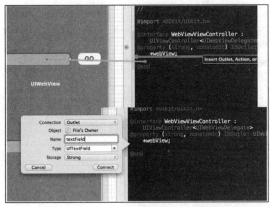

图 8-22　输入控件名字"textField"

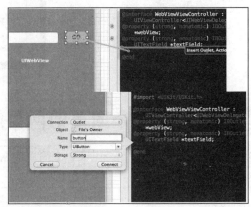

图 8-23　输入控件名字"Button"

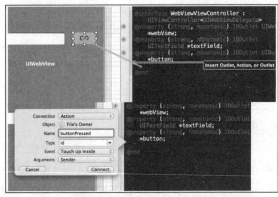

图 8-24　给 button 添加方法

在声明控件输出口的时候，系统也会自动生成一些代码，具体如下。

```
@synthesize textField;
@synthesize webView;
@synthesize button;
- (void)viewDidUnload
{
    [self setTextField:nil];
    [self setWebView:nil];
    [self setButton:nil];
    [super viewDidUnload];
}
- (IBAction)buttonPressed:(id)sender {
}
```

（5）viewDidLoad 方法的实现代码如下所示。

```
- (void)viewDidLoad
{
    [super viewDidLoad];
//自动缩放页面，以适应屏幕
    webView.scalesPageToFit = YES;
    webView.delegate = self;

//    指定进度轮大小
```

```objc
    self.activityIndicatorView = [[UIActivityIndicatorView alloc] initWithFrame:CGRectMake(0, 0, 32, 32)];
//  设置进度轮的中心也可以用[self.activityIndicatorView setCenter:CGPointMake(30, 30)];
    [self.activityIndicatorView setCenter:self.view.center];
//  设置 activityIndicatorView 风格
    [self.activityIndicatorView setActivityIndicatorViewStyle:UIActivityIndicatorViewStyleGray];
    [self.webView addSubview:self.activityIndicatorView];
    [self buttonPressed:nil];
}
```

（6）定义 loadWebPageWithString 方法用于加载指定的 URL，具体代码如下所示。

```objc
-(void)loadWebPageWithString:(NSString *)urlString
{
    if (self.textField.text != nil) {
//      追加一个字符串
        urlString = [@"http://" stringByAppendingFormat:urlString];
        NSURL *url = [NSURL URLWithString:urlString];
//      NSURLRequest 类方法用于获取 URL
        NSURLRequest *request = [NSURLRequest requestWithURL:url];
//      webView 加载 URL
        [webView loadRequest:request];
    }
}
```

（7）编写按钮处理事件，具体代码如下所示。

```objc
//按钮事件，点击按钮开始调用 loadWebPageWithString 方法
- (IBAction)buttonPressed:(id)sender {
    [textField resignFirstResponder];
    [self loadWebPageWithString:textField.text];
//  点击完 button 后隐藏 textField 和 button
    if (sender==button) {
        textField.hidden = YES;
        button.hidden = YES;
    }
}
```

（8）UIWebView 委托方法的实现代码如下所示。

```objc
//UIWebView 委托方法，开始加载一个 url 时候调用此方法
-(void)webViewDidStartLoad:(UIWebView *)webView
{
    [self.activityIndicatorView startAnimating];
}
//UIWebView 委托方法，url 加载完成的时候调用此方法
-(void)webViewDidFinishLoad:(UIWebView *)webView
{
    [self.activityIndicatorView stopAnimating];
}
//加载 url 出错的时候调用此方法
-(void)webView:(UIWebView *)webView didFailLoadWithError:(NSError *)error
{
//  判断 button 是否被触摸
    if (!self.button ) {
        UIAlertView *alert = [[UIAlertView alloc] initWithTitle:@""
                                                        message:[error localizedDescription]
                                                       delegate:nil
                                              cancelButtonTitle:@"OK"
                                              otherButtonTitles: nil];
        [alert show];
    }
}
```

最终的执行效果如图 8-25 所示，在输入框输入一个网址后回到整个页面，例如，输入"www.sohu.com"的效果如图 8-26 所示。

图 8-25　执行效果　　　　　　　　图 8-26　来到搜狐

实例 205　下载并显示远程 URL 地址的 JPEG 图片

实例 205	下载并显示远程 URL 地址的 JPEG 图片
源码路径	\daima\205\

实例说明

在本实例的功能是，下载指定 URL 地址的 JPEG 图片，并将图片显示在屏幕中。本实例需要用到如下所示的知识点。

- (void)connection:(NSURLConnection*)connection didReceiveResponse:(NSHTTPURLResponse*)response：表示要装载 URL 请求，这个 request 对象作为初始化进程的一部分，被深度复制（deep-copied）。在这个方法返回之后，再修改 request，将不会影响用在装载的过程中的 request。
- (void)connection:(NSURLConnection*)connection didReceiveData:(NSData*)data：表示输出，由服务器返回的 URL 响应。
- (void)connection:(NSURLConnection*)connection didFailWithError:(NSError*)error：表示输出，如果在处理请求的过程中发生错误就会使用，无错误时为 NULL。
- (void)connectionDidFinishLoading:(NSURLConnection *)connection：表示加载完毕。
- sendSynchronousRequest:returningResponse:error：用于同步加载一个 URL 请求。

具体实现

文件 MyDownloader.m 实现加载功能，具体代码如下所示。

```
#import "MyDownloader.h"
@implementation MyDownloader
@synthesize connection, receivedData, request;
- (id) initWithRequest: (NSURLRequest*) req {
    self = [super init];
    if (self) {
        self->request = [req copy];
        self->connection = [[NSURLConnection alloc] initWithRequest:req delegate:self startImmediately:NO];
        self->receivedData = [[NSMutableData alloc] init];
    }
    return self;
}
- (void) connection:(NSURLConnection *)connection didReceiveResponse:(NSURLResponse *)response {
    [receivedData setLength:0];
}
- (void) connection:(NSURLConnection *)connection didReceiveData:(NSData *)data {
    [receivedData appendData:data];
}
```

实例 206　解析指定的 XML 文件

```
- (void)connection:(NSURLConnection *)connection didFailWithError:(NSError *)error {
    [[NSNotificationCenter defaultCenter] postNotificationName:@"connectionFinished"
object:self userInfo:[NSDictionary dictionaryWithObject:error forKey:@"error"]];
}
- (void)connectionDidFinishLoading:(NSURLConnection *)connection {
    [[NSNotificationCenter defaultCenter] postNotificationName:@"connectionFinished"
object:self];
}

@end
```

文件 MyImageDownloader.m 用于下载指定 URL 的图片，具体代码如下所示。

```
#import "MyImageDownloader.h"
@implementation MyImageDownloader
@synthesize image;
- (UIImage*) image {
    if (image)
        return image;
    [self.connection start];
    return nil;
}
- (void)connection:(NSURLConnection *)connection didFailWithError:(NSError *)error {
    self.connection = [[NSURLConnection alloc] initWithRequest:self.request delegate:self startImmediately:NO];
}
- (void)connectionDidFinishLoading:(NSURLConnection *)connection {
    UIImage* im = [UIImage imageWithData:self->receivedData];
    if (im) {
        self.image = im;
        [[NSNotificationCenter defaultCenter] postNotificationName:@"imageDownloaded"
object:self];
    }
}
@end
```

执行后的效果如图 8-27 所示。

图 8-27　执行效果

实例 206　解析指定的 XML 文件

实例 206	解析指定的 XML 文件
源码路径	\daima\206\

实例说明

　　XML 是用于标记电子文件使其具有结构性的标记语言，可以用来标记数据、定义数据类型，是一种允许用户对自己的标记语言进行定义的源语言。在本实例的功能是，在屏幕中解析指定 URL 地址的 XML 文件。整个实例的原理比较简单，在具体实现时实现了网络数据请求（利用 ASIHTTPRequest），然后利用 NSXMLParser 解析 XML 数据，最后把数据存放到 table 中。本实例利用了开源框架资料，这些资料来自于 http://allseeing-i.com/ASIHTTPRequest。

具体实现

实例文件 testViewController.m 的具体代码如下所示。

```objc
#import "testViewController.h"
#import "ASIHTTPRequest.h"
@interface testViewController ()
@end
@implementation testViewController
@synthesize m_strCurrentElement;
@synthesize tempString;
- (id)initWithNibName:(NSString *)nibNameOrNil bundle:(NSBundle *)nibBundleOrNil
{
    self = [super initWithNibName:nibNameOrNil bundle:nibBundleOrNil];
    if (self) {
        // Custom initialization
    }
    return self;
}
-(void)dealloc
{
    [super dealloc];
}
- (void)ASIHttpRequestFailed:(ASIHTTPRequest *)request{
    if (request) {
        [request release];
    }
    NSError *error = [request error];
    NSLog(@"the error is %@",error);
}
//
- (void)ASIHttpRequestSuceed:(ASIHTTPRequest *)request{
    //成功，这里怎么写
    NSData *responseData = [request responseData];
    
    NSXMLParser *m_parser = [[NSXMLParser alloc] initWithData:responseData];
    [m_parser setDelegate:self];   //设置代理为本地
    BOOL flag = [m_parser parse];  //开始解析
    if(flag) {
        NSLog(@"ok");
    }else{
        NSLog(@"获取指定路径的 XML 文件失败");
    }
    [m_parser release];
}
- (void)parserDidStartDocument:(NSXMLParser *)parser {
    parserObjects = [[NSMutableArray alloc] init];
    //每一组信息都用数组来存，最后得到的数据即在此数组中
}
- (void)parser:(NSXMLParser *)parser didStartElement:(NSString *)elementName namespaceURI:(NSString *)namespaceURI qualifiedName:(NSString *)qualifiedName attributes:(NSDictionary *)attributeDict {
    
    NSArray *elementArray = [[NSArray alloc] initWithObjects:@"NewsId",@"NewsTitle",@"NewsUser",@"NewsContent",@"NewsDate",nil];
    if ([elementName isEqualToString:@"News"]) {   //开始解析 News 节点
        [dataDict release];
        dataDict = [[NSMutableDictionary alloc] initWithCapacity:0];
        //每一条信息都用字典来存储
        NSLog(@"%@",dataDict);
    }else {   //开始解析子节点
        for (NSString *e in elementArray) {
            if ([e isEqualToString:elementName]) {
                self.m_strCurrentElement = elementName;
                self.tempString = [NSMutableString string];
                break;
            }
        }
    }
```

```objc
}
- (void)parser:(NSXMLParser *)parser foundCharacters:(NSString *)string {
    //填充 string
    if (m_strCurrentElement) {
        [self.tempString appendString:string];
        [dataDict setObject:string forKey:m_strCurrentElement];
    }
}

- (void)parser:(NSXMLParser *)parser didEndElement:(NSString *)elementName namespaceURI:
(NSString *)namespaceURI qualifiedName:(NSString *)qName {
    //填充 dic
    if (m_strCurrentElement) {
        [dataDict setObject:self.tempString forKey:m_strCurrentElement];
        
        self.m_strCurrentElement = nil;
        self.tempString = nil;
    }
//结束解析 News 节点
    if ([elementName isEqualToString:@"News"]) {
        if (dataDict) {
            [parserObjects addObject:dataDict];
        }
    }
}

- (void)parserDidEndDocument:(NSXMLParser *)parser {
    [tablev reloadData];
}
- (void)viewDidLoad {
    [super viewDidLoad];
    NSURL *url = [NSURL URLWithString:@"http://www.vc111.cn/LSXH/GETNEWS2.ASP"];
    ASIHTTPRequest *request = [ASIHTTPRequest requestWithURL:url];
    [request setValidatesSecureCertificate:NO];
    [request setDelegate:self];
    [request setDidFailSelector:@selector(ASIHttpRequestFailed:)];
    [request setDidFinishSelector:@selector(ASIHttpRequestSuceed:)];
    [request startAsynchronous];
    [request setDefaultResponseEncoding:NSUTF8StringEncoding];
    tablev=[[UITableView alloc]initWithFrame:CGRectMake(200, 100, 400, 400) style:
UITableViewStylePlain];
    tablev.delegate=self;
    tablev.dataSource=self;
    [self.view addSubview:tablev];
    [tablev release];
}
- (NSInteger)tableView:(UITableView *)tableView numberOfRowsInSection:(NSInteger)section{
    return [parserObjects count];
    NSLog(@"num:%d",[parserObjects count]);
}
- (CGFloat)tableView:(UITableView *)tableView  heightForRowAtIndexPath:(NSIndexPath
*)indexPath
{
    return 40;
}

- (UITableViewCell *)tableView:(UITableView *)tableView cellForRowAtIndexPath:(NSIndexPath
*)indexPath{
    static NSString *CellIdentifier = @"Cell";
    
    UITableViewCell *cell = [tableView dequeueReusableCellWithIdentifier:CellIdentifier];
    if (cell == nil) {
        cell = [[[UITableViewCell alloc] initWithStyle:UITableViewCellStyleSubtitle
                            reuseIdentifier:CellIdentifier] autorelease];
    }
    if ([parserObjects count]!=0) {
        NSUInteger row = [indexPath row];
        dic = [parserObjects objectAtIndex:row];
        UILabel*lab=[[UILabel alloc]init];
        lab.frame=CGRectMake(0, 0, 350, 40);
        lab.text=[dic objectForKey:@"NewsTitle"];
```

```
            [cell addSubview:lab];
            UILabel*label=[[UILabel alloc]initWithFrame:CGRectMake(300, 20, 100, 20)];
            label.text=[dic objectForKey:@"NewsDate"];
            [cell addSubview:label];
            [lab release];
            [label release];
            //cell.textLabel.text = [dic objectForKey:@"NewsTitle"];
        }
        return cell;
}
- (void)tableView:(UITableView *)tableView didSelectRowAtIndexPath:(NSIndexPath*)indexPath{
        NSUInteger row = [indexPath row];
        dic = [parserObjects objectAtIndex:row];
        iview=[[UIView alloc]initWithFrame:tablev.frame];
        [self.view addSubview:iview];
        iview.backgroundColor=[UIColor whiteColor];
        UILabel*lab1=[[UILabel alloc]initWithFrame:CGRectMake(100, 20, 200, 50)];
        lab1.text=[dic objectForKey:@"NewsTitle"];
        
        UITextView*tview=[[UITextView alloc]initWithFrame:CGRectMake(20, 100, 360, 400)];
        tview.text=[dic objectForKey:@"NewsContent"];
        [iview addSubview:tview];
        [iview addSubview:lab1];
}
@end
```

执行后的效果如图 8-28 所示。

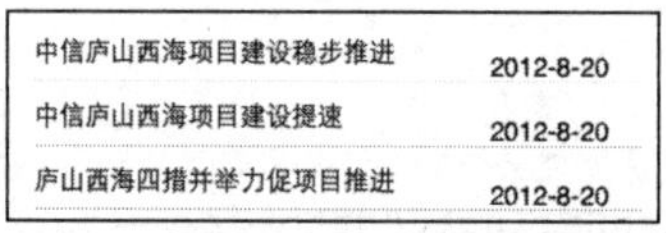

图 8-28 执行效果

实例 207　实时检测 Wi-Fi 状况

实例 207	实时检测 Wi-Fi 状况
源码路径	\daima\207\

实例说明

Wi-Fi 是一种可以将个人电脑、手持设备（如 PDA、手机）等终端以无线方式互相连接的技术。Wi-Fi 是一个无线网路通信技术的品牌，由 Wi-Fi 联盟（Wi-Fi Alliance）所持有，其目的是改善基于 IEEE 802.11 标准的无线网路产品之间的互通性。现时一般人会把 Wi-Fi 及 IEEE 802.11 混为一谈。甚至把 Wi-Fi 等同于无线网际网路。本实例的功能是，实现动态检测当前的 Wi-Fi 状况，不需要用户手动刷新，便可通知用户网络状态的变化。

具体实现

编写文件 AHReach.m，功能是判断当前网络的连接类型，此文件可以区分出无网络、Wi-Fi 和 wwan（2G&2.5G&3G）类型的网络连接类型。具体代码如下所示。

```
#import "AHReach.h"
#import <SystemConfiguration/SystemConfiguration.h>
#include <sys/socket.h>
#include <arpa/inet.h>
#include <ifaddrs.h>
#include <netdb.h>
enum {
    AHReachRouteNone = 0,
    AHReachRouteWiFi = 1,
    AHReachRouteWWAN = 2,
};
```

```objc
typedef NSInteger AHReachRoutes;

@interface AHReach ()
@property(nonatomic) SCNetworkReachabilityRef reachability;
@property(nonatomic, copy) AHReachChangedBlock changedBlock;
@end

void AHReachabilityCallback(SCNetworkReachabilityRef target, SCNetworkReachabilityFlags flags, void *info);

@implementation AHReach

@synthesize reachability, changedBlock;

#pragma mark - Factory methods

+ (AHReach *)reachForHost:(NSString *)host {
    SCNetworkReachabilityRef reachabilityRef = SCNetworkReachabilityCreateWithName(NULL, [host UTF8String]);
    return [[AHReach alloc] initWithReachability:reachabilityRef];
}

+ (AHReach *)reachForAddress:(const struct sockaddr_in *)addr {
    SCNetworkReachabilityRef reachabilityRef = SCNetworkReachabilityCreateWithAddress(NULL, (const struct sockaddr *)addr);
    return [[AHReach alloc] initWithReachability:reachabilityRef];
}

+ (AHReach *)reachForDefaultHost {
    return [self reachForHost:@kAHReachDefaultHost];
}

#pragma mark - Object lifetime

- (id)initWithReachability:(SCNetworkReachabilityRef)reachabilityRef {
    if((self = [super init]) && reachabilityRef) {
        reachability = reachabilityRef;
        return self;
    }

    return nil;
}

- (void)dealloc {
    [self stopUpdating];
    if(reachability) {
        CFRelease(reachability);
        reachability = NULL;
    }
}

#pragma mark - Reachability and notification methods

- (AHReachRoutes)availableRoutes {
    AHReachRoutes routes = AHReachRouteNone;
    SCNetworkReachabilityFlags flags = 0;
    SCNetworkReachabilityGetFlags(self.reachability, &flags);

    if(flags & kSCNetworkReachabilityFlagsReachable)
    {
        // 由于广域网可能需要连接,我们最初的假设路线没有连接所需的无线上网
        if(!(flags & kSCNetworkReachabilityFlagsConnectionRequired)) {
            routes |= AHReachRouteWiFi;
        }

        BOOL automatic = (flags & kSCNetworkReachabilityFlagsConnectionOnDemand) ||
                        (flags & kSCNetworkReachabilityFlagsConnectionOnTraffic);

        // 如果一个连接连接点没有干预,则可能是无线网络
        if(automatic && !(flags & kSCNetworkReachabilityFlagsInterventionRequired))
```

```objc
                    routes |= AHReachRouteWiFi;
            }
        }

                // 但是如果我们明确表示存在广域网, 则抛弃所有其他网络
                if(flags & kSCNetworkReachabilityFlagsIsWWAN) {
                    routes &= ~AHReachRouteWiFi;
                    routes |= AHReachRouteWWAN;
                }
        }

        return routes;
}

- (BOOL)isReachable {
    return [self availableRoutes] != AHReachRouteNone;
}

- (BOOL)isReachableViaWWAN {
    return [self availableRoutes] & AHReachRouteWWAN;
}

- (BOOL)isReachableViaWiFi {
    return [self availableRoutes] & AHReachRouteWiFi;
}

- (void)startUpdatingWithBlock:(AHReachChangedBlock)block {
    if(block && self.reachability) {
        self.changedBlock = block;
        SCNetworkReachabilityContext context = { 0, (__bridge void *)self, NULL, NULL, NULL };
        SCNetworkReachabilitySetCallback(self.reachability, AHReachabilityCallback, &context);
        SCNetworkReachabilityScheduleWithRunLoop(self.reachability, CFRunLoopGetCurrent(), kCFRunLoopDefaultMode);
    } else {
        [self stopUpdating];
    }
}

- (void)reachabilityDidChange {
    if(self.changedBlock)
        self.changedBlock(self);
}

- (void)stopUpdating {
    self.changedBlock = nil;

    if(self.reachability) {
        SCNetworkReachabilityUnscheduleFromRunLoop(self.reachability, CFRunLoopGetCurrent(), kCFRunLoopDefaultMode);
        SCNetworkReachabilitySetCallback(self.reachability, NULL, NULL);
    }
}

@end

#pragma mark - Reachability callback function

void AHReachabilityCallback(SCNetworkReachabilityRef target, SCNetworkReachabilityFlags flags, void *info)
{
    AHReach *reach = (__bridge AHReach *)info;
    [reach reachabilityDidChange];
}
```

编写文件 **AHViewController.m** 输出探测结果, 具体实现代码如下所示。

```objc
#import "AHViewController.h"
#import "AHReach.h"
@interface AHViewController ()
@property(nonatomic, strong) IBOutlet UITextField *defaultHostField;
```

实例 207 实时检测 Wi-Fi 状况

```objc
@property(nonatomic, strong) IBOutlet UITextField *hostField;
@property(nonatomic, strong) IBOutlet UITextField *addressField;
@property(nonatomic, strong) NSArray *reaches;
@end
@implementation AHViewController
@synthesize defaultHostField, hostField, addressField, reaches;
- (void)viewDidLoad {
    [super viewDidLoad];
    defaultHostField.text = @"<No updates yet>";
    hostField.text = @"<No updates yet>";
    addressField.text = @"<No updates yet>";
    AHReach *defaultHostReach = [AHReach reachForDefaultHost];
    [defaultHostReach startUpdatingWithBlock:^(AHReach *reach) {
        [self updateAvailabilityField:self.defaultHostField withReach:reach];
    }];
    [self updateAvailabilityField:self.defaultHostField withReach:defaultHostReach];

    AHReach *hostReach = [AHReach reachForHost:@"auerhaus.com"];
    [hostReach startUpdatingWithBlock:^(AHReach *reach) {
        [self updateAvailabilityField:self.hostField withReach:reach];
    }];
    [self updateAvailabilityField:self.hostField withReach:hostReach];
    struct sockaddr_in addr;
    memset(&addr, 0, sizeof(struct sockaddr_in));
    addr.sin_len = sizeof(struct sockaddr_in);
    addr.sin_family = AF_INET;
    addr.sin_port = htons(80);
    inet_aton("173.194.43.0", &addr.sin_addr);
    AHReach *addressReach = [AHReach reachForAddress:&addr];
    [addressReach startUpdatingWithBlock:^(AHReach *reach) {
        [self updateAvailabilityField:self.addressField withReach:reach];
    }];
    [self updateAvailabilityField:self.addressField withReach:addressReach];
    self.reaches = [NSArray arrayWithObjects:defaultHostReach, hostReach, addressReach, nil];
}
- (BOOL)shouldAutorotateToInterfaceOrientation:(UIInterfaceOrientation)interfaceOrientation {
    return (interfaceOrientation != UIInterfaceOrientationPortraitUpsideDown);
}
- (void)updateAvailabilityField:(UITextField *)field withReach:(AHReach *)reach {
    field.text = @"不可用";

    if([reach isReachableViaWWAN])
        field.text = @"可通过广域网";
    if([reach isReachableViaWiFi])
        field.text = @"可通过 WiFi";
}
@end
```

因为本实例用的是无线网络，所以执行后的效果如图 8-29 所示。

图 8-29 执行效果

实例 208 断点续传下载后实现播放

实例 208	断点续传下载后实现播放
源码路径	\daima\208\

实例说明

本实例的功能比较强大，使用最新的库 AFNetworking 和 AFDownloadRequestOperation 实现了断点续传下载功能。本实例的核心有两个，一个是网络应用方面的断点下载，另一个是流媒体播放，这两个功能都用到了开源的库。

具体实现

先看流媒体播放模块，其中实现文件 AudioStreamer.h 的具体代码如下所示。

```
#if TARGET_OS_IPHONE
#import <UIKit/UIKit.h>
#else
#import <Cocoa/Cocoa.h>
#endif

#include <pthread.h>
#include <AudioToolbox/AudioToolbox.h>

#define LOG_QUEUED_BUFFERS 0

#define kNumAQBufs 16              //缓冲区队列中的分配音频数

#define kAQDefaultBufSize 2048     //在每个音频缓冲区队列的的字节数

#define kAQMaxPacketDescs 512      //描述数组中的包
typedef enum
{
    AS_INITIALIZED = 0,
    AS_STARTING_FILE_THREAD = 1,        // 启动线程
    AS_WAITING_FOR_DATA = 2,            // 准备数据
    AS_FLUSHING_EOF = 3,                // 数据准备完毕
    AS_WAITING_FOR_QUEUE_TO_START = 4,  // 排队播放
    AS_PLAYING = 5,                     // 正在播放
    AS_BUFFERING = 6,                   // 网络不好,自动缓冲
    AS_PAUSED = 7,                      // 手动暂停
    AS_STOPPING = 8,                    // 即将停止,自动提醒
    AS_STOPPED = 9,                     // 已停止播放
} AudioStreamerState;

typedef enum
{
    AS_NO_STOP = 0,
    AS_STOPPING_EOF,
    AS_STOPPING_USER_ACTION,
    AS_STOPPING_ERROR,
    AS_STOPPING_TEMPORARILY
} AudioStreamerStopReason;

typedef enum
{
    AS_NO_ERROR = 0,
    AS_NETWORK_CONNECTION_FAILED,
    AS_FILE_STREAM_GET_PROPERTY_FAILED,
    AS_FILE_STREAM_SEEK_FAILED,
    AS_FILE_STREAM_PARSE_BYTES_FAILED,
```

```objc
    AS_FILE_STREAM_OPEN_FAILED,
    AS_FILE_STREAM_CLOSE_FAILED,
    AS_AUDIO_DATA_NOT_FOUND,
    AS_AUDIO_QUEUE_CREATION_FAILED,
    AS_AUDIO_QUEUE_BUFFER_ALLOCATION_FAILED,
    AS_AUDIO_QUEUE_ENQUEUE_FAILED,
    AS_AUDIO_QUEUE_ADD_LISTENER_FAILED,
    AS_AUDIO_QUEUE_REMOVE_LISTENER_FAILED,
    AS_AUDIO_QUEUE_START_FAILED,
    AS_AUDIO_QUEUE_PAUSE_FAILED,
    AS_AUDIO_QUEUE_BUFFER_MISMATCH,
    AS_AUDIO_QUEUE_DISPOSE_FAILED,
    AS_AUDIO_QUEUE_STOP_FAILED,
    AS_AUDIO_QUEUE_FLUSH_FAILED,
    AS_AUDIO_STREAMER_FAILED,
    AS_GET_AUDIO_TIME_FAILED,
    AS_AUDIO_BUFFER_TOO_SMALL
} AudioStreamerErrorCode;

extern NSString * const ASStatusChangedNotification;
@interface AudioStreamer : NSObject
{
    NSURL *url;
    AudioQueueRef audioQueue;
    AudioFileStreamID audioFileStream;    // 音频文件流分析器
    AudioStreamBasicDescription asbd;     // 音频描述
    NSThread *internalThread;             // 线程下载后的解析音频文件流
    AudioQueueBufferRef audioQueueBuffer[kNumAQBufs];        // 音频缓冲区队列
    AudioStreamPacketDescription packetDescs[kAQMaxPacketDescs]; // 分组说明入队
    unsigned int fillBufferIndex;         // audioQueueBuffer 被填充
    UInt32 packetBufferSize;
    size_t bytesFilled;                   // 有多少字节填充
    size_t packetsFilled;                 // 有多少数据包已被填补
    bool inuse[kNumAQBufs];               // 一个标示,说明缓冲区仍在使用
    NSInteger buffersUsed;
    NSDictionary *httpHeaders;

    AudioStreamerState state;
    AudioStreamerStopReason stopReason;
    AudioStreamerErrorCode errorCode;
    OSStatus err;
    bool discontinuous;                   // 标示
    pthread_mutex_t queueBuffersMutex;    //一个互斥保护的一种标志
    pthread_cond_t queueBufferReadyCondition; // 条件变量处理

    CFReadStreamRef stream;
    NSNotificationCenter *notificationCenter;

    UInt32 bitRate;                       // 比特每秒的文件
    NSInteger dataOffset;                 // 第一音频数据包流的偏移量
    NSInteger fileLength;                 // 该文件中的字节长度
    NSInteger seekByteOffset;             // 寻求在文件中的字节偏移量
    UInt64 audioDataByteCount;            // 使用时的实际音频字节数
                                          // 该文件是已知的(更准确地估计整个文件音频)

    UInt64 processedPacketsCount;         // 估计数据包的数量积累的比特率
    UInt64 processedPacketsSizeTotal;     // 估计数据包累积的字节大小

    double seekTime;
    BOOL seekWasRequested;
    double requestedSeekTime;
    double sampleRate;                    // 采样率的文件(用来比较样品的队列当前播放时间)
    double packetDuration;                // 帧包采样率
    double lastProgress;                  // 计算进展
#if TARGET_OS_IPHONE
```

```
        BOOL pausedByInterruption;
#endif
}

@property AudioStreamerErrorCode errorCode;
@property (readonly) AudioStreamerState state;
@property (readonly) double progress;
@property (readonly) double duration;
@property (readwrite) UInt32 bitRate;
@property (readonly) NSDictionary *httpHeaders;

- (id)initWithURL:(NSURL *)aURL;
- (void)start;
- (void)stop;
- (void)pause;
- (BOOL)isFinishing;
- (BOOL)isPlaying;
- (BOOL)isPaused;
- (BOOL)isWaiting;
- (BOOL)isIdle;
- (void)seekToTime:(double)newSeekTime;
- (double)calculatedBitRate;
- (NSString *)currentTime;
- (NSString *)totalTime;
@end
```

文件 AudioStreamer.m 是文件 AudioStreamer.h 的实现，读者可以参考 Mac 官网的帮助文档来学习这个文件。此文件使用 CFReadStreamRef 为文件创建读操作流，一旦操作流被创建，它就可以被打开。打开一个操作流会导致这个流占用它所需要的任何系统资源，比如用于打开文件的文件描述符。例如下面的代码说明了如何打开读操作流。

```
if (!CFReadStreamOpen(myReadStream)) {
CFStreamError myErr = CFReadStreamGetError(myReadStream);
if (myErr.domain == kCFStreamErrorDomainPOSIX) {
} else if (myErr.domain == kCFStreamErrorDomainMacOSStatus) {
OSStatus macError = (OSStatus)myErr.error;
}
}
```

如果操作流打开成功，那么 CFReadStreamOpen 函数返回 TRUE，如果因为某种原因打开失败就会返回 FALSE。如果 CFReadStreamOpen 返回 FALSE，示例程序调用了 CFReadStreamGetError 函数，它返回了一个 CFStreamError 类型的结构，其中包含两个值：一个域代码和一个错误代码。域代码决定了错误代码将被怎样解释。比如，如果域代码是 kCFStreamErrorDomainPOSIX，错误代码是一个 UNIX error 值。其他的错误域比如 kCFStreamErrorDomainMacOSStatus，这说明错误代码是 MacErrors.h 中定义的一个 OSStatus 值，如果是 kCFStreamErrorDomainHTTP，这说明错误代码是枚举对象 CFStreamErrorHTTP 中定义的一个值。

打开一个操作流可能会耗费比较长的时间，因此 CFReadStreamOpen 和 CFWriteStreamOpen 两个函数都不会被阻塞，它们返回 TRUE 表示操作流的打开过程已经开始。如果想要检查打开过程的状态，可以调用函数 CFReadStreamGetStatus 和 CFWriteStreamGetStatus，如果返回 kCFStreamStatus-Opening 说明打开过程仍然在进行中，如果返回 kCFStreamStatusOpen 说明打开过程已经完成，而返回 kCFStreamStatusErrorOccurred 说明打开过程已经完成，但是失败了。大部分情况下，打开过程是否完成并不重要，因为 CFStream 中负责读写操作的函数在操作流打开以前会被阻塞。

想要从读操作流中读取数据的话，需要调用函数 CFReadStreamRead，它类似于 UNIX 的 read() 系统调用。二者的相同之处包括：都需要缓冲区和缓冲区大小作为参数，都会返回读取的字节数，如果到了文件末尾会返回 0，如果遇到错误就会返回-1。另外，二者都会在至少一个字节可以被读取之前被阻塞，并且如果没有遇到阻塞都会继续读取。

为了节省篇幅，在此不再详细讲解文件 AudioStreamer.m。

再看下载模块，文件 AFDownloadRequestOperation.m 实现了断点下载，具体代码如下所示。

```objc
#import "AFDownloadRequestOperation.h"
#import "AFURLConnectionOperation.h"
#import <CommonCrypto/CommonDigest.h>
#include <fcntl.h>
#include <unistd.h>
@interface AFURLConnectionOperation (AFInternal)
@property (nonatomic, strong) NSURLRequest *request;
@property (readonly, nonatomic, assign) long long totalBytesRead;
@end
typedef void (^AFURLConnectionProgressiveOperationProgressBlock)(NSInteger bytes, long long totalBytes, long long totalBytesExpected, long long totalBytesReadForFile, long long totalBytesExpectedToReadForFile);
@interface AFDownloadRequestOperation() {
    NSError *_fileError;
}
@property (nonatomic, strong) NSString *tempPath;
@property (assign) long long totalContentLength;
@property (nonatomic, assign) long long totalBytesReadPerDownload;
@property (assign) long long offsetContentLength;
@property (nonatomic, copy) AFURLConnectionProgressiveOperationProgressBlock progressiveDownloadProgress;
@end
@implementation AFDownloadRequestOperation
@synthesize targetPath = _targetPath;
@synthesize tempPath = _tempPath;
@synthesize totalContentLength = _totalContentLength;
@synthesize offsetContentLength = _offsetContentLength;
@synthesize shouldResume = _shouldResume;
@synthesize deleteTempFileOnCancel = _deleteTempFileOnCancel;
@synthesize progressiveDownloadProgress = _progressiveDownloadProgress;
@synthesize totalBytesReadPerDownload;
#pragma mark - Static
+ (NSString *)cacheFolder {
    static NSString *cacheFolder;
    static dispatch_once_t onceToken;
    dispatch_once(&onceToken, ^{
        NSString *cacheDir = NSTemporaryDirectory();
        cacheFolder = [cacheDir stringByAppendingPathComponent:kAFNetworkingIncompleteDownloadFolderName];

        // 确保所有的缓存目录存在（只需要一次）
        NSError *error = nil;
        if(!([[NSFileManager new] createDirectoryAtPath:cacheFolder withIntermediateDirectories:YES attributes:nil error:&error])) {
            NSLog(@"Failed to create cache directory at %@", cacheFolder);
        }
    });
    return cacheFolder;
}

// 计算校验散列键
+ (NSString *)md5StringForString:(NSString *)string {
    const char *str = [string UTF8String];
    unsigned char r[CC_MD5_DIGEST_LENGTH];
    CC_MD5(str, strlen(str), r);
    return [NSString stringWithFormat:@"%02x%02x%02x%02x%02x%02x%02x%02x%02x%02x%02x%02x%02x%02x%02x%02x",
            r[0], r[1], r[2], r[3], r[4], r[5], r[6], r[7], r[8], r[9], r[10], r[11], r[12], r[13], r[14], r[15]];
}
#pragma mark - Private
- (unsigned long long)fileSizeForPath:(NSString *)path {
    signed long long fileSize = 0;
    NSFileManager *fileManager = [NSFileManager new]; //不是线程安全的
    if ([fileManager fileExistsAtPath:path]) {
        NSError *error = nil;
        NSDictionary *fileDict = [fileManager attributesOfItemAtPath:path error:&error];
        if (!error && fileDict) {
            fileSize = [fileDict fileSize];
        }
```

```objc
    }
    return fileSize;
}
#pragma mark - NSObject
- (id)initWithRequest:(NSURLRequest *)urlRequest targetPath:(NSString *)targetPath
shouldResume:(BOOL)shouldResume {
    if ((self = [super initWithRequest:urlRequest])) {
        NSParameterAssert(targetPath != nil && urlRequest != nil);
        _shouldResume = shouldResume;

        self.runLoopModes = [NSSet setWithObject:NSRunLoopCommonModes];

        // 假定至少目录已经存在于目标路径
        BOOL isDirectory;
        if(![[NSFileManager defaultManager] fileExistsAtPath:targetPath isDirectory:
        &isDirectory]) {
            isDirectory = NO;
        }
        // 如果目标路径是一个目录,从我们得到的 url request 使用文件名
        if (isDirectory) {
            NSString *fileName = [urlRequest.URL lastPathComponent];
            _targetPath = [NSString pathWithComponents:[NSArray arrayWithObjects:
            targetPath, fileName, nil]];
        }else {
            _targetPath = targetPath;
        }

        // 下载保存到一个文件
        NSString *tempPath = [self tempPath];

        // 是否需要恢复的文件
        BOOL isResuming = NO;
        if (shouldResume) {
            unsigned long long downloadedBytes = [self fileSizeForPath:tempPath];
            if (downloadedBytes > 0) {
                NSMutableURLRequest *mutableURLRequest = [urlRequest mutableCopy];
                NSString *requestRange = [NSString stringWithFormat:@"bytes=%llu-",
                downloadedBytes];
                [mutableURLRequest setValue:requestRange forHTTPHeaderField:@"Range"];
                self.request = mutableURLRequest;
                isResuming = YES;
            }
        }

        // 尝试在目标位置创建/打开一个文件
        if (!isResuming) {
            int fileDescriptor = open([tempPath UTF8String], O_CREAT | O_EXCL | O_RDWR,
            0666);
            if (fileDescriptor > 0) {
                close(fileDescriptor);
            }
        }

        self.outputStream = [NSOutputStream outputStreamToFileAtPath:tempPath append:
        isResuming];

        // 如果输出流不能创造,则立即销毁对象
        if (!self.outputStream) {
            return nil;
        }
    }
    return self;
}

#pragma mark - Public

- (BOOL)deleteTempFileWithError:(NSError **)error {
    NSFileManager *fileManager = [NSFileManager new];
    BOOL success = YES;
    @synchronized(self) {
```

```objc
        NSString *tempPath = [self tempPath];
        if ([fileManager fileExistsAtPath:tempPath]) {
            success = [fileManager removeItemAtPath:[self tempPath] error:error];
        }
    }
    return success;
}

- (NSString *)tempPath {
    NSString *tempPath = nil;
    if (self.targetPath) {
        NSString *md5URLString = [[self class] md5StringForString:self.targetPath];
        tempPath = [[[self class] cacheFolder] stringByAppendingPathComponent:md5URLString];
    }
    return tempPath;
}
- (void)setProgressiveDownloadProgressBlock:(void (^)(NSInteger bytesRead, long long totalBytesRead, long long totalBytesExpected, long long totalBytesReadForFile, long long totalBytesExpectedToReadForFile))block {
    self.progressiveDownloadProgress = block;
}

#pragma mark - AFURLRequestOperation
- (void)setCompletionBlockWithSuccess:(void (^)(AFHTTPRequestOperation *operation, id responseObject))success
                              failure:(void (^)(AFHTTPRequestOperation *operation, NSError *error))failure
{
#pragma clang diagnostic push
#pragma clang diagnostic ignored "-Warc-retain-cycles"
    self.completionBlock = ^ {
        NSError *localError = nil;
        if([self isCancelled]) {
            //
            if (self.isDeletingTempFileOnCancel) {
                [self deleteTempFileWithError:&localError];
                if (localError) {
                    _fileError = localError;
                }
            }
            return;
        // 发生错误：失去网络连接
        }else if(!self.error) {
            // 将文件移动到最终位置并捕获错误
            @synchronized(self) {
                [[NSFileManager new] moveItemAtPath:[self tempPath] toPath:_targetPath error:&localError];
                if (localError) {
                    _fileError = localError;
                }
            }
        }

        if (self.error) {
            dispatch_async(self.failureCallbackQueue ?: dispatch_get_main_queue(), ^{
                failure(self, self.error);
            });
        } else {
            dispatch_async(self.successCallbackQueue ?: dispatch_get_main_queue(), ^{
                success(self, _targetPath);
            });
        }
    };
#pragma clang diagnostic pop
}

- (NSError *)error {
    if (_fileError) {
        return _fileError;
    } else {
        return [super error];
```

```objc
    }
}

#pragma mark - NSURLConnectionDelegate

- (void)connection:(NSURLConnection *)connection didReceiveResponse:(NSURLResponse *)response {
    [super connection:connection didReceiveResponse:response];

    // 检查是否有正确的反应
    NSHTTPURLResponse *httpResponse = (NSHTTPURLResponse *)response;
    if (![httpResponse isKindOfClass:[NSHTTPURLResponse class]]) {
        return;
    }

    // 检查有效的恢复下载响应
    long long totalContentLength = self.response.expectedContentLength;
    long long fileOffset = 0;
    if(httpResponse.statusCode == 206) {
        NSString *contentRange = [httpResponse.allHeaderFields valueForKey:@"Content-Range"];
        if ([contentRange hasPrefix:@"bytes"]) {
            NSArray *bytes = [contentRange componentsSeparatedByCharactersInSet:[NSCharacterSet characterSetWithCharactersInString:@" -/"]];
            if ([bytes count] == 4) {
                fileOffset = [[bytes objectAtIndex:1] longLongValue];
                totalContentLength = [[bytes objectAtIndex:2] longLongValue]; // if this is *, it's converted to 0
            }
        }
    }
    self.totalBytesReadPerDownload = 0;
    self.offsetContentLength = MAX(fileOffset, 0);
    self.totalContentLength = totalContentLength;
    [self.outputStream setProperty:[NSNumber numberWithLongLong:_offsetContentLength] forKey:NSStreamFileCurrentOffsetKey];
}
- (void)connection:(NSURLConnection *)connection didReceiveData:(NSData *)data {
    [super connection:connection didReceiveData:data];
    // 跟踪自定义的读字节
    self.totalBytesReadPerDownload += [data length];
    if (self.progressiveDownloadProgress) {
        self.progressiveDownloadProgress((long long)[data length], self.totalBytesRead, self.response.expectedContentLength,self.totalBytesReadPerDownload + self.offsetContentLength, self.totalContentLength);
    }
}
@end
```

执行后的效果如图 8-30 所示。

图 8-30 执行效果

第 9 章 地图定位应用实战

在当前智能手机系统应用中,地图导航已经成为了必不可少的功能之一。作为一款强大的智能设备系统,iOS 也具备了地图导航功能。本章将通过几个典型实例的实现过程,详细介绍在 iOS 系统中使用地图服务的基本知识。

实例 209 获得当前所在位置和苹果公司总部的距离

实例 209	获得当前所在位置和苹果公司总部的距离
源码路径	\daima\209\

实例说明

在 iOS 应用中,可以使用 Core Location 实现地图定位处理。Core Location 是 iOS SDK 中一个提供设备位置的框架。根据设备的当前状态(在服务区、在大楼内等),可以使用如下 3 种技术之一。

(1)使用 GPS 定位系统,可以精确地定位用户当前所在的地理位置,但由于 GPS 接收机需要对准天空才能工作,因此在室内环境基本无用。

(2)另一个找到自己所在位置的有效方法是使用手机基站,手机开机时,它会与周围的基站保持联系,如果知道这些基站的身份,就可以使用各种数据库(包含基站的身份和它们的确切地理位置)计算出手机的物理位置。基站不需要卫星,和 GPS 不同,它在室内环境同样可用。但使用基站没有 GPS 那样精确,它的精度取决于基站的密度,它在基站密集型区域的准确度最高。

(3)第三种方法是依赖 Wi-Fi,使用这种方法时,设备连接到 Wi-Fi 网络,通过检查服务提供商的数据确定位置,它既不依赖卫星,也不依赖基站,因此这个方法对于可以连接到 Wi-Fi 网络的区域有效,但它的精确度也是这 3 种方法中最差的。

在上述技术中 GPS 是精准的,如果有 GPS 硬件,则 Core Location 将优先使用它。如果设备没有 GPS 硬件(如 Wi-Fi iPad)或使用 GPS 获取当前位置时失败,Core Location 将退而求其次,选择使用蜂窝或 Wi-Fi。

本实例的功能是,得到当前位置距离 Apple 公司总部的距离。在创建该应用程序时,将分两步进行:首先使用 Core Location 指出当前位置离 Apple 公司总部有多少英里;然后,使用设备指南针显示一个箭头,在用户偏离轨道时指明正确方向。在具体实现时,先创建一个位置管理器实例,并使用其方法计算当前位置离 Apple 公司总部有多远。在计算距离期间,将显示一条消息,让用户耐心等待。如果用户位于 Apple 公司总部,应表示祝贺,否则以英里为单位显示与 Apple 公司总部的距离。

具体实现

(1)在 Xcode 中,使用模板 Single View Application 新建一个项目,并将其命名为 "juli",如图 9-1 所示。

(2)添加 Core Location 框架。

在默认情况下,没有链接 Core Location 框架,因此需要添加它。选择项目 Cupertino 的顶级编组,并确保编辑器中当前显示的是 Summary 选项卡。接下来在该选项卡中向下滚动到 Linked Libraries and Frameworks 部分,单击列表下方的 "+" 按钮,在出现的列表中选择 CoreLocation.framework,再单击 Add 按钮,如图 9-2 所示。

图 9-1 创建工程

（3）添加背景图像资源。

为确保用户牢记要去哪里，将一张漂亮的照片用作这个应用程序的背景图像。将文件夹 Image（它包含 apple.png）拖曳到项目导航器中的项目代码编组中。在 Xcode 提示时，务必选择复制文件并创建编组，如图 9-3 所示。

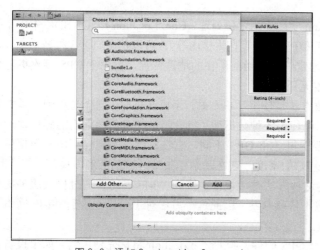

图 9-2　添加 CoreLocation.framework

图 9-3　工程组

（4）规划变量和连接。

ViewController 将充当位置管理器委托，它接收位置更新，并更新用户界面以指出当前位置。在这个视图控制器中，需要一个实例变量/属性（但不需要相应的输出口），它指向位置管理器实例。将把这个属性命名为 locMan。

在界面中，需要一个标签（distanceLabel）和两个子视图（distanceView 和 waitView）。其中，标签将显示到 Apple 总部的距离；子视图包含标签 distanceLabel，仅当获取了当前位置并计算出距离后才显示；而子视图 waitView 将在 iOS 设备获取航向时显示。

（5）添加表示 Apple 总部位置的常量。

要计算到 Apple 总部的距离，显然需要知道 Apple 总部的位置，以便将其与用户的当前位置进行比较。根据 http://gpsvisualizer.com/geocode 提供的信息，Apple 总部的纬度为 37.3229978，经度为 -122.0321823。在实现文件 ViewController.m 中的 #import 代码行后面，添加两个表示这些值的常量（kCupertinoLatitude 和 kCupertinoLongitude）。

```
#define kCupertinoLatitude 37.3229978
#define kCupertinoLongitude -122.0321823
```

（6）设计视图。

这个应用程序的用户界面很简单：不能执行任何操作来改变位置，因此只需更新屏幕，显示有

实例 209 获得当前所在位置和苹果公司总部的距离

关当前位置的信息即可。打开文件 MainStoryboard.storyboard，打开对象库（View→Utilities→Show Object Library），并开始设计界面。

首先，将一个图像视图（UIImageView）拖曳到视图中，使其居中并覆盖整个视图，它将用作应用程序的背景图像。在选择了该图像视图的情况下，按 Option+Command+4 打开 Attributes Inspector，并从下拉列表 Image 中选择 apple.png。

接下来，将一个视图（UIView）拖曳到图像视图底部。这个视图将充当主要的信息显示器，因此应将其高度设置为能显示大概两行文本。将 Alpha 设置为 0.75 并选中复选框 Hidden。

将一个标签（UILabel）拖曳到信息视图中，调整标签使其与全部 4 条边缘参考线对齐，并将其文本设置为"距离有多远"。使用 Attributes Inspector 将文本颜色改为白色，让文本居中，并根据需要调整字号，UI 视图如图 9-4 所示。

然后再添加一个半透明的视图，其属性与前一个视图相同，但不隐藏且高度大约为 1 英寸。拖曳这个视图，使其在背景中垂直居中，在设备定位时，这个视图将显示让用户耐心等待的消息。在这个视图中添加一个标签，将其文本设置为"检查距离"。调整该标签的大小，使其占据该视图的右边大约 2/3。然后从对象库拖曳一个活动指示器（UIActivityIndicatorView）到第二个视图中，并使其与标签左边缘对齐。指示器显示一个纺锤图标，它与标签 Checking the Distance 同时显示。使用 Attributes Inspector 选中属性 Animated 的复选框，让纺锤旋转，最终的视图应如图 9-5 所示。

（7）创建并连接输出口。

在本应用程序中，只需根据位置管理器提供的信息更新 UI。也就是说不需要连接操作。需要连接我们添加的两个视图，还需连接用于显示离 Apple 总部有多远的标签。切换到助手编辑器模式，按住 Control 键，从标签"距离有多远"拖曳到 ViewController.h 中代码行@interface 下方。在 Xcode 提示时，新建一个名为"distanceLabel"的输出口。然后对两个视图做同样的处理，将包含活动指示器的视图连接到输出口 waitView，将包含距离的视图连接到输出口 distanceView。

（8）准备位置管理器。

根据刚才设计的界面可知，应用程序将在启动时显示一条消息和转盘，让用户知道应用程序正在等待 Core Location 提供初始位置读数。将在加载视图后立即在视图控制器的 viewDidLoad 方法

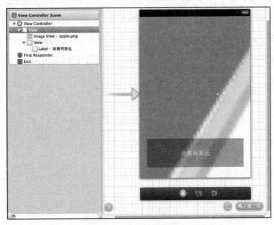

图 9-4 初始 UI 视图

图 9-5 最终 UI 视图

中请求这种读数。位置管理器委托获得读数后，将立即计算到 Apple 总部的距离，更新标签，隐藏活动指示器视图并显示距离视图。

首先，在文件 ViewController.h 中导入框架 Core Location 的头文件，然后在代码行@interface 中添加协议 CLLocationManagerDelegate。这能够创建位置管理器实例以及实现委托方法，但还需要一个指向位置管理器的实例变量/属性（locMan）。

完成上述修改后,文件 ViewController.h 的代码如下所示。

```
#import <UIKit/UIKit.h>
#import <CoreLocation/CoreLocation.h>

@interface ViewController : UIViewController <CLLocationManagerDelegate>

@property (strong, nonatomic) CLLocationManager *locMan;
@property (strong, nonatomic) IBOutlet UILabel *distanceLabel;
@property (strong, nonatomic) IBOutlet UIView *waitView;
@property (strong, nonatomic) IBOutlet UIView *distanceView;
@end
```

当声明属性 locMan 后,还需修改文件 ViewController.h,在其中添加配套的编译指令@synthesize:

```
@synthesize locMan;
```

并在方法 viewDidUnload 中将该实例变量设置为 nil:

```
[self setLocMan:nil];
```

现在该实现位置管理器并编写距离计算代码了。

(9)创建位置管理器实例。

在文件 ViewController.m 的方法 viewDidLoad 中,实例化一个位置管理器,将视图控制器指定为委托,将属性 desiredAccuracy 和 distanceFilter 分别设置为 kCLLocationAccuracyThreeKilometers 和 1609 米(1 英里)。使用方法 startUpdatingLocation 启动更新。具体实现代码如下所示。

```
- (void)viewDidLoad
{
    locMan = [[CLLocationManager alloc] init];
    locMan.delegate = self;
    locMan.desiredAccuracy = kCLLocationAccuracyThreeKilometers;
    locMan.distanceFilter = 1609; // a mile
    [locMan startUpdatingLocation];

    [super viewDidLoad];
    // Do any additional setup after loading the view, typically from a nib.
}
```

(10)实现位置管理器委托。

现在需要实现位置管理器委托协议的两个方法,将首先处理错误状态——locationManager:didFailWithError。对于获取当前位置失败的情形,标签 distanceLabel 包含默认消息,因此只需隐藏包含活动指示器的 waitView 视图,并显示视图 distanceView。如果用户禁止访问 Core Location 更新,还将清理位置管理器请求。文件 ViewController.m 中方法 locationManager:did FailWithError 的实现代码如下所示。

```
- (void)locationManager:(CLLocationManager *)manager
    didFailWithError:(NSError *)error {
    if (error.code == kCLErrorDenied) {
        // Turn off the location manager updates
        [self.locMan stopUpdatingLocation];
        [self setLocMan:nil];
    }
    self.waitView.hidden = YES;
    self.distanceView.hidden = NO;
}
```

在上述错误处理程序中,只考虑了位置管理器不能提供数据的情形。第 4 行检查错误编码,判断是否是用户禁止访问。如果是,则停止位置管理器(第 6 行)并将其设置为 nil(第 7 行)。第 9 行隐藏 waitView 视图,而第 10 行显示视图 distanceView(它包含默认文本"距离有多远")。

最后一个方法(locationManager:didUpdateToLocation:fromLocation)计算离 Apple 总部有多远,这需要使用 CLLocation 的另一个功能。不需要编写根据经度和纬度计算距离的代码,因为可使用 distanceFromLocation 计算两个 CLLocation 之间的距离。在 locationManager:didUpdate-Location: fromLocation 的实现中,将创建一个表示 Apple 总部的 CLLocation 实例,并将其与从 Core Location

获得的 CLLocation 实例进行比较，以获得以米为单位表示的距离，然后将米转换为英里。如果距离超过 3 英里，则显示它，并使用 NSNumberFormatter 在超过 1000 英里的距离中添加逗号；如果小于 3 英里，则停止位置更新，并输出祝贺用户信息"欢迎成为我们的一员"。LocationManager: didUpdateLo cation:fromLocation 的完整实现如下所示。

```
- (void)locationManager:(CLLocationManager *)manager
    didUpdateToLocation:(CLLocation *)newLocation
           fromLocation:(CLLocation *)oldLocation {

    if (newLocation.horizontalAccuracy >= 0) {
        CLLocation *Cupertino = [[CLLocation alloc]
                        initWithLatitude:kCupertinoLatitude
                               longitude:kCupertinoLongitude];
        CLLocationDistance delta = [Cupertino
                        distanceFromLocation:newLocation];
        long miles = (delta * 0.000621371) + 0.5; // meters to rounded miles
        if (miles < 3) {
            // Stop updating the location
            [self.locMan stopUpdatingLocation];
            // Congratulate the user
            self.distanceLabel.text = @"欢迎你\n 成为我们的一员！";
        } else {
            NSNumberFormatter *commaDelimited = [[NSNumberFormatter alloc]
                                    init];
            [commaDelimited setNumberStyle:NSNumberFormatterDecimalStyle];
            self.distanceLabel.text = [NSString stringWithFormat:
                            @"%@ 英里\n 到 Apple",
                            [commaDelimited stringFromNumber:
                             [NSNumber numberWithLong:miles]]];
        }
        self.waitView.hidden = YES;
        self.distanceView.hidden = NO;
    }
}
```

到此为止，整个实例设计完毕。单击 Run 并查看结果。确定当前位置后，应用程序将显示离加州 Apple 总部有多远，执行效果如图 9-6 所示。

图 9-6　执行效果

实例 210　使用磁性指南针

实例 210	使用磁性指南针
源码路径	\daima\210\

实例说明

　　iPhone 3GS 是第一款装备了磁性指南针的 iOS 设备，随后 iPad 也装备了磁性指南针。Apple 应用程序 Compass（指南针）和 Maps（地图，它使用指南针让地图的方向与用户面向的方向一致）

都使用了指南针。另外，还可使用 iOS 以编程方式访问指南针。本实例演示了使用指南针的过程，向用户提供一个指向左边、右边或前方的箭头，该箭头将引导用户到达 Apple 总部。与距离指示器一样，这里也只介绍了数字指南针潜在用途的很少一部分。本实例以上一个实例为基础，重新建立一个名为"zhen"的工程文件。

具体实现

（1）添加方向图像资源。

在文件夹 Images 中包含了 3 个箭头图像——arrow_up.png、arrow_right.png 和 arrow_lefi.png。为了实现新的方向指示器，需要在 ViewController 中添加一个输出口，它对应于显示合适箭头的 UIImageView；还需添加一个实例变量/属性，用于存储最新的位置。我们将把它们分别命名为"directionArrow"和"recentLocation"。

之所以需要存储最新位置，是因为每次收到航向更新时都需要使用最新位置进行计算。我们将在一个名为"headingToLocation:current"的方法中执行这种计算。

（2）添加用于在弧度和度之间进行转换的常量。

在计算方向时会涉及一些相当复杂的数学知识，借助于专用的公式可以在弧度和度之间进行转换。在文件 ViewController.m 中，在表示 Apple 总部的经度和纬度的常量后面添加两个常量，通过与这些常量相乘，可轻松地完成弧度和度之间的转换。

```
#define kDeg2Rad 0.0174532925
#define kRad2Deg 57.2957795
```

（3）修改用户界面。

为在这个应用程序中使用指南针，需要在界面中新增一个图像视图。为此，打开文件 MainStoryboard.storyboard 和对象库。将一个图像视图（UIImageView）拖曳到视图中，将其放在视图 waitView 上方。使用 Attributes Inspector（Option+Command+4）将其图像设置为 up_arrow.png。使用代码动态地设置图像，但指定默认图像有助于设计界面。接下来，使用 Attributes Inspector 将该图像视图配置成隐藏的，为此选中 View→Drawing 部分的复选框 Hidden。之所以这样做，是因为不想在计算前显示方向。

现在，使用 Size Inspector（Option+Command+5）将图像视图的宽度和高度都设置为 150 点。最后，让图像视图在屏幕上居中，不与视图 waitView 重叠。开发者可根据喜好调整元素的位置。最终的 UI 视图如图 9-7 所示。

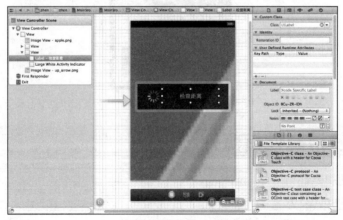

图 9-7　UI 视图

（4）创建并连接输出口。

设计好界面后，切换到助手编辑器模式。只需为刚添加的图像视图建立连接。按住 Control 键，

实例 210　使用磁性指南针

从图像视图拖曳到 ViewController.h 中最后一个@property 编译指令下方。在 Xcode 提示时，新建一个名为"directionArrow"的输出口。此时就可以实现航向更新了，切换到标准编辑器模式，并打开实现文件 ViewController.m。

（5）启动航向更新。

接下来开始修改应用程序逻辑。为了完成这个项目，还需要做如下 4 件事情。

- 需要让位置管理器实例获悉航向变化时都启动更新。
- 每当从 Core Location 获悉新位置时，都需要存储它，以便可以在航向计算中使用最新的位置。
- 必须计算从当前位置前往 Apple 总部的航向。
- 获得航向更新后，需要将其与计算得到的前往 Apple 总部的航向进行比较，如果需要调整航向，则修改 UI 中的箭头。

在请求航向更新前，需要使用位置管理器的方法 headingAvailable 检查它是否提供航向更新。如果没有航向更新，将不会显示箭头图像，而应用程序 Cupertino 将像以前那样运行。如果 headingAvailable 返回 YES，则将航向过滤器设置为 10，并使用 startUpdatingHeading 开始更新。将文件 ViewController.m 中的方法 viewDidLoad 修改成如下所示的代码。

```
- (void)viewDidLoad
{
    locMan = [[CLLocationManager alloc] init];
    locMan.delegate = self;
    locMan.desiredAccuracy = kCLLocationAccuracyThreeKilometers;
    locMan.distanceFilter = 1609; // a mile
    [locMan startUpdatingLocation];

    if ([CLLocationManager headingAvailable]) {
        locMan.headingFilter = 10; // 10 degrees
        [locMan startUpdatingHeading];
    }
    [super viewDidLoad];
}
```

在上述代码中，第 9 行检查是否有航向，如果有，要求仅当航向变化超过 10 度时才更新（第 10 行）。第 11 行请求位置管理器在航向发生变化时启动更新。读者可能会问，这里为何不设置委托，因为第 4 行已经给位置管理器设置了委托，这意味着 ViewController 必须处理位置更新和航向更新。

（6）存储最新的位置。

为存储最新的位置，需要声明一个实例变量/属性，以便在方法中使用它。其类型必须是 CLLocation，因此，在 ViewController.n 中添加合适的属性声明：

```
@property (strong, nonatomic) CLLocation *recentLocation;
```

在文件 ViewController.m 开头，在现有编译指令@synrhesize 后面添加配套的编译指令@synthesize：

```
@synthesize recentLocation;
```

最后，在方法 viewDidUnload 中将该属性设置为 nil。为此，在清理其他实例变量的代码后面添加如下代码行：

```
[self setRecentLocation:nil];
```

为了存储最新位置，需要在方法 locationManager:didUpdateLocation:fromLocation 中添加一行代码，将属性 recentLocation 设置为 newLocation。还需在离目的地不超过 3 英里时停止航向更新，就像停止位置更新一样。通过如下代码存储最近收到的位置供以后使用。

```
- (void)locationManager:(CLLocationManager *)manager
    didUpdateToLocation:(CLLocation *)newLocation
           fromLocation:(CLLocation *)oldLocation {
    if (newLocation.horizontalAccuracy >= 0) {
        // Store the location for use during heading updates
        self.recentLocation = newLocation;
        CLLocation *Cupertino = [[CLLocation alloc]
                          initWithLatitude:kCupertinoLatitude
```

```
                            longitude:kCupertinoLongitude];
        CLLocationDistance delta = [Cupertino
                          distanceFromLocation:newLocation];
        long miles = (delta * 0.000621371) + 0.5; // meters to rounded miles
        if (miles < 3) {
            // Stop updating the location and heading
            [self.locMan stopUpdatingLocation];
            [self.locMan stopUpdatingHeading];
            // Congratulate the user
            self.distanceLabel.text = @"欢迎你\n 成为我们的一员!";
        } else {
            NSNumberFormatter *commaDelimited = [[NSNumberFormatter alloc]
                                        init];
            [commaDelimited setNumberStyle:NSNumberFormatterDecimalStyle];
            self.distanceLabel.text = [NSString stringWithFormat:
                               @"%@ 英里\n 到 Apple",
                               [commaDelimited stringFromNumber:
                                 [NSNumber numberWithLong:miles]]];
        }
        self.waitView.hidden = YES;
        self.distanceView.hidden = NO;
    }
}
```

在上述代码中,只有第 8 和第 19 行是新增的。第 8 行将传入的位置存储到 recentLocation 中,而第 19 行在用户已经在 Apple 总部时停止航向更新。

(7) 计算前往 Apple 总部的航向。

在前两节中,避免了根据经度和纬度进行计算,但在这里,为得到前往 Apple 总部的航向,开发者必须自己做些计算,然后判断航向是向前还是向右或向左转。给定两个位置(如用户的当前位置以及 Apple 总部的位置),可使用一些基本的立体几何知识计算前往 Apple 总部的初始航向。通过在网上搜索,很快找到了这种算法的 JavaScript 代码(这里将其复制到了注释中)。根据这些 JavaScript 代码,很容易使用 Objective-C 语言实现该算法,并计算出航向。将这些代码放在一个新方法中——headingToLocation:current。这个方法接收两个位置作为参数,并返回从当前位置前往目的地的航向。

首先,在文件 ViewController.h 中添加这个方法的原型。这并非必需的,但是一个不错的习惯,有助于避免 Xcode 发出警告。为此,在属性声明后面添加如下代码行:

```
- (double) headingToLocation: (CLLocationCoordinate2D)desired
current: (CLLocationCoordinate2D) current;
```

接下来在文件 ViewController.m 中添加方法 headingToLocation:current,其代码如下所示。

```
-(double)headingToLocation:(CLLocationCoordinate2D)desired
              current:(CLLocationCoordinate2D)current {
    // Gather the variables needed by the heading algorithm
    double lat1 = current.latitude*kDeg2Rad;
    double lat2 = desired.latitude*kDeg2Rad;
    double lon1 = current.longitude;
    double lon2 = desired.longitude;
    double dlon = (lon2-lon1)*kDeg2Rad;
    double y = sin(dlon)*cos(lat2);
    double x = cos(lat1)*sin(lat2) - sin(lat1)*cos(lat2)*cos(dlon);
    double heading=atan2(y,x);
    heading=heading*kRad2Deg;
    heading=heading+360.0;
    heading=fmod(heading,360.0);
    return heading;
}
```

读者无需用担心其中的数学知识,没有必要理解它们。只需知道,给定两个位置(当前位置和目的地),这个方法返回一个浮点数,其单位为度。如果返回的值为零,则需要朝北走以前往目的地,如果是 180,则需要朝南走,依此类推。

(8) 处理航向更新。

ViewController 类遵守了 CLLocationManagerDelegate 协议。正如前面指出的,该协议有一个可

选方法 locationManager:didUpdateHeading，它在航向变化超过 headingFilter 指定的度数时提供航向更新。每当委托收到航向更新时，都应使用用户的当前位置计算前往 Apple 总部的航向，再将其与当前航向进行比较，并显示正确的箭头图像——向左、向右或向前。

为了确保航向计算有意义，需要知道当前位置并确保当前航向在一定的精度范围内，因此在执行计算前使用 if 语句检查这两个条件。如果未能通过这种检查，则隐藏 directionArrow。

由于这种航向不太现实，这里在当前航向与前往 Apple 总部的航向相差不超过 10 度时，将显示向前的箭头；如果超过 10 度，则根据转向正确航向的方向显示左箭头或右箭头。在文件 ViewController.m 中，方法 locationManager:didUpdateHeading 的实现代码如下所示。

```objectivec
(void)locationManager:(CLLocationManager *)manager
    didUpdateHeading:(CLHeading *)newHeading {
    if (self.recentLocation != nil && newHeading.headingAccuracy >= 0) {
        CLLocation *cupertino = [[CLLocation alloc]
                                initWithLatitude:kCupertinoLatitude
                                longitude:kCupertinoLongitude];
        double course = [self headingToLocation:cupertino.coordinate
                                current:recentLocation.coordinate];
        double delta = newHeading.trueHeading - course;
        if (abs(delta) <= 10) {
            self.directionArrow.image = [UIImage imageNamed:
                                @"up_arrow.png"];
        }
        else
        {
            if (delta > 180) {
                self.directionArrow.image = [UIImage imageNamed:
                                @"right_arrow.png"];
            }
            else if (delta > 0) {
                self.directionArrow.image = [UIImage imageNamed:
                                @"left_arrow.png"];
            }
            else if (delta > -180) {
                self.directionArrow.image = [UIImage imageNamed:
                                @"right_arrow.png"];
            }
            else {
                self.directionArrow.image = [UIImage imageNamed:
                                @"left_arrow.png"];
            }
        }
        self.directionArrow.hidden = NO;
    } else {
        self.directionArrow.hidden = YES;
    }
}
```

在上述代码中，第 4 行检查 recentLocation 包含有效的信息且航向精度有效。如果这些条件都不满足，则隐藏图像视图 directionArrow（第 36 行）。第 5～第 7 行新建一个 CLLocation 对象，它包含 Apple 总部的位置。第 8～第 9 行使用该对象计算从当前位置（recentLocation）前往 Apple 总部的航向，并将结果作为浮点数存储在变量 course 中。第 10 行执行简单的减法运算，是整个方法的核心。这里将从 Core Location 获悉的航向之一（newHeading.trueHeading）减去计算得到的航向，并将结果作为浮点数存储到变量 delta 中。

到此为止，整个项目实现完毕。运行该项目。如果用户的设备装备了磁性指南针，便可在办公椅上旋转，并看到箭头图像。如果用 iOS 模拟器运行这个项目，则不会看到箭头。

第 9 章 地图定位应用实战

实例 211 在屏幕中实现一个定位系统

实例 211	在屏幕中实现一个定位系统
源码路径	\daima\211\

实例说明

在本实例中，将通过一个定位系统的具体实现过程，来详细讲解开发这类项目的基本知识。本实例的源码来源于网络中的开源项目，功能是定位当前移动设备的位置。下载获取开源代码后，其目录结构如图 9-8 所示。

MainWindow.xib 是本项目的主窗口，默认的 Cocoa 程序都有这个窗口，启动主程序时会读取这个文件，根据这个文件配置的信息会启动对应的根控制器，其界面如图 9-9 所示。

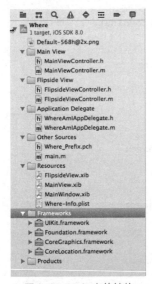

图 9-8　Xcode 中的结构

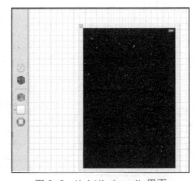

图 9-9　MainWindow.xib 界面

MainView.xib 是主视图的 nib 文件，是连接 MainViewController 和 MainView 的纽带。在此界面中，以表单的样式显示定位信息，如图 9-10 所示。

FlipsideView.xib 是主视图的 nib 文件，是 FlipsideViewController 和 FlipsideView 的纽带。在此界面中，以文本的样式显示当前系统的描述性信息，如图 9-11 所示。

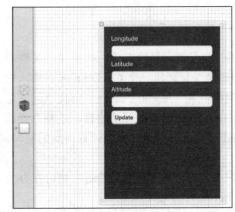

图 9-10　MainView.xib

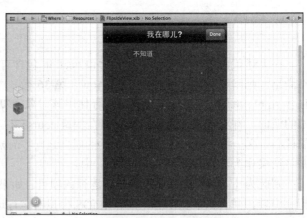

图 9-11　FlipsideView.xib

实例 211 在屏幕中实现一个定位系统

具体实现

在本实例中,文件 MainViewController.h 实现了主视图的关系映射,对应代码如下所示。

```objc
#import <UIKit/UIKit.h>
#import <CoreLocation/CoreLocation.h>
#import <CoreLocation/CLLocationManagerDelegate.h>
#import "FlipsideViewController.h"
@interface MainViewController : UIViewController <FlipsideViewControllerDelegate,
CLLocationManagerDelegate> {
      IBOutlet UITextField *altitude;
    IBOutlet UITextField *latitude;
    IBOutlet UITextField *longitude;

    CLLocationManager    *locmanager;
    BOOL                 wasFound;
}
@property (nonatomic,retain) UITextField *altitude;
@property (nonatomic,retain) UITextField *latitude;
@property (nonatomic,retain) UITextField *longitude;
@property (nonatomic,retain) CLLocationManager *locmanager;
- (IBAction)showInfo:(id)sender;
- (IBAction)update;
@end
```

在文件 MainViewController.m 中,通过- (IBAction)showInfo:方法处理单击图标后显示另一个视图界面。通过- (IBAction)update 方法响应单击"update"按钮后的事件,通过- (void)locationManager 方法实现定位管理功能。此文件的具体实现代码如下所示。

```objc
#import "MainViewController.h"
@implementation MainViewController
@synthesize altitude,latitude,longitude,locmanager;
- (IBAction)update {
      locmanager = [[CLLocationManager alloc] init];
      [locmanager setDelegate:self];
      [locmanager setDesiredAccuracy:kCLLocationAccuracyBest];
      [locmanager startUpdatingLocation];
}
// Implement viewDidLoad to do additional setup after loading the view, typically from a nib.
- (void)viewDidLoad {
      [self update];
}
- (void)locationManager:(CLLocationManager *)manager didUpdateToLocation:(CLLocation *)newLocation fromLocation:(CLLocation *)oldLocation
{
      if (wasFound) return;
      wasFound = YES;
          CLLocationCoordinate2D loc = [newLocation coordinate];
          latitude.text = [NSString stringWithFormat: @"%f", loc.latitude];
      longitude.text  = [NSString stringWithFormat: @"%f", loc.longitude];
      altitude.text = [NSString stringWithFormat: @"%f", newLocation.altitude];
      //   NSString *mapUrl = [NSString stringWithFormat: @"http://maps.google.com/maps?q=%f,%f", loc.latitude, loc.longitude];
      //   NSURL *url = [NSURL URLWithString:mapUrl];
      //   [[UIApplication sharedApplication] openURL:url];
}
- (void)locationManager:(CLLocationManager *)manager didFailWithError:(NSError *)error
{

    UIAlertView *alert = [[UIAlertView alloc] initWithTitle:@"错误通知"
                   message:[error description]
         delegate:nil cancelButtonTitle:@"OK" otherButtonTitles:nil];
      [alert show];
      [alert release];
}
- (void)flipsideViewControllerDidFinish:(FlipsideViewController *)controller {

    [self dismissModalViewControllerAnimated:YES];
}
- (IBAction)showInfo:(id)sender {
```

```objc
        FlipsideViewController *controller = [[FlipsideViewController alloc] initWithNib
Name:@"FlipsideView" bundle:nil];
        controller.delegate = self;
        controller.modalTransitionStyle = UIModalTransitionStyleFlipHorizontal;
        [self presentModalViewController:controller animated:YES];
        [controller release];
}
- (void)didReceiveMemoryWarning {
    // Releases the view if it doesn't have a superview.
    [super didReceiveMemoryWarning];

    // Release any cached data, images, etc. that aren't in use.
}
- (void)viewDidUnload {
    [locmanager stopUpdatingLocation];
}
/*
// Override to allow orientations other than the default portrait orientation.
- (BOOL)shouldAutorotateToInterfaceOrientation:(UIInterfaceOrientation)
interfaceOrientation {
    // Return YES for supported orientations.
    return (interfaceOrientation == UIInterfaceOrientationPortrait);
}
*/
- (void)dealloc {
    [altitude release];
    [latitude release];
    [longitude release];
    [locmanager release];
    [super dealloc];
}
@end
```

在上述代码中，方法(void)flipsideViewControllerDidFinish:是在委托协议 FlipsideViewController Delegate 中定义的方法，作为 FlipsideViewControllerDelegate 协议的实现者，MainViewController 视图控制器必须实现这个方法，此方法的作用是调用[self dismissModalViewControllerAnimated: YES]语句关闭模态视图控制器。

在本实例中，FlipsideView 视图显示系统的说明信息。在前面的文件 MainViewController.m 中，通过调用 - (IBAction)showInfo:方法显示 FlipsideView 视图。其中实现文件 FlipsideViewController.h 的代码如下所示。

```objc
#import <UIKit/UIKit.h>
@protocol FlipsideViewControllerDelegate;
@interface FlipsideViewController : UIViewController {
  id <FlipsideViewControllerDelegate> delegate;
}
@property (nonatomic, assign) id <FlipsideViewControllerDelegate> delegate;
- (IBAction)done:(id)sender;
@end
@protocol FlipsideViewControllerDelegate
- (void)flipsideViewControllerDidFinish:(FlipsideViewController *)controller;
@end
```

在上述文件 FlipsideViewController.h 中，不但定义了 FlipsideViewController 类，而且还定义了委托协议 FlipsideViewControllerDelegate。

其中，实例文件 FlipsideViewController.m 的代码如下所示。

```objc
#import "FlipsideViewController.h"
@implementation FlipsideViewController
@synthesize delegate;
- (void)viewDidLoad {
    [super viewDidLoad];
    self.view.backgroundColor = [UIColor viewFlipsideBackgroundColor];
}
- (IBAction)done:(id)sender {
    [self.delegate flipsideViewControllerDidFinish:self];
}
```

```
- (void)didReceiveMemoryWarning {
    // Releases the view if it doesn't have a superview.
    [super didReceiveMemoryWarning];
    // Release any cached data, images, etc that aren't in use.
}
- (void)viewDidUnload {
    // Release any retained subviews of the main view.
    // e.g. self.myOutlet = nil;
}
/*
// Override to allow orientations other than the default portrait orientation.
- (BOOL)shouldAutorotateToInterfaceOrientation:(UIInterfaceOrientation)interfaceOrientation {
    // Return YES for supported orientations
    return (interfaceOrientation == UIInterfaceOrientationPortrait);
}
*/
- (void)dealloc {
    [super dealloc];
}
@end
```

在上述代码中，通过-(void)viewDidLoad方法实现初始化处理。当单击"Done"按钮时会调用-(IBAction)done：方法，通过此方法关闭模态视图控制器。

到此为止，整个实例的主要功能就介绍完毕了，主视图的执行效果如图9-12所示，可以实现定位功能。

图9-12　主视图的执行效果

实例212　在屏幕中使用谷歌地图

实例212	在屏幕中使用谷歌地图
源码路径	\daima\212\

实例说明

在iOS应用中，通过Google Maps实现向用户提供了一个地图应用程序，它响应速度快，使用起来很有趣。通过使用Map Kit，开发者的应用程序也能提供这样的用户体验。Map Kit让开发者能够将地图嵌入到视图中，并提供显示该地图所需的所有图块（图像）。它在需要时处理滚动、缩放和图块加载。Map Kit还能执行反向地理编码（reverse geocoding），即根据坐标获取位置信息（国家、州、城市、地址）。程序员无需编写任何代码就可使用Map Kit，只需将Map Kit框架加入到项目中，并使用 Interface Builder将一个MKMapView实例加入到视图中。添加地图视图后，便可在Attributes Inspector中设置多个属性，以进一步定制它。

在本章上一个实例中，我们的定位系统只有数字。在本实例中，将对整个实例进行升级，最大变化是引入了谷歌地图，通过地图的样式实现定位功能，这样整个界面将更加直观。

> **注意**　Map Kit 图块（map tile）来自 Google Maps/Google Earth API，虽然我们不能直接调用该 API，但 Map Kit 代表开发都进行这些调用，因此使用 Map Kit 的地图数据时，开发者和其应用程序必须遵守 Google Maps/Google Earth API 服务条款。

具体实现

（1）添加打开地图功能。

在 Main View 中添加打开地图功能的触发点，本实例中添加了一个"open web map"按钮。此时升级后文件 MainViewController.h 的实现代码如下所示。

```
#import <UIKit/UIKit.h>
#import <CoreLocation/CoreLocation.h>
#import <CoreLocation/CLLocationManagerDelegate.h>
#import "FlipsideViewController.h"
@interface MainViewController : UIViewController <FlipsideViewControllerDelegate,
CLLocationManagerDelegate> {
    IBOutlet UITextField *altitude;
    IBOutlet UITextField *latitude;
    IBOutlet UITextField *longitude;
    CLLocationManager   *locmanager;
    BOOL wasFound;
}
@property (nonatomic,retain) UITextField *altitude;
@property (nonatomic,retain) UITextField *latitude;
@property (nonatomic,retain) UITextField *longitude;
@property (nonatomic,retain) CLLocationManager *locmanager;
- (IBAction)showInfo:(id)sender;
- (IBAction)update;
- (IBAction)openWebMap;
@end
```

而在文件 MainViewController.m 中添加了一个地图界面的方法- (IBAction)openWebMap，其他代码没有变化。文件 MainViewController.m 的具体代码如下所示。

```
#import "MainViewController.h"
@implementation MainViewController
@synthesize altitude,latitude,longitude,locmanager;
- (IBAction)openWebMap {
NSString *urlString = [NSString stringWithFormat:
                      @"http://maps.google.com/maps?q=%f,%f",
                      [latitude.text floatValue],
                      [longitude.text floatValue]];
    NSURL *url = [NSURL URLWithString:urlString];
[[UIApplication sharedApplication] openURL:url];
}
- (IBAction)update {
    locmanager = [[CLLocationManager alloc] init];
    [locmanager setDelegate:self];
    [locmanager setDesiredAccuracy:kCLLocationAccuracyBest];
    [locmanager startUpdatingLocation];
}
// Implement viewDidLoad to do additional setup after loading the view, typically from a nib.
- (void)viewDidLoad {
    [self update];
}
- (void)locationManager:(CLLocationManager *)manager didUpdateToLocation:(CLLocation *)newLocation fromLocation:(CLLocation *)oldLocation
{
    if (wasFound) return;
    wasFound = YES;
    
    CLLocationCoordinate2D loc = [newLocation coordinate];
        latitude.text = [NSString stringWithFormat: @"%f", loc.latitude];
    longitude.text  = [NSString stringWithFormat: @"%f", loc.longitude];
    altitude.text = [NSString stringWithFormat: @"%f", newLocation.altitude];
}
```

```objc
- (void)locationManager:(CLLocationManager *)manager didFailWithError:(NSError *)error
{
    UIAlertView *alert = [[UIAlertView alloc] initWithTitle:@"错误通知"
            message:[error description]
delegate:nil cancelButtonTitle:@"OK"
                                            otherButtonTitles:nil];
    [alert show];
    [alert release];
}
- (void)flipsideViewControllerDidFinish:(FlipsideViewController *)controller {
    [self dismissModalViewControllerAnimated:YES];
}
- (IBAction)showInfo:(id)sender {
    CLLocation *lastLocation = [locmanager location];
  if(!lastLocation) {
    UIAlertView *alert;
    alert = [[UIAlertView alloc]
            initWithTitle:@"系统错误"
            message:@"还没有接收到数据！"
            delegate:nil cancelButtonTitle:nil
            otherButtonTitles:@"OK", nil];
[alert show];
    [alert release];
    return;
  }
    FlipsideViewController *controller = [[FlipsideViewController alloc] initWithNib
Name:@"FlipsideView" bundle:nil];
    controller.delegate = self;
    controller.lastLocation = lastLocation;
    controller.modalTransitionStyle = UIModalTransitionStyleFlipHorizontal;
    [self presentModalViewController:controller animated:YES];
    [controller release];
}
- (void)didReceiveMemoryWarning {
    [super didReceiveMemoryWarning];
}
- (void)viewDidUnload {
    [locmanager stopUpdatingLocation];
}
// Override to allow orientations other than the default portrait orientation.
- (BOOL)shouldAutorotateToInterfaceOrientation:(UIInterfaceOrientation)interfaceOrientation {
    // Return YES for supported orientations.
    return (interfaceOrientation == UIInterfaceOrientationPortrait);
}
- (void)dealloc {
    [altitude release];
    [latitude release];
    [longitude release];
    [locmanager release];
   [super dealloc];
}
@end
```

此时主界面执行后的效果如图 9-13 所示。

图 9-13　主界面的执行效果

(2)升级视图控制器。

视图控制器的变化也比较多,首先在-(void)viewDidLoad 方法中追加 mapView 代码,然后添加(IBAction)search:方法以在地图上标注我们的位置,并且添加了反编码查询方法(void)-reverseGeocoder:,具体代码如下所示。

```objc
#import "FlipsideViewController.h"
@implementation FlipsideViewController
@synthesize delegate;
@synthesize lastLocation;
@synthesize mapView;
- (void)viewDidLoad {
    [super viewDidLoad];
    self.view.backgroundColor = [UIColor viewFlipsideBackgroundColor];
        mapView.mapType = MKMapTypeStandard;
        // mapView.mapType = MKMapTypeSatellite;
        //mapView.mapType = MKMapTypeHybrid;
        mapView.delegate = self;
}
- (IBAction)search:(id)sender {
        MKCoordinateRegion viewRegion = MKCoordinateRegionMakeWithDistance(lastLocation.coordinate, 2000, 2000);
        [mapView setRegion:viewRegion animated:YES];
    MKReverseGeocoder *geocoder = [[MKReverseGeocoder alloc] initWithCoordinate:lastLocation.coordinate];
        geocoder.delegate = self;
        [geocoder start];
}
- (IBAction)done:(id)sender {
        [self.delegate flipsideViewControllerDidFinish:self];
}
- (void)didReceiveMemoryWarning {
    [super didReceiveMemoryWarning];
}
- (void)viewDidUnload {
        // Release any retained subviews of the main view.
        // e.g. self.myOutlet = nil;
}
- (void)dealloc {
    [lastLocation release];
    [mapView release];
    [super dealloc];
}
#pragma mark -
#pragma mark Reverse Geocoder Delegate Methods
- (void)reverseGeocoder:(MKReverseGeocoder *)geocoder didFailWithError:(NSError *)error {
    UIAlertView *alert = [[UIAlertView alloc]
                        initWithTitle:@"地理解码错误息"
                        message: [error description]
                        delegate:nil
                        cancelButtonTitle:@"Ok"
                        otherButtonTitles:nil];
    [alert show];
    [alert release];
geocoder.delegate = nil;
    [geocoder autorelease];
}
- (void)reverseGeocoder:(MKReverseGeocoder *)geocoder didFindPlacemark:(MKPlacemark *)placemark {
    MapLocation *annotation = [[MapLocation alloc] init];
    annotation.streetAddress = placemark.thoroughfare;
    annotation.city = placemark.locality;
    annotation.state = placemark.administrativeArea;
    annotation.zip = placemark.postalCode;
    annotation.coordinate = geocoder.coordinate;
      [mapView removeAnnotations:mapView.annotations];
      [mapView addAnnotation:annotation];
      [annotation release];
    geocoder.delegate = nil;
      [geocoder autorelease];
}
```

```
#pragma mark -
#pragma mark Map View Delegate Methods
- (MKAnnotationView *) mapView:(MKMapView *)theMapView viewForAnnotation:(id <MKAnnotation>)
annotation {
    MKPinAnnotationView *annotationView
    = (MKPinAnnotationView *)[mapView dequeueReusableAnnotationViewWithIdentifier:
@"PIN_ANNOTATION"];
    if(annotationView == nil) {
        annotationView = [[[MKPinAnnotationView alloc] initWithAnnotation:annotation
        reuseIdentifier:@"PIN_ANNOTATION"] autorelease];
    }
    annotationView.pinColor = MKPinAnnotationColorPurple;
    annotationView.animatesDrop = YES;
    annotationView.canShowCallout = YES;
    return annotationView;
}
- (void)mapViewDidFailLoadingMap:(MKMapView *)theMapView withError:(NSError *)error {
    UIAlertView *alert = [[UIAlertView alloc]
                    initWithTitle:@"地图加载错误"
                    message:[error localizedDescription]
                    delegate:nil
                    cancelButtonTitle:@"Ok"
                    otherButtonTitles:nil];
    [alert show];
    [alert release];
}
@end
```

（3）添加自定义地图标注对象。

接下来需要添加自定义地图标注对象，在此需要实现 MKAnnotation 协议和 NSCoding 协议。文件 MapLocation.h 的实现代码如下所示。

```
#import <Foundation/Foundation.h>
#import <MapKit/MapKit.h>

@interface MapLocation : NSObject <MKAnnotation, NSCoding> {
    NSString *streetAddress;
    NSString *city;
    NSString *state;
    NSString *zip;
    CLLocationCoordinate2D coordinate;
}
@property (nonatomic, copy) NSString *streetAddress;
@property (nonatomic, copy) NSString *city;
@property (nonatomic, copy) NSString *state;
@property (nonatomic, copy) NSString *zip;
@property (nonatomic, readwrite) CLLocationCoordinate2D coordinate;
@end
```

在文件 MapLocation.m 中，通过-(NSString *) title 方法获取标题，通过-(NSString *) subtitle 方法获取子标题，通过-(void) encodeWithCoder: 将对象的状态写入到文件中。文件 MapLocation.m 的具体代码如下所示。

```
#import "MapLocation.h"
#import <MapKit/MapKit.h>
@implementation MapLocation
@synthesize streetAddress;
@synthesize city;
@synthesize state;
@synthesize zip;
@synthesize coordinate;
#pragma mark -
- (NSString *)title {
    return @"您的位置!";
}
- (NSString *)subtitle {
    NSMutableString *ret = [NSMutableString string];
    if (streetAddress)
        [ret appendString:streetAddress];
```

第9章 地图定位应用实战

```
        if (streetAddress && (city || state || zip))
            [ret appendString:@" • "];
        if (city)
            [ret appendString:city];
        if (city && state)
            [ret appendString:@", "];
        if (state)
            [ret appendString:state];
        if (zip)
            [ret appendFormat:@", %@", zip];
        return ret;
    }
    #pragma mark -
    - (void)dealloc {
        [streetAddress release];
        [city release];
        [state release];
        [zip release];
        [super dealloc];
    }
    #pragma mark -
    #pragma mark NSCoding Methods
    - (void) encodeWithCoder: (NSCoder *)encoder {
        [encoder encodeObject: [self streetAddress] forKey: @"streetAddress"];
        [encoder encodeObject: [self city] forKey: @"city"];
        [encoder encodeObject: [self state] forKey: @"state"];
        [encoder encodeObject: [self zip] forKey: @"zip"];
    }
    - (id) initWithCoder: (NSCoder *)decoder {
        if (self = [super init]) {
            [self setStreetAddress: [decoder decodeObjectForKey: @"streetAddress"]];
            [self setCity: [decoder decodeObjectForKey: @"city"]];
            [self setState: [decoder decodeObjectForKey: @"state"]];
            [self setZip: [decoder decodeObjectForKey: @"zip"]];
        }
        return self;
    }
@end
```

到此为止，就成功地在项目中添加了谷歌地图功能。单击"open web map"按钮后的效果如图 9-14 所示。

图 9-14　显示谷歌地图

实例 213　在收集地图中实现定位和位置标示

实例 213	在收集地图中实现定位和位置标示
源码路径	\daima\213\

实例说明

在本实例中，通过 Google Maps 在 iOS 屏幕中显示了一个地图。然后在屏幕中设置了如下所示的 3 个按钮。
- 现在的位置：触摸后可以在地图中快速定位到当前的所在位置。
- 转换位置：触摸后可以在地图中快速定位到设置的位置。
- 找出北京和上海：触摸后可以在地图中快速标示出北京和上海的位置。

具体实现

视图文件 ViewController.xib 的界面效果如图 9-15 所示。

在实例文件 ViewController.m 中定义了触摸 3 个按钮的事件处理程序，具体实现代码如下所示。

图 9-15　UI 视图界面

```objc
#import "ViewController.h"
@implementation ViewController
- (void)didReceiveMemoryWarning
{
    [super didReceiveMemoryWarning];
    // Release any cached data, images, etc that aren't in use.
}
-(IBAction)changeMapType:(id)segcontrol {
    UISegmentedControl *ctrl = (UISegmentedControl*) segcontrol;
    NSInteger temp = ctrl.selectedSegmentIndex;//获取分段控件上的选择：0/1/2
    mv.mapType=temp;//设置地图的显示类型（也是使用 0/1/2 来代表 3 种显示）
}
- (IBAction)addPin
{
    CLLocationCoordinate2D coordinate1 = {31.240948,121.485958};//上海的经纬度
    NSDictionary *address = [NSDictionary dictionaryWithObjectsAndKeys:@"中国",
@"Country",@"上海",@"Locality", nil];//上述位置的地址信息
    MKPlacemark *shanghai = [[MKPlacemark alloc] initWithCoordinate:coordinate1
addressDictionary:address]; //创建 MKPlacemark
    [mv addAnnotation:shanghai]; //在地图上标识
    //按照同样的方法，在地图上标记北京
    CLLocationCoordinate2D c = {39.908605,116.398019};
    address = [NSDictionary dictionaryWithObjectsAndKeys:@"中国", @"Country",@"北京
",@"Locality", nil];
    MKPlacemark *mysteryspot = [[MKPlacemark alloc] initWithCoordinate:c address
Dictionary:address];
    [mv addAnnotation:mysteryspot];
}
-(IBAction)reverseGeoTest{
    CLLocationCoordinate2D c = {39.908605,116.398019};//一个位置
    //调用 MKReverseGeocoder 来查询上述位置的地址名称
    geo=[[MKReverseGeocoder alloc] initWithCoordinate:c];
    geo.delegate=self; //设置回调（找到后的回调方法和找不到的回调方法）
    [geo start]; //开始转换
}
//找不到地址信息，就调用下述方法
-(void) reverseGeocoder:(MKReverseGeocoder*)geocoder didFailwithError:(NSError*)error{
```

```objc
        NSLog(@"reverseGeoCoder error");
}
//找到了地址信息，就标识在地图上
-(void)reverseGeocoder:(MKReverseGeocoder*)geocoder
didFindPlacemark:(MKPlacemark*)placemark{
    MKPlacemark *mysteryspot = [[MKPlacemark alloc] initWithCoordinate:placemark.coordinate addressDictionary:placemark.addressDictionary];
    [mv addAnnotation:mysteryspot];//标记在地图上
    [v setCenterCoordinate:placemark.coordinate animated:YES];
}
- (IBAction)currentLocation
{
    mv.showsUserLocation = YES;//允许显示用户当前位置（缺省为否）
    MKUserLocation *userLocation = mv.userLocation;//获取当前位置
    CLLocationCoordinate2D coordinate = userLocation.location.coordinate;//经纬度
    if (!geo)
    {//使用MKReverseGeocoder，获取当前位置的地址信息，并在地图上标记
        geo = [[MKReverseGeocoder alloc] initWithCoordinate:coordinate];
        geo.delegate = self;
        [geo start];
    }
}
#pragma mark - View lifecycle

- (void)viewDidLoad
{
   [super viewDidLoad];
}
- (void)viewDidUnload
{
   [super viewDidUnload];
}
- (void)viewWillAppear:(BOOL)animated
{
   [super viewWillAppear:animated];
}
- (void)viewDidAppear:(BOOL)animated
{
   [super viewDidAppear:animated];
}
- (void)viewWillDisappear:(BOOL)animated
{
  [super viewWillDisappear:animated];
}
- (void)viewDidDisappear:(BOOL)animated
{
  [super viewDidDisappear:animated];
}
- (BOOL)shouldAutorotateToInterfaceOrientation:(UIInterfaceOrientation)interfaceOrientation
{
   // Return YES for supported orientations
   return (interfaceOrientation != UIInterfaceOrientationPortraitUpsideDown);
}
@end
```

执行后的效果如图 9-16 所示。

图 9-16 执行效果

实例 214　在地图中实现标注

实例 214	在地图中实现标注
源码路径	\daima\214\

实例说明

在 iOS 应用中，当通过 Google Maps 在 iOS 屏幕中显示地图后，我们可以在地图上做一些具有提醒功能和解释功能的标注信息。这一功能是通过 MKAnnotationView 和 MKPinAnnotationView 实现的，其中，MKAnnotation 协议能够实现最基本的标注，此标注协议有 3 个属性——coordinate、title 和 subtitle，其中必须设置 coordinate 属性。设置好 Annotation 后就可以用其将标注在地图上显示出来。而 MKPinAnnotationView 能够以大头针的方式显示标注，此协议继承自 MKAnnotationView，同时添加了如下两个属性。

- @property (nonatomic) MKPinAnnotationColor pinColor：设置大头针的颜色，有红、绿、紫 3 种颜色可选择。
- @property (nonatomic) BOOL animatesDrop：设置大头针是否以掉下来的动画方式显示。

如果想创建以静态图片作为大头针图片的话，可以通过创建 MKAnnotationView 实现。如果想使用 Apple 自带的大头针则创建 MKPinAnnotationView。

下面是在地图上添加 Annotation 的步骤。

（1）创建一个实现 MKAnnotation 协议的类，在该类的初始化函数中进行 coordinate 属性设置。

（2）用上述方法创建 Annotation。

（3）把创建的 Annotation 用 addAnnotation 的方法添加到 MapView 中。

（4）实现 MKMapViewDelegate 代理，在代理函数 - (MKAnnotationView *)mapView:(MKMapView *)mView viewForAnnotation:(id <MKAnnotation>)annotation 中把 Annotation 以 MKPinAnnotationView 或 MKAnnotationView 的方式标注在地图上显示。

在本实例中，演示了使用 MKAnnotation 和 MKPinAnnotationView 实现标注的方法。

具体实现

在实例文件 RootViewController.m 中，首先使用 switch 语句设置了界面初始载入的视图，其中 "case7" 使用 MKOverlay 协议在地图中绘制了一个覆盖三角形；然后使用方法 mapView：viewForAnnotation:判断绘制的标注视图是否可用。此文件的具体实现代码如下所示。

```
#define which 4 // try 2, 3, 4, 5, 6, 7, 8
-(void)viewDidAppear:(BOOL)animated {
    [super viewDidAppear: animated];
    switch (which) {
        case 1: // figure 15-4
        {
            CLLocationCoordinate2D loc = CLLocationCoordinate2DMake(34.923964,-120.219558);
            MKCoordinateRegion reg = MKCoordinateRegionMakeWithDistance(loc, 1000, 1000);
            self->map.region = reg;
            self->map.hidden = NO;
            break;
        }
        case 2: // figure 15-23
        {
            CLLocationCoordinate2D loc = CLLocationCoordinate2DMake(34.923964,-120.219558);
            MKCoordinateRegion reg = MKCoordinateRegionMakeWithDistance(loc, 1000, 1000);
            self->map.region = reg;
            MKPointAnnotation* ann = [[MKPointAnnotation alloc] init];
            ann.coordinate = loc;
            ann.title = @"Park here";
```

```objc
            ann.subtitle = @"Fun awaits down the road!";
            [self->map addAnnotation:ann];
            self->map.hidden = NO;
            break;
        }
        case 3: // set delegate to make pin green, get our own annotation view, etc.
        case 4:
        case 5:
        {
            self->map.delegate = self;
            CLLocationCoordinate2D loc = CLLocationCoordinate2DMake(34.923964,-120.219558);
            MKCoordinateRegion reg = MKCoordinateRegionMakeWithDistance(loc, 1000, 1000);
            self->map.region = reg;
            MKPointAnnotation* ann = [[MKPointAnnotation alloc] init];
            ann.coordinate = loc;
            ann.title = @"Park here";
            ann.subtitle = @"Fun awaits down the road!";
            [self->map addAnnotation:ann];
            self->map.hidden = NO;
            break;
        }
        case 6: // use our own annotation class too
        {
            self->map.delegate = self;
            CLLocationCoordinate2D loc = CLLocationCoordinate2DMake(34.923964,-120.219558);
            MKCoordinateRegion reg = MKCoordinateRegionMakeWithDistance(loc, 1000, 1000);
            self->map.region = reg;
            MyAnnotation* ann = [[MyAnnotation alloc] initWithLocation:loc];
            ann.title = @"Park here";
            ann.subtitle = @"Fun awaits down the road!";
            [self->map addAnnotation:ann];
            self->map.hidden = NO;
            break;
        }
        case 7: // overlay,figure 15-25
        {
            self->map.delegate = self;
            CLLocationCoordinate2D loc = CLLocationCoordinate2DMake(34.923964,-120.219558);
            MKCoordinateRegion reg = MKCoordinateRegionMakeWithDistance(loc, 1000, 1000);
            self->map.region = reg;
            MyAnnotation* ann = [[MyAnnotation alloc] initWithLocation:loc];
            ann.title = @"Park here";
            ann.subtitle = @"Fun awaits down the road!";
            [self->map addAnnotation:ann];
            loc = self->map.region.center;
            CGFloat lat = loc.latitude;
            CLLocationDistance metersPerPoint = MKMetersPerMapPointAtLatitude(lat);
            MKMapPoint c = MKMapPointForCoordinate(loc);
            c.x += 150/metersPerPoint;
            c.y -= 50/metersPerPoint;
            MKMapPoint p1 = MKMapPointMake(c.x, c.y);
            p1.y -= 100/metersPerPoint;
            MKMapPoint p2 = MKMapPointMake(c.x, c.y);
            p2.x += 100/metersPerPoint;
            MKMapPoint p3 = MKMapPointMake(c.x, c.y);
            p3.x += 300/metersPerPoint;
            p3.y -= 400/metersPerPoint;
            MKMapPoint pts[3] = {
                p1, p2, p3
            };
            MKPolygon* tri = [MKPolygon polygonWithPoints:pts count:3];
            [self->map addOverlay:tri];
            
            self->map.hidden = NO;
            break;
        }
        case 8: // nicer overly,
        {
            self->map.delegate = self;
            CLLocationCoordinate2D loc = CLLocationCoordinate2DMake(34.923964,-120.219558);
            MKCoordinateRegion reg = MKCoordinateRegionMakeWithDistance(loc, 1000, 1000);
            self->map.region = reg;
```

```objc
                MyAnnotation* ann = [[MyAnnotation alloc] initWithLocation:loc];
                ann.title = @"Park here";
                ann.subtitle = @"Fun awaits down the road!";
                [self->map addAnnotation:ann];
                // start with our position and derive a nice unit for drawing
                loc = self->map.region.center;
                CGFloat lat = loc.latitude;
                CLLocationDistance metersPerPoint = MKMetersPerMapPointAtLatitude(lat);
                MKMapPoint c = MKMapPointForCoordinate(loc);
                CGFloat unit = 75.0/metersPerPoint;
                // size and position the overlay bounds on the earth
                CGSize sz = CGSizeMake(4*unit, 4*unit);
                MKMapRect mr = MKMapRectMake(c.x + 2*unit, c.y - 4.5*unit, sz.width, sz.height);
                // describe the arrow as a CGPath
                CGMutablePathRef p = CGPathCreateMutable();
                CGPoint start = CGPointMake(0, unit*1.5);
                CGPoint p1 = CGPointMake(start.x+2*unit, start.y);
                CGPoint p2 = CGPointMake(p1.x, p1.y-unit);
                CGPoint p3 = CGPointMake(p2.x+unit*2, p2.y+unit*1.5);
                CGPoint p4 = CGPointMake(p2.x, p2.y+unit*3);
                CGPoint p5 = CGPointMake(p4.x, p4.y-unit);
                CGPoint p6 = CGPointMake(p5.x-2*unit, p5.y);
                CGPoint points[] = {
                    start, p1, p2, p3, p4, p5, p6
                };
                // rotate the arrow around its center
                CGAffineTransform t1 = CGAffineTransformMakeTranslation(unit*2, unit*2);
                CGAffineTransform t2 = CGAffineTransformRotate(t1, -M_PI/3.5);
                CGAffineTransform t3 = CGAffineTransformTranslate(t2, -unit*2, -unit*2);
                CGPathAddLines(p, &t3, points, 7);
                CGPathCloseSubpath(p);
                // create the overlay and give it the path
                MyOverlay* over = [[MyOverlay alloc] initWithRect:mr];
                over.path = [UIBezierPath bezierPathWithCGPath:p];
                CGPathRelease(p);
                // add the overlay to the map
                [self->map addOverlay:over];

                self->map.hidden = NO;
                break;
            }
        }
    }

- (MKAnnotationView *)mapView:(MKMapView *)mapView
         viewForAnnotation:(id <MKAnnotation>)annotation {
    switch (which) {
        case 3:
        {
            MKAnnotationView* v = nil;
            if ([annotation.title isEqualToString:@"Park here"]) {
                static NSString* ident = @"greenPin";
                v = [mapView dequeueReusableAnnotationViewWithIdentifier:ident];
                if (v == nil) {
                    v = [[MKPinAnnotationView alloc] initWithAnnotation:annotation
                                                reuseIdentifier:ident];
                    ((MKPinAnnotationView*)v).pinColor = MKPinAnnotationColorGreen;
                    v.canShowCallout = YES;
                } else {
                    v.annotation = annotation;
                }
            }
            return v;
            break;
        }
        case 4: // figure 15-24
        {
            MKAnnotationView* v = nil;
            if ([annotation.title isEqualToString:@"Park here"]) {
                static NSString* ident = @"greenPin";
                v = [mapView dequeueReusableAnnotationViewWithIdentifier:ident];
```

```objc
                if (v == nil) {
                    v = [[MKAnnotationView alloc] initWithAnnotation:annotation
                                                      reuseIdentifier:ident];
                    v.image = [UIImage imageNamed:@"clipartdirtbike.gif"];
                    CGRect f = v.bounds;
                    f.size.height /= 3.0;
                    f.size.width /= 3.0;
                    v.bounds = f;
                    v.centerOffset = CGPointMake(0,-20);
                    v.canShowCallout = YES;
                } else {
                    v.annotation = annotation;
                }
            }
            return v;
            break;
        }
        case 5: // use custom MKAnnotationView subclass instead
        {
            MKAnnotationView* v = nil;
            if ([annotation.title isEqualToString:@"Park here"]) {
                static NSString* ident = @"bike";
                v = [mapView dequeueReusableAnnotationViewWithIdentifier:ident];
                if (v == nil) {
                    v = [[MyAnnotationView alloc] initWithAnnotation:annotation
                                                     reuseIdentifier:ident];
                    v.canShowCallout = YES;
                } else {
                    v.annotation = annotation;
                }
            }
            return v;
            break;
        }
        case 6: // use custom MKAnnotation too
        case 7:
        case 8:
        {
            MKAnnotationView* v = nil;
            if ([annotation isKindOfClass:[MyAnnotation class]]) { // much better test
                static NSString* ident = @"bike";
                v = [mapView dequeueReusableAnnotationViewWithIdentifier:ident];
                if (v == nil) {
                    v = [[MyAnnotationView alloc] initWithAnnotation:annotation
                                                     reuseIdentifier:ident];
                    v.canShowCallout = YES;
                } else {
                    v.annotation = annotation;
                }
            }
            return v;
            break;
        }
    }
    return nil; // shut the compiler up
}

- (MKOverlayView *)mapView:(MKMapView *)mapView
          viewForOverlay:(id <MKOverlay>)overlay {
    switch (which) {
        case 7: {
            MKPolygonView* v = nil;
            if ([overlay isKindOfClass:[MKPolygon class]]) {
                v = [[MKPolygonView alloc] initWithPolygon:(MKPolygon*)overlay];
                v.fillColor = [[UIColor redColor] colorWithAlphaComponent:0.1];
                v.strokeColor = [[UIColor redColor] colorWithAlphaComponent:0.8];
                v.lineWidth = 2;
            }
            return v;
            break;
        }
        case 8:
```

```
        {
            MKOverlayView* v = nil;
            if ([overlay isKindOfClass: [MyOverlay class]]) {
                v = [[MKOverlayPathView alloc] initWithOverlay:overlay];
                MKOverlayPathView* vv = (MKOverlayPathView*)v; // typecast for simplicity
                vv.path = ((MyOverlay*)overlay).path.CGPath;
                vv.strokeColor = [UIColor blackColor];
                vv.fillColor = [[UIColor redColor] colorWithAlphaComponent:0.2];
                vv.lineWidth = 2;
            }
            return v;
            break;
        }
    }
    return nil; // shut the compiler up
}
@end
```

编写文件 MyAnnotation.m 实现自己的标注类，具体代码如下所示。

```
#import "MyAnnotation.h"
@implementation MyAnnotation
@synthesize coordinate, title, subtitle;
- (id)initWithLocation: (CLLocationCoordinate2D) coord {
    self = [super init];
    if (self) {
        self->coordinate = coord;
    }
    return self;
}
@end
```

在文件 MyAnnotationView.m 中创建了自己的 MyAnnotationView 子类，并实现了自己绘制功能。具体实现代码如下所示。

```
#import "MyAnnotationView.h"
@implementation MyAnnotationView
- (id)initWithAnnotation:(id <MKAnnotation>)annotation
        reuseIdentifier:(NSString *)reuseIdentifier {
    self = [super initWithAnnotation:annotation reuseIdentifier:reuseIdentifier];
    if (self) {
        UIImage* im = [UIImage imageNamed:@"clipartdirtbike.gif"];
        self.frame =
        CGRectMake(0, 0, im.size.width / 3.0 + 5, im.size.height / 3.0 + 5);
        self.centerOffset = CGPointMake(0,-20);
        self.opaque = NO;
    }
    return self;
}
- (void) drawRect: (CGRect) rect {
    UIImage* im = [UIImage imageNamed:@"clipartdirtbike.gif"];
    [im drawInRect:CGRectInset(self.bounds, 5, 5)];
}
@end
```

执行后的效果如图 9-17 所示。

图 9-17 执行效果

实例 215　在地图中灵活标注

实例 215	在地图中灵活标注
源码路径	\daima\215\

实例说明

在本实例中，使用 UIGestureRecognizer 获取了触摸地图的点坐标，然后使用方法-（CLLocation Coordinate2D）convertPoint：（CGPoint）point toCoordinateFromView：（UIView *）view 将点坐标转化为经纬度方法。并且在屏幕的地图界面中，在长时触摸的经纬度处加入了大头针标示。点击大头针后会弹出泡泡图标按钮，点击泡泡按钮会导航到指定页面的效果。

具体实现

实例文件 NotificationViewController.m 的具体实现代码如下所示。

```
#import "NotificationViewController.h"
@implementation NotificationViewController
- (void)dealloc
{
    [super dealloc];
}
- (void)didReceiveMemoryWarning
{
    [super didReceiveMemoryWarning];
}
-(void)viewWillAppear:(BOOL)animated{
    //隐藏导航栏
    [self.navigationController setNavigationBarHidden:YES];
}
/*********************
 使用谷歌地图需要导入 MapKit.framework,同时在.h 内 #import <MapKit/MapKit.h>
 *********************/
- (void)viewDidLoad
{
    [super viewDidLoad];
    //创建地图视图，初始化参数
    m_mapview = [[MKMapView alloc] initWithFrame:CGRectMake(0, 0, 320, 480)];
    //设置地图的代理
    m_mapview.delegate = self;
    [self.view addSubview:m_mapview];
        //地图的类型：MKMapTypeStandard 显示街道和道路，MKMapTypeSatellite 显示卫星, MKMapType Hybrid 显示混合地图
    [m_mapview setMapType:MKMapTypeStandard];

        //显示用户当前的坐标，打开地图有相应的提示
    //    m_mapview.showsUserLocation=YES;
        //定义经纬坐标
    CLLocationCoordinate2D theCoordinate;
    theCoordinate.latitude = 32.05000;
    theCoordinate.longitude = 118.78333;

        //定义显示的范围
    MKCoordinateSpan theSpan;
    theSpan.latitudeDelta=0.1;
    theSpan.longitudeDelta=0.1;
        //定义一个区域（用定义的经纬度和范围来大小来定义）
    MKCoordinateRegion theRegion;
    theRegion.center=theCoordinate;
    theRegion.span=theSpan;
        //在地图上显示此区域
    [m_mapview setRegion:theRegion animated:YES];

        //长按事件
```

```objc
    UILongPressGestureRecognizer    *lpress   =   [[UILongPressGestureRecognizer  alloc]
initWithTarget:self action:@selector(longPress:)];
    lpress.minimumPressDuration = 0.5;//按0.5秒响应longPress方法
    lpress.allowableMovement = 10.0;
    [m_mapview addGestureRecognizer:lpress];
    [lpress release];
}
- (void)longPress:(UIGestureRecognizer*)gestureRecognizer{
    if (gestureRecognizer.state == UIGestureRecognizerStateBegan){
        tagNum++;
            //取地图上的长按的点坐标
        CGPoint touchPoint = [gestureRecognizer locationInView:m_mapview];
            //将点坐标转换为经纬度坐标
        CLLocationCoordinate2D touchMapCoordinate =
        [m_mapview convertPoint:touchPoint toCoordinateFromView:m_mapview];
        NSLog(@"%f    %f",touchMapCoordinate.latitude,touchMapCoordinate.longitude);
            //初始化详情弹出框对象
        MKPointAnnotation *pointAnnotation = [[MKPointAnnotation alloc] init];
            //位置
        pointAnnotation.coordinate = touchMapCoordinate;
            //显示标题
        pointAnnotation.title = [NSString stringWithFormat:@"位置%d",tagNum];
        [m_mapview addAnnotation:pointAnnotation];
        [pointAnnotation release];
    }
}
/*******************************************
函数名称 : viewForAnnotation
函数描述 : 在地图上加入大头针及其动画。
输入参数 : mapView, theMapView, annotation。
输出参数 : N/A
返回值   : N/A
*******************************************/
- (MKAnnotationView *)mapView:(MKMapView *)mV viewForAnnotation:(id <MKAnnotation>)annotation{
/*
定义中用到的标注属性
image 标注图片
pinColor 颜色//MKPinAnnotationColorRed ,MKPinAnnotationColorGreen,MKPinAnnotationColorPurple
canShowCallout / /是否弹出
animatesDrop / /落下动画
centerOffset / /大头针偏移量
annotationView.calloutOffset / /标注偏移量
rightCalloutAccessoryView / /右边点击按钮
leftCalloutAccessoryView / /左边点击按钮
*/
    static NSString *AnnotationIdentifier = @"AnnotationIdentifier";
    MKPinAnnotationView *customPinView = (MKPinAnnotationView *)[mV
dequeueReusableAnnotationViewWithIdentifier:AnnotationIdentifier];
        //初始化大头针对象
    if (!customPinView) {
        customPinView = [[[MKPinAnnotationView alloc]  initWithAnnotation:annotation
reuseIdentifier:AnnotationIdentifier] autorelease];

        customPinView.pinColor = MKPinAnnotationColorRed;//设置大头针的颜色
        customPinView.animatesDrop = YES;                //坠落动画
        customPinView.canShowCallout = YES;              //显示详情

            //添加导航按钮
        UIButton *rightButton = [UIButton buttonWithType:UIButtonTypeDetailDisclosure];
        [rightButton addTarget:self action:@selector(showDetails:) forControlEvents:
UIControlEventTouchUpInside];
        customPinView.rightCalloutAccessoryView = rightButton;
    }else{
        customPinView.annotation = annotation;
    }
    return customPinView;
}

- (void)showDetails:(UIButton*)sender
{
    OneViewController *oneController = [[OneViewController alloc] init];
```

```objc
        //给Controller的标题赋值
    oneController.title = currentPin;
    [self.navigationController pushViewController:oneController animated:YES];
        //给Controller的label赋值
    [oneController setLabelText:currentPin];
    [oneController release];
}

/***********************************************
函数名称   : didSelectAnnotationView
函数描述   : 点击大头针时调用此方法
输入参数   : mapView, view。
输出参数   : N/A
返回值     : N/A
***********************************************/
- (void)mapView:(MKMapView *)mapView didSelectAnnotationView:(MKAnnotationView *)view
{
        //点击大头针时，取出其详情信息
    MKPointAnnotation *currentAnnotation = (MKPointAnnotation *)view.annotation;
    currentPin = currentAnnotation.title;
}
    //拖动地图，改变地图比例时调用此方法
/***********************************************
函数名称   : regionWillChangeAnimated
函数描述   : 拖动地图，改变地图比例时调用此方法
输入参数   : mapView, animated。
输出参数   : N/A
返回值     : N/A
***********************************************/
- (void)mapView:(MKMapView *)mapView regionWillChangeAnimated:(BOOL)animated
{

}

- (void)viewDidUnload
{
    [super viewDidUnload];
    // Release any retained subviews of the main view.
    // e.g. self.myOutlet = nil;
}

-
(BOOL)shouldAutorotateToInterfaceOrientation:(UIInterfaceOrientation)interfaceOrientation
{
    // Return YES for supported orientations
    return (interfaceOrientation == UIInterfaceOrientationPortrait);
}

@end
```

实例文件 OneViewController.m 的具体实现代码如下所示。

```objc
#import "OneViewController.h"
@implementation OneViewController
@synthesize m_label;
- (id)initWithNibName:(NSString *)nibNameOrNil bundle:(NSBundle *)nibBundleOrNil
{
    self = [super initWithNibName:nibNameOrNil bundle:nibBundleOrNil];
    if (self) {
    }
    return self;
}
- (void)didReceiveMemoryWarning
{
    [super didReceiveMemoryWarning];

}
#pragma mark - View lifecycle
- (void)viewDidLoad
{
    [super viewDidLoad];
        //显示导航栏
    [self.navigationController setNavigationBarHidden:NO];
```

```
        //初始化 label
    UILabel *label = [[UILabel alloc] initWithFrame:CGRectMake(0, 100, 320, 100)];
    [label setBackgroundColor:[UIColor blackColor]];
    label.font = [UIFont boldSystemFontOfSize:24];
    label.textColor = [UIColor whiteColor];
    label.textAlignment = UITextAlignmentCenter;
    self.m_label = label;
    [self.view addSubview:label];
    [label release];
}
-(void)setLabelText:(NSString *)str{
    m_label.text = str;
}
- (void)viewDidUnload
{
    [super viewDidUnload];
}
- (BOOL)shouldAutorotateToInterfaceOrientation:(UIInterfaceOrientation)interfaceOrientation
{
    return (interfaceOrientation == UIInterfaceOrientationPortrait);
}
@end
```

执行后的效果如图 9-18 所示。

图 9-18 执行效果

实例 216 实现复杂的地图标注

实例 216	实现复杂的地图标注
源码路径	\daima\216\

实例说明

本实例是一个 iPad 项目，本实例将通过一个地图系统的具体实现过程，来详细讲解在地图中实现复杂标注的基本知识。本实例是对前面实例的总结运用，是对本章前面内容的一个综合运用。

具体实现

（1）界面视图。

本项目的主界面视图是 MainWindow.xib，在主界面显示已经标注的信息，如图 9-19 所示。

地图界面视图是 GIKMapView.xib，能够显示地图，如图 9-20 所示。

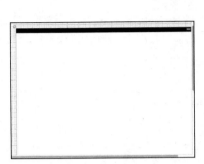

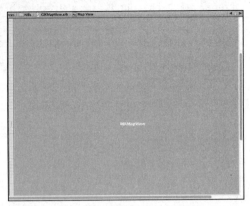

图 9-19　MainWindow.xib 视图　　　　　图 9-20　GIKMapView.xib 视图

酒店详情视图是 HotelDetailTableView.xib，功能是显示标注酒店的详细信息，如图 9-21 所示。
动画标注视图是 AnimatedCalloutViewController.xib，功能是在标注信息时实现动画效果，如图 9-22 所示。

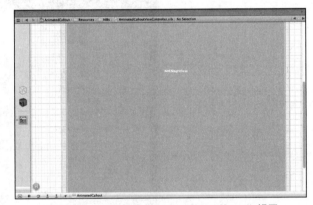

图 9-21　HotelDetailTableView.xib 视图　　　图 9-22　AnimatedCalloutViewController.xib 视图

（2）实现地图视图。

在文件 MapViewController.h 中定义了类 MapViewController，具体代码如下所示。

```
#import <UIKit/UIKit.h>
#import "GIKMapViewController.h"
@interface MapViewController : GIKMapViewController <GIKCalloutDetailDataSource> {
    NSArray *hotels;
}
@end
```

文件 MapViewController.m 是文件 MapViewController.h 的实现，在此文件中定义了如下方法。

- （void）viewDidLoad：载入初始视图，设置初始坐标位置是三藩市。
- （BOOL）shouldAutorotateToInterfaceOrientation：返回设备支持的方向。
- （void）showAnnotations：显示酒店的详细信息。
- （void）detailController：实现数据对象的详细的标注。

文件 MapViewController.m 的具体实现代码如下所示。

```
#import "MapViewController.h"
#import "Hotel.h"
#import "HotelAnnotation.h"
#import "HotelDetailViewController.h"

@interface MapViewController ()
```

实例216 实现复杂的地图标注

```objc
@property (nonatomic, retain) NSArray *hotels;
- (void)showAnnotations;
@end
@implementation MapViewController
@synthesize hotels;
- (void)dealloc {
  [hotels release];
  [super dealloc];
}
- (id)init {
    if (!(self = [super initWithNibName:@"GIKMapView" bundle:nil])) {
        return nil;
    }
    return self;
}
#pragma mark -
#pragma mark View management

- (void)viewDidLoad {
    [super viewDidLoad];
    MKCoordinateRegion startupRegion;
    // 初始坐标为三藩市区周围的莫斯克尼西区域
    startupRegion.center = CLLocationCoordinate2DMake(37.785334, -122.406964);
    startupRegion.span = MKCoordinateSpanMake(0.003515, 0.007129);
    [self.mapView setRegion:startupRegion animated:YES];
    [self.mapView setShowsUserLocation:NO];
    // 基类需要访问自定义标注数据,无需知道具体实现细节
    self.detailDataSource = self;
    HotelDetailViewController *controller = [[HotelDetailViewController alloc] initWithNibName:@"HotelDetailTableView" bundle:nil];
    self.calloutDetailController = controller;
    [controller release];
    [self showAnnotations];
}
- (BOOL)shouldAutorotateToInterfaceOrientation:(UIInterfaceOrientation)interfaceOrientation {
    // 返回支持的方向
    return YES;
}
- (void)showAnnotations {
    NSMutableArray *hotelAnnotations = [NSMutableArray arrayWithCapacity:4];
    for (NSDictionary *hotel in self.hotels) {
        Hotel *theHotel = [[Hotel alloc] init];
        theHotel.name = [hotel objectForKey:@"name"];
        theHotel.street = [hotel objectForKey:@"street"];
        theHotel.city = [hotel objectForKey:@"city"];
        theHotel.state = [hotel objectForKey:@"state"];
        theHotel.zip = [hotel objectForKey:@"zip"];
        theHotel.phone = [hotel objectForKey:@"phone"];
        theHotel.url = [hotel objectForKey:@"url"];
        theHotel.latitude = [[hotel objectForKey:@"latitude"] doubleValue];
        theHotel.longitude = [[hotel objectForKey:@"longitude"] doubleValue];
        HotelAnnotation *annotation = [[HotelAnnotation alloc] initWithLatitude:theHotel.latitude longitude:theHotel.longitude];
        annotation.hotel = theHotel;
        [hotelAnnotations addObject:annotation];
        [annotation release];
        [theHotel release];
    }
    [self.mapView addAnnotations:hotelAnnotations];
}
#pragma mark -
#pragma mark GIKCalloutDetailDataSource
// 数据对象的详细的标注
- (void)detailController:(UIViewController *)detailController detailForAnnotation:(id)annotation {
    [(HotelDetailViewController *)detailController setHotel:[(HotelAnnotation *)annotation hotel]];
}
#pragma mark -
```

```
#pragma mark Memory management
- (void)didReceiveMemoryWarning {
     // 如果没有视图则释放
    [super didReceiveMemoryWarning];
}
- (void)viewDidUnload {
  [super viewDidUnload];
}

#pragma mark Accessors
- (NSArray *)hotels {
     if (hotels == nil) {
         hotels = [[NSArray alloc] initWithContentsOfFile:[[NSBundle mainBundle] pathForResource:@"Hotels" ofType:@"plist"]];
     }
     return hotels;
}
@end
```

（3）酒店详情视图。

在文件 HotelDetailViewController.h 中，在此界面可以显示标注酒店的网址、地址、电话和线路图链接。具体代码如下所示。

```
#import <UIKit/UIKit.h>
@class Hotel;
@interface HotelDetailViewController : UIViewController <UITableViewDelegate, UITableViewDataSource, UIGestureRecognizerDelegate> {
     UITableView *table;
     Hotel *hotel;
@private
     NSArray *directions;
}
@property (nonatomic, retain) IBOutlet UITableView *table;
@property (nonatomic, retain) Hotel *hotel;
@end
```

文件 HotelDetailViewController.m 是文件 HotelDetailViewController.h 的实现，在此文件中定义了如下方法。

- (void)viewDidLoad：载入初始视图，设置了背景图片 CalloutTableBackground.png。
- (UITableViewCell*)tableView:(UITableView*)tableViewcellForRowAtIndexPath:(NSIndexPath *)indexPath：此方法的核心是使用 switch 语句获取用户的选择，根据用户的选择来到对应的界面，实现对应的功能。

文件 MapViewController.m 的具体实现代码如下所示。

```
#import "HotelDetailViewController.h"
#import "Hotel.h"
typedef enum {
     kDirections,
     kPhone,
     kURL,
     kAddress,
     NUMBER_OF_SECTIONS
} TableSections;

@interface HotelDetailViewController ()
@property (nonatomic, readonly) NSArray *directions;
@end
@implementation HotelDetailViewController
@synthesize table, hotel;
@synthesize directions;
- (void)dealloc {
     [directions release];
     [table release];
     [hotel release];
     [super dealloc];
}
```

实例216 实现复杂的地图标注

```objc
- (id)initWithNibName:(NSString *)nibNameOrNil bundle:(NSBundle *)nibBundleOrNil {
    if (!(self = [super initWithNibName:nibNameOrNil bundle:nibBundleOrNil])) {
        return nil;
    }
    return self;
}
- (void)viewDidLoad {
  [super viewDidLoad];
    self.table.backgroundColor = [UIColor clearColor];
    UIImage *backgroundImage = [[UIImage imageNamed:@"CalloutTableBackground.png"] stretchableImageWithLeftCapWidth:0 topCapHeight:6];
    UIImageView *backgroundImageView = [[UIImageView alloc] initWithImage:backgroundImage];
    backgroundImageView.frame = self.view.bounds;
    self.table.backgroundView = backgroundImageView;
    [backgroundImageView release];
}
- (void)viewWillAppear:(BOOL)animated {
  [super viewWillAppear:animated];
  [self.table reloadData];
}
- (void)didReceiveMemoryWarning {
    // Releases the view if it doesn't have a superview.
    [super didReceiveMemoryWarning];
}
- (void)viewDidUnload {
    [super viewDidUnload];
  self.table = nil;
}
#pragma mark -
#pragma mark UITableViewDelegate methods
- (CGFloat)tableView:(UITableView *)tableView heightForRowAtIndexPath:(NSIndexPath *)indexPath {
    if (indexPath.section == kAddress) {
        return 80.0f;
    }
    return 44.0f;
}
#pragma mark -
#pragma mark UITableViewDataSource methods
- (UITableViewCell *)tableView:(UITableView *)tableView cellForRowAtIndexPath:(NSIndexPath *)indexPath {
    static NSString *kDirectionCellIdentifier = @"DirectionCellIdentifier";
    static NSString *kOtherCellIdentifier = @"OtherCellIdentifier";
    NSString *workingCellIdentifier = (indexPath.section == kDirections) ? kDirectionCellIdentifier : kOtherCellIdentifier;
    UITableViewCell *cell = [tableView dequeueReusableCellWithIdentifier:workingCellIdentifier];
    if (cell == nil) {
        if ([workingCellIdentifier isEqualToString:kDirectionCellIdentifier]) {
            cell = [[[UITableViewCell alloc] initWithStyle:UITableViewCellStyleDefault reuseIdentifier:workingCellIdentifier] autorelease];        }
        else {
            cell = [[[UITableViewCell alloc] initWithStyle:UITableViewCellStyleValue2 reuseIdentifier:workingCellIdentifier] autorelease];
        }
    }
    switch (indexPath.section) {
        case kDirections:
            cell.textLabel.textAlignment = UITextAlignmentCenter;
            cell.textLabel.text = [self.directions objectAtIndex:indexPath.row];
            cell.textLabel.textColor = [UIColor colorWithRed:82.0f/255.0f green:102.0f/255.0f blue:145.0f/255.0f alpha:1.0f];
            break;
        case kPhone:
            cell.textLabel.text = @"phone";
            cell.detailTextLabel.text = self.hotel.phone;
            break;
        case kURL:
```

```objc
                cell.textLabel.text = @"home page";
                cell.detailTextLabel.text = self.hotel.url;
                break;
            case kAddress:
                cell.textLabel.text = @"address";
                cell.detailTextLabel.lineBreakMode = UILineBreakModeWordWrap;
                cell.detailTextLabel.numberOfLines = 4;
                cell.detailTextLabel.text = [NSString stringWithFormat:@"%@\n%@ %@\n%@",
                                             self.hotel.street,
           self.hotel.state, self.hotel.zip], self.hotel.zip];
                break;
            default:
                break;
        }
        return cell;
}
- (NSInteger)numberOfSectionsInTableView:(UITableView *)tableView {
        return NUMBER_OF_SECTIONS;
}
- (NSInteger)tableView:(UITableView *)tableView numberOfRowsInSection:(NSInteger)section {
        NSInteger rows = 0;
        switch (section) {
            case kDirections:
                rows = [self.directions count];
                break;
            case kPhone:
            case kURL:
            case kAddress:
                rows = 1;
                break;
            default:
                break;
        }
        return rows;
}
- (void)tableView:(UITableView *)tableView didSelectRowAtIndexPath:(NSIndexPath *)indexPath {
        [self.table.superview performSelector:@selector(disableMapSelections)];
        [tableView deselectRowAtIndexPath:indexPath animated:YES];
        if (indexPath.section == kURL) {
            [[UIApplication sharedApplication] openURL:[NSURL URLWithString:self.hotel.url]];
        }
}
#pragma mark -
#pragma mark Accessors
- (void)setHotel:(Hotel *)newHotel {
        if (hotel != newHotel) {
            [hotel release];
            hotel = [newHotel retain];
        }
        [self.table reloadData];
}
- (NSArray *)directions {
        if (directions == nil) {
            directions = [[NSArray alloc] initWithObjects:@"Directions To Here",
     @"Directions From Here", nil];
        }
        return directions;
}
@end
```

(4) 酒店信息。

文件 Hotel.h 定义了酒店信息变量，包括名字、街道、城市、电话、邮编和主页等，具体代码如下所示。

```objc
#import <Foundation/Foundation.h>
#import <MapKit/MapKit.h>
@interface Hotel : NSObject {
        NSString *name;
```

```
        NSString *street;
        NSString *city;
        NSString *state;
        NSString *zip;
        NSString *phone;
        NSString *url;
        CLLocationDegrees latitude;
        CLLocationDegrees longitude;
}
@property (nonatomic, copy) NSString *name;
@property (nonatomic, copy) NSString *street;
@property (nonatomic, copy) NSString *city;
@property (nonatomic, copy) NSString *state;
@property (nonatomic, copy) NSString *zip;
@property (nonatomic, copy) NSString *phone;
@property (nonatomic, copy) NSString *url;
@property (nonatomic, assign) CLLocationDegrees latitude;
@property (nonatomic, assign) CLLocationDegrees longitude;
@end
```

文件 Hotel.m 是文件 Hotel.h 的实现,功能是释放酒店信息变量的内存,具体代码如下所示。

```
#import "Hotel.h"
@implementation Hotel
@synthesize name, street, city, state, zip, phone, url;
@synthesize latitude, longitude;
- (void)dealloc {
        [name release];
        [street release];
        [city release];
        [state release];
        [zip release];
        [phone release];
        [url release];
        [super dealloc];
}
@end
```

(5)动画效果标注。

在文件 AnimatedCalloutAppDelegate.h 中定义类 AnimatedCalloutAppDelegate,具体代码如下所示。

```
#import <UIKit/UIKit.h>
@class MapViewController;
@interface AnimatedCalloutAppDelegate : NSObject <UIApplicationDelegate> {
    UIWindow *window;
    MapViewController *viewController;
}
@property (nonatomic, retain) IBOutlet UIWindow *window;
@property (nonatomic, retain) MapViewController *viewController;
@end
```

文件 AnimatedCalloutAppDelegate.m 是文件 AnimatedCalloutAppDelegate.h 的实现,主要包含了如下方法。

- (BOOL)application:(UIApplication*)application didFinishLaunchingWithOptions:(NSDictionary*)launchOptions:设置应用程序启动后使用控制点进行定制操作。

- (void)applicationWillResignActive:(UIApplication *)application:设置应用程序从活动变为不活动状态。此功能可以对某些类型的临时中断(如来电或短信)或当用户退出应用程序并开始过渡到国家的背景起作用。通过使用这个方法,可以暂停正在进行的任务,禁用定时器和节流中胚胎帧速率。游戏应该使用这种方法暂停游戏。

- (void)applicationDidBecomeActive:(UIApplication *)application:重新启动的任何任务,这些任务包括被暂停(或尚未开始)的申请无效的任务。

文件 AnimatedCalloutAppDelegate.m 的具体代码如下所示。

```
#import "AnimatedCalloutAppDelegate.h"
#import "MapViewController.h"
```

```objc
@implementation AnimatedCalloutAppDelegate
@synthesize window;
@synthesize viewController;

A#pragma mark -
#pragma mark Application lifecycle
- (BOOL)application:(UIApplication *)application didFinishLaunchingWithOptions:
(NSDictionary *)launchOptions {
    // 应用程序启动后使用控制点定制
    MapViewController *controller = [[MapViewController alloc] init];
    self.viewController = controller;
    [controller release];
    [self.window addSubview:self.viewController.view];
    [self.window makeKeyAndVisible];
    return YES;
}
- (void)applicationWillResignActive:(UIApplication *)application {
}
- (void)applicationDidBecomeActive:(UIApplication *)application {
}
- (void)applicationWillTerminate:(UIApplication *)application {
    /*
    当应用程序要终止：
    */
}
#pragma mark -
#pragma mark Memory management
- (void)applicationDidReceiveMemoryWarning:(UIApplication *)application {
    /*
    尽可能清除缓存的数据对象
    */
}
- (void)dealloc {
    [viewController release];
    [window release];
    [super dealloc];
}
@end
```

到此为止，本项目的主要代码介绍完毕。执行后的初始界面效果如图 9-23 所示。某个酒店详细信息界面如图 9-24 所示。

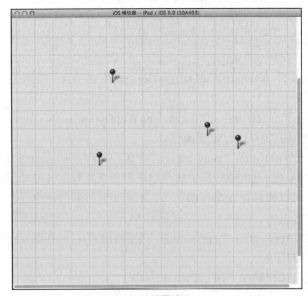

图 9-23　初始界面效果

图 9-24　酒店详情界面

前往线路界面如图 9-25 所示。

图 9-25　前往线路界面

第 10 章 传感器、触摸和交互

在当前智能手机系统应用中，通常使用触摸的方式操控设备。在操控的过程中，通常使用传感器来实现用户需要的功能。触摸操作设备的最终目的是实现和设备之间的交互。本章将通过几个典型实例的实现过程，详细介绍在 iOS 系统中使用传感器等技术实现触摸交互的基本知识。

实例 217 实现一个可触摸识别程序

实例 217	实现一个可触摸识别程序
源码路径	\daima\217\

实例说明

iOS 系统在推出之时，最吸引用户的便是多点触摸功能，通过对屏幕的触摸实现了良好的用户体验。通过使用多点触摸屏技术，让用户能够使用大量的自然手势来完成原本只能通过菜单、按钮和文本来完成的操作。另外，iOS 系统还提供了高级手势识别功能，我们可以在应用程序中轻松实现它们。

在本实例中将实现 5 种手势识别器（轻按、轻扫、张合、旋转和摇动）以及这些手势的反馈。每种手势都会更新标签，指出有关该手势的信息。在张合、旋转和摇动的基础上更进一步。当用户执行这些手势时，将缩放、旋转或重置一个图像视图。

为了给手势输入提供空间，这个应用程序显示的屏幕中包含 4 个嵌套的视图（UIView），在故事板场景中，直接给每个嵌套视图指定了一个手势识别器。当您在视图中执行操作时，将调用视图控制器中相应的方法，在标签中显示有关手势的信息；另外，根据执行的手势，还可能更新屏幕上的一个图像视图（UIImageView）。

具体实现

（1）启动 Xcode，使用模板 Single View Application 创建一个名为"shoushi"的应用程序，如图 10-1 所示。

（2）添加图像资源。

这个应用程序的界面包含一幅可旋转或缩放的图像，这旨在根据用户的手势提供视觉反馈。在本章的项目文件夹中，子文件夹 Images 包含一幅名为"flower.png"的图像。将文件夹 Images 拖放到项目的代码编组中，并选择必要时复制资源并创建编组。

图 10-1 创建工程

(3) 规划变量和连接。

对于我们要检测的每个触摸手势，都需要提供让其能够得以发生的视图。通常，这可使用主视图，但出于演示目的，我们将在主视图中添加 4 个 UIView，每个 UIView 都与一个手势识别器相关联。令人惊讶的是，这些 UIView 都不需要输出口，因为我们将在 Interface Builder 编辑器中直接将它们连接到手势识别器。

但是我们需要两个输出口/属性——outputLabel 和 imageView，它们分别连接到一个 UILabel 和一个 UIImageView。其中，标签用于向用户提供文本反馈，而图像视图在用户执行张合和旋转手势时提供视觉反馈。

在这 4 个视图中检测到手势时，应用程序需要调用一个操作方法，以便与标签和图像交互。我们把手势识别器 UI 连接到方法 foundTap、foundSwipe、foundPinch、foulldRotation。

(4) 添加表示默认图像大小的常量。

当手势识别器对 UI 中的图像视图调整大小或旋转时，我们希望能够恢复到默认大小和位置。为此，需要在代码中记录默认大小和位置。这里选择将 UIImageView 的大小和位置存储在 4 个常量中，而这些常量的值是这样确定的：将图像视图放到所需的位置，然后从 Interface Builder Size Inspector 读取其框架值。

如果我们决定自行设计该项目的界面，则需要自己记录图像视图的框架值。对于 iPhone 版本，可以在文件 ViewController.m 的代码行#import 后面输入如下代码。

```
#define kOriginWidth 125.0
#define kOriginHeight 115.0
#define kOriginX 100.0
#define kOriginY 330.0
```

如果创建的是 iPad 应用程序，应该按照下面的代码定义这些常量。

```
#define kOriginWidth 265.0
#define kOriginHeight 250.0
#define kOriginX 250.0
#define kOriginY 750.0
```

使用这些常量可以快速记录 UIImageView 的位置和大小，但这并非唯一的解决方案。其可以在应用程序启动时读取并存储图像视图的 frame 属性，并在以后恢复它们。然而我们的目的是帮助我们理解工作原理，而不是过度考虑解决方案是否巧妙。

(5) 设计界面。

打开文件 MainStoryboard.storyboard，并在工作区中腾出一些空间。为了创建界面，首先拖曳 4 个 UIView 实例到主视图中。将第一个视图调整为小型矩形，并位于屏幕的左上角，它将捕获轻按手势；将第二个视图放在第一个视图右边，它用于检测轻扫手势；将其他两个视图放在前两个视图下方，且与这两个视图等宽，它们分别用于检测张合手势和旋转手势。使用 Attributes Inspector（Option+Command+4）将每个视图的背景设置为不同的颜色。

接下来，在每个视图中添加一个标签，这些标签的文本应分别为"Tap 我"、"Swipe 我"、"Pinch 我"和"Rotate 我"。然后再拖放一个 UILabel 实例到主视图中，让其位于屏幕顶端并居中；使用 Attributes Inspector 将其设置为居中对齐。这个标签将用于向用户提供反馈，请将其默认文本设置为"动起来"。最后，在屏幕底部中央添加一个 UIImageView。使用 Attributes Inspector（Option+Command+4）和 Size Inspector（Option+Command+5）将图像设置为 flower.png，并按如下设置其大小和位置：X 为 100.0、Y 为 330.0、W 为 125.0、H 为 115.0（对于 iPhone 应用程序），或 X 为 250.0、Y 为 750.0、W 为 265.0、H 为 250.0（对于 iPad 应用程序），如图 10-2 所示。这些值与前面定义的常量值一致。

在大多数项目中，设计好视图后，我们都通过输出口和操作将界面连接到代码，但这里不是这样的。要建立连接，必须先将手势识别器加入到故事板中。

第 10 章　传感器、触摸和交互

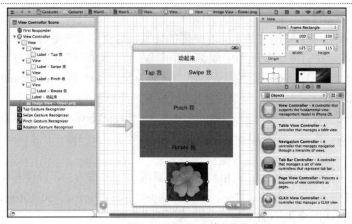

图 10-2　设置 UIImageView 的大小和位置

（6）轻按手势识别器。

接下来开始给视图添加手势识别器，添加手势识别器的方式之一是使用代码。对于初始化要使用的识别器，配置其参数，再将其加入 MainStoryboard.storyboard 时将调用的方法中。另一种方式是，从对象库中将手势识别器拖放到视图中，这几乎不需要编写任何代码。下面开始就这样做。首先确保打开了文件 MainStoryboard.storyboard，并且文档大纲可见。

第一步是在项目中添加一个 UITapGestureRecognizer 实例。为此，在对象库中找到轻按手势识别器，将其拖放到包含标签"Tap 我"的 UIView 实例中，如图 10-3 所示。识别器将作为一个对象出现在文档大纲底部，而无论将其放在哪里。

通过将轻按手势识别器拖放到视图中，就创建了一个手势识别器对象，并将其关联到了该视图（可根据需要将任意数目的手势识别器加入到同一个视图中）。

接下来需要配置该识别器，让其知道要检测哪种手势。轻按手势识别器有如下两个属性。

- Taps：需要轻按对象多少次才能识别出轻按手势。
- Touches：需要有多少个手指在屏幕上才能识别出轻按手势。

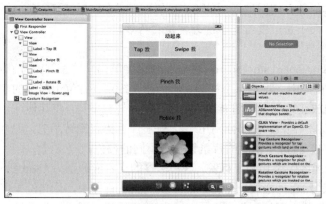

图 10-3　将识别器拖放到将使用它的视图上

在本实例中，我们将轻按手势定义为用一个手指轻按屏幕一次，因此指定一次轻按和一个触点。选择轻按手势识别器，再打开 Attributes Inspector(Option+Command+4)，如图 10-4 所示。

将文本框 Taps 和 Touches 都设置为 1，这是试验识别器属性的绝佳时机。这样就在项目中添加了第一个手势识别器，并对其进行了配置。稍后还需将其连接到一个操作，但是需要先添加其他的识别器。

（7）轻扫手势识别器。

轻扫手势识别器的实现方式几乎与轻按手势识别器完全相同，然而，不是指定轻按次数，而是指定轻扫的方向（上、下、左、右），还需指定多少个手指触摸屏幕（触点数）时才能视为轻扫手势。同样，在对象库中找到轻扫手势识别器（UISwipeGestureRecognizer），并将其拖放到包含标签"Swipe 我"的视图上。接下来，选择该识别器，并打开 Attributes Inspector 以便配置它，如图 10-5 所示。这里对轻扫手势识别器进行配置，使其监控用一个手指向右轻扫的手势。

图 10-4　使用 Attributes Inspector 配置手势识别器

（8）张合手势识别器。

在视图中将两个手指并拢或张开时，将触发张合手势；这两种操作常用于缩放对象。与轻按手势识别器和轻扫手势识别器相比，添加张合手势识别器需要做的配置更少，因为这种手势已经有明确的定义。然而，实现响应张合手势的方法更困难些，因为除了知道发生了张合手势外，还需考虑张合的程度（缩放比例）和速度。稍后将更详细地阐述这一点。

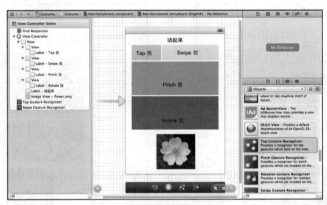

图 10-5　配置轻扫方向和触点数

在对象库中找到张合手势识别器（UIPinGestureRecognizer），并将其拖放到包含标签"Pinch 我"的视图上。不用做其他的配置工作。

（9）旋转手势识别器。

旋转手势指的是两个手指沿圆圈移动。想想使用两个手指旋转门把手的情形，就能知道 iOS 将哪些操作视为有效的旋转手势了。与张合手势识别器一样，旋转手势识别器也无需做任何配置，只需诠释结果——旋转的角度（单位为弧度）和速度。

在对象库中找到旋转手势识别器（UIRotationGestureRecognizer），并将其拖放到包含标签"Rotate 我"的视图上。这样就在故事板中添加了最后一个对象。

（10）添加输出口。

接下来开始创建并连接输出口和操作。为了在主视图控制器中响应手势并访问反馈对象，需要创建前面确定的输出口和操作。需要的输出口如下所示。

- 图像视图（UIImageView）：imageView。
- 提供反馈的标签（UILabel）：outputLabel。

需要的操作如下所示。
- 响应轻按手势：foundTap。
- 响应轻扫手势：foundSwipe。
- 响应张合手势：foundPinch。
- 响应旋转手势：foundRotation。

为了建立连接准备好工作区，打开文件 MainStoryboard.storyboard，并切换到助手编辑器模式。由于将从场景中的手势识别器开始拖曳，请确保要么文档大纲可见（Editor→Show Document Outline），要么能够在视图下方的对象栏中区分不同的识别器。按住 Control 键，并从标签 Do Something!拖曳到文件 ViewController.h 中代码行@interface 下方。在 Xcode 提示时，新建一个名为 outputLabel 的输出口，如图 10-6 所示。对图像视图重复上述操作，并将输出口命名为"imageView"。

（11）添加操作。

要将手势识别器连接到前面确定的操作方法，可以采取和前面一样的方式，唯一的不同之处在于将对象连接到操作时，实际上连接的是该对象的特定事件，如按钮的 To Up Inside 事件。而将手势识别器连接到操作时，实际上建立的是从识别器的选择器（selector）到方法的连接。还记得前面的代码示例吗？选择器用于指定检测到特定手势时应该调用的方法。

要将手势识别器连接到操作方法，只需按住 Control 键，并从文档大纲中的手势识别器拖曳到文件 ViewController.h。现在就对轻按手势识别器进行上述操作，并拖曳到前面定义的属性下方。在 Xcode 提示时，将连接类型指定为操作，并将名称指定为"foundTap"，如图 10-7 所示。

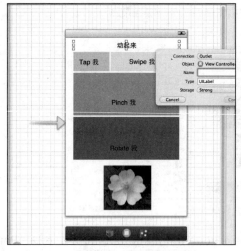
图 10-6　将标签和图像视图连接到输出口

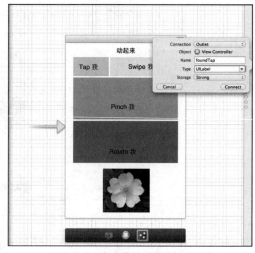
图 10-7　将手势识别器连接到操作

对于其他每个手势识别器重复上述操作，将轻扫手势识别器连接到 foundSwipe，将张合手势识别器连接到 foundPinch，将旋转手势识别器连接到 foundRotation。为了检查建立的连接，选择识别器之一（这里是轻按手势识别器），并查看 Connections Inspector（Option+ Command+6），将看到 Sent Actions 部分指定了操作，而 Referencing Outlet Collection 部分引用了使用识别器的视图。

（12）响应轻按手势识别器。

接下来开始实现应用程序逻辑，看实现手势识别器逻辑的过程。首先实现轻按手势识别器。实现一个识别器后，将发现其他识别器的实现方式极其类似。唯一不同的是摇动手势，这就是将它留在最后的原因。

切换到标准编辑器模式，并打开视图控制器实现文件 ViewController.m。要响应轻按手势识别器，只需实现方法 foundTap。修改这个方法的存根，使其实现代码如下所示。

```
- (IBAction)foundTap:(id)sender {
    self.outputLabel.text=@"Tapped";
}
```

这个方法不需要处理输入，除指出自己被执行外，它什么也不做。将标签 outPutLabel 的属性 text 设置为 Tapped 就足够了。

这样第一个手势识别器就实现好了。对其他 4 个手势识别器重复上述过程就可以很快地完成。

（13）响应轻扫手势识别器。

要想响应轻扫手势识别器，方式与响应轻按手势识别器相同：更新输出标签，指出检测到了轻扫手势。为此按如下代码实现方法 foundSwipe。

```
- (IBAction)foundSwipe:(id)sender {
    self.outputLabel.text=@"Swiped";
}
```

到目前为止，一切都很顺利，下面将实现张合手势识别器。这需要做的工作稍微多些，因为将利用张合手势与图像视图交互。

（14）响应张合手势识别器。

轻按和轻扫都是简单手势，它们只存在发不发生的问题；而张合手势和旋转手势更加复杂一些，它们返回更多的值，让开发者能够更好地控制用户界面。例如，张合手势包含属性 velocity（张合手势发生的速度）和 scale（与手指间距离变化呈正比的小数）。例如，如果手指间距离缩小了 50%，则缩放比例（scale）将为 0.5；如果手指间距离为原来的两倍，则缩放比例为 2。

接下来使用方法 foundPinch 完成多项工作。它重置 UIImageView 的旋转角度（以免受旋转手势带来的影响），使用张合手势识别器返回的缩放比例和速度值创建一个反馈字符串，并缩放图像视图，以便立即向用户提供可视化反馈。

方法 foundPinch 的实现代码如下所示。

```
- (IBAction)foundPinch:(id)sender {
    UIPinchGestureRecognizer *recognizer;
    NSString *feedback;
    double scale;
    recognizer=(UIPinchGestureRecognizer *)sender;
    scale=recognizer.scale;
    self.imageView.transform = CGAffineTransformMakeRotation(0.0);
    feedback=[[NSString alloc]
              initWithFormat:@"Pinched, Scale:%1.2f, Velocity:%1.2f",
              recognizer.scale,recognizer.velocity];
    self.outputLabel.text=feedback;
    self.imageView.frame=CGRectMake(kOriginX,
                                    kOriginY,
                                    kOriginWidth*scale,
                                    kOriginHeight*scale);
}
```

如果现在生成并运行该应用程序，开发者将能够在 pinchView 视图中使用张合手势缩放图像，甚至可以将图像放大到超越屏幕边界，如图 10-8 所示。

（15）响应旋转手势识别器。

我们将添加的最后一个多点触摸手势识别器是旋转手势识别器。与张合手势一样，旋转手势也返回一些有用的信息，其中最著名的是速度和旋转角度，可以使用它们来调整屏幕对象的视觉效果。返回的旋转角度是一个弧度值，表示用户沿着顺时针或逆时针方向旋转了多少弧度。

在文件 ViewController.m 中，foundRotation 方法的实现代码如下所示。

图 10-8　使用张合手势缩放图像

```
- (IBAction)foundRotation:(id)sender {
    UIRotationGestureRecognizer *recognizer;
    NSString *feedback;
    double rotation;
    recognizer=(UIRotationGestureRecognizer *)sender;
    rotation=recognizer.rotation;
    feedback=[[NSString alloc]
            initWithFormat:@"Rotated, Radians:%1.2f, Velocity:%1.2f",
            recognizer.rotation,recognizer.velocity];
    self.outputLabel.text=feedback;
    self.imageView.transform = CGAffineTransformMakeRotation(rotation);
}
```

(16)实现摇动识别器。

摇动的处理方式与其他手势稍有不同,必须拦截一个类型为 UIEventTypeMotion 的 UIEvent。为此,视图或视图控制器必须是响应者链中的第一响应者,还必须实现方法 motionEnded:withEvent。

第 1 步:成为第一响应者。

要让视图控制器成为第一响应者,必须通过方法 canBecomeFirstResponder 允许它成为第一响应者,这个方法除了返回 YES 外什么都不做;然后在视图控制器加载视图时要求它成为第一响应者。首先,在实现文件 ViewController.m 中添加方法 canBecomeFirstResponder,具体代码如下所示。

```
- (BOOL)canBecomeFirstResponder{
    return YES;
}
```

通过上述代码,可以让视图控制器能够成为第一响应者。

接下来需要在视图控制器加载其视图后立即发送消息 becomeFirstResponder,让视图控制器成为第一响应者。为此可以修改文件 ViewController.m 中的方法 viewDidAppear,具体代码如下所示。

```
- (void)viewDidAppear:(BOOL)animated
{
    [self becomeFirstResponder];
    [super viewDidAppear:animated];
}
```

至此,视图控制器为成为第一响应者并接收摇动事件做好了准备,我们只需要实现 motionEnded:withEvent 以捕获并响应摇动手势即可。

第 2 步:响应摇动手势。

为了响应摇动手势,motionEnded:withEvent 方法的实现代码如下所示。

```
- (void)motionEnded:(UIEventSubtype)motion withEvent:(UIEvent *)event {
    if (motion==UIEventSubtypeMotionShake) {
        self.outputLabel.text=@"Shaking things up!";
        self.imageView.transform = CGAffineTransformMakeRotation(0.0);
        self.imageView.frame=CGRectMake(kOriginX,
                                        kOriginY,
                                        kOriginWidth,
                                        kOriginHeight);
    }
}
```

图 10-9 执行效果

在上述代码中,通过检查确保收到的 motion 值(一个类型为 UIEventSubtype 的对象)确实是一个运动事件。为此,将其与常量 UIEventSubtypeMotionShake 进行比较,如果它们相同,说明用户刚摇动过设备。并且设置输出标签的文本,将图像视图旋转到默认朝向,并将图像视图的框架重置为视图大小常量指定的原始大小。换句话说,摇动设备将把图像重置到默认状态。

此时就可以运行该应用程序并使用本章实现的所有手势了。尝试使用张合手势缩放图像:摇动设备将图像恢复到原始大小;缩放和旋转图像、轻按、轻扫——一切都按开发者预期的那样进行,而令人惊讶的是,需要编写的代码很少。执行后的效果如图 10-9 所示。

实例 218　触摸按钮

实例 218	触摸按钮
源码路径	\daima\218\

实例说明

在 iOS 应用中，最常见的触摸操作是通过 UIButton 按钮实现的，这也是最简单的一种方式。iOS 中包含如下所示的操作手势。
- 点击（Tap）：点击作为最常用手势，用于按下或选择一个控件或条目（类似于普通的鼠标点击）。
- 拖动（Drag）：拖动用于实现一些页面的滚动，以及对控件的移动功能。
- 滑动（Flick）：滑动用于实现页面的快速滚动和翻页的功能。
- 横扫（Swipe）：横扫手势用于激活列表项的快捷操作菜单。

双击（Double Tap）：双击放大并居中显示图片，或恢复原大小（如果当前已经放大）。同时，双击能够激活针对文字编辑菜单。
- 放大（Pinch open）：放大手势可以实现以下功能——打开订阅源，打开文章的详情。在照片查看的时候，放大手势也可实现放大图片的功能。
- 缩小（Pinch close）：缩小手势，可以实现与放大手势相反且对应的功能——关闭订阅源退出到首页，关闭文章退出至索引页。在查看照片的时候，缩小手势也可实现缩小图片的功能。
- 长按（Touch &Hold）：如果针对文字长按，将出现放大镜辅助功能，松开后，则出现编辑菜单。针对图片长按，将出现编辑菜单。
- 摇晃（Shake） ：摇晃手势，将出现撤销与重做菜单，主要是针对用户文本输入的。

在本实例中，在屏幕中央设置了一个"触摸我"按钮，当触摸此按钮时会调用 buttonDidPush:(id)sender 方法，在屏幕中显示一个提示框效果。

具体实现

实例文件 UIKitPrjButton.m 的具体实现代码如下所示。

```
#import "UIKitPrjButton.h"
@implementation UIKitPrjButton
- (void)viewDidLoad {
  [super viewDidLoad];
  self.title = @"UIButton";
  self.view.backgroundColor = [UIColor whiteColor];
  // 创建按钮
  UIButton* button = [UIButton buttonWithType:UIButtonTypeRoundedRect];
  // 设置按钮标题
  [button setTitle:@"触摸我!" forState:UIControlStateNormal];
  // 根据标题长度自动决定按钮尺寸
  [button sizeToFit];
  // 将按钮布置在中心位置
  button.center = self.view.center;
  // 画面变化时按钮位置自动调整
  button.autoresizingMask = UIViewAutoresizingFlexibleWidth |
                            UIViewAutoresizingFlexibleHeight |
                            UIViewAutoresizingFlexibleLeftMargin |
                            UIViewAutoresizingFlexibleRightMargin |
                            UIViewAutoresizingFlexibleTopMargin |
                            UIViewAutoresizingFlexibleBottomMargin;

  // 设置按钮被触摸时响应方法
  [button addTarget:self
             action:@selector(buttonDidPush:)
   forControlEvents:UIControlEventTouchUpInside];
  // 将按钮追加到画面 view 中
```

```
    [self.view addSubview:button];
}
// 按钮被触碰时调用的方法
- (void)buttonDidPush:(id)sender {
  if ( [sender isKindOfClass:[UIButton class]] ) {
    UIButton* button = sender;
    UIAlertView* alert = [[[UIAlertView alloc] initWithTitle:nil
                                           message:button.currentTitle
                                           delegate:nil
                                           cancelButtonTitle:nil
                                           otherButtonTitles:@"OK", nil] autorelease];
    [alert show];
  }
}
@end
```

执行后首先显示一个"触摸我！"按钮，触摸此按钮后会弹出一个对话框，如图 10-10 所示。

图 10-10 执行效果

实例 219　同时滑动两个滑块

实例 219	同时滑动两个滑块
源码路径	\daima\219\

实例说明

在 iOS 应用中，除了 UIButton 按钮意外，最常见的触摸操作是通过 UISlider 滑块控件实现的。在本实例中预先设置了两个滑块，当使用触摸方式滑动一个滑块时，另一个滑块会以同样的进度进行同步滑动。

具体实现

实例文件 UIKitPrjSlider.m 定义了两个滑块的最小值和最大值，并且指定了滑块变化时调用方法 sliderDidChange，通过此方法设置两个滑块的值保持同步。文件 UIKitPrjSlider.m 的具体实现代码如下所示。

```
#import "UIKitPrjSlider.h"
@implementation UIKitPrjSlider
// 对象释放方法
- (void)dealloc {
  [sliderCopy_ release];
  [super dealloc];
}
- (void)viewDidLoad {
  [super viewDidLoad];
  self.title = @"UISlider 滑块";
  self.view.backgroundColor = [UIColor whiteColor];
  // 创建滑块控件
  UISlider* slider = [[[UISlider alloc] init] autorelease];
  slider.frame = CGRectMake( 0, 0, 200, 50 );
```

```
    slider.minimumValue = 0.0; //设置滑块最小值
    slider.maximumValue = 1.0; //设置滑块最大值
    slider.center = self.view.center;
    // 指定滑块变化时被调用的方法
    [slider addTarget:self
            action:@selector(sliderDidChange:)
     forControlEvents:UIControlEventValueChanged];
    // 拷贝滑块
    sliderCopy_ = [[UISlider alloc] init];
    sliderCopy_.frame = slider.frame;
    sliderCopy_.minimumValue = slider.minimumValue;
    sliderCopy_.maximumValue = slider.maximumValue;
    CGPoint point = slider.center;
    point.y += 50;
    sliderCopy_.center = point;
    // 在画面中追加两个滑块
    [self.view addSubview:slider];
    [self.view addSubview:sliderCopy_];
}
// 滑块变化时调用
- (void)sliderDidChange:(id)sender {
    if ( [sender isKindOfClass:[UISlider class]] ) {
      UISlider* slider = sender;
      // 将 sliderCopy_ 的值保持与 slider 相同
      sliderCopy_.value = slider.value;
    }
}
@end
```

执行后的效果如图 10-11 所示。

图 10-11 执行效果

实例 220　触摸屏幕检测

实例 220	触摸屏幕检测
源码路径	\daima\220\

实例说明

在本章前面的实例中，都是通过控件来完成触摸检测功能的。在本实例中，将不使用第三方控件来实现触摸功能。本实例比较简单，只有一个 UIViewController，当触摸屏幕后会调用 touchesBegan:withEvent 方法。

具体实现

实例文件 UIKitPrjTouchesBegan.m 的具体实现代码如下所示。

```
#import "UIKitPrjTouchesBegan.h"
@implementation UIKitPrjTouchesBegan
- (void)viewDidLoad {
  [super viewDidLoad];
  // 设置背景色（必需）
```

```
    // 默认状态下为透明，无法完成触摸
    self.view.backgroundColor = [UIColor whiteColor];
}
// 此方法为接受触摸事件的方法
- (void)touchesBegan:(NSSet*)touches withEvent:(UIEvent*)event {
    UIAlertView* alert = [[[UIAlertView alloc] initWithTitle:nil
                                          message:@"这是一个viewController!"
                                          delegate:nil
                                          cancelButtonTitle:nil
                                          otherButtonTitles:@"OK", nil] autorelease];
    [alert show];
}
@end
```

执行后触摸屏幕会弹出一个对话框，如图 10-12 所示。

图 10-12　执行效果

实例 221　触摸屏幕中的文字标签

实例 221	触摸屏幕中的文字标签
源码路径	\daima\221\

实例说明

在 iOS 应用中，除了可以触摸屏幕中的按钮和滑动块之外，还可以触摸其他的控件元素。在本实例中，在屏幕中设置了一个 UILable 控件，在屏幕中显示文字"触摸我吧!"。触摸这段文字后，可以弹出一个"这是一段文字"对话框。

具体实现

实例文件 UIKitPrjTouchesTheLabel.m 的具体实现代码如下所示。

```
#import "UIKitPrjTouchesTheLabel.h"
#pragma mark ----- TouchableLabel -----
// 定义 UILabel 的子类
@interface TouchableLabel : UILabel
@end
// 只追加 touchesBegan:withEvent:方法
@implementation TouchableLabel
- (void)touchesBegan:(NSSet*)touches withEvent:(UIEvent*)event {
    UIAlertView* alert = [[[UIAlertView alloc] initWithTitle:nil
                                          message:@"这是一段文字!"
                                          delegate:nil
                                          cancelButtonTitle:nil
                                          otherButtonTitles:@"OK", nil] autorelease];
    [alert show];
}
@end
#pragma mark ----- UIKitPrjTouchesTheLabel -----
@implementation UIKitPrjTouchesTheLabel
- (void)viewDidLoad {
```

```
    [super viewDidLoad];
    self.view.backgroundColor = [UIColor whiteColor];
    // 将新创建的标签布置在画面上
    TouchableLabel* label = [[[TouchableLabel alloc] init] autorelease];
    label.frame = CGRectMake( 60, 100, 200, 50 );
    label.text = @"触摸我吧!";
    label.textAlignment = UITextAlignmentCenter;
    label.backgroundColor = [UIColor grayColor];
    // 必须将 userInteractionEnabled 属性设置成 YES，默认为 NO
    label.userInteractionEnabled = YES;
    [self.view addSubview:label];
}
- (void)touchesBegan:(NSSet*)touches withEvent:(UIEvent*)event {
    UIAlertView* alert = [[[UIAlertView alloc] initWithTitle:nil
                                   message:@"这是一个viewController!"
                                   delegate:nil
                         cancelButtonTitle:nil
                         otherButtonTitles:@"OK", nil] autorelease];
    [alert show];
}
@end
```

执行后的效果如图 10-13 所示。

图 10-13 执行效果

实例 222 演示一次触摸和两次触摸

实例 222	演示一次触摸和两次触摸
源码路径	\daima\222\

实例说明

在 iOS 应用中，除了可以一次触摸屏幕外，还可以响应两次触摸的处理事件。在本实例中，首先根据 tapCount 属性判断到底是单次触碰还是多次触碰。如果是单次触摸则执行 singleTap 方法，通过此方法输出"这是一次触摸!"提示框。如果是两次触摸则执行 doubleTap 方法，通过此方法输出"这是两次触摸!"提示框。

具体实现

实例文件 UIKitPrjDoubleTap.m 的具体实现代码如下所示。

```
#import "UIKitPrjDoubleTap.h"
@implementation UIKitPrjDoubleTap
- (void)viewDidLoad {
    [super viewDidLoad];
    self.view.backgroundColor = [UIColor whiteColor];
}
- (void)touchesBegan:(NSSet*)touches withEvent:(UIEvent*)event {
    // 画面被触摸后解除单点触摸标志
    singleTapReady_ = NO;
}
- (void)touchesEnded:(NSSet*)touches withEvent:(UIEvent*)event {
    //参照 touches 中对象的 tapCount 属性
    //根据 tapCount 属性判断单次触碰或者多次触碰
    NSInteger tapCount = [[touches anyObject] tapCount];
```

```
        if ( 2 > tapCount ) {
            // tapCount 小于 2 时将单点触摸标志设置成 YES
            singleTapReady_ = YES;
            // 单点触碰方法延迟 0.3 秒执行
            [self performSelector:@selector(singleTap)
                       withObject:nil
                       afterDelay:0.3f];
        } else {
            // 执行双击确认方法（douleTap）
            [self performSelector:@selector(doubleTap)];
        }
    }
- (void)singleTap {
    // 如果有其他的 touchesBegan 被调用则方法退出
    if ( !singleTapReady_ ) return;
    UIAlertView* alert = [[[UIAlertView alloc] initWithTitle:nil
                                                    message:@"这是一次触摸!"
                                                   delegate:nil
                                          cancelButtonTitle:nil
                                          otherButtonTitles:@"OK", nil] autorelease];
    [alert show];
}
- (void)doubleTap {
    UIAlertView* alert = [[[UIAlertView alloc] initWithTitle:nil
                                                    message:@"这是两次触摸!!"
                                                   delegate:nil
                                          cancelButtonTitle:nil
                                          otherButtonTitles:@"OK", nil] autorelease];
    [alert show];
}
@end
```

执行后的效果如图 10-14 所示。

图 10-14 执行效果

实例 223 演示 3 次触摸

实例 223	演示 3 次触摸
源码路径	\daima\223\

实例说明

在 iOS 应用中，除了可以一次触摸和两次触摸屏幕外，还可以响应 3 次触摸的处理事件。在本实例中，首先根据 tapCount 属性判断到底是单次触碰还是多次触碰。如果是单次触摸则执行 singleTap 方法，通过此方法输出"这是一次触摸!"提示框。如果是两次触摸则执行 doubleTap 方法，通过此方法输出"这是两次触摸!!"提示框。如果是 3 次触摸则执行 tripleTap 方法，通过此方法输出"这是 3 次触摸!!!"提示框。

具体实现

实例文件 UIKitPrjTripleTap.m 的具体实现代码如下所示。

```
#import "UIKitPrjTripleTap.h"
@implementation UIKitPrjTripleTap

- (void)viewDidLoad {
    [super viewDidLoad];
    self.view.backgroundColor = [UIColor whiteColor];
```

```objc
}
- (void)touchesBegan:(NSSet*)touches withEvent:(UIEvent*)event {
  singleTapReady_ = NO;
  doubleTapReady_ = NO;
}
- (void)touchesEnded:(NSSet*)touches withEvent:(UIEvent*)event {
  NSInteger tapCount = [[touches anyObject] tapCount];
  if ( 2 > tapCount ) {
    singleTapReady_ = YES;
    [self performSelector:@selector(singleTap)
             withObject:nil
             afterDelay:0.3f];
  } else if ( 3 > tapCount ) {
    doubleTapReady_ = YES;
    [self performSelector:@selector(doubleTap)
             withObject:nil
             afterDelay:0.3f];
  } else {
    [self performSelector:@selector(tripleTap)];
  }
}
- (void)singleTap {
  if ( !singleTapReady_ ) return;
  UIAlertView* alert = [[[UIAlertView alloc] initWithTitle:nil
                                            message:@"这是一次触摸!"
                                            delegate:nil
                                            cancelButtonTitle:nil
                                            otherButtonTitles:@"OK", nil] autorelease];
  [alert show];
}
- (void)doubleTap {
  if ( !doubleTapReady_ ) return;
  UIAlertView* alert = [[[UIAlertView alloc] initWithTitle:nil
                                            message:@"这是两次触摸!!"
                                            delegate:nil
                                            cancelButtonTitle:nil
                                            otherButtonTitles:@"OK", nil] autorelease];
  [alert show];
}
- (void)tripleTap {
  UIAlertView* alert = [[[UIAlertView alloc] initWithTitle:nil
                                            message:@"这是3次触摸!!!"
                                            delegate:nil
                                            cancelButtonTitle:nil
                                            otherButtonTitles:@"OK", nil] autorelease];
  [alert show];
}
@end
```

执行后的效果如图10-15所示。

图10-15　3次触摸

实例224　拖曳方式移动屏幕中的图片

实例224	拖曳方式移动屏幕中的图片
源码路径	\daima\224\

实例说明

在iOS应用中，除了可以在屏幕中显示指定图片外，还可以使用拖曳的方式移动屏幕中的图片。在本实例中，通过调用touchesMoved:withEvent:方法移动了屏幕中的图片。此方法能够检测手指在屏幕中的位置，通过向 locationInView 方法中传递 UIView 后，可以获取手指按住的位置。

具体实现

实例文件 UIKitPrjDrag.m 的具体实现代码如下所示。

```objectivec
#import "UIKitPrjDrag.h"
#pragma mark ----- Private Methods Definition -----
@interface UIKitPrjDrag ()
- (void)theCharacterWillWalk;
@end
#pragma mark ----- Start Implementation For Methods -----
@implementation UIKitPrjDrag
- (void)dealloc {
  shouldWalk_ = NO;
  [character_ release];
  [super dealloc];
}
- (void)viewDidLoad {
  [super viewDidLoad];
  self.view.backgroundColor = [UIColor whiteColor];

  // 为实现动画导入两枚图片
  UIImage* image1 = [UIImage imageNamed:@"chara1.png"];
  UIImage* image2 = [UIImage imageNamed:@"chara2.png"];
  NSArray* images = [[NSArray alloc] initWithObjects:image1, image2, nil];
  // 创建 UIImageView 实例并初始化
  character_ = [[UIImageView alloc] initWithImage:image1];
  // 将动画图片以数值形式设置到 animationImages 属性中
  character_.animationImages = images;
  character_.animationDuration = 0.3;
  [images release];
  [self.view addSubview:character_];
}
- (void)viewWillDisappear:(BOOL)animated {
  [super viewWillDisappear:animated];
  shouldWalk_ = NO;
}
#pragma mark ----- Responder -----
// 手指在画面上触摸的瞬间动画开始，调用人物移动的方法
// 当 shouldWalk_ 为 YES 时持续调用人物移动方法
- (void)touchesBegan:(NSSet*)touches withEvent:(UIEvent*)event {
  shouldWalk_ = YES;
  [character_ startAnimating];
  targetPoint_ = [[touches anyObject] locationInView:self.view];
  [self theCharacterWillWalk];
}
// 画面拖动时保持当时位置
- (void)touchesMoved:(NSSet*)touches withEvent:(UIEvent*)event {
  targetPoint_ = [[touches anyObject] locationInView:self.view];
}
// 手指从画面离开时动画停止，并将 shouldWalk_ 设置成 NO
- (void)touchesEnded:(NSSet*)touches withEvent:(UIEvent*)event {
  shouldWalk_ = NO;
  [character_ stopAnimating];
}
// 放弃触摸状态后动画停止，并将 shouldWalk_ 设置成 NO
- (void)touchesCancelled:(NSSet*)touches withEvent:(UIEvent*)event {
  shouldWalk_ = NO;
  [character_ stopAnimating];
}
#pragma mark ----- Private Methods -----
// 向手指触摸位置一点点移动人物位置的方法
// shouldWalk_ 为 YES 时以 0.3 秒为间隔递归调用本方法
- (void)theCharacterWillWalk {
  if ( !shouldWalk_ ) {
    return;
  }
  static const NSInteger kMaximumSteps = 8;
  CGPoint newPoint = character_.center;
  if ( kMaximumSteps < abs( targetPoint_.x - newPoint.x ) ) {
    if ( targetPoint_.x > newPoint.x ) {
      newPoint.x += kMaximumSteps;
```

```
        } else {
          newPoint.x -= kMaximumSteps;
        }
      } else {
        newPoint.x = targetPoint_.x;
      }
      if ( kMaximumSteps < abs( targetPoint_.y - newPoint.y ) ) {
        if ( targetPoint_.y > newPoint.y ) {
          newPoint.y += kMaximumSteps;
        } else {
          newPoint.y -= kMaximumSteps;
        }
      } else {
        newPoint.y = targetPoint_.y;
      }
      character_.center = newPoint;

      [self performSelector:@selector(theCharacterWillWalk)
              withObject:nil
              afterDelay:0.3];
}
@end
```

执行后的效果如图 10-16 所示。

图 10-16　执行效果

实例 225　可以检测上、下、左、右 4 个方向的触摸

实例 225	可以检测上、下、左、右 4 个方向的触摸
源码路径	\daima\225\

实例说明

在 iOS 应用中，我们可以拖动设备的屏幕，可以让整个屏幕画面向左或向右滑动显示。在本实例中，定义了枚举 DirectionForSlide 来保存移动位置，可以检测上、下、左、右 4 个方向的触摸操作。

```
//标示滑动方向的枚举类型定义
typedef enum
{
  kSlideNone,//原状态
  kSlideHorizontal,//横向滑动
  kSlideVertical,//纵向滑动
} DirectionForSlide;
```

具体实现

实例文件 UIKitPrjSlide.m 的具体实现代码如下所示。

```
#import "UIKitPrjSlide.h"
@implementation UIKitPrjSlide
- (void)dealloc {
  [label_ release];
  [super dealloc];
}
```

```objc
- (void)viewDidLoad {
  [super viewDidLoad];

  self.view.backgroundColor = [UIColor blackColor];
  label_ = [[UILabel alloc] init];
  label_.frame = self.view.bounds;
  label_.backgroundColor = [UIColor whiteColor];
  label_.textAlignment = UITextAlignmentCenter;
  label_.text = @"可以上下左右滑动";
  label_.autoresizingMask =
  UIViewAutoresizingFlexibleWidth | UIViewAutoresizingFlexibleHeight;
  [self.view addSubview:label_];
  [[NSNotificationCenter defaultCenter] addObserver:self
                                    selector:@selector(suspend)
                                        name:UIApplicationWillResignActiveNotification
                                      object:nil];
}
#pragma mark ----- Responder -----
- (void)touchesBegan:(NSSet*)touches withEvent:(UIEvent*)event {
  // 保存触摸位置
  touchBegan_ = [[touches anyObject] locationInView:self.view];
  // 保存标签原位置
  labelOrigin_ = label_.center;
  // 初始化运动方向
  direction_ = kSlideNone;
}
- (void)touchesMoved:(NSSet*)touches withEvent:(UIEvent*)event {
  static const NSInteger kNeedMove = 10;
  CGPoint point = [[touches anyObject] locationInView:self.view];
  // 计算初始触摸位置与当前位置的坐标差
  NSInteger distanceHorizontal = point.x - touchBegan_.x;
  NSInteger distanceVertical = point.y - touchBegan_.y;
  if ( kSlideNone == direction_ ) {
    // 判断运动方向
    if ( ABS( distanceHorizontal ) > ABS( distanceVertical ) ) {
      // 横向运动
      if ( kNeedMove <= ABS( distanceHorizontal ) ) {
        direction_ = kSlideHorizontal;
      }
    } else {
      // 纵向运动
      if ( kNeedMove <= ABS( distanceVertical ) ) {
        direction_ = kSlideVertical;
      }
    }
  }
  if ( kSlideNone != direction_ ) {
    // 判断运动距离
    CGPoint newPoint = labelOrigin_;
    if ( kSlideHorizontal == direction_ ) {
      newPoint.x += distanceHorizontal;
    } else {
      newPoint.y += distanceVertical;
    }
    // 移动的目标点
    label_.center = newPoint;
  }
}
- (void)touchesEnded:(NSSet*)touches withEvent:(UIEvent*)event {
  // 放开手指的话,标签返回原位置
  [UIView beginAnimations:nil context:nil];
  label_.center = self.view.center;
  [UIView commitAnimations];
}
- (void)touchesCancelled:(NSSet*)touches withEvent:(UIEvent*)event {
  [self touchesEnded:touches withEvent:event];
}
- (void)suspend {
  [self touchesCancelled:nil withEvent:nil];
}
@end
```

执行后的效果如图 10-17 所示。

实例 226 检测触摸滑动的方向

图 10-17 执行效果

实例 226 检测触摸滑动的方向

实例 226	检测触摸滑动的方向
源码路径	\daima\226\

实例说明

在 iOS 应用中，通过滑动（Flick）可以实现页面的快速滚动和翻页的功能。在本实例中，通过方法 touchesMoved:withEvent 实现了快速滑动处理，通过使用 if 语句检测在哪个方向滑动。

具体实现

实例文件 UIKitPrjFlick.m 的具体实现代码如下所示。

```
#import "UIKitPrjFlick.h"
@implementation UIKitPrjFlick
- (void)viewDidLoad {
    [super viewDidLoad];
    self.view.backgroundColor = [UIColor whiteColor];
}
#pragma mark ----- Responder -----
- (void)touchesBegan:(NSSet*)touches withEvent:(UIEvent*)event {
    // 保存触摸时间与位置
    UITouch* touch = [touches anyObject];
    timestampBegan_ = event.timestamp;
    pointBegan_ = [touch locationInView:self.view];
}
- (void)touchesEnded:(NSSet*)touches withEvent:(UIEvent*)event {
    static const NSTimeInterval kFlickJudgeTimeInterval = 0.3;
    static const NSInteger kFlickMinimumDistance = 10;
    UITouch* touchEnded = [touches anyObject];
    CGPoint pointEnded = [touchEnded locationInView:self.view];
    NSInteger distanceHorizontal = ABS( pointEnded.x - pointBegan_.x );
    NSInteger distanceVertical = ABS( pointEnded.y - pointBegan_.y );
    if ( kFlickMinimumDistance > distanceHorizontal && kFlickMinimumDistance > distanceVertical )
{
        // 纵向与横向如果几乎都没有移动的话，方法返回（return）
        return;
    }
    NSTimeInterval timeBeganToEnded = event.timestamp - timestampBegan_;
    if ( kFlickJudgeTimeInterval > timeBeganToEnded ) {
        // 快速滑动处理
        NSString* message;
        // 判断向哪个方向滑动
        if ( distanceHorizontal > distanceVertical ) {
            if ( pointEnded.x > pointBegan_.x ) {
                message = @"正在向右滑动！";
            } else {
                message = @"正在向左滑动！";
            }
        } else {
            if ( pointEnded.y > pointBegan_.y ) {
                message = @"正在向下滑动！";
            } else {
                message = @"正在向上滑动！";
```

```
        }
      }
      UIAlertView* alert = [[[UIAlertView alloc] initWithTitle:nil
                                    message:message
                                    delegate:nil
                              cancelButtonTitle:nil
                              otherButtonTitles:@"OK", nil] autorelease];

      [alert show];
  }
}
@end
```

执行后的效果如图 10-18 所示。

图 10-18 执行效果

实例 227 实现屏幕的多点触摸

实例 227	实现屏幕的多点触摸
源码路径	\daima\227\

实例说明

在 iOS 应用中,我们可以多点触摸设备屏幕。iOS 默认是不能多点触摸的,如果将 UIView 的属性 multipleTouchEnabled 设置为 YES,则可以允许多点触摸屏幕。在本实例中通过设置此属性实现了多点触摸,并且触摸后会提示几个手指触摸了屏幕。

具体实现

实例文件 UIKitPrjDoubleTouch.m 的具体实现代码如下所示。

```
#import "UIKitPrjDoubleTouch.h"
@implementation UIKitPrjDoubleTouch
- (void)viewDidLoad {
  [super viewDidLoad];
  self.view.backgroundColor = [UIColor whiteColor];
  // 允许多点触摸时必须将此属性设置成 YES
  self.view.multipleTouchEnabled = YES;
}
#pragma mark ----- Responder -----
- (void)touchesBegan:(NSSet*)touches withEvent:(UIEvent*)event {
  NSString* message = [NSString stringWithFormat:@"共有%d 个手指同时触摸了屏幕", [touches count]];
  UIAlertView* alert = [[[UIAlertView alloc] initWithTitle:nil
                                    message:message
                                    delegate:nil
                              cancelButtonTitle:nil
                              otherButtonTitles:@"OK", nil] autorelease];

  [alert show];
}
@end
```

执行效果如图 10-19 所示。

图 10-19 执行效果

实例 228 检测双指滑动

实例 228	检测双指滑动
源码路径	\daima\228\

实例说明

在本实例中演示了一个多点触摸的实现过程，此处可以通过两个手指让标签上下滑动。在方法 touchesMoved：withEvent:中，通过 if 语句检测参数 touches 值是否为 2，如果是 2 则判断向哪个方向移动，并计算各自纵向的移动距离。并且设置双方都向下移动的话，标签也向下移动。双方都向上移动的话，标签也向上移动。

具体实现

实例文件 UIKitPrjDoubleSlide.m 的具体实现代码如下所示。

```
#import "UIKitPrjDoubleSlide.h"
@implementation UIKitPrjDoubleSlide
- (void)dealloc {
  [label_ release];
  [super dealloc];
}
- (void)viewDidLoad {
  [super viewDidLoad];
  self.view.backgroundColor = [UIColor blackColor];
  self.view.multipleTouchEnabled = YES;
  label_ = [[UILabel alloc] init];
  label_.frame = self.view.bounds;
  label_.autoresizingMask =
    UIViewAutoresizingFlexibleWidth | UIViewAutoresizingFlexibleHeight;
  label_.textAlignment = UITextAlignmentCenter;
  label_.text = @"通过2个手指进行滑动让屏幕上下移动！";
  [self.view addSubview:label_];
}
#pragma mark ----- Responder -----

- (void)touchesMoved:(NSSet*)touches withEvent:(UIEvent*)event {
  if ( 2 == [touches count] ) {
    // 两个手指触摸的情况下计算各自纵向的移动距离
    NSInteger distance[2];
    int i = 0;
    for ( UITouch* touch in touches ) {
      CGPoint before = [touch previousLocationInView:self.view];
      CGPoint now = [touch locationInView:self.view];
      distance[i] = now.y - before.y;
      ++i;
    }
    CGPoint newPoint = label_.center;
    if ( 0 < distance[0] && 0 < distance[1] ) {
      // 双方都向下移动的话，标签也向下移动
      newPoint.y += MAX( distance[0], distance[1] );
    } else if ( 0 > distance[0] && 0 > distance[1] ) {
      // 双方都向上移动的话，标签也向上移动
      newPoint.y += MAX( distance[0], distance[1] );
    }
    label_.center = newPoint;
  }
}
```

@end

执行后的效果如图 10-20 所示。

图 10-20 执行效果

实例 229　通过触摸方式放大或缩小屏幕中的图片

实例 229	通过触摸方式放大或缩小屏幕中的图片
源码路径	\daima\229\

实例说明

本实例是基于本章上一个实例改造而来的，在方法 touchesMoved:withEvent 中，通过 if 语句检测参数 touches 值是否为 2，并将当前的距离与上次的距离进行比较，以判断距离是缩短了还是扩大了。如果距离缩短了则进行缩小处理，如果距离扩大了则进行扩大处理。

具体实现

实例文件 UIKitPrjPinch.m 的具体实现代码如下所示。

```
#import "UIKitPrjPinch.h"
#import <math.h>
#pragma mark ----- Private Methods Definition -----
@interface UIKitPrjPinch ()
- (CGFloat)distanceWithPointA:(CGPoint)pointA pointB:(CGPoint)pointB;
@end
#pragma mark ----- Start Implementation For Methods -----
@implementation UIKitPrjPinch
- (void)dealloc {
 [imageView_ release];
 [super dealloc];
}
- (void)viewDidLoad {
 [super viewDidLoad];
 self.view.backgroundColor = [UIColor blackColor];
 self.view.multipleTouchEnabled = YES;
 UIImage* image = [UIImage imageNamed:@"dog.jpg"];
 imageView_ = [[UIImageView alloc] initWithImage:image];
 imageView_.center = self.view.center;
 imageView_.contentMode = UIViewContentModeScaleAspectFill;
 imageView_.autoresizingMask = UIViewAutoresizingFlexibleLeftMargin |
                               UIViewAutoresizingFlexibleRightMargin |
                               UIViewAutoresizingFlexibleTopMargin |
                               UIViewAutoresizingFlexibleBottomMargin;
 [self.view addSubview:imageView_];
}
#pragma mark ----- Responder -----
- (void)touchesMoved:(NSSet*)touches withEvent:(UIEvent*)event {
 if ( 2 == [touches count] ) {
  // 如果两个手指触摸的情况下，计算 2 点间的距离
  // 将当前的距离与上次的距离进行比较，以判断距离是缩短了还是扩大了
  NSArray* twoFingers = [touches allObjects];
  UITouch* touch1 = [twoFingers objectAtIndex:0];
  UITouch* touch2 = [twoFingers objectAtIndex:1];
```

```
    CGPoint previous1 = [touch1 previousLocationInView:self.view];
    CGPoint previous2 = [touch2 previousLocationInView:self.view];
    CGPoint now1 = [touch1 locationInView:self.view];
    CGPoint now2 = [touch2 locationInView:self.view];
    CGFloat previousDistance = [self distanceWithPointA:previous1 pointB:previous2];
    CGFloat distance = [self distanceWithPointA:now1 pointB:now2];
    CGFloat scale = 1.0;
    if ( previousDistance > distance ) {
      // 距离缩短了的话，进行缩小处理
      scale -= ( previousDistance - distance ) / 300.0;
    } else if ( distance > previousDistance ) {
      // 距离扩大了的话，进行扩大处理
      scale += ( distance - previousDistance ) / 300.0;
    }
    CGAffineTransform newTransform =
      CGAffineTransformScale( imageView_.transform, scale, scale );
    imageView_.transform = newTransform;
    imageView_.center = self.view.center;
  }
}
#pragma mark ----- Private Methods -----
- (CGFloat)distanceWithPointA:(CGPoint)pointA pointB:(CGPoint)pointB {
  CGFloat dx = fabs( pointB.x - pointA.x );
  CGFloat dy = fabs( pointB.y - pointA.y );
  return sqrt( dx * dx + dy * dy );
}
@end
```

执行效果如图 10-21 所示。

图 10-21　执行效果

实例 230　通过触摸方式放大或缩小屏幕中的图片

实例 230	通过触摸方式放大或缩小屏幕中的图片
源码路径	\daima\230\

实例说明

从 iOS 3.0 开始支持 Motion（震动）事件，特别是在摇动设备时，包括如下 3 个检测振动的方法。

- motionBegan:withEvent：运动开始时执行。
- motionEnded:withEvent：运动结束时执行。
- motionCancelled:withEvent：运动被取消时执行。

在本实例中，在屏幕中央设置了一个提示文本"摇动我吧！"。当开始运动时调用 motionBegan:withEvent:方法，当运动结束时调用 motionEnded:withEvent 方法。

具体实现

实例文件 UIKitPrjMotion.m 的具体实现代码如下所示。

```objc
#import "UIKitPrjMotion.h"
//实现UILabel的子类
@implementation LabelForMotion
- (BOOL)canBecomeFirstResponder {
  return YES;
}
@end
//实现UIViewController的子类
@implementation UIKitPrjMotion
- (void)viewDidLoad {
  [super viewDidLoad];
  //创建新标签并追加到画面中
  LabelForMotion* label = [[[LabelForMotion alloc] init] autorelease];
  label.frame = self.view.bounds;
  label.autoresizingMask =
    UIViewAutoresizingFlexibleWidth | UIViewAutoresizingFlexibleHeight;
  label.textAlignment = UITextAlignmentCenter;
  label.text = @"摇动我吧!";
  [self.view addSubview:label];
  //将标签设置为第一响应者
  [label becomeFirstResponder];
}
- (void)motionBegan:(UIEventSubtype)motion withEvent:(UIEvent*)event {
  NSLog( @"开始了!" );
}
//振动结束时调用的方法
- (void)motionEnded:(UIEventSubtype)motion withEvent:(UIEvent*)event {
  NSLog( @"结束了" );
  UIAlertView* alert = [[[UIAlertView alloc] initWithTitle:nil
                                      message:@"地震了!!"
                                      delegate:nil
                                cancelButtonTitle:nil
                              otherButtonTitles:@"OK", nil] autorelease];
  [alert show];
}
- (void)motionCancelled:(UIEventSubtype)motion withEvent:(UIEvent*)event {
  NSLog( @"取消吧" );
}
@end
```

执行效果如图10-22所示。

图10-22 执行效果

实例231 使用加速传感器

实例231	使用加速传感器
源码路径	\daima\231\

实例说明

加速度传感器俗称加速规和 G-Sensor，可以感应物体的加速度性。事实上加速度传感器的实现方式也有许多种，MEMS 只是手法之一，用 MEMS 实现加速度传感器确实是目前的趋势。加速度传感器一般有「X、Y 两轴」与「X、Y、Z 三轴」两种，两轴多用于车、船等平面移动，三轴多用于飞弹、飞机等飞行物。iPhone 可以感应实体翻转而自动对应翻转画面，也是靠这个传感器。

在 iOS 4 之前，加速度计由 UIAccelerometer 类来负责采集工作，而电子罗盘则由 Core Location 接管。而随着 iPhone 4 等后续版本的推出，由于加速度计的升级（有消息说使用的是这款芯片）和陀螺仪的引入，与 motion 相关的编程成为重头戏，所以，苹果在 iOS 4 中增加一个专门负责该方面处理的框架，就是 Core Motion Framework。这个 Core Motion 有什么好处呢？简单来说，它不仅提供给开发者获得实时的加速度值和旋转速度值，更重要的是，苹果在其中集成了很多算法，可以直接输出把重力加速度分量剥离的加速度，省去高通滤波操作，以及提供一个专门设备的三维 attitude 信息。

在公开的 API 中，通过 UIAccelerometer 类实现相应的功能。加速计（UIAccelerometer）是一个单例模式的类，所以需要通过方法 sharedAccelerometer 获取其唯一的实例。加速计需要设置如下两点。

- 设置其代理：用以执行获取加速计信息的方法。
- 设置加速计获取信息的频率：最高支持每秒 100 次。

例如下面的代码：

```
UIAccelerometer *accelerometer = [UIAccelerometer sharedAccelerometer];
accelerometer.delegate = self;
accelerometer.updateInterval = 1.0/30.0f;
下面是加速计的代理方法，需要符合协议<UIAccelerometerDelegate>。
-(void)accelerometer:(UIAccelerometer  *)accelerometer  didAccelerate:(UIAcceleration
*)acceleration
{
//    NSString *str = [NSString stringWithFormat:@"x:%g\ty:%g\tz:%g",acceleration.x,
acceleration.y,acceleration.z];
//    NSLog(@"%@",str);

    // 检测摇动，1.5 为轻摇，2.0 为重摇
//    if (fabsf(acceleration.x)>1.8||
//        fabsf(acceleration.y)>1.8||
//        fabsf(acceleration.z)>1.8)) {
//        NSLog(@"你摇动我了~");
//    }
    static NSInteger shakeCount = 0;
    static NSDate *shakeStart;
    NSDate *now = [[NSDate alloc]init];
    NSDate *checkDate = [[NSDate alloc]initWithTimeInterval:1.5f sinceDate:shakeStart];
    if ([now compare:checkDate] == NSOrderedDescending || shakeStart == nil) {
        shakeCount = 0;
        [shakeStart release];
        shakeStart = [[NSDate alloc]init];
    }
    [now release];
    [checkDate release];
    if (fabsf(acceleration.x)>1.7||
        fabsf(acceleration.y)>1.7||
        fabsf(acceleration.z)>1.7) {
        shakeCount ++;
        if (shakeCount >4) {
            NSLog(@"你摇动我了~");
            shakeCount = 0;
            [shakeStart release];
            shakeStart = [[NSDate alloc]init];
        }
    }
}
```

在本实例中，通过 Accelerometer 测试了屏幕中滚动的小球。

具体实现

实例文件 UIKitPrjAccelerometer.m 的具体实现代码如下所示。

```objc
#import "UIKitPrjAccelerometer.h"
#pragma mark ----- Private Methods Definition -----
@interface UIKitPrjAccelerometer ()
- (CGFloat)lowpassFilter:(CGFloat)accel before:(CGFloat)before;
- (CGFloat)highpassFilter:(CGFloat)accel before:(CGFloat)before;
@end
#pragma mark ----- Start Implementation For Methods -----
@implementation UIKitPrjAccelerometer
- (void)dealloc {
  [imageView_ release];
  [super dealloc];
}
- (void)viewDidLoad {
  [super viewDidLoad];
  self.view.backgroundColor = [UIColor whiteColor];
  // 追加球体的 UIImageView
  UIImage* image = [UIImage imageNamed:@"metal.png"];
  imageView_ = [[UIImageView alloc] initWithImage:image];
  imageView_.center = self.view.center;
  imageView_.autoresizingMask = UIViewAutoresizingFlexibleLeftMargin |
                                UIViewAutoresizingFlexibleRightMargin |
                                UIViewAutoresizingFlexibleTopMargin |
                                UIViewAutoresizingFlexibleBottomMargin;
  [self.view addSubview:imageView_];
}

- (void)viewWillAppear:(BOOL)animated {
  [super viewWillAppear:animated];
  // 开始获取加速度传感器传过来的值
  UIAccelerometer* accelerometer = [UIAccelerometer sharedAccelerometer];
  accelerometer.updateInterval = 1.0 / 60.0; //< 60Hz
  accelerometer.delegate = self;
}
- (void)viewWillDisappear:(BOOL)animated {
  [super viewWillDisappear:animated];
  speedX_ = speedY_ = 0.0;
  // 结束从加速度传感器获取值
  UIAccelerometer* accelerometer = [UIAccelerometer sharedAccelerometer];
  accelerometer.delegate = nil;
}
// 处理从加速度传感器来的通知
- (void)accelerometer:(UIAccelerometer*)accelerometer
    didAccelerate:(UIAcceleration*)acceleration
{
  speedX_ += acceleration.x;//在 x 轴方向速度上附加 x 轴方向加速度
  speedY_ += acceleration.y;//在 y 轴方向速度上附加 y 轴方向加速度
  CGFloat posX = imageView_.center.x + speedX_;//根据速度调整球体位置坐标
  CGFloat posY = imageView_.center.y - speedY_;//根据速度调整球体位置坐标
  // 碰到边框后反弹的处理
  if ( posX < 0.0 ) {
    posX = 0.0;
    speedX_ *= -0.4; //碰到左边的边框后以 0.4 倍的速度反弹
  } else if ( posX > self.view.bounds.size.width ) {
    posX = self.view.bounds.size.width;
    speedX_ *= -0.4; //碰到右边的边框后以 0.4 倍的速度反弹
  }
  if ( posY < 0.0 ) {
    posY = 0.0;
    speedY_ = 0.0; //碰到上边的边框不反弹
  } else if ( posY > self.view.bounds.size.height ) {
    posY = self.view.bounds.size.height;
    speedY_ *= -1.5; //碰到下边的边框以 1.5 倍的速度反弹
  }
  imageView_.center = CGPointMake( posX, posY );
}
// 低通滤波器
- (CGFloat)lowpassFilter:(CGFloat)accel before:(CGFloat)before {
```

```
    static const CGFloat kFilteringFactor = 0.1; //常量
    return ( accel * kFilteringFactor ) + ( before * ( 1.0 - kFilteringFactor ) );
}
// 高通滤波器
- (CGFloat)highpassFilter:(CGFloat)accel before:(CGFloat)before {
    return accel - [self lowpassFilter:accel before:before];
}
@end
```

执行后的效果如图 10-23 所示。

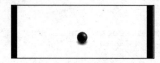

图 10-23　执行效果

实例 232　触摸屏幕后插入一幅图片

实例 232	触摸屏幕后插入一幅图片
源码路径	\daima\232\

实例说明

本实例演示轻击手势的用法，当在设备屏幕中触摸某一点时，会在这个位置插入一幅指定的图片。本实例是一个 iPad 项目，使用的素材图片是 "123.png"。

具体实现

实例文件 ViewController.m 的具体实现代码如下所示。

```
#import "ViewController.h"
@implementation ViewController
- (void)didReceiveMemoryWarning
{
    [super didReceiveMemoryWarning];
    // Release any cached data, images, etc that aren't in use.
}
- (void)handleTapFrom:(UITapGestureRecognizer *)recognizer{
    CGPoint location = [recognizer locationInView:self.view];
    CGRect rect = CGRectMake(location.x - 40, location.y - 40, 80.0f, 80.0f);
    UIImageView *image = [[UIImageView alloc] initWithFrame:rect];
    [image setImage:[UIImage imageNamed:@"flower 123.png" ]];
    [self.view addSubview:image];
}
#pragma mark - View lifecycle
- (void)viewDidLoad
{
    [super viewDidLoad];
    UITapGestureRecognizer *tapRecognizer =
    [[UITapGestureRecognizer alloc]initWithTarget:self action:@selector(handleTapFrom:)];
    [tapRecognizer setNumberOfTapsRequired:1];
    [self.view addGestureRecognizer:tapRecognizer];
}
- (void)viewDidUnload
{
    [super viewDidUnload];
    // Release any retained subviews of the main view.
    // e.g. self.myOutlet = nil;
}
- (void)viewWillAppear:(BOOL)animated
{
    [super viewWillAppear:animated];
}

- (void)viewDidAppear:(BOOL)animated
{
```

```
        [super viewDidAppear:animated];
}
- (void)viewWillDisappear:(BOOL)animated
{
        [super viewWillDisappear:animated];
}
- (void)viewDidDisappear:(BOOL)animated
{
        [super viewDidDisappear:animated];
}
- (BOOL)shouldAutorotateToInterfaceOrientation:(UIInterfaceOrientation)
interfaceOrientation
{
    // Return YES for supported orientations
    return YES;
}
@end
```

执行后的效果如图 10-24 所示。

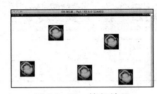

图 10-24　执行效果

实例 233　触摸后实现开花效果

实例 233	触摸后实现开花效果
源码路径	\daima\233\

实例说明

本实例演示了多次触摸手势处理的用法，当在设备屏幕中触摸某一点时，会在这个位置插入一幅指定的图片。如果继续触摸这个插入的图片，则用另外一幅指定的图片替换，实现开花的效果。本实例是一个 iPad 项目，使用的素材图片是 "123.png" 和 "456.png"。

具体实现

实例文件 ViewController.m 的主要实现代码如下所示。

```
#import "ViewController.h"
@implementation ViewController
- (void)didReceiveMemoryWarning
{
    [super didReceiveMemoryWarning];
    // Release any cached data, images, etc that aren't in use.
}
#pragma mark - View lifecycle
- (void)viewDidLoad
{
    [super viewDidLoad];
    UITapGestureRecognizer *tapRecognizer =
    [[UITapGestureRecognizer alloc]
     initWithTarget:self
     action:@selector(handleTapFrom:)];
    [tapRecognizer setNumberOfTapsRequired:1];
    [self.view addGestureRecognizer:tapRecognizer];
}
- (void)handleTapFrom:(UITapGestureRecognizer *)recognizer {
    CGPoint location = [recognizer locationInView:self.view];
    UIView *hitView = [self.view hitTest:location withEvent:nil];
    if ([hitView isKindOfClass:[UIImageView class]]){
        [(UIImageView *)hitView setImage:[UIImage imageNamed:@"123.png" ]];
    }
```

```
    else
    {
        CGRect rect = CGRectMake(location.x - 40,location.y - 40, 80.0f, 80.0f);
        UIImageView *image =[[UIImageView alloc] initWithFrame:rect];
        [image setImage:[UIImage imageNamed:@"456.png" ]];
        [image setUserInteractionEnabled: YES];
        [self.view addSubview:image];
    }
}
```

执行后的效果如图 10-25 所示。

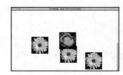

图 10-25　执行效果

实例 234　使用轻扫手势触摸

实例 234	使用轻扫手势触摸
源码路径	\daima\234\

实例说明

本实例演示了使用轻扫手势触摸处理的用法，当在设备屏幕中触摸某一点时，会在这个位置插入一幅指定的图片。如果继续触摸这个插入的图片，则用另外一幅指定的图片替换，实现开花的效果。如果使用清扫手势触摸屏幕，整个屏幕的界面会发生变化。本实例是一个 iPad 项目，使用的素材图片是"123.png"和"456.png"。

具体实现

实例文件 ViewController.m 的主要实现代码如下所示。

```
#import "ViewController.h"
@implementation ViewController
- (void)didReceiveMemoryWarning
{
    [super didReceiveMemoryWarning];
    // Release any cached data, images, etc that aren't in use.
}
#pragma mark - View lifecycle
- (void)viewDidLoad
{
    [super viewDidLoad];
    UITapGestureRecognizer *tapRecognizer =
    [[UITapGestureRecognizer alloc]
     initWithTarget:self
     action:@selector(handleTapFrom:)];
    [tapRecognizer setNumberOfTapsRequired:1];
    [self.view addGestureRecognizer:tapRecognizer];
        UISwipeGestureRecognizer *swipeRecognizer =
    [[UISwipeGestureRecognizer alloc]
     initWithTarget:self
     action:@selector(handleSwipeFrom:)];
    swipeRecognizer.direction = UISwipeGestureRecognizerDirectionRight|
    UISwipeGestureRecognizerDirectionLeft;
    [self.view addGestureRecognizer:swipeRecognizer];
}
- (void)handleTapFrom:(UITapGestureRecognizer *)recognizer {
    CGPoint location = [recognizer locationInView:self.view];
    UIView *hitView = [self.view hitTest:location withEvent:nil];
    if ([hitView isKindOfClass:[UIImageView class]]){
        [(UIImageView *)hitView  setImage:[UIImage imageNamed:@"123.png" ]];
```

```objc
        }
        else
        {
            CGRect rect = CGRectMake(location.x - 40,location.y - 40, 80.0f, 80.0f);
            UIImageView *image =[[UIImageView alloc] initWithFrame:rect];
            [image setImage:[UIImage imageNamed:@"456.png" ]];
            [image setUserInteractionEnabled: YES];
            [self.view addSubview:image];
        }
    }
- (void)handleSwipeFrom:(UISwipeGestureRecognizer *)recognizer {
    for (UIView *subview in [self.view subviews]){
        [subview removeFromSuperview];
    }
    [UIView beginAnimations:nil context:nil];
    [UIView setAnimationDuration:60];
    [UIView setAnimationBeginsFromCurrentState:YES];
    [UIView setAnimationTransition:UIViewAnimationTransitionFlipFromLeft
                    forView:self.view
                        cache:YES];
    [UIView commitAnimations];
}
```

实例 235 双指触摸放大或缩小屏幕中图片

实例 235	双指触摸放大或缩小屏幕中图片
源码路径	\daima\235\

实例说明

本实例演示了使用轻扫手势触摸处理的用法，当在设备屏幕中触摸某一点时，会在这个位置插入一幅指定的图片。如果继续触摸这个插入的图片，则用另外一幅指定的图片替换，实现开花的效果。使用双指触摸手势触摸屏幕中的图片时，会放大或缩小屏幕中图片。本实例是一个 iPad 项目，使用的素材图片是 "123.png" 和 "456.png"。

具体实现

实例文件 ViewController.m 的主要实现代码如下所示。

```objc
#import "ViewController.h"
@implementation ViewController
- (void)didReceiveMemoryWarning
{
    [super didReceiveMemoryWarning];
}
#pragma mark - View lifecycle
- (void)viewDidLoad
{
    [super viewDidLoad];
    UITapGestureRecognizer *tapRecognizer =
    [[UITapGestureRecognizer alloc]
     initWithTarget:self
     action:@selector(handleTapFrom:)];
    [tapRecognizer setNumberOfTapsRequired:1];
    [self.view addGestureRecognizer:tapRecognizer];
}
- (void)handleTapFrom:(UITapGestureRecognizer *)recognizer {
    CGPoint location = [recognizer locationInView:self.view];
    UIView *hitView = [self.view hitTest:location withEvent:nil];
    if ([hitView isKindOfClass:[UIImageView class]]){
        [(UIImageView *)hitView  setImage:[UIImage imageNamed:@"123.png" ]];
    }
    else
    {
        CGRect rect=CGRectMake(location.x - 40, location.y - 40, 80.0f, 80.0f);
        UIImageView *image = [[UIImageView alloc] initWithFrame:rect];
```

```
            [image setImage:[UIImage imageNamed:@"456.png" ]];
            [image setUserInteractionEnabled: YES];
            UIPinchGestureRecognizer *pinchRecognizer =
            [[UIPinchGestureRecognizer alloc]
             initWithTarget:self
             action:@selector(handlePinchFrom:)];
            [image addGestureRecognizer:pinchRecognizer];
            [self.view addSubview:image];

        }
}
- (void)handlePinchFrom:(UIPinchGestureRecognizer *)recognizer{
    CGFloat scale = [recognizer scale];
    CGAffineTransform transform= CGAffineTransformMakeScale(scale, scale);
    recognizer.view.transform = transform;
}
```

执行后的效果如图 10-26 所示。

图 10-26　执行效果

实例 236　自定义触摸手势删除屏幕中的图片

实例 236	自定义触摸手势删除屏幕中的图片
源码路径	\daima\236\

实例说明

除了可以使用 iOS 提供的内置触摸手势外，还可以自定义手势。在本实例中自定义了一个删除手势，当在设备屏幕中触摸某一点时，会在这个位置插入一幅指定的图片。如果继续触摸这个插入的图片，则用另外一幅指定的图片替换，实现开花的效果。使用自定义的删除手势触摸屏幕中的图片时，会删除选中的图片。本实例是一个 iPad 项目，使用的素材图片是"123.png"和"456.png"。

具体实现

在文件 DeleteGestureRecognizer.m 中自定义了删除手势，具体代码如下所示。

```
#import "DeleteGestureRecognizer.h"
@implementation DeleteGestureRecognizer
@synthesize viewToDelete;
- (void)reset {
    [super reset];
    strokeMovingUp = YES;
    touchChangedDirection = 0;
    self.viewToDelete = nil;
}
- (void)touchesBegan:(NSSet *)touches withEvent:(UIEvent *)event {
    [super touchesBegan:touches withEvent:event];
    if ([touches count] != 1) {
        self.state = UIGestureRecognizerStateFailed;
        return;
    }
}
- (void)touchesMoved:(NSSet *)touches withEvent:(UIEvent *)event {
    [super touchesMoved:touches withEvent:event];
```

```objc
        if (self.state == UIGestureRecognizerStateFailed) return;
        CGPoint nowPoint = [[touches anyObject] locationInView:self.view];
        CGPoint prevPoint = [[touches anyObject] previousLocationInView:self.view];
        if (strokeMovingUp == YES) {
            if (nowPoint.y < prevPoint.y ){
                strokeMovingUp = NO;
                touchChangedDirection++;
            }
        } else if (nowPoint.y > prevPoint.y ) {
            strokeMovingUp = YES;
            touchChangedDirection++;
        }
        if (viewToDelete == nil) {
            UIView *hit = [self.view hitTest:nowPoint withEvent:nil];
            if (hit != nil && hit != self.view){
                self.viewToDelete = hit;
            }
        }
    }
    - (void)touchesEnded:(NSSet *)touches withEvent:(UIEvent *)event {
        [super touchesEnded:touches withEvent:event];
        if (self.state == UIGestureRecognizerStatePossible) {
            if (touchChangedDirection >= 3){
                self.state = UIGestureRecognizerStateRecognized;
            }
            else
            {
                self.state = UIGestureRecognizerStateFailed;
            }
        }
    }
    -(void)touchesCancelled:(NSSet *)touches withEvent:(UIEvent *)event {
        [super touchesCancelled:touches withEvent:event];
        [self reset];
        self.state = UIGestureRecognizerStateFailed;
    }
@end
```

实例文件 ViewController.m 调用了自定义的删除手势，主要实现代码如下所示。

```objc
- (void)viewDidLoad
{
    [super viewDidLoad];

    UITapGestureRecognizer *tapRecognizer = [[UITapGestureRecognizer alloc] initWithTarget:self action:@selector(handleTapFrom:)];
    [tapRecognizer setNumberOfTapsRequired:1];
    [self.view addGestureRecognizer:tapRecognizer];
    UISwipeGestureRecognizer *swipeRecognizer =
        [[UISwipeGestureRecognizer alloc]
         initWithTarget:self
         action:@selector(handleSwipeFrom:)];
    swipeRecognizer.direction = UISwipeGestureRecognizerDirectionRight |
    UISwipeGestureRecognizerDirectionLeft;
    [self.view addGestureRecognizer:swipeRecognizer];
    DeleteGestureRecognizer *deleteRecognizer =
    [[DeleteGestureRecognizer alloc]
     initWithTarget:self
     action:@selector(handleDeleteFrom:)];
    [self.view addGestureRecognizer:deleteRecognizer];

}

- (void)handleDeleteFrom:(DeleteGestureRecognizer *)recognizer {
    if (recognizer.state == UIGestureRecognizerStateRecognized){
        UIView *viewToDelete = [recognizer viewToDelete];
        [viewToDelete removeFromSuperview];
    }
}
- (void)handleTapFrom:(UITapGestureRecognizer *)recognizer {
    CGPoint location = [recognizer locationInView:self.view];
    UIView *hitView = [self.view hitTest:location withEvent:nil];
    if ([hitView isKindOfClass:[UIImageView class]]){
```

```
             [(UIImageView *)hitView
              setImage:[UIImage imageNamed:@"123.png" ]];
             //[self makePopSound];
         } else{
             CGRect rect = CGRectMake(location.x - 40,
                                     location.y - 40, 120.0f, 120.0f);
             UIImageView *image = [[UIImageView alloc]
                                   initWithFrame:rect];
             [image setImage:[UIImage imageNamed:@"456.png"]];
             [image setUserInteractionEnabled:YES];
             
             UIPinchGestureRecognizer *pinchRecognizer =
             [[UIPinchGestureRecognizer alloc]
              initWithTarget:self
              action:@selector(handlePinchFrom:)];
             [image addGestureRecognizer:pinchRecognizer];
             
             [self.view addSubview:image];
             
         }
}
- (void)handleSwipeFrom:(UISwipeGestureRecognizer *)recognizer {
    for (UIView *subview in [self.view subviews]) {
        [subview removeFromSuperview];
    }
    [UIView beginAnimations:nil context:nil];
    [UIView setAnimationDuration:.75];
    [UIView setAnimationBeginsFromCurrentState:YES];
    [UIView
     setAnimationTransition:UIViewAnimationTransitionFlipFromLeft
     forView:self.view
     cache:YES];
    [UIView commitAnimations];
}
```

执行本实例后，当在设备屏幕中触摸某一点时，会在这个位置插入一幅指定的图片。如果继续触摸这个插入的图片，则用另外一幅指定的图片替换，实现开花的效果。如果上下触摸某开花图片，会调用自定义的删除手势删除这一幅图片。执行效果如图10-27所示。

图10-27　执行效果

实例 237　触摸屏幕时发出声音

实例 237	触摸屏幕时发出声音
源码路径	\daima\237\

实例说明

在本实例中，当在设备屏幕中触摸某一点时，会在这个位置插入一幅指定的图片。如果继续触摸这个插入的图片，则用另外一幅指定的图片替换，实现开花的效果，在开花的同时会发出一个声音。本实例是一个 iPad 项目，使用的素材图片是"123.png"和"456.png"，声音素材文件是"789.mp3"。

第 10 章 传感器、触摸和交互

具体实现

实例文件 ViewController.m 的主要代码如下所示。

```objc
- (void)didReceiveMemoryWarning
{
    [super didReceiveMemoryWarning];
    // Release any cached data, images, etc that aren't in use.
}
#pragma mark - View lifecycle
- (void)viewDidLoad
{
    [super viewDidLoad];

    NSURL *url = [NSURL fileURLWithPath:
                  [NSString stringWithFormat:@"%@/789.mp3" ,
                   [[NSBundle mainBundle] resourcePath]]];
    NSError *error;
    player = [[AVAudioPlayer alloc] initWithContentsOfURL:url error:&error];
    player.numberOfLoops = 0;

    UITapGestureRecognizer *tapRecognizer =
      [[UITapGestureRecognizer alloc]
       initWithTarget:self
       action:@selector(handleTapFrom:)];
    [tapRecognizer setNumberOfTapsRequired:1];
    [self.view addGestureRecognizer:tapRecognizer];

}
-(void)makePopSound {
    [player play];
}
- (void)handleTapFrom:(UITapGestureRecognizer *)recognizer {
    CGPoint location = [recognizer locationInView:self.view];
    UIView *hitView = [self.view hitTest:location withEvent:nil];
    if ([hitView isKindOfClass:[UIImageView class]]){
        [(UIImageView *)hitView
          setImage:[UIImage imageNamed:@"123.png" ]];
        [self makePopSound];
    } else {
        CGRect rect = CGRectMake(location.x - 40,
                                 location.y - 40, 120.0f, 120.0f);
        UIImageView *image =
        [[UIImageView alloc] initWithFrame:rect];
        [image setImage:[UIImage imageNamed:@"456.png" ]];
        [image setUserInteractionEnabled: YES];
        [self.view addSubview:image];

    }
}
```

执行后的效果如图 10-28 所示。

图 10-28　执行效果

第 11 章 和设备之间的操作实战

在 iOS 应用中,经常需要和硬件设备进行交互,通过获取这些设备的信息可以实现更好的服务。例如及时显示剩余电量、获取硬件配置信息和休眠处理等。在本章的内容中,将通过具体实例的实现过程来详细讲解和设备操作相关的知识,为读者步入本书后面知识的学习打下基础。

实例 238　在屏幕中添加标记

实例 238	在屏幕中添加标记
源码路径	\daima\238\

实例说明

在使用 iOS 设备时,这个功能比较常见。例如,iPone 屏幕中有很多程序图标,iPhone 通常会用一个标记标示出未接电话之类的未读信息。在 iOS 应用中,我们可以使用 applicationIconBadgeNumber 来标示某个程序,当将此属性设置为 1 时,将通过红字来标示指定的程序,这样起到了提醒的效果。如果取消提醒效果,可以将此属性设置为 0。在本实例中,分别设置了添加标记和删除标记的过程。

具体实现

实例文件 UIKitPrjBadge.m 的主要代码如下所示。

```
#import "UIKitPrjBadge.h"
#pragma mark ----- Private Methods Definition -----
@interface UIKitPrjBadge ()
- (void)updateLabel:(UILabel*)label withNumber:(NSInteger)number;
@end
#pragma mark ----- Start Implementation For Methods -----
@implementation UIKitPrjBadge
- (void)dealloc {
  [label_ release];
  [super dealloc];
}
- (void)viewDidLoad {
  [super viewDidLoad];
  label_ = [[UILabel alloc] init];
  label_.frame = self.view.bounds;
  label_.autoresizingMask =
    UIViewAutoresizingFlexibleWidth | UIViewAutoresizingFlexibleHeight;
  label_.textAlignment = UITextAlignmentCenter;
  label_.backgroundColor = [UIColor blackColor];
  label_.textColor = [UIColor whiteColor];
  label_.font = [UIFont systemFontOfSize:128];
  [self.view addSubview:label_];
}
- (void)viewWillAppear:(BOOL)animated {
  [super viewWillAppear:animated];
  badgeNumber_ = [UIApplication sharedApplication].applicationIconBadgeNumber;
  [self updateLabel:label_ withNumber:badgeNumber_];
}
- (void)viewWillDisappear:(BOOL)animated {
  [super viewWillDisappear:animated];
  [UIApplication sharedApplication].applicationIconBadgeNumber = badgeNumber_;
}
- (void)touchesEnded:(NSSet*)touches withEvent:(UIEvent*)event {
  if ( 1 < [[touches anyObject] tapCount] ) {
    badgeNumber_ = 0;
```

```
    } else {
      ++badgeNumber_;
    }
    [self updateLabel:label_ withNumber:badgeNumber_];
}
#pragma mark ----- Private Methods -----
- (void)updateLabel:(UILabel*)label withNumber:(NSInteger)number {
    label.text = [NSString stringWithFormat:@"%d", number];
}
@end
```

执行效果如图 11-1 所示。

图 11-1　执行效果

实例 239　调用外部程序

实例 239	调用外部程序
源码路径	\daima\239\

实例说明

在 iOS 应用中，经常需要调用外部应用程序，例如谷歌地图、邮件系统和拨号处理等。在本实例中，使用 UIApplication 中的方法 openURL 调用了如下外部应用程序：

- Web Links；
- Mail Links；
- Phone Links；
- Text Links；
- Map Links；
- iTunes Links；
- Customize Links。

具体实现

实例文件 UIKitPrjLinks.m 的主要代码如下所示。

```
#import "UIKitPrjLinks.h"
@implementation UIKitPrjLinks
- (void)dealloc {
    [dataSource_ release];
    [urls_ release];
    [super dealloc];
}
- (void)viewDidLoad {
    [super viewDidLoad];
    dataSource_ = [[NSArray alloc] initWithObjects:
                    @"Web Links",
                    @"Mail Links",
                    @"Phone Links",
                    @"Text Links",
                    @"Map Links",
                    @"iTunes Links",
                    @"Customize Links",
```

```objc
                        nil ];
  urls_ = [[NSArray alloc] initWithObjects:
                        @"http://www.apple.com/",
@"mailto:monster@n-i-l-x.example.com?subject=hello&body=hello%20mail!",
                        @"tel:150-0000-0000",
                        @"sms:150-0000-0000",
@"http://maps.google.com/maps?ll=31.143289%2C121.347239&z=15",
//@"http://www.tudou.com/v/jO7ei5KqCeA/&resourceId=0_04_05_99/v.swf",
@"http://phobos.apple.com/WebObjects/MZStore.woa/wa/viewSoftware?id=441898703&mt=8",
                        @"sampleapp://sampleapp.yourcompany.com/?Hello",
                        nil ];
}
- (NSInteger)tableView:(UITableView*)tableView numberOfRowsInSection:(NSInteger)section {
  return [dataSource_ count];
}
- (UITableViewCell*)tableView:(UITableView*)tableView
  cellForRowAtIndexPath:(NSIndexPath*)indexPath
{
  static NSString* identifier = @"basis-cell";
  UITableViewCell* cell = [tableView dequeueReusableCellWithIdentifier:identifier];
  if ( nil == cell ) {
    cell = [[UITableViewCell alloc] initWithStyle:UITableViewCellStyleDefault
                        reuseIdentifier:identifier];
    [cell autorelease];
  }
  cell.textLabel.text = [dataSource_ objectAtIndex:indexPath.row];
  return cell;
}
- (void)tableView:(UITableView*)tableView didSelectRowAtIndexPath:
(NSIndexPath*)indexPath {
  NSString* urlString = [urls_ objectAtIndex:indexPath.row];
  NSURL* url = [NSURL URLWithString:urlString];
  UIApplication* application = [UIApplication sharedApplication];
  if ( [application canOpenURL:url] ) {
    [application openURL:url];
  }else {
      //URL 不正确时的处理
  }
}
@end
```

执行效果如图 11-2 所示，触摸屏幕中的某个链接后会调用对应的应用程序。

图 11-2　执行效果

实例 240　使用接近传感器

实例 240	使用接近传感器
源码路径	\daima\240\

实例说明

在 iOS 应用中，靠近传感器是代替限位开关等接触式检测方式，以无需接触检测对象进行检测为目的的传感器的总称。例如在通话过程中,当 iPhone 放在耳边时,手机屏幕中的画面会关闭,这样起到了省电的作用。当 iPhone 离开耳边时,画面会再打开。这个过程就是用到了接近传感器。在 iOS 应用中,在默认情况下,接近传感器是关闭的,这一功能是通过将类 UIDevice 的属性 proximityMonitoringEnabled 设置为 NO 实现的。在本实例中,将此属性设置为了 YES,演示了使用接近传感器的方法。

具体实现

实例文件 UIKitPrjProximityMonitoring.m 的主要代码如下所示。

```objc
#import "UIKitPrjProximityMonitoring.h"
#pragma mark ----- Private Methods Definition -----
@interface UIKitPrjProximityMonitoring ()
- (void)onDidPush;
- (void)offDidPush;
- (void)proximityStateDidChange;
@end
#pragma mark ----- Start Implementation For Methods -----
@implementation UIKitPrjProximityMonitoring
- (void)dealloc {
  [label_ release];
  [super dealloc];
}
- (void)viewDidLoad {
  [super viewDidLoad];
  label_ = [[UILabel alloc] init];
  label_.frame = self.view.bounds;
  label_.textAlignment = UITextAlignmentCenter;
  label_.autoresizingMask =
    UIViewAutoresizingFlexibleWidth | UIViewAutoresizingFlexibleHeight;
  label_.backgroundColor = [UIColor whiteColor];
  [self.view addSubview:label_];
}
//在画面显示后向 NSNotificationCenter 中注册检测到接近传感器变成 ON / OFF 时的处理方法
- (void)viewWillAppear:(BOOL)animated {
  [super viewWillAppear:animated];
  [self onDidPush];
  [[NSNotificationCenter defaultCenter] addObserver:self
                            selector:@selector(proximityStateDidChange)
                                name:UIDeviceProximityStateDidChangeNotification
                              object:nil];
}
#pragma mark ----- Private Methods -----

- (void)onDidPush {
  UIBarButtonItem* button =
    [[[UIBarButtonItem alloc] initWithTitle:@"接近 OFF"
                                  style:UIBarButtonItemStyleBordered
                                 target:self
                                 action:@selector(offDidPush)] autorelease];
  self.navigationItem.rightBarButtonItem = button;
  // 接近传感器设置为 ON
  [UIDevice currentDevice].proximityMonitoringEnabled = NO;
  label_.text = @"";
}
- (void)offDidPush {
  UIBarButtonItem* button =
    [[[UIBarButtonItem alloc] initWithTitle:@"接近 ON"
                                  style:UIBarButtonItemStyleDone
                                 target:self
                                 action:@selector(onDidPush)] autorelease];
  self.navigationItem.rightBarButtonItem = button;
  // 接近传感器 ON
  [UIDevice currentDevice].proximityMonitoringEnabled = YES;
}
- (void)proximityStateDidChange {
  if ( [UIDevice currentDevice].proximityState ) {
```

```
         //label_为预先在画面中追加的 UILabel 实例
         label_.text = @"已经变暗了一";
    }
}
@end
```

执行后的效果如图 11-3 所示。

图 11-3 执行效果

实例 241 获取电池的状态

实例 241	获取电池的状态
源码路径	\daima\241\

实例说明

在 iOS 应用中,可以使用类 UIDevice 获取电池的状态,当需要实现这一功能时,需要将属性 batteryMonitoringEnabled 设置为 YES。并且可以使用属性 batteryLevel 获取电池的剩余电量,可以使用属性 batteryState 获取充电状态。在本实例中,演示了获取电池状态的基本方法。

具体实现

实例文件 UIKitPrjBatteryMonitor.m 的主要代码如下所示。

```
#import "UIKitPrjBatteryMonitor.h"
#pragma mark ----- Private Methods Definition -----
@interface UIKitPrjBatteryMonitor ()
- (void)refreshDidPush;
- (NSString*)batteryStateToString:(UIDeviceBatteryState)state;
@end
#pragma mark ----- Start Implementation For Methods -----
@implementation UIKitPrjBatteryMonitor
- (void)dealloc {
  [textView_ release];
  [super dealloc];
}
- (void)viewDidLoad {
  [super viewDidLoad];
  textView_ = [[UITextView alloc] init];
  textView_.editable = NO;
  textView_.frame = self.view.bounds;
  textView_.autoresizingMask =
    UIViewAutoresizingFlexibleWidth | UIViewAutoresizingFlexibleHeight;
  textView_.backgroundColor = [UIColor blackColor];
  textView_.textColor = [UIColor whiteColor];
  textView_.font = [UIFont systemFontOfSize:16];
  [self.view addSubview:textView_];
  UIBarButtonItem* button =
    [[UIBarButtonItem alloc] initWithBarButtonSystemItem:UIBarButtonSystemItemRefresh
                                       target:self
                                       action:@selector(refreshDidPush)];
  self.navigationItem.rightBarButtonItem = button;
  [button release];
  [UIDevice currentDevice].batteryMonitoringEnabled = YES;
  [self refreshDidPush];
  [[NSNotificationCenter defaultCenter] addObserver:self
                                       selector:@selector(refreshDidPush)
name:UIDeviceBatteryLevelDidChangeNotification
                                       object:nil];
  [[NSNotificationCenter defaultCenter] addObserver:self
```

```
                       selector:@selector(refreshDidPush)
name:UIDeviceBatteryStateDidChangeNotificationobject:nil];
}
- (void)refreshDidPush {
  UIDevice* device = [UIDevice currentDevice];
  NSMutableString* text = [[NSMutableString alloc] initWithCapacity:1024];
  [text appendFormat:@"batteryState: %@\n", [self batteryStateToString:device.
batteryState]];
  [text appendFormat:@"batteryLevel: %f\n", device.batteryLevel];
  textView_.text = text;
}
- (NSString*)batteryStateToString:(UIDeviceBatteryState)state {
  switch ( state ) {
    case UIDeviceBatteryStateUnplugged: return @"UIDeviceBatteryStateUnplugged";
    case UIDeviceBatteryStateCharging: return @"UIDeviceBatteryStateCharging";
    case UIDeviceBatteryStateFull: return @"UIDeviceBatteryStateFull";
    default: return @"UIDeviceBatteryStateUnknown";
  }
}
@end
```

因为是在模拟器中运行的，所以执行效果如图 11-4 所示。

图 11-4 执行效果

实例 242 获取系统信息

实例 242	获取系统信息
源码路径	\daima\242\

实例说明

在 iOS 应用中，可以使用类 UIDevice 获取当前系统的各种中断信息，例如系统版本和信号等信息。这些功能是通过如下方法实现的。

- localizedModel：本地型号名。
- model：型号名。
- name：用户自定义终端名。
- systemName：OS 名。
- systemVersion：OS 的版本。

在本实例中，演示了使用上述方法获取系统信息的基本方法。

具体实现

实例文件 UIKitPrjDeviceInfo.m 的主要代码如下所示。

```
#import "UIKitPrjDeviceInfo.h"
@implementation UIKitPrjDeviceInfo
- (void)dealloc {
  [titles_ release];
  [datas_ release];
  [super dealloc];
}
- (void)viewDidLoad {
  [super viewDidLoad];
  titles_ = [[NSArray alloc] initWithObjects:
                 @"本地型号名",
```

```
                    @"型号名",
                    @"用户自定义终端名",
                    @"OS 名",
                    @"OS 的版本",
                    nil];
    UIDevice* device = [UIDevice currentDevice];
    datas_ = [[NSArray alloc] initWithObjects:
                    device.localizedModel,
                    device.model,
                    device.name,
                    device.systemName,
                    device.systemVersion,
                    nil];
}
- (NSInteger)tableView:(UITableView*)tableView numberOfRowsInSection:(NSInteger)section {
    return [titles_ count];
}
- (UITableViewCell*)tableView:(UITableView*)tableView
    cellForRowAtIndexPath:(NSIndexPath*)indexPath
{
    static NSString* identifier = @"basis-cell";
    UITableViewCell* cell = [tableView dequeueReusableCellWithIdentifier:identifier];
    if ( nil == cell ) {
        cell = [[UITableViewCell alloc] initWithStyle:UITableViewCellStyleValue1
                            reuseIdentifier:identifier];
        [cell autorelease];
    }
    cell.textLabel.text = [titles_ objectAtIndex:indexPath.row];
    cell.detailTextLabel.text = [datas_ objectAtIndex:indexPath.row];
    return cell;
}
@end
```

执行后的效果如图 11-5 所示。

图 11-5 执行效果

实例 243　获取设备的终端识别符

实例 243	获取设备的终端识别符
源码路径	\daima\243\

实例说明

在 iOS 应用中，通常使用 iPhone 的终端识别符作为代替用户 ID 的元素来使用。在当今市面中的每一台 iPhone、iPad 等设备都有一个唯一的识别符。在获取识别符时，可以通过属性 uniqueIdentifier 来实现。在本实例中，演示了通过属性 uniqueIdentifier 获取设备的终端识别符的基本方法。

具体实现

实例文件 UIKitPrjUniqueIdentifier.m 的主要代码如下所示。

```
#import "UIKitPrjUniqueIdentifier.h"
@implementation UIKitPrjUniqueIdentifier
- (void)dealloc {
```

```objectivec
    [super dealloc];
}
- (void)viewDidLoad {
    [super viewDidLoad];
    UILabel* label = [[[UILabel alloc] init] autorelease];
    label.frame = self.view.bounds;
    label.autoresizingMask =
      UIViewAutoresizingFlexibleWidth | UIViewAutoresizingFlexibleHeight;
    label.numberOfLines = 4;
    NSString* identifier = [UIDevice currentDevice].uniqueIdentifier;
    label.text = [NSString stringWithFormat:@"识别符是: %@", identifier];
    label.font = [UIFont systemFontOfSize:36];
    [self.view addSubview:label];
}
- (void)copy:(id)sender {
    [UIPasteboard generalPasteboard].string = [UIDevice currentDevice].uniqueIdentifier;
}
- (void)touchesBegan:(NSSet*)touches withEvent:(UIEvent*)event {
    if ( [self becomeFirstResponder] ) {
        UITouch* touch = [touches anyObject];
        CGPoint point = [touch locationInView:self.view];
        UIMenuController* menu = [UIMenuController sharedMenuController];
        CGRect minRect;
        minRect.origin = point;
        [menu setTargetRect:minRect inView:self.view];
        [menu setMenuVisible:YES animated:YES];
    }
}
- (BOOL)canBecomeFirstResponder {
    return YES;
}
- (BOOL)canPerformAction:(SEL)action withSender:(id)sender {
    if ( @selector(copy:) == action ) {
        return YES;
    }
    return [super canPerformAction:action withSender:sender];
}
@end
```

执行效果如图 11-6 所示。

图 11-6　执行效果

实例 244　设置一个复制菜单

实例 244	设置一个复制菜单
源码路径	\daima\244\

实例说明

　　在 iOS 应用中有两个系统粘贴板，其中通用系统粘贴板用于复制/粘贴操作，而查找粘贴板用于保存上一次搜寻用的字符串。另外，应用程序还可以创建自己的粘贴板，甚至可以被其他应用程序使用。iOS 中的所有粘贴板都是 UIPasteboard 的子类，最常用的粘贴板操作包括获取/设置字符串、图像、URL 和颜色。Apple 提供了如下实现复制、粘贴功能的方法。

```
NSString *string = pasteboard.string;
UIImage *image = pasteboard.image;
NSURL *url = pasteboard.URL;
UIColor *color = pasteboard.color;
pasteboard.string = @"paste me somewhere";
```

实例 244 设置一个复制菜单

在本实例中,演示了通过 UIPasteboard 复制文本的基本方法,并且自定义了一个"复制"菜单。

具体实现

实例文件 UIKitPrjCopyBase.m 的主要代码如下所示。

```objc
#import "UIKitPrjCopyBase.h"
@implementation UIKitPrjCopyBase
- (void)viewDidLoad {
    [super viewDidLoad];
    label_ = [[UILabel alloc] init];
    label_.frame = [[UIScreen mainScreen] bounds];
    label_.font = [UIFont boldSystemFontOfSize:48];
    label_.numberOfLines = 3;
    label_.text = @"将这些字符串复制到粘贴板中";
    [self.view addSubview:label_];
}
//不成为第一响应者无法显示编辑菜单
- (BOOL)canBecomeFirstResponder {
    return YES;
}
- (void)touchesEnded:(NSSet*)touches withEvent:(UIEvent*)event {
    UITouch* touch = [touches anyObject];
    if ( [self becomeFirstResponder] && 1 < [touch tapCount] ) {
        // 连续两次触碰后显示编辑菜单
        UIMenuController* menu = [UIMenuController sharedMenuController];
        CGPoint touchPoint_ = [touch locationInView:self.view];
        CGRect minRect;
        minRect.origin = touchPoint_;
        [menu setTargetRect:minRect inView:self.view];
        [menu setMenuVisible:YES animated:YES];
    }
}
//使复制命令可被使用
- (BOOL)canPerformAction:(SEL)action withSender:(id)sender {
    if ( @selector(copy:) == action ) {
        return YES;
    }
    return [super canPerformAction:action withSender:sender];
}
//执行复制命令时被调用
//label_为 UILabel 的实例
- (void)copy:(id)sender {
    [UIPasteboard generalPasteboard].string = label_.text;
}
- (void)didReceiveMemoryWarning {
  // Releases the view if it doesn't have a superview.
  [super didReceiveMemoryWarning];

  // Release any cached data, images, etc. that aren't in use.
}
- (void)viewDidUnload {
  [super viewDidUnload];
  // Release any retained subviews of the main view.
  // e.g. self.myOutlet = nil;
}
- (void)dealloc {
  [super dealloc];
}
@end
```

执行后的效果如图 11-7 所示。

图 11-7 执行效果

实例 245 复制/剪切/粘贴屏幕中的图片

实例 245	复制/剪切/粘贴屏幕中的图片
源码路径	\daima\245\

实例说明

在 iOS 应用中除了可以复制屏幕中的文本外，还可以复制/粘贴屏幕中的图片。在本实例中，首先设置了 3 幅素材图片："bug1.png"、"bug2.png" 和 "bug3.png"，然后使用方法 touchesEnded:withEvent:自定义了复制和粘贴菜单，最后使用 copy 方法复制屏幕中的图片，使用 cut 方法剪切屏幕中的图片，使用 paste 方法粘贴屏幕中的图片。

具体实现

实例文件 UIKitPrjCopyAndPaste.m 的主要代码如下所示。

```objc
#import "UIKitPrjCopyAndPaste.h"
//定义私有方法
@interface UIKitPrjCopyAndPaste ()
- (UIImageView*)imageContainsPoint:(CGPoint)point;
@end
@implementation UIKitPrjCopyAndPaste
- (void)viewWillAppear:(BOOL)animated {
  [super viewWillAppear:animated];
  self.view.backgroundColor = [UIColor whiteColor];
  UIImage* image;
  image = [UIImage imageNamed:@"bug1.png"];
  UIImageView* bug1 = [[UIImageView alloc] initWithImage:image];
  bug1.center = self.view.center;
  [self.view addSubview:bug1];
  [bug1 release];
  image = [UIImage imageNamed:@"bug2.png"];
  UIImageView* bug2 = [[UIImageView alloc] initWithImage:image];
  bug2.center = CGPointMake( 200, 300 );
  [self.view addSubview:bug2];
  [bug2 release];
  image = [UIImage imageNamed:@"bug3.png"];
  UIImageView* bug3 = [[UIImageView alloc] initWithImage:image];
  bug3.center = CGPointMake( 50, 80 );
  [self.view addSubview:bug3];
  [bug3 release];
}
//不成为第一响应者无法显示编辑菜单
- (BOOL)canBecomeFirstResponder {
  return YES;
}
- (void)touchesEnded:(NSSet*)touches withEvent:(UIEvent*)event {
  UITouch* touch = [touches anyObject];
  if ( [self becomeFirstResponder] && 1 < [touch tapCount] ) {
    // 连续两次触碰后显示编辑菜单
    UIMenuController* menu = [UIMenuController sharedMenuController];
    touchPoint_ = [touch locationInView:self.view];
    CGRect minRect;
    minRect.origin = touchPoint_;
    [menu setTargetRect:minRect inView:self.view];
    [menu setMenuVisible:YES animated:YES];
  }
}
- (BOOL)canPerformAction:(SEL)action withSender:(id)sender {
  if ( @selector(copy:) == action ) {
    if ( [self imageContainsPoint:touchPoint_] ) {
      return YES;
    }
  } else if ( @selector(cut:) == action ) {
    if ( [self imageContainsPoint:touchPoint_] ) {
```

```objc
      return YES;
    }
  } else if ( @selector(paste:) == action ) {
    return ( nil != [UIPasteboard generalPasteboard].image );
  }
  return NO;
}
- (UIImageView*)imageContainsPoint:(CGPoint)point {
  for ( UIView* view in self.view.subviews ) {
    if ( CGRectContainsPoint( view.frame, point ) ) {
      if ( [view isKindOfClass:[UIImageView class]] ) {
        return (UIImageView*)view;
      }
    }
  }
  return nil;
}
- (void)copy:(id)sender {
  UIImageView* imageView = [self imageContainsPoint:touchPoint_];
  if ( imageView ) {
    [UIPasteboard generalPasteboard].image = imageView.image;
  }
}
- (void)cut:(id)sender {
  UIImageView* imageView = [self imageContainsPoint:touchPoint_];
  if ( imageView ) {
    [UIPasteboard generalPasteboard].image = imageView.image;
    [imageView removeFromSuperview];
  }
}
- (void)paste:(id)sender {
  UIPasteboard* pasteBoard = [UIPasteboard generalPasteboard];
  if ( pasteBoard.image ) {
    UIImageView* bug =
      [[[UIImageView alloc] initWithImage:pasteBoard.image] autorelease];
    bug.center = touchPoint_;
    [self.view addSubview:bug];
    pasteBoard.image = bug.image;
  }
}
@end
```

执行结果如图 11-8 所示。

图 11-8 执行效果

实例 246　在粘贴板中保存自定义类

实例 246	在粘贴板中保存自定义类
源码路径	\daima\246\

实例说明

　　在 iOS 应用中，因为可以使用类 UIPasteboard 的方法 setData:forPasteboardType:在粘贴板中保存 NSData 类型的数据。所以在本实例中特意自定义了一个类，然后将此自定义类转换成了 NSData 类型，并在粘贴板中进行了保存。

具体实现

实例文件 UIKitPrjSaveAnyClass.m 的主要代码如下所示。

```objc
#import "UIKitPrjSaveAnyClass.h"
#import "TitleAndBody.h"
#define kPasteboardTypeSample @"sample"
@implementation UIKitPrjSaveAnyClass
- (void)viewDidLoad {
    [super viewDidLoad];
    UIPasteboard *pasteboard = [UIPasteboard generalPasteboard];
    TitleAndBody *object = [[TitleAndBody alloc] init];
    object.title = @"tit";
    object.body  = @"bod";
    //将自定义转换成 NSData
    NSData *data = [NSKeyedArchiver archivedDataWithRootObject:object];
    [object release];
    //指定类型名将 NSData 保存于剪贴板
    [pasteboard setData:data forPasteboardType:kPasteboardTypeSample];
    //指定类型名,从剪贴板中读取 NSData
    if ([pasteboard containsPasteboardTypes:[NSArray arrayWithObject:kPasteboardTypeSample]]) {
        NSData *pasteData = [pasteboard dataForPasteboardType:kPasteboardTypeSample];
        //将 NSData 复原成自定义类
        TitleAndBody *pasteObj = [NSKeyedUnarchiver unarchiveObjectWithData:pasteData];
        NSLog(pasteObj.title);
        NSLog(pasteObj.body);
    }
}
- (void)didReceiveMemoryWarning {
    [super didReceiveMemoryWarning];
}
- (void)viewDidUnload {
    [super viewDidUnload];
}
- (void)dealloc {
    [super dealloc];
}
@end
```

实例 247　获取电池的详细信息

实例 247	获取电池的详细信息
源码路径	\daima\247\

实例说明

在本章前面的实例 241 中,已经演示了获取电池信息的基本方法。在本实例中,将进一步讲解获取 iOS 设备电池信息的方法,本实例将用比较美观的界面显示获取的信息,视图界面效果如图 11-9 所示。

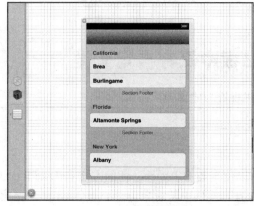

图 11-9　视图界面效果

具体实现

实例文件 RootViewController.m 的主要代码如下所示。

```objc
#import "RootViewController.h"
@implementation RootViewController
enum ControlTableSections
{
    kMonitoringSection = 0,
    kLevelSection,
    kBatteryStateSection
};
- (void)dealloc
{
    [numberFormatter release];
    [super dealloc];
}
- (void)viewDidLoad
{
    [super viewDidLoad];
    // 此标题将显示在导航栏
    self.title = NSLocalizedString(@"电池状态", @"");
    // 注册为电池的电量和状态更改通知
    [[NSNotificationCenter defaultCenter] addObserver:self selector:@selector(batteryLevelDidChange:)
        name:UIDeviceBatteryLevelDidChangeNotification object:nil];

    [[NSNotificationCenter defaultCenter] addObserver:self selector:@selector(batteryStateDidChange:)
        name:UIDeviceBatteryStateDidChangeNotification object:nil];
}
- (void)didReceiveMemoryWarning
{
    // 释放视图
    [super didReceiveMemoryWarning];
    //释放数据
}
- (void)viewDidUnload
{
}
- (NSNumberFormatter *)numberFormatter
{
    if (numberFormatter == nil)
    {
        numberFormatter = [[NSNumberFormatter alloc] init];
        [numberFormatter setNumberStyle:NSNumberFormatterPercentStyle];
        [numberFormatter setMaximumFractionDigits:1];
    }
    return numberFormatter;
}
#pragma mark - Switch action handler

- (void)switchAction:(id)sender
{
    if ([sender isOn])
    {
        [UIDevice currentDevice].batteryMonitoringEnabled = YES;
        //更新用户界面
    }
    else {
        [UIDevice currentDevice].batteryMonitoringEnabled = NO;
        [self.tableView reloadData];
    }
}
#pragma mark - Battery notifications
- (void)batteryLevelDidChange:(NSNotification *)notification
{
    [self.tableView reloadData];
}
- (void)batteryStateDidChange:(NSNotification *)notification
```

```objc
{
    [self.tableView reloadData];
}
#pragma mark - UITableView delegates
- (NSInteger)numberOfSectionsInTableView:(UITableView *)tableView
{
    return 3;
}
- (NSString *)tableView:(UITableView *)tableView titleForHeaderInSection:(NSInteger)section
{
    NSString *title = nil;
    switch (section)
    {
        case kBatteryStateSection:
        {
            title = NSLocalizedString(@"电池状态", @"");
            break;
        }
    }
    return title;
}
- (NSInteger)tableView:(UITableView *)tableView numberOfRowsInSection:(NSInteger)section
{
    NSInteger rowCount = 1;
    
    if (section == kBatteryStateSection)
    {
        rowCount = 4;
    }
    
    return rowCount;
}
static NSInteger kLevelTag = 2;
// Customize the appearance of table view cells.
- (UITableViewCell *)tableView:(UITableView *)tableView cellForRowAtIndexPath:(NSIndexPath *)indexPath
{
    static NSString *kMonitoringCellIdentifier = @"Monitoring";
    static NSString *kLevelCellIdentifier = @"Level";
    static NSString *kStateCellIdentifier = @"State";
    
    UITableViewCell *cell = nil;
    
    switch (indexPath.section)
    {
        case kMonitoringSection:
        {
            cell = [tableView dequeueReusableCellWithIdentifier:kMonitoringCellIdentifier];
            if (cell == nil)
            {
                cell = [[[UITableViewCell alloc] initWithStyle:UITableViewCellStyleDefault reuseIdentifier:kMonitoringCellIdentifier] autorelease];
                cell.textLabel.text = NSLocalizedString(@"Monitoring", @"");
                
                UISwitch *switchCtl = [[[UISwitch alloc] initWithFrame:CGRectMake(197, 8, 94, 27)] autorelease];
                [switchCtl addTarget:self action:@selector(switchAction:)
                    forControlEvents:UIControlEventValueChanged];
                switchCtl.backgroundColor = [UIColor clearColor];
                
                [cell.contentView addSubview:switchCtl];
            }
            
            break;
        }
        case kLevelSection:
        {
            cell = [tableView dequeueReusableCellWithIdentifier:kLevelCellIdentifier];
            UILabel *levelLabel = nil;
            if (cell == nil)
```

```objc
            {
                cell = [[[UITableViewCell alloc] initWithStyle:UITableViewCell Style-
Default reuseIdentifier:kLevelCellIdentifier] autorelease];
                cell.selectionStyle = UITableViewCellSelectionStyleNone;
                cell.textLabel.text = NSLocalizedString(@"Level", @"");

                levelLabel = [[[UILabel alloc] initWithFrame:CGRectMake(171, 11,
                120, 21)] autorelease];
                levelLabel.tag = kLevelTag;
                levelLabel.textAlignment = UITextAlignmentRight;
                [cell.contentView addSubview:levelLabel];
                levelLabel.backgroundColor = [UIColor clearColor];
            }
            else {
                levelLabel = (UILabel *) [cell.contentView viewWithTag:kLevelTag];
            }

            float batteryLevel = [UIDevice currentDevice].batteryLevel;
            if (batteryLevel < 0.0)
            {
                // -1.0 表示点知状态未知
                levelLabel.text = NSLocalizedString(@"Unknown", @"");
            }
            else {
                NSNumber *levelObj = [NSNumber numberWithFloat:batteryLevel];

                //当第一次使用时，使用 numberFormatter 属性创建对象
                levelLabel.text = [self.numberFormatter stringFromNumber:levelObj];
            }
            break;
        }
        case kBatteryStateSection:
        {
            cell = [tableView dequeueReusableCellWithIdentifier:kStateCellIdentifier];
            if (cell == nil)
            {
                cell = [[[UITableViewCell alloc] initWithStyle:UITableViewCell
                StyleDefault reuseIdentifier:kStateCellIdentifier] autorelease];
                cell.selectionStyle = UITableViewCellSelectionStyleNone;

                cell.accessoryView = [[[UIImageView alloc] initWithImage:[UIImage
                imageNamed:@"StatusClear.png"]] autorelease];
            }

            switch (indexPath.row)
            {
                case 0:
                {
                    cell.textLabel.text = NSLocalizedString(@"Unknown", @"");
                    break;
                }
                case 1:
                {
                    cell.textLabel.text = NSLocalizedString(@"Unplugged", @"");
                    break;
                }
                case 2:
                {
                    cell.textLabel.text = NSLocalizedString(@"Charging", @"");
                    break;
                }
                case 3:
                {
                    cell.textLabel.text = NSLocalizedString(@"Full", @"");
                    break;
                }
            }

            UIImageView *statusImageView = (UIImageView *) cell.accessoryView;
            if (indexPath.row + UIDeviceBatteryStateUnknown == [UIDevice current
            Device].batteryState)
            {
```

```
                statusImageView.image = [UIImage imageNamed:@"StatusGreen.png"];
            }
            else {
                statusImageView.image = [UIImage imageNamed:@"StatusClear.png"];
            }
            break;
    }
    // 设置属性
    cell.selectionStyle = UITableViewCellSelectionStyleNone;
    return cell;
}
@end
```

因为是在模拟器中运行的，所以执行效果如图 11-10 所示。

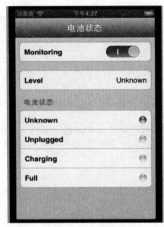

图 11-10　执行效果

实例 248　获取 iPhone 的硬件版本以及系统信息

实例 248	获取 iPhone 的硬件版本以及系统信息
源码路径	\daima\248\

实例说明

在本章前面的实例 242 中，已经演示了获取系统信息的基本方法。在本实例中，将进一步讲解获取 iOS 设备的硬件版本以及系统信息的方法，本实例的视图界面效果如图 11-11 所示。

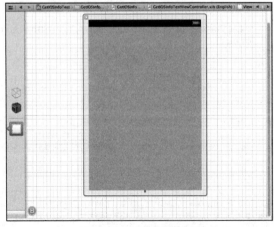

图 11-11　视图界面效果

具体实现

实例文件 UIDeviceHardware.m 用于获取硬件信息和系统信息，具体实现代码如下所示。

```objc
#import "UIDeviceHardware.h"
#include <sys/types.h>
#include <sys/sysctl.h>
@implementation UIDeviceHardware
- (NSString *) platform{
    size_t size;
    sysctlbyname("hw.machine", NULL, &size, NULL, 0);
    char *machine = malloc(size);
    sysctlbyname("hw.machine", machine, &size, NULL, 0);
    NSString *platform = [NSString stringWithCString:machine encoding:NSUTF8StringEncoding];
    free(machine);
    return platform;
}
- (NSString *) platformString{
    NSString *platform = [self platform];
    if ([platform isEqualToString:@"iPhone1,1"])    return @"iPhone 1G";
    if ([platform isEqualToString:@"iPhone1,2"])    return @"iPhone 3G";
    if ([platform isEqualToString:@"iPhone2,1"])    return @"iPhone 3GS";

    if ([platform isEqualToString:@"iPhone3,1"])    return @"iPhone 4";
    if ([platform isEqualToString:@"iPhone4,1"])    return @"iPhone 4S";
    if ([platform isEqualToString:@"iPhone5,1"])    return @"iPhone 4S";
    if ([platform isEqualToString:@"iPhone6,1"])    return @"iPhone 5";
    if ([platform isEqualToString:@"iPod1,1"])      return @"iPod Touch 1G";
    if ([platform isEqualToString:@"iPod2,1"])      return @"iPod Touch 2G";
    if ([platform isEqualToString:@"iPod3,1"])      return @"iPod Touch 3G";
    if ([platform isEqualToString:@"iPod4,1"])      return @"iPod Touch 4G";
    if ([platform isEqualToString:@"iPad1,1"])      return @"iPad";
    if ([platform isEqualToString:@"i386"] || [platform isEqualToString:@"x86_64"])  return @"iPhone Simulator";
    return platform;
}
@end
```

实例文件 GetIOSInfoTestViewController.m 用于在屏幕中输出当前设备的硬件信息和系统信息，具体实现代码如下所示。

```objc
#import "GetIOSInfoTestViewController.h"
#import "UIDeviceHardware.h"
@implementation GetIOSInfoTestViewController
- (void)dealloc
{
    [super dealloc];
}
- (void)didReceiveMemoryWarning
{
    // 释放视图资源
    [super didReceiveMemoryWarning];
    // 释放数据
}
#pragma mark - View lifecycle
// Implement viewDidLoad to do additional setup after loading the view, typically from a nib.
- (void)viewDidLoad
{
    [super viewDidLoad];
    UILabel *label = [[[UILabel alloc] initWithFrame:CGRectMake(20, 10, 280, 20)] autorelease];
    label.text = [@"systemName : " stringByAppendingString:[[UIDevice currentDevice] systemName]];
    [self.view addSubview:label];
    label = [[[UILabel alloc] initWithFrame:CGRectMake(20, 40, 280, 20)] autorelease];
    label.text = [@"systemVersion : " stringByAppendingString:[[UIDevice currentDevice] systemVersion]];
    [self.view addSubview:label];

    label = [[[UILabel alloc] initWithFrame:CGRectMake(20, 70, 280, 20)] autorelease];
```

```objc
    label.text = [@"model : " stringByAppendingString:[[UIDevice currentDevice] model]];
    [self.view addSubview:label];
    
    label = [[[UILabel alloc] initWithFrame:CGRectMake(20, 100, 280, 20)] autorelease];
    label.text = [@"id : " stringByAppendingString:[[UIDevice currentDevice] uniqueIdentifier]];
    label.font = [UIFont systemFontOfSize:10];
    [self.view addSubview:label];
    
    label = [[[UILabel alloc] initWithFrame:CGRectMake(20, 130, 280, 20)] autorelease];
    label.text = [@"name : " stringByAppendingString:[[UIDevice currentDevice] name]];
    [self.view addSubview:label];
    
    label = [[[UILabel alloc] initWithFrame:CGRectMake(20, 160, 280, 20)] autorelease];
    label.text = [@"localizedModel : " stringByAppendingString:[[UIDevice currentDevice] localizedModel]];
    [self.view addSubview:label];
    UIDeviceHardware *d = [[[UIDeviceHardware alloc] init] autorelease];
    label = [[[UILabel alloc] initWithFrame:CGRectMake(20, 190, 280, 20)] autorelease];
    label.text = [@"platform : " stringByAppendingString:[d platform]];
    [self.view addSubview:label];
    label = [[[UILabel alloc] initWithFrame:CGRectMake(20, 220, 280, 20)] autorelease];
    label.text = [@"platformString : " stringByAppendingString:[d platformString]];
    [self.view addSubview:label];
}
- (void)viewDidUnload
{
    [super viewDidUnload];
    
}
- (BOOL)shouldAutorotateToInterfaceOrientation:(UIInterfaceOrientation)interfaceOrientation
{
    // 返回支持的方向
    return (interfaceOrientation == UIInterfaceOrientationPortrait);
}
@end
```

执行效果如图 11-12 所示。

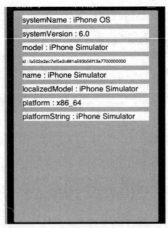

图 11-12 执行效果

第 12 章　游戏应用实战

自从手持设备诞生以来，游戏就成为了最重要的应用之一。无论是在旅行和上下班的路上，还是躺在家中的床上，用户都可以用手机游戏来打发无聊的时间。本章将通过几个典型实例的实现过程，详细介绍在 iOS 系统中实现游戏项目的基本知识。

实例 249　实现一个连连看游戏

实例 249	实现一个连连看游戏
源码路径	\daima\249\

实例说明

《连连看》游戏的规则比较简单，只要将相同的两张牌用 3 条以内的直线连在一起就可以消除。整个游戏速度节奏快，画面清晰可爱，适合细心的玩家。丰富的道具和公共模式的加入，增强了游戏的竞争性。多样式的地图使玩家在各个游戏水平都可以寻找到挑战的目标，长期地保持游戏的新鲜感。在本实例的内容中，将通过一个连连看游戏的实现过程，详细讲解使用 Xcode 集成开发环境，并使用 Objective-C 语言开发 iPhone 手机游戏项目的基本流程。在讲解过程中，首先统一规划了整个项目需要的类，然后进行了具体编码工作。希望读者仔细品味每一段代码，为自己在以后的开发应用工作打好基础。

本 iPhone 连连看游戏实例具备如下功能。

（1）用户通过触摸屏幕可以选定相同的两张图片并根据游戏规则消除它们，第一张图片选定后会变成暗黄色。

（2）在用户游戏的过程中进行计时，并将用户所有时间的长短排列在分数排行榜中。

（3）游戏分为 5 个难度等级，并提供游戏重新启动的功能，此时游戏界面会重新排列并重新计时。

（4）用户退出时会保存用户游戏所用的时间长短。

具体实现

（1）设计类。

在本实例中有如下 5 个类。

- **main.m 类**：类 main.m 为 iPhone 应用程序的启动类，由 iPhone OS 系统底层调用并传递相应的应用程序"沙盒"路径信息等给 iPhone 应用程序，之后 iPhone OS 调用 UIApplicationMain( ) 方法初始化 iPhone 应用程序的 UIApplication 等资源。

- **LianAppDelegate 类**：类 LianAppDelegate 是 iPhone 应用程序的委托类，iPhone OS 底层把相应的 iPhone 应用程序委托给 LianAppDelegate 类负责，LianAppDelegate 类负责 iPhone 应用程序的生命周期和系统事件响应，在 applicationDidFinishLaunching:方法中生成了一个 LianScreen 视图控件类对象，然后通过 addSubView:方法加载到当前主窗口上。

- **HelpScreen 类**：类 HelpScreen 内部封装一个 UIWebView 对象，负责从 help.html 文件中加载 HTML 格式的帮助信息并显示到 UIWebView 视图上。

- **ScoresScreen 类**：类 ScoresScreen 继承自 UIView，它内部封装了一个 UIScrollView 对象，用来滚动地显示 40 个游戏时间记录。这些游戏时间记录都是 UILabel 控件，它们通过 addSubview:

方法添加到 UIScrollView 视图对象上。
- SoundEffect 类：类 SoundEffect 主要为游戏音频播放，使用 System Sound Services 音频技术控制进行音频文件的播放。

（2）设计界面。

在 iPhone 系统中，所有的视图都派生自 UIView 类。而复合视图是包含了其他视图的视图，被包含的视图一般被称为子视图。子视图可以是简单控件，也可以是复合视图，一个复合视图可以包含一个或多个子视图。

在 Xcode 中创建基于 Window-based Application 模板的项目并命名为"SwitchViewProj"，在创建 SwitchViewProj 项目之后，Xcode 会自动为开发人员生成模板代码，编译并运行项目。模板代码默认为开发人员生成了一个 UIWindow 窗口，应用程序启动时会从 MainWindow.xib 文件加载这个窗口。

接下来创建 SwitchViewController 视图控件类，在 Xcode 左边 Groups&Files 窗口的 Classes 文件夹下单击右键选择 Add→New File…，选择 Cocoa Touch Class 下的 UIViewController subclass 并命名为"SwitchViewController"。把 Xcode 生成的 SwitchViewController.xib 文件拖入到 Resources 文件夹下，默认情况下 SwitchViewController.xib 文件在 Classes 目录下。

然后使用相同的方法为 SwitchViewProj 项目分别添加视图控件类 FirstViewController 和 SecondViewController。

最后双击文件 FirstViewController.xib 打开 Interface Builder，然后依次选择 Tools→Library 菜单（或使用快捷键 Command+Shift+L）打开控件库窗口，拖动一个 UILabel 和一个 UIButton 到 FirstViewController 的窗口上。然后使用同样的方法，为 SecondViewController 视图控件类添加一个 UILabel 和一个 UIButton。到此为止，通过 Interface Builder 已经快速建立了视图控件类 First ViewController 和 SecondViewController 的界面，接下来就需要在 Xcode 中声明相应的 IBAction 以及 IBOut 来操作这些 Interface Builder 中的控件元素。

视图控件类 SwitchViewController 用于管理 FirstViewController 视图控件以及 SecondViewController 视图控件，通过方法 showFirstView()以及方法 showSecondView()在两个视图控件类之间切换。其中头文件的声明代码如下所示。

```
#import <UIKit/UIKit.h>

@class FirstViewController;
@class SecondViewController;

@interface SwitchViewController : UIViewController {
    FirstViewController* firstviewcontroller;
    SecondViewController* secondviewcontroller;
}

@property (nonatomic,retain) FirstViewController* firstviewcontroller;
@property (nonatomic,retain) SecondViewController* secondviewcontroller;

-(void)initView;
-(void)showFirstView;
-(void)showSecondView;
-(void)removeAllViews;

@end
```

属性中的 nonatomic 关键字以及 retain 关键字的实现代码如下所示。

```
#import "SwitchViewController.h"
#import "FirstViewController.h"
#import "SecondViewController.h"

@implementation SwitchViewController

@synthesize firstviewcontroller;
@synthesize secondviewcontroller;
```

```objc
-(void)initView
{
    if(self.firstviewcontroller == nil)
    {
        firstviewcontroller = [[FirstViewController alloc]
initWithNibName:@"FirstView" bundle:nil];
    }
    [self removeAllViews];
    [self.view insertSubview:firstviewcontroller.view atIndex:0];
}

-(void)showFirstView
{
    if(self.firstviewcontroller == nil)
    {
        firstviewcontroller = [[FirstViewController alloc]
initWithNibName:@"FirstView" bundle:nil];
    }
    [self removeAllViews];
    [self.view insertSubview:firstviewcontroller.view atIndex:0];
}

-(void)showSecondView
{
    if(self.secondviewcontroller == nil)
    {
        secondviewcontroller = [[SecondViewController alloc]
initWithNibName:@"SecondView" bundle:nil];
    }
    [self removeAllViews];
    [self.view insertSubview:secondviewcontroller.view atIndex:0];
}

-(void)removeAllViews
{
    for(NSInteger i = 0;i < [self.view.subviews count];i++)
    {
        [[self.view.subviews objectAtIndex:i] removeFromSuperview];
    }
}

/*
// The designated initializer.  Override if you create the controller
programmatically and want to perform customization that is not
appropriate for viewDidLoad.
- (id)initWithNibName:(NSString *)nibNameOrNil
bundle:(NSBundle *)nibBundleOrNil
{
if (self = [super initWithNibName:nibNameOrNil bundle:nibBundleOrNil]) {
// Custom initialization
}
return self;
}
*/

/*
typically from a nib.
- (void)viewDidLoad {
[super viewDidLoad];
}
// Override to allow orientations other than the default portrait
orientation.
- (BOOL)shouldAutorotateToInterfaceOrientation:
(UIInterfaceOrientation)interfaceOrientation
{
// Return YES for supported orientations
return (interfaceOrientation == UIInterfaceOrientationPortrait);
}
*/
- (void)didReceiveMemoryWarning {
    // Releases the view if it doesn't have a superview.
[super didReceiveMemoryWarning];
```

```objc
}
- (void)viewDidUnload {
    // Release any retained subviews of the main view.
    // e.g. self.myOutlet = nil;
}

- (void)dealloc {
[super dealloc];
    [firstviewcontroller release];
    [secondviewcontroller release];
}
@end
```

在上述代码中，注释部分代码是 Xcode 为 Window-based Application 模板自动生成的代码，开发人员应该熟悉这些接口的作用并根据需要实现相应的接口。

接下来将 SwitchViewController 视图控件类对象添加到 SwitchViewProj 项目的主窗口中，首先在 SwitchViewProjAppDelegate.h 中声明，代码如下所示。

```objc
#import <UIKit/UIKit.h>
@class SwitchViewController;
@interface SwitchViewProjAppDelegate : NSObject <UIApplicationDelegate> {
UIWindow *window;
    SwitchViewController* switchviewcontroller;
}
@property (nonatomic, retain) IBOutlet UIWindow *window;
@property (nonatomic, retain) IBOutlet SwitchViewController* switchviewcontroller;
@end
```

然后在头文件 SwitchViewProjAppDelegate.h 中声明了一个 SwitchViewController 视图控件类的对象 switchviewcontroller。这个对象负责管理 FirstViewController 视图控件以及 SecondViewController 视图控件，具体代码如下所示。

```objc
#import "SwitchViewProjAppDelegate.h"
#import "SwitchViewController.h"
@implementation SwitchViewProjAppDelegate
@synthesize window;
@synthesize switchviewcontroller;
- (void)applicationDidFinishLaunching:(UIApplication *)application {
switchviewcontroller = [[SwitchViewController alloc]
            initWithNibName:@"SwitchViewController" bundle:nil];
[window addSubview:switchviewcontroller.view];
[switchviewcontroller initView];
// Override point for customization after application launch
[window makeKeyAndVisible];
}
- (void)dealloc {
[window release];
[switchviewcontroller release];
[super dealloc];
}
@end
```

在上述 applicationDidFinishLaunching:方法中，对 switchviewcontroller 对象进行了初始化操作，接着调用 switchviewcontroller 对象的 initView()方法初始化并加载 First ViewController 视图控件类，最后把 switchviewcontroller 对象的视图加载到当前窗口上。

单击 Show SecondView UIButton 控件时，程序可以跳到 SecondViewController 视图控件上，为了响应 Interface Builder 中的控件元素，需要在头文件 FirstViewController.h 中声明一个 IBAction 方法，具体代码如下所示。

```objc
#import <UIKit/UIKit.h>
@interface FirstViewController : UIViewController {
}
-(IBAction) buttonClick:(id)sender;
@end
```

关键字 IBAction 为开发人员响应 Interface Builder 中的控件元素操作提供了方便，这里通过 buttonClick:方法响应按钮操作，具体实现代码如下所示。

```
#import "FirstViewController.h"
#import "SwitchViewProjAppDelegate.h"
@implementation FirstViewController
-(IBAction)buttonClick:(id)sender
{
    [((SwitchViewProjAppDelegate*)[[UIApplication sharedApplication]
delegate]).switchviewcontroller showSecondView];
}
@end
```

接下来，程序需要能够响应 ShowFirstView 按钮并跳转回 FirstViewController 视图控件上，同样需要在 SecondViewController.h 头文件中声明一个 IBAction 方法，具体代码如下所示。

```
#import <UIKit/UIKit.h>
@interface SecondViewController : UIViewController {
}
-(IBAction)buttonClick:(id)sender;
@end
```

和 FirstViewController.h 头文件中的声明一样，在此声明了一个 IBAction 属性的 buttonClick:方法来响应按钮 ShowFirstView，具体实现代码如下所示。

```
#import "SecondViewController.h"
#import "SwitchViewProjAppDelegate.h"
@implementation SecondViewController
-(IBAction)buttonClick:(id)sender
{
    [((SwitchViewProjAppDelegate*)[[UIApplication sharedApplication]
            delegate]).switchviewcontroller showFirstView];
}
```

最后需要在 Interface Builder 中关联 ShowFirstView 按钮和 buttonClick:方法，否则 ShowFirstView 按钮不知道如何响应用户操作。

（3）按键处理。

在 iPhone 程序中，当手指触摸屏幕时，系统会自动调用方法 touchesBegan:withEvent:，通过参数 touches 可以获取 UITouch 对象，声明 UITouch 类的代码如下所示。

```
UIKIT_EXTERN_CLASS @interface UITouch : NSObject
{
NSTimeInterval _timestamp;
UITouchPhase _phase;
UITouchPhase _savedPhase;
NSUInteger _tapCount;

UIWindow* _window;
UIView *_view;
UIView *_gestureView;
NSMutableArray *_gestureRecognizers;

CGPoint _locationInWindow;
CGPoint _previousLocationInWindow;
struct {
unsigned int _firstTouchForView:1;
unsigned int _isTap:1;
unsigned int _isWarped:1;
unsigned int _isDelayed:1;
unsigned int _sentTouchesEnded:1;
} _touchFlags;
}

@property(nonatomic,readonly) NSTimeInterval timestamp;
@property(nonatomic,readonly) UITouchPhasephase;
@property(nonatomic,readonly) NSUInteger tapCount;
@property(nonatomic,readonly,retain) UIWindow*window;
@property(nonatomic,readonly,retain) UIView *view;
```

```
- (CGPoint)locationInView:(UIView *)view;
- (CGPoint)previousLocationInView:(UIView *)view;

@end
```

- 属性 tapCount：用来记录轻击屏幕的次数，它只对连续的轻击进行计数，如果两次轻击的间隔时间过长，或者屏幕检测到多点触摸，tapCount 的值会被重置为 0。
- 属性 timestamp：代表触摸事件发生的时间点。
- 属性 phase：用来指定当前触摸事件是触摸事件的哪一个阶段。
- 属性 window：代表触摸事件发生时所在的 UIWindow 对象。
- 属性 view：代表触摸事件发生时所在的 View 对象。

在大多数情况下，方法 touchesBegan:withEvent:和方法 touchesEnded:withEvent:可以实现相同的功能，只是一个是在触摸开始时触发；另一个是在触摸结束时触发。

如果要实现拖曳的功能，那么必须在方法 touchesMoved:withEvent:中响应用户的拖曳操作，例如在本项目中可以在用户拖曳图片的时候让图片的中心始终处于用户手指所在的位置，具体代码如下所示。

```
-(void)touchesMoved:(NSSet *)touches withEvent:(UIEvent *)event
{
    UITouch *touch = [touches anyObject];
    for (intX=0; intX<GridSize-2;intX++)
    {
        for(intY=0; intY<GridSize-2;intY++)
        {
            {
                if ([touch view] ==theImage[intX][intY]&&
theImage[intX][intY].tag==1)
                {
                    CGPoint location = [touch locationInView:self.view];
                    [theImage[intX][intY] setCenter:
CGPointMake(location.x , location.y)];
                    theImage[intX][intY].image=pimage;
                }
            }
        }
    }
}
```

在上述代码中，首先通过[touch view]获取当前手指触摸到的视图，然后判断手指触摸的视图是否和连连看视图数组中的某个视图相同，接着通过[touch locationInView:self.view]方法获取当前手指的位置信息，最后通过[theImage[intX][intY] setCenter:CGPointMake(location.x,location.y)]方法设置连连看视图的中心为当前手指触摸的位置。

当 iPhone 系统探测到用户动作时，例如手指触摸或滑过表面时，系统会不断地发送事件对象给应用程序处理。当前的 iPhone 支持两种类型的事件，分别是触摸事件和运动事件。

类 UIEven 在 iPhone OS 3.0 中得到了扩展，不仅可以支持这两种事件类型而且还可以容纳将来更多的事件类型，下面列出了已声明的枚举常量。

```
typedef enum {
UIEventTypeTouches,
UIEventTypeMotion,
}UIEventType;
Typedef enum {
UIEventSubtypeNone = 0,
UIEventSubtypeMotionShake = 1,
}UIEventSubtype;
```

此处的每一个事件都具备 UIEventType 事件类型和相关的 UIEventSubtype 子类型，开发人员可以通过 UIEvent 的 type 和 subtype 属性来访问它们。

在 iPhone 应用程序中，当手指触摸屏幕时，系统至少需要调用下面方法中的一个来告诉 UIResp。nder 对象接收和处理这些按键消息。

```objc
-(void)touchesBegan:(NSSet *)touches withEvent:(UIEvent *)event;
-(void)touchesMoved:(NSSet *)touches withEvent:(UIEvent *)event;
-(void)touchesEnded:(NSSet *)touches withEvent:(UIEvent *)event;
```

上述这3个事件处理函数的框架基本是一样的，具体过程如下所示。

第1步：获取所有触摸信息，可以通过event参数获取，具体代码如下所示。

`NSSet *allTouches = [event allTouches];`

第2步：依次处理每一个触摸点。通过[allTouches count]来判断是多点触摸还是单点触摸，获取第一个触摸点的方法如下。

`UITouch *touch = [[allTouches allObjects] objectAtIndex:0];`

获取第二个触摸点的方法如下。

`UITouch *touch2 = [[allTouches allObjects] objectAtIndex:1];`

获取第三个、第四个触摸点的方法依此类推。

第3步：针对每个触摸点的处理。通过[touch tapCount]判断每个触摸点是单击还是双击。

在实际的开发过程中，开发人员可以综合以上3种方法来判断用户的触摸手势，以正确地响应用户操作，下面列出了一些演示代码。

判断单击、双击触摸的代码如下：

```objc
-(void)touchesEnded:(NSSet *)touches withEvent:(UIEvent *)event
{
NSSet *allTouches = [event allTouches];
switch ([allTouches count])
{
 case 1:
 {
UITouch *touch = [[allTouches allObjects] objectAtIndex:0];
switch([touch tapCount])
{
case 1:
//单击触摸事件
break;
  case 2:
//双击触摸事件
break;
  }
 }
  break;
}
}
```

判断两个手指的分开、合拢手势。首先，封装一个计算两个点之间的距离的函数，代码如下。

```objc
-(CGFloat)distanceBetweenTwoPoints:(CGPoint)fromPoint
toPoint:(CGPoint)toPoint
{
float x = toPoint.x-fromPoint.x;
float y = toPoint.y-fromPoint.y;
return sqrt(x * x + y * y);
}
```

然后需要先在touchesBegan:withEvent:方法里记录多触点之间的初始距离，作为判断手指分开还是合拢的依据，具体代码如下所示。

```objc
-(void)touchesBegan:(NSSet *)touches withEvent:(UIEvent *)event
{
NSSet *allTouches = [event allTouches];
switch ([allTouches count])
{
case 1:
break;
case 2:
UITouch *touch1 = [[allTouches allObjects] objectAtIndex:0];
UITouch *touch2 = [[allTouches allObjects] objectAtIndex:1];
initialDistance = [self distanceBetweenTwoPoints:
```

```
[touch1 locationInView:[self view]] toPoint:
[touch2 locationInView:[self view]]];
break;
default:
break;
}
}
```

最后在 touchesMoved:withEvent:方法中判断手指移动时两个手指之间的距离，通过和 touchesBegan:withEvent:方法中手指之间的初始距离做比较来判断两个手指的手势是分开还是合拢的。具体代码如下所示。

```
-(void)touchesMoved:(NSSet *)touches withEvent:(UIEvent *)event
{
NSSet *allTouches = [event allTouches];
switch ([allTouches count])
{
case 1:
break;
case 2:
UITouch *touch1 = [[allTouches allObjects] objectAtIndex:0];
UITouch *touch2 = [[allTouches allObjects] objectAtIndex:1];

CGFloat finalDistance = [self distanceBetweenTwoPoints:
[touch1 locationInView:[self view]] toPoint:
[touch2 locationInView:[self view]]];

if(initialDistance > finalDistance) {//合拢手势
}
else {//分开手势
}
break;
}
}
```

通过上面的代码，就可以处理 iPhone 应用程序中的单击、双击和多触点手势操作。

（4）响应屏幕刷新事件。

类 NSTimer 是 iPhone 应用中比较常用的事件触发定时器类，基本操作如下。

```
-(void)handleTimer:(NSTimer*)timer
{
    //…
}
NSTimer* timer;
timer = [NSTimer scheduledTimerWithTimeInterval:0.5
target:self
selector:@selector(handleTimer:)
userInof:nil
repeats:YES];
```

上述定时器类对象 timer 会每隔 0.5 秒调用一次 handleTimer:方法并执行里面的操作，接下来演示了 NSTimer 类的使用方法。

第 1 步：在 Xcode 中创建基于 Window-based Application 模板的项目并命名为"NSTimerTest"。

第 2 步：创建 NSTimerTest 视图控件类，在 Xcode 左边 Groups&Files 窗口的 Classes 文件夹下单击鼠标右键，选择 Add→New File…，选择 Cocoa Touch Class 下的 UIViewController subclass 并命名为"NSTimerTest"。

在 NSTimerTestAppDelegate.h 头文件中声明 NSTimerTest 视图控件类对象，具体代码如下所示。

```
#import <UIKit/UIKit.h>
@class NSTimerTest;
@interface NSTimerTestAppDelegate : NSObject <UIApplicationDelegate> {
UIWindow *window;
NSTimerTest *viewController;
}
@property (nonatomic, retain) IBOutlet UIWindow *window;
@property (nonatomic, retain) IBOutlet NSTimerTest *viewController;
@end
```

在委托类 NSTimerTestAppDelegate 中声明了一个 NSTimerTest 类的对象 viewController，接下来需要从 XIB 文件初始化这个视图控件类对象，具体代码如下所示。

```
#import "NSTimerTestAppDelegate.h"
#import "NSTimerTest.h"
@implementation NSTimerTestAppDelegate

@synthesize window;
@synthesize viewController;
- (void)applicationDidFinishLaunching:(UIApplication *)application
{
    viewController = [[NSTimerTest alloc] initWithNibName:
@"NSTimerTest" bundle:nil];
[window addSubview:viewController.view];
    [window makeKeyAndVisible];
}
- (void)dealloc
{
[viewController release];
[window release];
[super dealloc];
}

@end
```

在委托类的方法 applicationDidFinishLaunching:中，初始化了 viewController 视图控件类对象，并把它的视图添加到当前主窗口上，声明头文件 NSTimerTest.h 的代码如下所示。

```
#import <UIKit/UIKit.h>
@interface NSTimerTest : UIViewController {
    NSTimer *myTimer;
}
@end
```

在视图控件类 NSTimerTest 中声明了一个定时器类对象 myTimer，myTimer 会定时触发事件调用 moveACar( )方法，具体代码如下所示。

```
#import "NSTimerTest.h"
@implementation NSTimerTest
- (void)viewDidLoad {
[super viewDidLoad];
CGRect workingFrame;
workingFrame.origin.x = 15;
workingFrame.origin.y = 400;
workingFrame.size.width = 40;
workingFrame.size.height = 40;
for(int i = 0; i < 6; i++)
{
      UIView *myView = [[UIView alloc] initWithFrame:workingFrame];
      [myView setTag:i];
      [myView setBackgroundColor:[UIColor redColor]];
      workingFrameworkingFrame.origin.x =
workingFrame.origin.x + workingFrame.size.width + 10;
      [self.view addSubview:myView];
}
myTimer = [NSTimer scheduledTimerWithTimeInterval:0.5
target:self
selector:@selector(handleTimer)
userInfo:nil
repeats:YES];
}
- (void)handleTimer
{
int r = rand( ) % 6;
for(UIView *aView in [self.view subviews])
    {
        if([aView tag] == r)
        {
            int movement = rand( ) % 100;
            CGRect workingFrame = aView.frame;
            workingFrameworkingFrame.origin.y = workingFrame.origin.y - movement;
```

```
            [UIView beginAnimations:nil context:NULL];
            [UIView setAnimationDuration:.2];
            [aView setFrame:workingFrame];
            [UIView commitAnimations];
            if(workingFrame.origin.y < 0)
            {
                [myTimer invalidate];
            }
        }
    }
}
- (void)didReceiveMemoryWarning {
[super didReceiveMemoryWarning];
}
- (void)viewDidUnload {
}
- (void)dealloc {
[super dealloc];
}
@end
```

在上述代码中，先在方法 viewDidLoad:中初始化了 6 个 UIView 对象，并把它们作为寄宿视图添加到当前窗口的视图中，然后初始化了一个定时器对象 myTimer，myTimer 定时器会每隔 0.5 秒触发事件调用一次 handleTimer()方法，方法 handlerTimer()会每次选择一个子视图进行移动。

（5）具体编码。

界面设计完毕之后，接下来开始进行具体编码工作。在此阶段的任务是编写响应事件的处理程序，让整个项目"动"起来。本项目的具体编码过程如下所示。

第 1 步：打开 Xcode，创建一个名为"Lian"的项目，在类 LianAppDelegate 的方法 applicationDidFinishLaunching:中隐藏状态栏，这样可以实现全屏显示效果。具体代码如下所示。

```
- (void)applicationDidFinishLaunching:(UIApplication *)application {
    [[UIApplication sharedApplication]
                    setStatusBarHidden:YES animated:YES];
}
```

第 2 步：在类 LianAppDelegate 的 applicationDidFinishLaunching:方法中初始化 LianScreen 视图控件类对象并添加到当前窗口上，具体代码如下所示。

```
- (void)applicationDidFinishLaunching:(UIApplication *)application {
    [[UIApplication sharedApplication]
setStatusBarHidden:YES animated:YES];
    LianScreen *aLian =[[LianScreen alloc] initWithNibName:
                    @"LianScreen" bundle:nil];
self.lianScreen=aLian;
[aLian release];
UIView *aview=[lianScreen view];
[window addSubview:aview];
[window makeKeyAndVisible];
}
```

第 3 步：双击文件 LianScreen.xib，打开 Interface Builder，然后依次选择 Tools→Library 菜单（或者使用快捷键 Command+ Shift+L）打开控件库，在 Data Views 下拖动一个 UIButton 控件到窗口上，选择 Document 窗口上的 UIButton 控件，单击 Tools→Attributes Inspector（可以使用快捷键 Command+1 打开）可以看到 Type 选项框，选择 Type 下的 Info Light 选项。在头文件 LianScreen.h 中声明 UIButton 属性 helpButton 并设置为 IBOutlet，在 Interface Builder 中关联 helpButton 和图 5-28 所示的 Info Light 类型的 UIButton 按钮，具体代码如下所示。

```
@interface LianScreen : UIViewController
{
UIButton *startButton,*restartButton,*undoButton,
*soundButton,*backButton,*helpButton,*splashBackButton;
}
@property (nonatomic,retain) UIButton *startButton,
*restartButton,*undoButton,*soundButton,*backButton;
```

@property (nonatomic,retain) IBOutlet UIButton *helpButton, *scoreButton;

第 4 步：在头文件 LianScreen.h 中声明如下 7 个 UIButton 对象。

- startButton：开始游戏按钮。
- restartButton：重启游戏按钮。
- undoButton：取消按钮。
- soundButton：音效按钮。
- backButton：返回按钮。
- helpButton：帮助信息按钮。
- splashBackButton：Splash 视图上的按钮。

初始化上述按钮的代码如下所示。

```
#import "LianScreen.h"
@implementation LianScreen
#define DefaultTest @"DefaultTest.plist"
#define DefaultTestOne @"DefaultTestOne.plist"
@synthesize startButton,restartButton,helpButton,
undoButton,soundButton,backButton,scoreButton;
-(void)viewWillAppear:(BOOL)animated
{
    {
        startButton = [[UIButton alloc]initWithFrame:
CGRectMake(102, 445, 30, 30)];
        [startButton setImage:[UIImage imageNamed:@"Start.png"]
forState:UIControlStateNormal];
        [startButton addTarget:nil action:@selector(startGame:)
forControlEvents:UIControlEventTouchUpInside];
        [self.view addSubview:startButton];
        [startButton release];
    }
    {
        undoButton = [[UIButton alloc]initWithFrame:
CGRectMake(189, 443, 33, 33)];
        [undoButton setImage:[UIImage imageNamed:@"Undo.png"]
forState:UIControlStateNormal];
        [undoButton addTarget:self action:@selector(backLast)
forControlEvents:UIControlEventTouchUpInside];
        [self.view addSubview:undoButton];
        [undoButton release];
    }
    {
        restartButton = [[UIButton alloc]initWithFrame:
CGRectMake(143, 443, 33, 33)];
        [restartButton setImage:[UIImage imageNamed:
@"RestartGame.png"] forState:UIControlStateNormal];
        [restartButton addTarget:self action:@selector(restartGame)
forControlEvents:UIControlEventTouchUpInside];
        [self.view addSubview:restartButton];
        [restartButton release];
    }
    {
        soundButton = [[UIButton alloc]initWithFrame:
CGRectMake(239, 443, 32, 32)];
        [soundButton setTitle:@"  " forState:UIControlStateNormal];
        [soundButton setImage:[UIImage imageNamed:@"SoundOn.png"]
forState:UIControlStateNormal];
        [soundButton addTarget:self
action:@selector(pauseBackgroundSound)
forControlEvents:UIControlEventTouchUpInside];
        [self.view addSubview:soundButton];
        [soundButton release];
    }
    {
        [helpButton addTarget:self
action:@selector(showHelp)
forControlEvents:UIControlEventTouchUpInside];
```

```objc
        scoreButton = [[UIButton alloc]initWithFrame:
CGRectMake(105, 420, 20, 20)];
        [scoreButton setImage:[UIImage imageNamed:@"Score.png"]
forState:UIControlStateNormal];
        [scoreButton setTitle:@"-" forState:UIControlStateNormal];
        [self.view addSubview:scoreButton];
        [scoreButton addTarget:self
action:@selector(initScoresScreen)
forControlEvents:UIControlEventTouchUpInside];
        [scoreButton release];
    }
        splashBackButton = [[UIButton alloc] initWithFrame:
CGRectMake(0,0,320,480)];
        splashBackButton.backgroundColor = [UIColor clearColor];
        splashBackButton.enabled=YES;
        [splashBackButton addTarget:self
action:@selector(removeSplash)
forControlEvents:UIControlEventTouchUpInside];
        [self.view addSubview:splashBackButton];
    }
}
```

通过上述代码初始化了 6 个按钮对象，并且分别指定了各自的响应方法。

- startButton 按钮：响应方法为 startGame。
- undoButton 按钮：响应方法为 backLast( )方法。
- restartButton 按钮：响应方法为 restartGame( )方法。
- soundButton 按钮：响应方法为 pauseBackgroundSound( )方法。
- scoreButton 按钮：响应方法为 initScoresScreen( )方法。
- splashBackButton 按钮：响应方法为 removeSplash( )方法。

当用户触摸响应的按钮时，按钮就会调用相应的方法来响应用户的操作，下面是方法 startGame: 的具体实现代码。

```objc
-(void)startGame:(BOOL)a
{
    [allArray removeAllObjects];
    for (intX=0; intX<GridSize-2;intX++)
    {
        for(intY=0; intY<GridSize-2;intY++)
        {
[self disappearAlien:intX :intY];
        [self disappearDisk:intX :intY];
        }
    }
    {   acurrentTime=0;
        ctrlcurrentime =YES;
}
    intN=[jumpesNumber.text intValue];
    if(intN<1){
        intN=1;
    jumpesNumber.text=[NSString stringWithFormat:@"%d", intN];
    }
    if(intN>41){
    intN=41;
    jumpesNumber.text=[NSString stringWithFormat:@"%d", intN];
    }
    intC=intN+2;
    if(intC>40)
    {intC=41;}
    if(intC<3)
    {intC=3;}
    if(numberC==41)
    {
        UIAlertView *alert=[[UIAlertView alloc]
initWithTitle:@"Congratulations!"
message:@"You have passed all level!"
delegate:self
cancelButtonTitle:@"Replay"
```

```
otherButtonTitles:nil];
        [alert show];
        intC=3;
    }
    [self levelCreate];
    [self saveImage];
    [self saveLast];
}
```

第 5 步：方法 startGame:先调用方法 disappearAlien:disy:把数组 theImage 中的图像设置为 nil，然后进行初始化操作。如果用户当前的等级是 41，则提示用户已经通过了游戏所有的关卡。调用方法 levelCreate()创建游戏视图，此方法的实现代码如下所示。

```
-(void)levelCreate  {
    [self level];
}
-(void)level
{
    int count1;
    count1=0;
    for (int i=0; i<GridSize-2;i++)
    {
        for(int j=0; j<GridSize-2;j++)
        {
            iintX = i;
            intY = j;
            [self randomCreate];
            if(theImage[i][j].image!=nil)
            {
                count1++;
                intValued=count1;
            }
        }
    }
}
```

第 6 步：方法 level 初始化了一个 8×8 的视图数组，视图数组中的每一个元素都代表一个 Jump Attribute 对象，这个对象通过 randomCreate()方法进行初始化，方法 randomCreate()的实现代码如下所示。

```
-(void)randomCreate
{
    int N =arc4random( )%20+1;

    if(theImage[intX][intY].image == nil)
    {
        [self disappearDisk:intX :intY];
        [self createAlien:N];
    }
}
```

在上述代码中，方法 randomCreate( )首先获取一个随机数 N，N 为一个不大于 20 的整数，它代表连连看中的图片索引，然后通过 createAlien:方法指定 theImage 数组中 JumpAttribute 对象的图片索引。方法 createAlien:的实现代码如下所示。

```
-(void)createAlien:(int)imageIndex  {
    [self createDisk];
    theImage[intX][intY].image =
[UIImage imageNamed:[NSString stringWithFormat:@"%d.png",
imageIndex]];
    theImage[intX][intY].tag = imageIndex;
}
```

数组 theImage 被初始化为一个 8×8 的数组 JumpAttribute，功能是保存连连游戏中的图片数据和属性，头文件 JumpAttribute.h 的声明代码如下所示。

```
#import <UIKit/UIKit.h>
@interface JumpAttribute : UIImageView {
BOOL north,south,east,west;
BOOL inite,initw,inits,initn ;
CGRect rt;
```

```
}
@property(readwrite) BOOL north,south,east,west,
inite,initw,inits,initn ;
@end
```

类 JumpAttribute 继承自 UIImageView，并且封装了一些初始化信息，例如方向和有效范围等信息。类 JumpAttribute 的实现代码如下所示。

```
#import "JumpAttribute.h"
@implementation JumpAttribute
@synthesize  north,south,east,west, inite,initw,inits ,initn;
- (id)initWithFrame:(CGRect)frame
{
if ([super initWithFrame:frame] == nil) {
return nil;
}
    rt = frame;
    north=NO;
    south=NO;
    east=NO;
    west=NO;

    inite=NO;
    initw =NO;
    inits =NO;
    initn =NO;
return self;
}
       //…
- (void)dealloc {
[super dealloc];
}
@end
```

第 7 步：接下来开始编写用户操作响应代码。本游戏的用户操作响应代码位于 touchesBegan:withEvent:方法中，具体代码如下所示。

```
-(void)touchesBegan:(NSSet*)touches withEvent:(UIEvent*)event
{
    UITouch *touch = [touches anyObject];
    lastBestTime=9999;

    for (intX=0; intX<GridSize-2;intX++)
    {
        for(intY=0; intY<GridSize-2;intY++) {
            if ([touch view] == theImage[intX][intY]) {
                currentJumpAttribute = theImage[intX][intY];

                if(lastJumpAttribute == nil)
                {
                    currentJumpAttribute.backgroundColor=
[UIColor yellowColor];
                    currentJumpAttribute.alpha =0.5;
                    lastJumpAttribute = currentJumpAttribute;
                }else
                {
                    if(lastJumpAttribute == currentJumpAttribute)
                    {
                        return;
                    }else
                    {
                        if(lastJumpAttribute.tag ==
currentJumpAttribute.tag)
                        {
                            lastJumpAttribute.image = nil;
                            currentJumpAttribute.image = nil;

                            lastJumpAttribute.backgroundColor =
[UIColor clearColor];
                            lastJumpAttribute.alpha = 1.0;
                            lastJumpAttribute = nil;
                            currentJumpAttribute = nil;
```

```
                        }else
                        {
                            lastJumpAttribute.backgroundColor =
[UIColor clearColor];
                            lastJumpAttribute.alpha = 1.0;
                            currentJumpAttribute.backgroundColor=
[UIColor yellowColor];
                            currentJumpAttribute.alpha =0.5;
                            lastJumpAttribute = currentJumpAttribute;
                        }
                    }
                    [self.view bringSubviewToFront:theImage[intX][intY]];
                }
            }
        }
    }
```

在上述代码中，先用[touches anyObject]获取用户的第一次触摸操作，然后遍历界面上所有的图片元素判断用户的触摸是否为图片元素中的某一个，如果是就判断当前的触摸是否是第一次触摸。这主要通过判断 lastJumpAttribute 是否为空来实现，如果为空就为第一次触摸，如果不为空就为第二次触摸。接下来就判断第二次触摸的图片索引是否和第一次触摸的图片索引相同，如果相同则消失，如果不同则把第二次触摸的图片赋值给 lastJumpAttribute 作为第一次触摸的图片。

第8步：接下来开始编写音效控制按钮的处理代码。当用户单击 soundButton 按钮时会响应方法 pauseBackgroundSound()。在上面初始化 soundButton 按钮的代码中把按钮 soundButton 的初始标题设置为@" "，这代表按钮 soundButton 的默认标题为两个空格，方法 pauseBackgroundSound( )能够根据标题内容来判断背景音效是否处于打开状态。具体实现代码如下所示。

```
-(void)pauseBackgroundSound
{
    if([soundButton.currentTitle isEqualToString:@" "])
    {
        soundPlay =YES;
        [soundButton setTitle:@"  " forState:UIControlStateNormal];
        [soundButton setImage:[UIImage imageNamed:@"SoundOn.png"]
forState:UIControlStateNormal];
        [self.player play];
        AudioSessionSetActive(YES);
    }
    else {
        [soundButton setTitle:@" " forState:UIControlStateNormal];
        [self.player pause];
        soundPlay =NO;
        AudioSessionSetActive(NO);
        [soundButton setImage:[UIImage imageNamed:@"SoundOff.png"]
forState:UIControlStateNormal];
    }
}
```

通过上述代码，如果按钮 soundButton 的标题是两个空格，则代表背景音效处于打开状态，那么上面的程序会重新设置 soundButton 按钮的标题为@" "，并且调用方法[self.player pause]暂时播放背景音效，最后通过方法[soundButton setImage:forState]重新设置按钮的背景图片和显示状态。

第9步：接下来开始编写排行榜界面的处理代码。当用户单击按钮 scoreButton 时会响应方法 initScoresScreen()，此方法会初始化一个 ScoresScreen 视图类对象 tempScreen，并且赋值给 scoresScreen 属性，最后把 ScoresScreen 视图作为子视图添加到当前窗口。具体代码如下所示。

```
-(void)initScoresScreen
{
    ScoresScreen * tempScreen = [[ScoresScreen alloc] initWithFrame:
CGRectMake(0.0,0.0,320,480)];
    [tempScreen.backButton addTarget:self
action:@selector(scoresScreenShow:)
forControlEvents:UIControlEventTouchUpInside];
    self.scoresScreen = tempScreen;
```

```
    [tempScreen release];
    [self.view addSubview:scoresScreen];
}
```

类 ScoresScreen 的功能是维护 UIScrollView 视图对象 scrollView，此视图对象通过本游戏"沙盒"文档目录下的 DefaultTest.plist 文件加载排行榜信息，具体代码如下所示。

```
for (int i=1; i<40; i++)
{
UILabel * labelLevel = [[UILabel alloc] initWithFrame:
CGRectMake(0,3+30*i-30,80 ,40)];
labelLevel.textAlignment = UITextAlignmentLeft;
labelLevel.backgroundColor = [UIColor clearColor];
labelLevel.font = [UIFont fontWithName:@"Arial" size:20.0f];
labelLevel.text = [NSString stringWithFormat:@"%d",i];
labelLevel.textColor = [UIColor yellowColor];
[scrollView addSubview:labelLevel];
[labelLevel release];
}
NSArray *paths = NSSearchPathForDirectoriesInDomains(NSDocumentDirectory, NSUserDomainMask, YES);
NSString *documentsPath = [paths objectAtIndex:0];
NSMutableArray *defaultArray = [[[NSMutableArray alloc]
initWithContentsOfFile:
[documentsPath stringByAppendingPathComponent:
DefaultTest]] autorelease];
for(int i=1; i<40; i++)
{
{
UILabel * tempLevelsLable = [[UILabel alloc] initWithFrame:
CGRectMake(60,3+30*i-30,80 ,40)];
tempLevelsLable.textAlignment = UITextAlignmentLeft;
tempLevelsLable.backgroundColor = [UIColor clearColor];
tempLevelsLable.font = [UIFont fontWithName:@"Arial" size:20.0f];
[self secondChangeShow:[defaultArray objectAtIndex:3*(i -1)+1]]
int trans =[[defaultArray objectAtIndex:3*(i-1)+2] intValue]/10;
NSString *astring = [NSString stringWithFormat:@"%d",trans];
tempLevelsLable.text = [self secondChangeShow:[astring intValue]];
tempLevelsLable.textColor = [UIColor greenColor];
levelsLable = tempLevelsLable;
[scrollView addSubview:levelsLable];
[tempLevelsLable release];
}

{
UILabel * tempScoresLable = [[UILabel alloc] initWithFrame:
CGRectMake(140,3+30*i-30,80 ,40)];
tempScoresLable.textAlignment = UITextAlignmentLeft;
tempScoresLable.backgroundColor = [UIColor clearColor];
tempScoresLable.font = [UIFont fontWithName:@"Arial" size:20.0f];
int transl =[[defaultArray objectAtIndex:3*(i-1)+1] intValue]/10;
NSString *string =[NSString stringWithFormat:@"%d",transl];
tempScoresLable.text = [self secondChangeShow:[string intValue]];
tempScoresLable.textColor = [UIColor whiteColor];
lastBestTimeLable = tempScoresLable;
[scrollView addSubview:lastBestTimeLable];
[tempScoresLable release];
}
}
```

通过上述代码，在对象 scrollView 中初始化了 40 行排行内容，每一行的内容由 3 个 UILabel 组成，这 3 个 UILabel 相对于屏幕左边的偏移量分别为 0、60、140，偏移量为 0 的 UILabel 显示索引 1～39；偏移量为 60 的 UILabel 显示 DefaultTest.plist 文件中起始索引为 2、偏移量为 3 的内容，比如 DefaultTest.plist 文件中索引依次为 2、5、8、11 等对象的内容；偏移量为 140 的 UILabel 显示 DefaultTest.plist 文件中起始索引为 1、偏移量为 3 的内容，比如 DefaultTest.plist 文件中索引依次为 1、4、7、10 等对象的内容。

第 10 步：接下来开始编写帮助信息界面的处理代码。当用户单击按钮 helpButton 时会响应方法 showHelp()，此方法会初始化一个 HelpScreen 视图类对象 tempScreen，并且赋值给 helpScreen 属性，最后把 HelpScreen 视图作为子视图添加到当前窗口上。具体实现代码如下所示。

```
-(void)showHelp
{
HelpScreen * tempScreen =[[HelpScreen alloc] initWithNibName:
@"HelpScreen"
bundle:nil];
[tempScreen.hbackButton addTarget:self
action:@selector(backMenu)
forControlEvents:UIControlEventTouchUpInside];
self.helpScreen = tempScreen;
[tempScreen release];
[self.view addSubview:helpScreen.view];
}
```

在视图类 HelpScreen 中维护了一个 UIWebView 对象 webView，类 UIWebView 继承自 UIView，开发人员可以使用视图 UIWebView 在 iPhone 应用程序中嵌入 Web 页面。

接下来需要在头文件 MyWebNavAppDelegate.h 中声明一个 IBOutlet 的 MyWebNavVeiwController 视图控件类对象，具体代码如下所示。

```
#import <UIKit/UIKit.h>
@class MyWebNavViewController;
@interface MyWebNavAppDelegate : NSObject <UIApplicationDelegate>
{
UIWindow *window;
MyWebNavViewController *viewController;
}

@property (nonatomic, retain) IBOutlet UIWindow *window;
@property (nonatomic, retain) IBOutlet MyWebNavViewController *viewController;

@end
```

然后双击文件 MainWindow.xib 打开 Interface Builder，选择 Document 窗口中的 MyWebNavAppDelegate 项，然后依次选择 Tools→Connections Inspector（或者使用快捷键 Command+2）菜单打开 Connections Inspector 窗口，单击 viewController 后面的圆圈并拖动到 MyWebNavViewController 项上。接下来需要在头文件 MyWebNavViewController.h 中分别声明 IBOutlet 变量关联 Interface Buidler 中的 UIWebView 视图元素以及 IBAction 方法响应 Interface Buidler 中的 google 按钮、myBlog 按钮、apple 按钮。具体代码如下所示。

```
#import <UIKit/UIKit.h>
@interface MyWebNavViewController : UIViewController {
    IBOutlet UIWebView *myWevView;
}
@property (nonatomic,retain) UIWebView *myWevView;
-(IBAction) NavURL:(id)sender;
@end
```

关联 myWebView 对象到 Interface Builder 中的 UIWebView 视图上，指定 Interface Buidler 中 google 按钮、myBlog 按钮以及 apple 按钮的响应方法为 NavURL:。视图控件类 MyWebNavView Controller 的具体实现代码如下。

```
#import "MyWebNavViewController.h"
@implementation MyWebNavViewController
@synthesize myWevView;
-(IBAction) NavURL:(id)sender
{
 switch ([sender tag]) {
    case 1:
        [myWevView loadRequest:[NSURLRequest requestWithURL:
[NSURL URLWithString:@"http://www.google.com"]]];
        break;
    case 2:
        [myWevView loadRequest:[NSURLRequest requestWithURL:
```

```
[NSURL URLWithString:@"http://blog.csdn.net/dongfengsun"]]];
            break;
        case 3:
            [myWevView loadRequest:[NSURLRequest requestWithURL:
[NSURL URLWithString:@"http://www.apple.com"]]];
            break;
    }
}
- (void)didReceiveMemoryWarning
{
    [super didReceiveMemoryWarning];
}
- (void)dealloc
{
    [super dealloc];
}
@end
```

第 11 步：同样在视图控件类 HelpScreen 中也采用了解析 HTML 文件的方法显示帮助信息，只不过帮助说明文件 help.html 存放在程序本地。因为帮助文件 help.html 采用 UTF-8 格式存放帮助信息，所以在处理上和 MyWebNav 项目中会略有不同。头文件 HelpScreen.h 的声明代码如下所示。

```
#import <UIKit/UIKit.h>

@interface HelpScreen : UIViewController {
    IBOutlet UIWebView *webView;
    UIButton *hbackButton;
}
@property (nonatomic,retain)UIButton *hbackButton;
@property (nonatomic,retain)UIWebView *webView;
-(void)backMenu;
@end
```

其中按钮 hbackButton 对象的功能是响应用户的返回操作，在用户单击返回按钮时会返回到连连看游戏界面，具体实现代码如下所示。

```
#import "HelpScreen.h"
@implementation HelpScreen
@synthesize hbackButton;
@synthesize webView;
-(void)viewDidLoad
{
    self.view.backgroundColor=[UIColor clearColor];
    {
    UIImageView *imageView =[[UIImageView alloc]initWithFrame:
CGRectMake(0, 0, 320, 480)];
    UIImage *image=[UIImage imageNamed:@"HelpAlien.png"];
    imageimageView.image=image;
    [self.view addSubview:imageView];
    }
    {
        UIWebView *tempwebvew=[[UIWebView alloc] initWithFrame:
CGRectMake(15, 100, 300, 330)];
        self.webView=tempwebvew;
        [tempwebvew release];
        NSString *fullPath = [NSBundle 
pathForResource:@"help" 
ofType:@"html" 
inDirectory:[[NSBundle mainBundle] bundlePath]];
        NSData *htmlData = [NSData dataWithContentsOfFile:fullPath];
        if (htmlData) {
            [webView loadData:htmlData 
MIMEType:@"text/html" 
textEncodingName:@"UTF-8" 
baseURL:[NSURL fileURLWithPath:fullPath]];
        }

        webView.opaque = NO;
        webView.scalesPageToFit=YES;
        webView.backgroundColor = [UIColor clearColor];
```

```
    [self.view addSubview:webView];
}
    UIButton * tempBackButton = [[UIButton alloc] initWithFrame:
CGRectMake(40,440,74,37)];
    tempBackButton.backgroundColor = [UIColor clearColor];
    [tempBackButton setBackgroundImage:[UIImage imageNamed:@"Back.png"]
forState:UIControlStateNormal];
    tempBackButton.enabled=YES;
    self.hbackButton = tempBackButton;
    [self.view addSubview:hbackButton];
    [hbackButton addTarget:self action:@selector(backMenu)
forControlEvents:UIControlEventTouchUpInside];
    [tempBackButton release];
}
-(void)backMenu
{
    [self.view removeFromSuperview];
}
- (void)didReceiveMemoryWarning {
 [super didReceiveMemoryWarning];
}

- (void)viewDidUnload {
}
- (void)dealloc {
    [webView release];
[super dealloc];
    [hbackButton release];
}
@end
```

在上述代码中，先通过[NSData dataWithContentsOfFile:fullPath]获取 help.html 文件的文本信息，然后通过 loadData:MIMEType：textEncodingName:baseURL:方法指定 help.html 文本内容的编码格式为"UTF-8"格式。

到此为止，本章的连连看游戏全部讲解完毕。执行后的效果如图 12-1 所示。

图 12-1　执行效果

实例 250　实现一个移动老虎机游戏

实例 250	实现一个移动老虎机游戏
源码路径	\daima\250\

实例说明

老虎机（one-arm bandit）是一种用零钱赌博的机器，因为上面有老虎图案的筹码而得名。它有 3 个玻璃框，里面有不同的图案，投币之后拉下拉杆，就会开始转，如果出现特定的图形（比如 3 个相同）就会吐钱出来。本实例在 iPhone 系统中开发一个移动老虎机游戏。

具体实现

（1）打开 Xcode，新建一个名为"tigger"的项目，如图 12-2 所示。

（2）右键单击"Rsources"文件夹，然后使用"Add…"命令添加需要的素材文件。添加之后的结构如图 12-3 所示。

图 12-2 创建的工程　　　　　　　　　图 12-3 添加素材文件之后的结构

（3）在 UI 视图中添加一个 UIPickerView 控件，通过控件以表格的样式显示老虎机中的图标。

（4）分别定义 UIPickerView 对象、UILabel 和 NSArray 对象，并分别实现对应的 IBAction 方法。

（5）这是最后一个步骤，在此分别定义本实例控件属性值的操作方法。

接下来开始看整理的编码文件，其中视图文件 DayViewContro11er.h 的实现代码如下所示。

```
#import <UIKit/UIKit.h>
#import <Foundation/Foundation.h>

@interface DayViewController : UIViewController {
    IBOutlet  UILabel *winLabel;
    IBOutlet  UIPickerView *picker;
    NSArray *column1;
    NSArray *column2;
    NSArray *column3;
    NSArray *column4;
    NSArray *column5;
}
@property (nonatomic,retain) UIPickerView *picker;
@property (nonatomic,retain) UILabel *winLabel;
@property (nonatomic,retain) NSArray *column1;
@property (nonatomic,retain) NSArray *column2;
@property (nonatomic,retain) NSArray *column3;
@property (nonatomic,retain) NSArray *column4;
@property (nonatomic,retain) NSArray *column5;
-(IBAction)spin;
@end
```

文件 DayViewController.m 是文件 DayViewController.h 的实现，具体代码如下所示。

```
#import "DayViewController.h"
@implementation DayViewController
@synthesize picker;
@synthesize winLabel;
@synthesize column1;
@synthesize column2;
@synthesize column3;
@synthesize column4;
@synthesize column5;
// 界面初始化
- (void)viewDidLoad {
    [super viewDidLoad];
    UIImage *seven = [UIImage imageNamed:@"seven.png"];
    UIImage *bar = [UIImage imageNamed:@"bar.png"];
```

```objc
        UIImage *crown = [UIImage imageNamed:@"crown.png"];
        UIImage *cherry = [UIImage imageNamed:@"cherry.png"];
        UIImage *lemon = [UIImage imageNamed:@"lemon.png"];
        UIImage *apple = [UIImage imageNamed:@"apple.png"];
        for(int i =1; i<=5; i++)
        {
            UIImageView *sevenView = [[UIImageView alloc] initWithImage:seven];
            UIImageView *barView = [[UIImageView alloc] initWithImage:bar];
            UIImageView *crownView = [[UIImageView alloc] initWithImage:crown];
            UIImageView *cherryView = [[UIImageView alloc] initWithImage:cherry];
            UIImageView *lemonView = [[UIImageView alloc] initWithImage:lemon];
            UIImageView *appleView = [[UIImageView alloc] initWithImage:apple];
            NSArray *imageViewArray = [[NSArray alloc] initWithObjects:
            sevenView,barView, crownView,cherryView,lemonView,appleView,nil];

            NSString *fieldName = [[NSString alloc] initWithFormat:@"column%d",i];
            [self setValue:imageViewArray forKey:fieldName];
            [fieldName release];
            [imageViewArray release];

            [sevenView release];
            [barView release];
            [crownView release];
            [cherryView release];
            [lemonView release];
            [appleView release];
        }
        srandom(time(NULL));

}
-(IBAction)spin{
    BOOL win = NO;
    int numInRow = 1;
    int lastVal = -1;
    for(int i = 0; i<5 ;i++)
    {
        int newValue = random() % [self.column1 count];
        if(newValue == lastVal)
            numInRow++;
        else
            numInRow =1;
        lastVal = newValue;
        [picker selectRow:newValue inComponent:i animated:YES];
        [picker reloadComponent:i];
        if(numInRow >= 3)
            win = YES;
    }

    if(win)
        winLabel.text = @"胜利!!!";
    else
        winLabel.text = @"失败~";
}

-(NSInteger)numberOfComponentsInPickerView:(UIPickerView *)pickerView
{
    return 5;
}

-(NSInteger)pickerView:(UIPickerView *)pickerView
numberOfRowsInComponent:(NSInteger)component
{
    return [self.column1 count];
}

-(UIView *)pickerView:(UIPickerView *)pickerView
           viewForRow:(NSInteger)row
         forComponent:(NSInteger)component reusingView:(UIView *)view
{
    NSString *arrayName = [[NSString alloc] initWithFormat:@"column%d",component+1];
    NSArray *array = [self valueForKey:arrayName];
    return [array objectAtIndex:row];
```

第 12 章　游戏应用实战

```
}
- (void)didReceiveMemoryWarning {
    [super didReceiveMemoryWarning];
}

- (void)viewDidUnload {
}

- (void)dealloc {
    [super dealloc];

    [picker release];
    [winLabel release];

    [column1 release];
    [column2 release];
    [column3 release];
    [column4 release];
    [column5 release];
}
@end
```

到此为止，整个游戏实例介绍完毕，执行后的效果如图 12-4 所示。

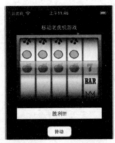

图 12-4　执行效果

实例 251　实现一个移动打砖块游戏

实例 251	实现一个移动打砖块游戏
源码路径	\daima\251\

实例说明

　　打砖块游戏是一种动作电子游戏的名称。玩家操作一根荧幕上水平的"棒子"，让一颗不断弹来弹去的"球"在撞击作为过关目标消去的"砖块"的途中不会落到荧幕底下。球碰到砖块、棒子与底下以外的三边会反弹，落到底下会失去一颗球，把砖块全部消去就可以破关。

　　打砖块游戏是史蒂夫•乔布斯与他的好友沃兹（苹果公司的另一位创始人）于 1975 年的夏末，花了 4 天时间设计完成的游戏《乒乓》。同时，美国英宝格公司于 1976 年推出的街机游戏"Breakout"，由该公司在 1972 年发行的"PONG"（世界上第一款电子游戏，类似台球）改良而来。相较于其前作，一个人就可以玩与变化丰富这两项特点让 Breakout 相当卖座，各家公司竞相模仿。因为规则简单与充满游戏性，现在许多移动电话都有内建打砖块游戏，也有许多因特网小游戏版本。打砖块游戏一般不大，甚至用软盘即可容下。

　　在本实例中，演示了在 iPhone 系统中开发一个打砖块游戏的实现过程。

具体实现

　　（1）打开 Xcode，创建一个名为"zhuan"的项目，如图 12-5 所示。
　　（2）因为本项目用到了素材图片和背景音乐素材文件，所以在开发之初需要添加这些素材文件。右键单击"Rsources"文件夹，然后使用"Add…"命令添加需要的素材文件。

添加之后的结构如图 12-6 所示。

图 12-5　创建工程

图 12-6　添加素材文件之后的结构

（3）在项目中添加框架文件 QuartzCore.framework 和 AudioToolbox.framework，如图 12-7 所示。

（4）本实例有两个 UI 视图界面，首先看第一个玩游戏界面 MainView.xib。在此 UI 视图中添加一个 UIImageView 控件，然后自定义 UIImageView 的视图检查器。

（5）在 MainView.xib 屏幕上方定义 3 个 UILabel 对象，分别表示"最高成绩"、"现在得分"和"游戏界别"。在 MainView.xib 屏幕下方定义 3 个 UIButton 对象，分别表示"左移"、"右移"和"开始游戏"3 个按钮，最终的 UI 视图界面效果如图 12-8 所示。

图 12-7　添加框架文件

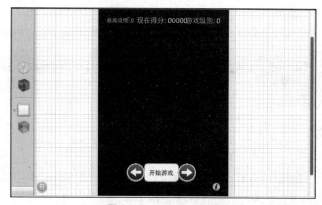

图 12-8　MainView.xib 视图

（6）接下来看第二个视图——FlipsideView.xib，在此视图中使用 UITextView 控件显示此游戏的说明信息。

接下来开始看整理的编码文件，其中玩游戏视图文件 MainViewController.h 的实现代码如下所示。

```
@interface MainViewController : UIViewController <FlipsideViewControllerDelegate> {
    IBOutlet UILabel *highestLabel;
    IBOutlet UILabel *levelLabel;
    IBOutlet UILabel *scoreLabel;
    NSTimer *timer;
    
    UIImageView *ball;
    CGPoint moveDis;
    BoardView *board;
    NSMutableArray *bricks;
    int level,numOfBricks,score,highest;
    double speed;
    NSString *soundFile;
}
```

第 12 章 游戏应用实战

```objc
@property int level,score,highest;
@property int numOfBricks;
@property double speed;
@property(nonatomic,retain)NSString *soundFile;
- (IBAction)onLeft:(id)sender;
- (IBAction)onRight:(id)sender;
- (IBAction)onStart:(id)sender;
- (void)levelMap:(int)inlevel;
-(void)playSound:(NSString*)soundAct;
- (IBAction)showInfo:(id)sender;
@end
```

文件是 MainViewController.h 是文件 MainViewController.h 的实现，具体代码如下所示。

```objc
#import "MainViewController.h"
@implementation MainViewController
@synthesize level;
@synthesize numOfBricks;
@synthesize speed,score,highest;
@synthesize soundFile;
-(void)playSound:(NSString*)soundAct{
    
    NSString *path = [NSString stringWithFormat:@"%@%@",
                      [[NSBundle mainBundle] resourcePath],
                      soundAct];
    
    NSLog(@"%@\n", path);
    SystemSoundID soundID;
    NSURL *filePath = [NSURL fileURLWithPath:path isDirectory:NO];
    AudioServicesCreateSystemSoundID((CFURLRef)filePath, &soundID);
    AudioServicesPlaySystemSound(soundID);
    [filePath release];
}
- (IBAction)onLeft:(id)sender {
    [UIImageView beginAnimations:@"animLeft" context:NULL];
    [UIImageView setAnimationDuration:0.2];
    board.center=CGPointMake(board.center.x-20, board.center.y);
    
    [UIImageView commitAnimations];
    soundFile = [NSString stringWithFormat:@"/button_press.caf"];
    [self playSound: soundFile];
}
- (IBAction)onRight:(id)sender {
    [UIImageView beginAnimations:@"animRight" context:NULL];
    [UIImageView setAnimationDuration:0.2];
    board.center=CGPointMake(board.center.x+20, board.center.y);
    [UIImageView commitAnimations];
    soundFile = [NSString stringWithFormat:@"/button_press.caf"];
    [self playSound: soundFile];
}
- (IBAction)onStart:(id)sender {
    if(!timer){
        timer=[NSTimer scheduledTimerWithTimeInterval:speed target:self selector:@selector(onTimer) userInfo:nil repeats:YES];
        ball.frame=CGRectMake(160, 328, 32, 32);
        [self.view addSubview:ball];
        board.frame=CGRectMake(160, 360, 48, 10);
    }
}
- (void)viewDidLoad {
    [super viewDidLoad];
    moveDis=CGPointMake(-3, -3)
    speed=0.03;
    board=[[BoardView alloc] initWithImage:[UIImage imageNamed:@"board.png"]];
    [board setUserInteractionEnabled:YES];
    board.frame=CGRectMake(160, 360, BOARDWIDTH, BOARDHIGHT);
    ball=[[UIImageView alloc] initWithImage:[UIImage imageNamed:@"ball.png"]];
    [self.view addSubview:board];
    level=1; score=0; highest=0;
    levelLabel.text=[NSString stringWithFormat:@"游戏级别: %i",level];
    scoreLabel.text=[NSString stringWithFormat:@"现时得分: %i",score];
```

```objc
        highestLabel.text=[NSString stringWithFormat:@"最高成绩: %i",highest];
        [self levelMap:level];
}
- (void)levelMap:(int)inlevel {
    UIImageView *brick;
    switch (inlevel) {
        case 1:
            bricks=[NSMutableArray arrayWithCapacity:20];
            numOfBricks=20;

            for (int i=0; i<3; i++) {
                for (int j=0;j<6;j++) {
                    brick=[[UIImageView alloc] initWithImage:[UIImage imageNamed:
                    @"brick.png"]];
                    brick.frame=CGRectMake(20+j*BRICKWIDTH+j*5,
                    TOP+10+BRICKHIGHT*i+5*i, BRICKWIDTH, BRICKHIGHT);
                    [self.view addSubview:brick];
                    [bricks addObject:brick];
                }
            }
            brick=[[UIImageView alloc] initWithImage:[UIImage imageNamed:@"brick.png"]];
            brick.frame=CGRectMake(20, TOP+10+20*2+5*4, BRICKWIDTH, BRICKHIGHT);
            [self.view addSubview:brick];
            [bricks addObject:brick];
            brick=[[UIImageView alloc] initWithImage:[UIImage imageNamed:@"brick.png"]];

            brick.frame=CGRectMake(20+5*BRICKWIDTH+5*5, TOP+10+20*2+5*4, BRICKWIDTH,
            BRICKHIGHT);
            [self.view addSubview:brick];
            [bricks addObject:brick];
            [bricks retain];
            break;
        case 2:
            bricks=[NSMutableArray arrayWithCapacity:28];
            numOfBricks=20;
            for (int i=0; i<7; i++) {
                (int j=0;j<2;j++) {
                    brick=[[UIImageView alloc] initWithImage:[UIImage imageNamed:
                    @"brick.png"]];
                    brick.frame=CGRectMake(20+j*BRICKWIDTH+j*5, TOP+10+
                    BRICKHIGHT*i+5*i, BRICKWIDTH, BRICKHIGHT);
                    [self.view addSubview:brick];
                    [bricks addObject:brick];
                }
            }
            for (int i=0; i<7; i++) {
                for (int j=0;j<2;j++) {
                    brick=[[UIImageView alloc] initWithImage:[UIImage imageNamed:@
                    "brick.png"]];
    brick.frame=CGRectMake(20+j*BRICKWIDTH+j*5+180, TOP+10+BRICKHIGHT*i+5*i,
BRICKWIDTH, BRICKHIGHT);
                    [self.view addSubview:brick];
                    [bricks addObject:brick];
                }
            }
            [bricks retain];
            break;
        default:
            break;
    }
}
- (void)onTimer{
    float posx,posy;
    posx=ball.center.x; posy=ball.center.y;
    ball.center = CGPointMake(posx+moveDis.x, posy+moveDis.y);
    if (ball.center.x>305 || ball.center.x<15 ) {
        moveDis.x=-moveDis.x;
    }
    if ( ball.center.y<TOP ) {
        moveDis.y=-moveDis.y;
    }
```

第12章 游戏应用实战

```objc
        int j=[bricks count];
        for (int i=0; i<j; i++) {
            UIImageView *brick=(UIImageView *)[bricks objectAtIndex:i];
            if (CGRectIntersectsRect(ball.frame, brick.frame)&&[brick superview]) {
                soundFile = [NSString stringWithFormat:@"/Brick_move.caf"];
                [self playSound: soundFile];
                score+=100;
                [brick removeFromSuperview];
                if (rand()%5==0) {
                    UIImageView* imageView = [[UIImageView alloc] initWithImage: [UIImage imageNamed:@"shorten.png"]];
                    imageView.frame = CGRectMake(brick.frame.origin.x, brick.frame.origin.y, 48, 48);
                    [self.view addSubview:imageView];
                    [UIView beginAnimations:nil context:imageView];
                    [UIView setAnimationDuration:5.0];
                    [UIView setAnimationCurve:UIViewAnimationCurveEaseOut];
                    imageView.frame = CGRectMake(brick.frame.origin.x, 380, 40, 40);
                    [UIView setAnimationDelegate:self];
                    [UIView setAnimationDidStopSelector:@selector(removeSmoke:finished:context:)];
                    [UIView commitAnimations];
                }
                numOfBricks--;
                if ((ball.center.y-16<brick.frame.origin.y+BRICKHIGHT || ball.center.y+16>brick.frame.origin.y)
                    && ball.center.x>brick.frame.origin.x && ball.center.x<brick.frame.origin.x+BRICKWIDTH) {
                    moveDis.y=-moveDis.y;
                }else if (ball.center.y>brick.frame.origin.y && ball.center.y<brick.frame.origin.y+BRICKHIGHT
                          && (ball.center.x+16>brick.frame.origin.x || ball.center.x-16<brick.frame.origin.x+BRICKWIDTH)){
                    moveDis.x=-moveDis.x;

                }else{
                    moveDis.x=-moveDis.x;
                    moveDis.y=-moveDis.y;
                }
                break;

            }
        }
        if (numOfBricks==0) {
            if (level<2) {
                [ball removeFromSuperview];
                [timer invalidate];
                level++; speed=speed-0.003;
                levelLabel.text=[NSString stringWithFormat:@"Level %i",level];
                [self levelMap:level];
            }else{
                UIAlertView *alert=[[UIAlertView alloc] initWithTitle:@"K.O."
                    message:@"恭喜! 你赢了"
                              delegate:self cancelButtonTitle:@"OK"
             otherButtonTitles:nil];
                [alert show];
                soundFile = [NSString stringWithFormat:@"/Win.caf"];
                [self playSound: soundFile];
            }
        }
        if (CGRectIntersectsRect(ball.frame, board.frame)) {
            soundFile = [NSString stringWithFormat:@"/Kick.caf"];
            [self playSound: soundFile];
            if (ball.center.x>board.frame.origin.x&&ball.center.x<board.frame.origin.x+BOARDWIDTH) {

                moveDis.y=-moveDis.y;
            }else {

                moveDis.x=-moveDis.x;
```

```
                    moveDis.y=-moveDis.y;
                }
        }else{
                if (ball.center.y>380){
                   [ball removeFromSuperview];
                    [timer invalidate];
                   timer=NULL;
                   UIAlertView *alert=[[UIAlertView alloc] initWithTitle:@"Game over"
                                        message:@"你输了,继续获取更好的成绩..."
                     delegate:self
                          cancelButtonTitle:@"确定"
                              otherButtonTitles:nil];
                    [alert show];
                   soundFile = [NSString stringWithFormat:@"/Lose.caf"];
                   [self playSound: soundFile];
                }
        }
        scoreLabel.text=[NSString stringWithFormat:@"现时得分: %i",score];
}
- (void)flipsideViewControllerDidFinish:(FlipsideViewController *)controller {
    [self dismissModalViewControllerAnimated:YES];
}
- (IBAction)showInfo:(id)sender {
    FlipsideViewController *controller = [[FlipsideViewController alloc] initWithNib
Name:@"FlipsideView" bundle:nil];
    controller.delegate = self;
    controller.modalTransitionStyle = UIModalTransitionStyleFlipHorizontal;
    [self presentModalViewController:controller animated:YES];
    [controller release];
}
- (void)didReceiveMemoryWarning {
   [super didReceiveMemoryWarning];
}
- (void)viewDidUnload {
}
- (void)dealloc {
    [soundFile release];
    [super dealloc];
}
@end
```

编写文件 BoardView.h,在此自定义 UIImageView 控件,具体代码如下所示。

```
#import <UIKit/UIKit.h>
#import <QuartzCore/QuartzCore.h>
@interface BoardView : UIImageView {
     CGPoint startLocation;
}
@end
```

文件是 BoardView.h 是文件 BoardView.m 的实现,具体代码如下所示。

```
#import "BoardView.h"
@implementation BoardView
- (void)touchesBegan:(NSSet *)touches withEvent:(UIEvent *)event{
 startLocation= [[touches anyObject] locationInView:self];
 [[self superview] bringSubviewToFront:self];
}
- (void)touchesMoved:(NSSet *)touches withEvent:(UIEvent *)event{
 CGPoint pt=[[touches anyObject] locationInView:self];
 CGRect frame=[self frame];
 frame.origin.x = frame.origin.x + (pt.x-startLocation.x);
 [self setFrame:frame];
}
- (void)didReceiveMemoryWarning {
   [super didReceiveMemoryWarning];
}
- (void)viewDidUnload {
   [super viewDidUnload];
}
- (void)dealloc {
    [super dealloc];
```

```
}
@end
```

再看游戏说明视图文件 FlipsideViewController.h，具体代码如下所示。

```
#import <UIKit/UIKit.h>
@protocol FlipsideViewControllerDelegate;
@interface FlipsideViewController : UIViewController {
    id <FlipsideViewControllerDelegate> delegate;
}
@property (nonatomic, assign) id <FlipsideViewControllerDelegate> delegate;
- (IBAction)done:(id)sender;
@end
@protocol FlipsideViewControllerDelegate
- (void)flipsideViewControllerDidFinish:(FlipsideViewController *)controller;
@end
```

文件是 FlipsideViewController.h 是文件 FlipsideViewController.h 的实现，具体代码如下所示。

```
#import "FlipsideViewController.h"
@implementation FlipsideViewController
@synthesize delegate;
- (void)viewDidLoad {
    [super viewDidLoad];
    self.view.backgroundColor = [UIColor viewFlipsideBackgroundColor];
}
- (IBAction)done:(id)sender {
    [self.delegate flipsideViewControllerDidFinish:self];
}
- (void)didReceiveMemoryWarning {
    [super didReceiveMemoryWarning];
}
- (void)viewDidUnload {
}
- (void)dealloc {
    [super dealloc];
}
@end
```

到此为止，整个实例介绍完毕，执行后玩游戏界面的效果如图 12-9 所示。

游戏说明界面效果如图 12-10 所示。

图 12-9 执行效果

图 12-10 游戏说明界面

第 13 章 移动 Web 实战

随着移动手机设备的不断升级，以及 Android、iOS、Windows Phone 等智能系统的市场占有率越来越高，智能手机已经步入了飞速发展的黄金时期。本章将通过几个典型实例的实现过程，详细介绍在 iOS 系统中开发移动 Web 项目的基本知识。

实例 252 实现页眉定位

实例 252	实现页眉定位
源码路径	\daima\252\

实例说明

在设计过程中，有如下 3 种样式可以用于定位页眉。

（1）Default（默认）：默认的页眉会在屏幕的顶部边缘显示，而且在屏幕滚动时，页眉将会滑到可视范围之外。

```
<div data-role="header">
<h1>Default Header</h1>
</div>
```

（2）Fixed（固定）：固定的页眉总是位于屏幕的顶部边缘位置，而且总是保持可见。但是在屏幕滚动的过程中，页眉是不可见的，当滚动结束之后才会显示页眉。通过添加 data-position="fixed" 属性的方式可以创建一个固定的页眉，例如：

```
<div data-role="header" data-position="fixed">
<h1>Fixed Header</h1>
</div>
```

（3）Responsive（响应式）：当创建一个全屏页面时会全屏显示页面中的内容，而页眉和页脚则基于触摸响应来出现或消失。全屏模式在显示照片和播放视频方面相当有用。

要想创建一个全屏的页面，需要在页面容器中添加如下代码。

```
data-fullscreen="true"
```

然后在页眉和页脚元素中添加如下所示的属性。

```
data-position="fixed"
```

在接下来的内容中，将通过一个具体实例的实现过程，详细讲解在 iOS 中实现页眉定位的方法。

具体实现

实例文件 position-full.html 的具体实现代码如下所示。

```
<!DOCTYPE html>
<html>
<head>
<meta charset="utf-8">
<title>Fullscreen Example</title>
<meta name="viewport" content="width=device-width, maximum-scale=1">
<link rel="stylesheet" href="http://code.jquery.com/mobile/1.0/jquery.mobile-1.0.min.css" />
<style>
    .detailimage { width: 100%; text-align: center; margin-right: 0; margin-left: 0; }
    .detailimage img { width: 100%; }
</style>
<script
```

```
  src="http://code.jquery.com/jquery-1.6.4.min.js"></script>
   <script
  src="http://code.jquery.com/mobile/1.0/jquery.mobile-1.0.min.js"></script>
  </head>
  <body>
  <div data-role="page" data-fullscreen="true">
    <div data-role="header" data-position="fixed">
        <h6>4/10</h6>
    </div>

    <div data-role="content">
        <div class="detailimage"><img src="images/1213.jpg" /></div>
    </div>

    <!-- toolbar with icons -->
    <div data-role="footer" data-position="fixed">
        <div data-role="navbar">
            <ul>
                <li><a href="#" data-icon="forward"></a></li>
                <li><a href="#" data-icon="arrow-l"></a></li>
                <li><a href="#" data-icon="arrow-r"></a></li>
                <li><a href="#" data-icon="delete"></a></li>
            </ul>
        </div>
    </div>
  </div>
  </body>
  </html>
```

执行上述代码后将首先显示一个有页眉的效果,如图 13-1 所示。

图 13-1　有页眉的效果

在图 13-1 所示的效果中有一个用来显示照片的全屏页面,如果用户轻敲屏幕,则页眉和页脚将会出现或消失,这样便形成了一个全屏显示效果,如图 13-2 所示。

图 13-2　页眉消失后全屏显示

在本实例中有一个照片查看器,而且其页眉显示照片的计数信息,页脚显示一个工具栏以辅助导航,发送电子邮件或删除照片。

实例 253　在页眉中使用按钮

实例 253	在页眉中使用按钮
源码路径	\daima\253\

实例说明

在移动 Web 设计应用中可能需要在页眉中添加控件，目的是管理屏幕中的内容。例如，在编辑数据时，保存和取消按钮是经常会用到的两个控件。可以添加到页眉中的按钮有 3 种类型，具体说明如下所示。

（1）只带有文本的按钮。

（2）只带有图标的按钮。只带有图标的按钮需要添加两个属性：data-icon 和 data-iconpos="notext"。

（3）既有文本又有图标的按钮，这种类型的按钮也需要 data-icon 属性。

每一种类型的按钮示例如下所示。

```
<!---只带有文本的按钮-->
<a href="#">Done</a>
<!--只带有图标的按钮-->
<a href="#" data-icon="plus" data-iconpos="notext"></a>
<!--既有文本又有图标的按钮-->
<a href="#" data-icon="check">Done</a>
```

在接下来的内容中，将通过一个具体实例的实现过程，详细讲解在页眉中实现既有文本又有图标的按钮效果的方法。

具体实现

实例文件 buttons.html 的具体实现代码如下所示。

```
<!DOCTYPE html>
<html>
  <head>
  <meta charset="utf-8">
  <title>Header Example</title>
  <meta name="viewport" content="width=device-width, initial-scale=1">
  <link rel="stylesheet" href="http://code.jquery.com/mobile/1.0/jquery.mobile-1.0.min.css" />
  <script src="http://code.jquery.com/jquery-1.6.4.min.js"></script>
  <script src="http://code.jquery.com/mobile/1.0/jquery.mobile-1.0.min.js"></script>
</head>
<body>
<div data-role="page" data-theme="b">
  <div data-role="header" data-position="inline">
      <a href="#" data-icon="delete">取消</a>
      <h1>发布评论</h1>
      <a href="#" data-icon="check">完成</a>
  </div>

  <div data-role="content">
    <fieldset data-role="controlgroup" data-theme="c">
        <legend>评分：</legend>
        <input type="radio" name="radio-choice-1" id="radio-choice-1" value="choice-1" data-theme="c"/>
        <label for="radio-choice-1">我去看看</label>

        <input type="radio" name="radio-choice-1" id="radio-choice-2" value="choice-2" data-theme="c" />
        <label for="radio-choice-2">不好看，不看了</label><br>
```

```
                <label for="comments">内容:</label>
                <textarea cols="40" rows="8" name="comments" id="comments" data-theme="d">
    </textarea>
            </fieldset>
        </div>
    </div>
    </body>
    </html>
```

图 13-3 执行效果

在上述实例代码中，按钮被设计为一个普通的链接。可以通过属性 data-icon 为每一个按钮附加一个图标。在页眉内部，按钮依据它们的语义顺序进行摆放。例如，第一个按钮是左对齐的，第二个按钮是右对齐的。如果页眉只包含一个按钮，可以通过将属性 class="ui-btn-right" 添加到按钮的标记中的方法来右对齐按钮。上述实例代码执行后的效果如图 13-3 所示。

在图 13-3 所示的执行效果中，页眉中带有一个"取消"按钮和一个"完成"按钮，用来帮助管理评论信息。

实例 254 在页眉中使用分段控件

实例 254	在页眉中使用分段控件
源码路径	\daima\254\

实例说明

分段控件是一组内联（inline）的控件，其中每一个控件可以显示一个不同的视图。在页眉中可以使用分段控件，具体使用时，建议将分段控件放置在主页眉内。如果将页眉作为一个固定控件来放置，则这种放置方式可以让分段控件与主页眉无缝集成。

接下来的内容中，将通过一个具体实例的实现过程，详细讲解在页眉中使用分段控件的方法。

具体实现

实例文件 fenduan.html 的具体实现代码如下所示。

```html
<!DOCTYPE html>
<html>
<head>
<meta charset="utf-8">
<title>Segmented Control Example</title>
<meta name="viewport" content="width=device-width, initial-scale=1">
<link rel="stylesheet" href="http://code.jquery.com/mobile/1.0/jquery.mobile-1.0.min.css" />
<style>
    .segmented-control { text-align:center;}
    .segmented-control .ui-controlgroup { margin: 0.2em; }
    .ui-control-active, .ui-control-inactive { border-style: solid; border-color: gray; }
    .ui-control-active { background: #BBB; }
    .ui-control-inactive { background: #DDD; }
</style>
<script src="http://code.jquery.com/jquery-1.6.4.min.js"></script>
<script src="http://code.jquery.com/mobile/1.0/jquery.mobile-1.0.min.js"></script>
</head>
<body>

<div data-role="page">
  <div data-role="header" data-position="fixed">
      <h1>精彩影视</h1>
      <div class="segmented-control ui-bar-d">
          <div data-role="controlgroup" data-type="horizontal">
              <a href="#" data-role="button" class="ui-control-active">剧院模式</a>
```

```
                <a href="#" data-role="button" class="ui-control-inactive">马上回来</a>
                <a href="#" data-role="button" class="ui-control-inactive">最受欢迎的</a>
            </div>
        </div>
    </div>

    <div data-role="content">
        <ul data-role="listview">
            <li>
                <a href="#">
                    <img src="images/111.jpg" />
                    <h3>变形金刚</h3>
                    <p>评论：PG</p>
                    <p>时长：95 min.</p>
                </a>
            </li>
            <li>
                <a href="#">
                    <img src="images/222.jpg" />
                    <h3>X 战警</h3>
                    <p>评论：PG-13</p>
                    <p>时长：137 min.</p>
                </a>
            </li>
            <li>
                <a href="#">
                    <img src="images/3313.jpg" />
                    <h3>雷雨</h3>
                    <p>评论 PG-13</p>
                    <p>时长：131 min.</p>
                </a>
            </li>
            <li>
                <a href="#">
                    <img src="images/444.jpg" />
                    <h3>小李飞刀</h3>
                    <p>评论：PG</p>
                    <p>时长：95 min.</p>
                </a>
            </li>
        </ul>
    </div>
</div>

</body>
</html>
```

本实例的执行效果如图 13-4 所示。

图 13-4　执行效果

在上述实例代码中，分段控件可以按照特定的分类来显示电影。该分段控件允许用户通过他们选择的分类（剧院模式、马上回来或最受欢迎的）来切换模式。

实例 255　在 iOS 网页中使用页脚

实例 255	在 iOS 网页中使用页脚
源码路径	\daima\255\

实例说明

页脚使用属性 data-role="footer"定义，是一些按照从左到右的顺序直线放置它的按钮。这种灵活性可以用来创建工具栏或标签栏。使用 data-theme 属性可以调整页脚的主题。如果不为页脚设置主题，则它会继承页面组件的主题。默认的主题是黑色的（data-theme="a"）。通过添加 data-position="fixed"属性的方式，可以固定页脚的位置。在默认情况下，所有的页脚级别(H1~H6)具有相同的风格，以维持视觉上的一致性。

在现实应用中，最简单的页脚形式如下面的代码所示。

```
<div data-role="footer">
<!--在此添加页脚文本或按钮-->
</div>
```

其中 data-role="footer"是唯一需要的属性。在页脚内可以包含任何语义 HTML。页脚通常包含工具栏和标签控件。工具栏提供了一组用户可以在当前环境中使用的动作。标签栏则可以允许用户在应用程序内的不同视图之间进行切换。

在接下来的内容中，将通过一个具体实例的实现过程，详细讲解在 iOS 系统中使用页脚的基本方法。

具体实现

实例文件 foot.html 的具体实现代码如下所示。

```
<!DOCTYPE html>
<html>
    <head>
    <meta charset="utf-8">
    <title>Default Header Footer Example</title>
    <meta name="viewport" content="width=device-width, initial-scale=1">
    <link rel="stylesheet" href="http://code.jquery.com/mobile/1.0/jquery.mobile-1.0.min.css" />
    <script src="http://code.jquery.com/jquery-1.6.4.min.js"></script>
    <script src="http://code.jquery.com/mobile/1.0/jquery.mobile-1.0.min.js"></script>
</head>
<body>
<div data-role="page">
    <div data-role="header">
        <h1>页头</h1>
    </div>

    <div data-role="content">
    在默认的底部位置时，内容不消耗整个装置的高度。
    </div>

    <div data-role="footer">
        <h3>页脚</h3>
    </div>
</div>
</body>
</html>
```

上述实例代码执行后的效果如图13-5所示。

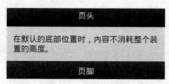

图13-5 执行效果

为了将页脚内容定位在屏幕的最底部显示，可以为页脚元素添加属性data-position="fixed"。在默认的情况下，页脚位于内容的后面，而不是位于屏幕底部。如果内容只是占据了一半的屏幕高度，则页脚会出现在屏幕的中央位置。

实例 256　在 iOS 系统中使用工具栏

实例 256	在 iOS 系统中使用工具栏
源码路径	\daima\256\

实例说明

工具栏可用来辅助管理当前屏幕中的内容。例如，邮件应用程序通常使用工具栏来管理电子邮件。当用户需要执行与当前屏幕中的对象相关联的动作时，工具栏会非常有用。在构建工具栏时，可以选择使用图标或文本。只有图标构成的工具栏的好处是，与文本构成的工具栏相比占据更少的屏幕空间。在选择图标时，建议选择能够表达正确含义的标准图标。请读者看如下所示的实例，演示了在 iOS 系统中使用工具栏的过程。

具体实现

实例文件 gongju.html 的具体实现代码如下所示。

```html
<!DOCTYPE html>
<html>
    <head>
        <meta charset="utf-8">
        <title>Toolbar example with icons</title>
        <meta name="viewport" content="width=device-width, initial-scale=1">
        <link rel="stylesheet" href="http://code.jquery.com/mobile/1.0/jquery.mobile-1.0.min.css" />
        <style>
            /* wrap the text for the movie review */
            .ui-li-desc { white-space: normal; margin-right: 20px; }
        </style>
        <script src="http://code.jquery.com/jquery-1.6.4.min.js"></script>
        <script src="http://code.jquery.com/mobile/1.0/jquery.mobile-1.0.min.js"></script>
    </head>
<body>

<div data-role="page">
    <div data-role="header">
        <h1>电影评论</h1>
    </div>

    <div data-role="content">
        <ul data-role="listview" data-inset="true" data-theme="e">
            <li data-role="list-divider">X-战警
              <p class="ui-li-aside">评级：<em>1,588</em></p></li>
            <li>
                <img src="images/thumbs-up.png" class="ui-li-icon">
                <p>去看看它！这部电影是好演员和特殊效果是令人置信的。值得的门票价格。</p>
```

```html
            </li>
        </ul>

        <ul data-role="listview" data-inset="true" data-theme="e">
          <li data-role="list-divider">评论</li>
                <li>
                    <img src="images/111-user.png" class="ui-li-icon">
                    <p>感谢评论，这周末我就去看。</p>
                    <span class="ui-li-count">1 天前</span>
                </li>
                <li>
                    <img src="images/111-user.png" class="ui-li-icon">
                    <p>你的评论非常有用！</p>
                    <span class="ui-li-count">3 天前</span>
                </li>
        </ul>
    </div>

    <!-- toolbar with icons -->
    <div data-role="footer" data-position="fixed">
        <div data-role="navbar">
            <ul>
                <li><a href="#" data-icon="arrow-l"></a></li>
                <li><a href="#" data-icon="back"></a></li>
                <li><a href="#" data-icon="star"></a></li>
                <li><a href="#" data-icon="plus"></a></li>
                <li><a href="#" data-icon="arrow-r"></a></li>
            </ul>
        </div>
    </div>
</div>
</body>
</html>
```

上述实例执行后的效果如图 13-6 所示。

图 13-6　执行效果

在上述执行效果中有一个显示电影评论的屏幕，为了帮助用户管理评论，可以利用一个由标准图标构成的工具栏，此工具栏允许用户执行如下所示的 5 种动作。

- 导航到前面的评论。
- 回复评论。
- 将评论标记为最喜欢的评论。
- 添加一条新的电影评论。

- 导航到后面的评论。

在创建工具栏时，仅需要最少的标记。在含有属性 data-role="navbF"的 div 中，只需要其中包含按钮的一个无序列表即可。工具栏按钮相当灵活，而且可以根据设备的宽度进行等间距排放。

实例 257　使用带有标准图标的标签栏

实例 257	使用带有标准图标的标签栏
源码路径	\daima\257\

实例说明

在移动 Web 设计应用中，可以将页脚设计为一个标签栏，通过标签栏可以以不同的视图来查看应用程序。其实，标签栏的行为与 Web 上可以见到的基于标签的导航相类似。标签栏通常作为一个永久的页脚出现在屏幕的底部边缘，而且用户可以在应用程序的任何位置访问它。标签栏通常包含同时显示图标和文本的按钮。

在接下来的内容中，将通过一个具体实例的实现过程，详细讲解在 iOS 中使用带有标准图标的标签栏的方法。

具体实现

实例文件 tabbar.html 的具体实现代码如下所示。

```html
<body>
<div data-role="page">
    <div data-role="header">
        <h1>精彩视频</h1>
    </div>

    <div data-role="content">
        <ul data-role="listview">
            <li>
                <a href="#">
                    <img src="images/111.jpg" />
                    <h3>变形金刚</h3>
                    <p>评级: PG</p>
                    <p>时长: 95 min.</p>
                </a>
            </li>
            <li>
                <a href="#">
                    <img src="images/222.jpg" />
                    <h3>X 战警</h3>
                    <p>评级: PG-13</p>
                    <p>时长: 137 min.</p>
                </a>
            </li>
            <li>
                <a href="#">
                    <img src="images/3313.jpg" />
                    <h3>雷雨</h3>
                    <p>评级: PG-13</p>
                    <p>时长: 131 min.</p>
                </a>
            </li>
            <li>
                <a href="#">
                    <img src="images/444.jpg" />
                    <h3>小李飞刀</h3>
                    <p>评级: PG</p>
                    <p>时长: 95 min.</p>
                </a>
            </li>
        </ul>
```

```
        </div>
        <!-- tab bar with standard icons -->
        <div data-role="footer" data-position="fixed">
            <div data-role="navbar">
                <ul>
                    <li><a href="#" data-icon="home">主页</a></li>
                    <li><a href="#" data-icon="star" class="ui-btn-active">电影</a></li>
                    <li><a href="#" data-icon="grid">剧场</a></li>
                </ul>
            </div>
        </div>
    </div>
</body>
```

上述实例执行后的效果如图 13-7 所示。

图 13-7　执行效果

实例 258　在 iOS 中使用链接按钮

实例 258	在 iOS 中使用链接按钮
源码路径	\daima\258\

实例说明

在 jQuery Mobile 中有多种形式的按钮，主要有链接按钮、表单按钮、图像按钮、只带有图标的按钮，以及同时带有文本和图标的按钮。在现实应用中，jQuery Mobile 按钮都具有一致的样式风格。无论使用链接按钮还是基于表单的按钮，jQuery Mobile 框架都会以完全相同的方式来处理这些按钮。

在 jQuery Mobile 应用中，链接按钮是最常使用的按钮类型。当需要将一个普通链接设计为按钮时，需要为链接添加如下所示的属性。

```
data-role="button"
```

在默认的情况下，页面中的内容区域内的按钮都被设计为块级元素，这样可以填充其外层容器（即内容区域）的整个宽度。但是，如果需要的是一个更为紧凑的按钮，使其宽度与按钮内部的文本和图标的宽度相同，则可以添加如下所示的属性。

```
data-inline="true"
```

在接下来的内容中，将通过一个具体实例的实现过程,详细讲解在 iOS 中使用链接按钮的方法。

具体实现

实例文件 link.html 的具体实现代码如下所示。

```
<!DOCTYPE html>
<html>
    <head>
        <meta charset="utf-8">
```

```html
        <title>按钮</title>
        <meta name="viewport" content="width=device-width, minimum-scale=1.0, maximum-scale=1.0;">
        <link rel="stylesheet" href="http://code.jquery.com/mobile/1.0/jquery.mobile-1.0.min.css" />
        <script src="http://code.jquery.com/jquery-1.6.14.min.js"></script>
        <script src="http://code.jquery.com/mobile/1.0/jquery.mobile-1.0.min.js"></script>
</head>
<body>

<div data-role="page" data-theme="b">
    <div data-role="header">
        <h1>演示按钮的用法</h1>
    </div>

    <div data-role="content">
        <p style="text-align:center;">
            <em>&lt;a href="#" <strong>data-role="button"</strong>&gt;链接按钮&lt;/a&gt;</em>链接按钮<a href="#" data-role="button"></a>

            <br><br>

            <em>&lt;a href="#" data-role="button" <strong>data-inline="true"</strong>&gt;同意&lt;/a&gt;</em>
            <a href="#" data-role="button" data-inline="true" data-rel="back" data-theme="a">不同意</a>
            <a href="#" data-role="button" data-inline="true" data-theme="c">同意</a>
        </p>
    </div>
</div>

</body>
</html>
```

在上述实例代码中,如果希望让按钮并排放置,并占据屏幕的整个宽度,则可以使用一个两列的网格。执行后的效果如图 13-8 所示。

图 13-8 执行效果

实例 259 在 iOS 中使用分组按钮

实例 259	在 iOS 中使用分组按钮
源码路径	\daima\259\

实例说明

在移动 Web 开发应用中,可以将按钮包含在一个控件组内。在 jQuery Mobile 应用中,要实现这一效果,可以使用如下所示的属性将一组按钮包装在容器中。

```
data-role="controlgroup"
```

在默认情况下,框架会对按钮进行垂直分组,并移除所有的页边空白(margin),以及在按钮

第13章 移动 Web 实战

之间添加边界。为了在视觉上增强分组，通常使用圆角设计第一个和最后一个元素。由于按钮在默认情况下是垂直摆放的，可以添加属性 data-type="horizontal"来水平摆放按钮。垂直摆放的按钮会占据其外层容器的整个宽度，而水平摆放的按钮的宽度则只与其内容一样宽。

在接下来的内容中，将通过一个具体实例的实现过程，详细讲解在 iOS 中使用分组按钮的方法。

具体实现

实例文件 fenduan.html 的具体实现代码如下所示。

```html
<!DOCTYPE html>
<html>
  <head>
    <meta charset="utf-8">
    <title>Segmented Control Example</title>
    <meta name="viewport" content="width=device-width, initial-scale=1">
    <link rel="stylesheet" href="http://code.jquery.com/mobile/1.0/jquery.mobile-1.0.min.css" />
    <style>
        .segmented-control { text-align:center;}
        .segmented-control .ui-controlgroup { margin: 0.2em; }
        .ui-control-active, .ui-control-inactive { border-style: solid; border-color: gray; }
        .ui-control-active { background: #BBB; }
        .ui-control-inactive { background: #DDD; }
    </style>
    <script src="http://code.jquery.com/jquery-1.6.14.min.js"></script>
    <script src="http://code.jquery.com/mobile/1.0/jquery.mobile-1.0.min.js"></script>
  </head>
  <body>

<div data-role="page">
  <div data-role="header" data-position="fixed">
      <h1>精彩影视</h1>
      <div class="segmented-control ui-bar-d">
          <div data-role="controlgroup" data-type="horizontal">
              <a href="#" data-role="button" class="ui-control-active">剧院模式</a>
              <a href="#" data-role="button" class="ui-control-inactive">马上回来</a>
              <a href="#" data-role="button" class="ui-control-inactive">最受欢迎的</a>
          </div>
      </div>
  </div>

  <div data-role="content">
      <ul data-role="listview">
          <li>
            <a href="#">
              <img src="images/111.jpg" />
              <h3>变形金刚</h3>
                <p>评论：PG</p>
                <p>时长：95 min.</p>
            </a>
          </li>
          <li>
            <a href="#">
              <img src="images/222.jpg" />
              <h3>X 战警</h3>
                <p>评论：PG-13</p>
                <p>时长：137 min.</p>
            </a>
          </li>
          <li>
            <a href="#">
              <img src="images/333.jpg" />
              <h3>雷雨</h3>
                <p>评论 PG-13</p>
```

```
            <p>时长：131 min.</p>
          </a>
        </li>
        <li>
          <a href="#">
            <img src="images/4414.jpg" />
            <h3>小李飞刀</h3>
            <p>评论：PG</p>
            <p>时长：95 min.</p>
          </a>
        </li>
      </ul>
    </div>
  </div>

  </body>
</html>
```

本实例的执行效果如图 13-9 所示。

图 13-9　执行效果

当对按钮进行水平分组时，控件组的宽度超出了屏幕宽度会发生重叠现象。

实例 260　在 iOS 中创建并使用动态按钮

实例 260	在 iOS 中创建并使用动态按钮
源码路径	\daima\260\

实例说明

在 jQuery Mobile 应用中，button 插件（plugin）是一个能自动增强本地按钮的微件（widget）。可以使用该插件动态创建、启用和禁用按钮。如果需要在代码中动态创建按钮，可以有两个选择：通过标记驱动的方法动态创建按钮；显式设置 button 插件的选项。在标记驱动的方法中可以为新按钮创建 jQuery Mobile 标记，然后将其添加到内容容器中，然后再进行增强处理。

在接下来的内容中，将通过具体实例的实现过程，详细讲解在 iOS 中实现动态按钮的方法。

具体实现

实例文件 d-buttons.html 的具体实现代码如下所示。

```
<!DOCTYPE html>
<html>
  <head>
  <meta charset="utf-8">
  <title>Buttons</title>
  <meta name="viewport" content="width=device-width, minimum-scale=1.0, maximum-scale
=1.0;">
```

```html
    <link rel="stylesheet" href="http://code.jquery.com/mobile/1.0/jquery.mobile-1.0.min.css" />
    <script src="http://code.jquery.com/jquery-1.6.14.min.js"></script>
    <script src="http://code.jquery.com/mobile/1.0/jquery.mobile-1.0.min.js"></script>
</head>
<body>

<div data-role="page" data-theme="b">
 <div data-role="header">
     <h1>创建动态按钮</h1>
 </div>

 <div data-role="content">
     <a href="#" data-role="button" id="create-button1">创建按钮 1</a>
     <a href="#" data-role="button" id="create-button2">创建按钮 2</a>

     <br><br>
     <a href="#" data-role="button" id="create-multiple-buttons">创建多个按钮</a>
     <a href="#" data-role="button" id="create-button5">创建按钮 5</a>
     <a href="#" data-role="button" id="create-button6">创建按钮 6</a>
     <a href="#" data-role="button" id="disable-button3" data-theme="d">禁用的按钮 3</a>
     <a href="#" data-role="button" id="enable-button3" data-theme="d">可用的按钮 3</a>
 </div>

 <script type="text/javascript">
 <!--使用标记驱动的方法来创建动态按钮-->
     $( "#create-button1" ).bind( "click", function() {
         $( '<a href="http://jquerymobile.com" id="button1" data-role="button" data-icon="star" data-inline="true" data-theme="a">Button1</a>' )
             .appendTo( ".ui-content" )
             .button();
     });
     <!--使用插件驱动的方法来创建动态按钮-->
     $( "#create-button2" ).bind( "click", function() {
         $( '<a href="http://jquerymobile.com" id="button2">Button2</a>' )
             .insertAfter( "#create-button2" )
             .button({
                 corners: true,
                 icon: "home",
                 inline: true,
                 shadow: true,
                 theme: 'a',
                 create: function(event) {
                     console.log( "Creating button..." );
                     for (prop in event) {
                         console.log(prop + ' = ' + event[prop]);
                     }
                 }
             })
     });

     $( "#create-button5" ).bind( "click", function() {
         $( '<input type="submit" id="button5" value="Button5" data-theme="a" />' )
             .insertAfter( "#create-button5" )
             .button();
     });

     $( "#create-button6" ).bind( "click", function() {
         $( '<input type="submit" id="button6" value="Button6" />' )
             .insertAfter( "#create-button6" )
             .button({
                 'icon': "home",
                 'inline': true,
                 'shadow': true,
                 'theme': 'a'
```

```
        })
    });

    $( "#create-multiple-buttons" ).bind( "click", function() {
        $( '<button id="button3" data-theme="a">Button3</button>' ).insertAfter( "#create-multiple-buttons" );
        $( '<button id="button4" data-theme="a">Button4</button>' ).insertAfter( "#button3" );
        $.mobile.pageContainer.trigger( "create" );
    });
    <!--创建按钮,并动态禁用/启动它们按钮-->
    $( "#disable-button3" ).bind( "click", function() {
        $( "#button3" ).button( "disable" );
    });

    $( "#enable-button3" ).bind( "click", function() {
        $( "#button3" ).button( "enable" );
    });
</script>
</div>

</body>
</html>
```

在上述示例代码中，JavaScript 语句是整个程序的核心，这段 JavaScript 语句的实现流程如下所示。

（1）使用标记驱动的方法来创建动态按钮。

在标记驱动的方法中，为新按钮创建 jQuery Mobile 标记，然后将其添加到内容容器中，然后再进行增强。

（2）使用插件驱动的方法来创建动态按钮。

对于插件驱动的方法而言，需要创建一个本地链接，将按钮插入到页面中，然后应用按钮增强。

（3）创建按钮，并动态禁用/启动它们。

创建多个表单按钮，但是不再为每个按钮分别调用 button 插件，而是通过一次触发页面容器的 create 方法，对所有的按钮进行增强。另外，也可以使用 button 插件的 enable 和 disable 方法动态启用或禁用按钮。

本实例执行后的初始效果如图 13-10 所示。

触摸单击图 13-10 中的某个按钮后，会动态创建对应的按钮。例如，摸单击"创建多个按钮"后，会在下方自动创建两个按钮："Button3"和"Button4"，如图 13-11 所示。

图 13-10 初始执行效果

图 13-11 动态自动创建两个按钮："Button3"和"Button4"

实例 261　在 iOS 中使用表单

实例 261	在 iOS 中使用表单
源码路径	\daima\261\

实例说明

在 jQuery Mobile 应用中，实现基于表单的应用程序的方法和传统构建 Web 表单的方法类似。在默认情况下，action 属性会默认为当前页面的相对路径，该路径可以通过$.mobile.path.get()找到，而未指定的 method 属性默认为"get"。在提交表单时，通过默认的"滑动"转换将当前的页面转换到后续页面。但是通过之前用来管理链接的属性可以配置表单的转换行为。

在接下来的内容中，将通过一个具体实例的实现过程，详细讲解在 iOS 中使用表单的方法。

具体实现

实例文件 form.html 的具体实现代码如下所示。

```html
<!DOCTYPE html>
<html>
    <head>
        <meta charset="utf-8">
        <title>Forms</title>
        <meta name="viewport" content="width=device-width, minimum-scale=1.0, maximum-scale=1.0;">
        <link rel="stylesheet" href="http://code.jquery.com/mobile/1.0/jquery.mobile-1.0.min.css" />
        <style>
            label {
                float: left;
                width: 5em;
            }

            input.ui-input-text {
                display: inline !important;
                width: 12em !important;
            }

            form p {
                clear: left;
                margin: 1px;
            }
        </style>
        <script src="http://code.jquery.com/jquery-1.6.4.min.js"></script>
        <script src="http://code.jquery.com/mobile/1.0/jquery.mobile-1.0.min.js"></script>
    </head>
    <body>
        <div data-role="page" data-theme="b">
            <div data-role="header">
                <h1>提交表单信息</h1>
            </div>
            <div data-role="content">
                <form name="test" id="test" action="form-response.php" method="post" data-transition="pop">
                    <p>
                        <label for="email">邮箱:</label>
                        <input type="email" name="email" id="email" value="" placeholder="Email" data-theme="d"/>
                    </p>
                    <p>
                        <button type="submit" data-theme="a" name="submit">提交</button>
                    </p>
                </form>
```

```
            </div>
        </div>
    </body>
</html>
```

在上述实例代码中，使用"form"标记简单实现了一个表单效果。执行后的效果如图 13-12 所示。

图 13-12　执行效果

可以继续在表单元素中添加如下所示的属性，以管理转换或禁用 Ajax。

```
data-transition="pop"
data-direction="reverse"
data-ajax="false"
```

在整个站点中，需要确保每一个表单的 id 属性都是唯一的。在进行表单转换时，jQuery Mobile 会同时将"from"页面和"to"页面载入到 DOM 中，以完成平滑的转换。为了避免任何冲突，所以要确保表单的 id 必须唯一。

实例 262　在 iOS 中使用选择菜单

实例 262	在 iOS 中使用选择菜单
源码路径	\daima\262\

实例说明

即使在无需添加额外标记的情况下，jQuery Mobile 框架也会自动增强所有本地的选择元素。在默认情况下，通过轻敲某个选择按钮的方式会为移动设备启动本地选择选择器，可以设置 jQuery Mobile 使其显示自定义的选择菜单。

在接下来的内容中，将通过一个具体实例的实现过程，详细讲解在 iOS 中使用选择菜单的方法。

具体实现

实例文件 select.html 的具体实现代码如下所示。

```
<div data-role="page" data-theme="b">
  <div data-role="header">
      <h1>使用选择菜单</h1>
  </div>

  <div data-role="content">
    <form id="test" id="test" action="#" method="post">

      <p>
          <label for="genre">属性:</label>
          <select name="genre" id="genre" multiple="multiple">
              <option value="action">Action</option>
              <option value="comedy">Comedy</option>
              <option value="drama">Drama</option>
              <option value="romance">Romance</option>
          </select>
      </p>
      <p>
          <label for="delivery">方式:</label>
          <select name="delivery" id="delivery">
            <option value="barcode">电子客票</option>
            <option value="nfc">NFC</option>
            <option value="overnight">晚上送</option>
```

第 13 章 移动 Web 实战

```
            <option value="express">快递</option>
            <option value="ground">地面</option>
            <option value="overnight">在晚上</option>
            <option value="express">快递</option>
            <option value="standard">地面</option>
                <optgroup label="Digital">
                    <option value="barcode" selected>E-Ticket</option>
                    <option value="nfc">NFC</option>
                </optgroup>
                <optgroup label="FedEx">
                    <option value="overnight">Overnight</option>
                    <option value="express">Express</option>
                    <option value="ground">Ground</option>
                </optgroup>
                <optgroup label="US Mail">
                    <option value="overnight">Overnight</option>
                    <option value="express">Express</option>
                    <option value="standard">Standard</option>
                </optgroup>
            </select>
        </p>
    </form>
  </div>
</div>
```

在用户选择之后，选择按钮会显示已选定选项的值。如果和按钮相比，文本值太长，则会截断文本，并在后面显示一个省略号。并且当用户选择了多个选项后，多选按钮会对已选中的选项显示计数泡或进行标记。

执行后的初始效果如图 13-13 所示。触摸按下某个选项后会自动弹出该选项下面的菜单，例如触摸按下"方式"后面的 ⌄ 后会弹出一个图 13-14 所示的菜单框。

图 13-13 初始效果

图 13-14 弹出选项下的菜单框

> **注意** 在使用 multiple="multiple" 属性创建选择菜单时，有些移动平台不支持多选特性。在需要使用多选菜单的时候，建议使用自定义菜单。

实例 263 在 iOS 中使用单选按钮

实例 263	在 iOS 中使用单选按钮
源码路径	\daima\263\

实例说明

在 jQuery Mobile 应用中，单选按钮只允许用户选择一个条目。在默认情况下，单选按钮会继承其父控件的主题。但是如果想为单选按钮应用其他主题，需要为相应单选按钮的标签添加 data-theme 属性。

在接下来的内容中，将通过一个具体实例的实现过程，详细讲解在 iOS 中使用单选按钮的方法。

具体实现

实例文件 radio.html 的具体实现代码如下所示。

```
<div data-role="page">
 <div data-role="header">
     <h1>使用单选按钮</h1>
 </div>

 <div data-role="content">
   <form id="test" id="test" action="#" method="post">

     <fieldset data-role="controlgroup">
     <legend>地图模式：</legend>
         <input type="radio" name="map" id="map1" value="Map" checked="checked" />
     <label for="map1" data-theme="b">街道</label>
         <input type="radio" name="map" id="map2" value="Satellite"  />
     <label for="map2" data-theme="b">卫星</label>
     <input type="radio" name="map" id="map3" value="Hybrid"  />
        <label for="map3" data-theme="b">鸟瞰</label></fieldset>

     <fieldset data-role="controlgroup" data-type="horizontal">
      <legend>观看模式：</legend>
         <input type="radio" name="map" id="map1" value="Map" checked="checked" />
     <label for="map1">城区</label>

     <input type="radio" name="map" id="map2" value="Satellite" />
     <label for="map2">卫星</label>

     <input type="radio" name="map" id="map3" value="Hybrid"  />
     <label for="map3">俯视</label></fieldset>

   </form>
  </div>
 </div>
```

在上述实例代码中添加了如下 3 个额外的属性，以帮助设计和放置单选按钮。

- 第一个属性 data-role="controlgroup"对按钮进行编组，而且编组后的按钮是圆角的。
- 第二个属性 data-type="horizontal"重写按钮默认的垂直定位，以水平方式显示按钮。
- 第三个属性用来对按钮进行主题化。

执行后的效果如图 13-15 所示。

如果水平放置的单选按钮的容器无法在一行内显示所有的单选按钮，则按钮会发生重叠现象。为了避免重叠，可以通过如下代码减小按钮的字体大小。

图 13-15　执行效果

```
ui- controlgroup- horizontal.ui- radio label{
font-size:13px !important;
}
```

实例 264　在 iOS 中水平放置复选框

实例 264	在 iOS 中水平放置复选框
源码路径	\daima\264\

实例说明

复选框允许用户从一系列选择中选择多个值,复选框和单选按钮相对,用于设计和定位复选框的标记与之前用于单选按钮的标记相同。在复选框中添加了如下所示的 3 个额外的属性。

- 第一个属性 data-role="controlgroup"将复选框元素进行编组,而且编组后的复选框是圆角的。
- 第二个属性 data-type="horizontal"重写按钮默认的垂直定位,以水平方式显示按钮。
- 第三个属性用来对按钮进行主题化。默认情况下,复选框会继承其父控件的主题。但是,如果想为复选框应用其他主题,需要为相应复选框的标签添加 data-theme 属性。

在 jQuery Mobile 应用中,checkboxradio 插件能够自动增强复选框和单选按钮。通过该插件可以动态创建、启用、禁用和刷新复选框。

在接下来的内容中,将通过一个具体实例的实现过程,详细讲解在 iOS 中水平放置复选框的方法。

具体实现

实例文件 check.html 的具体实现代码如下所示。

```html
<div data-role="page">
  <div data-role="header">
      <h1>使用复选框</h1>
  </div>

  <div data-role="content">
    <form id="test" id="test" action="#" method="post">

      <fieldset data-role="controlgroup">
      <legend>选择喜欢的类型:</legend>
          <input type="checkbox" name="genre" id="c1" />
      <label for="c1"data-theme="c">古装</label>

      <input type="checkbox" name="genre" id="c2" />
      <label for="c2" data-theme="c">言情</label>

      <input type="checkbox" name="genre" id="c3" />
       <label for="c3" data-theme="c">警匪</label>

      </fieldset>

      <fieldset data-role="controlgroup" data-type="horizontal">
      <legend>类型:</legend>
          <input type="checkbox" name="genre" id="c1" />
      <label for="c1" data-theme="b">古装</label>

      <input type="checkbox" name="genre" id="c2" />
      <label for="c2" data-theme="b">言情</label>

      <input type="checkbox" name="genre" id="c3" />
      <label for="c3" data-theme="b">警匪</label>
      </fieldset>

    </form>
  </div>
</div>
```

执行后的效果如图 13-16 所示。

图 13-16　执行效果

如果水平放置的复选框的容器无法在一行内显示所有的复选框,则复选框会发生重叠现象。为了避免重叠,可以通过如下代码减小复选框的字体大小:

```
ui- controlgroup- horizontal.ui-checkbox label{
font-size:11px !important;
}
```

实例 265 在 iOS 中使用列表

实例 265	在 iOS 中使用列表
源码路径	\daima\265\

实例说明

在 jQuery Mobile 应用中,列表的代码表示一个包含了 data-role="listview" 属性无序列表 ul。jQuery Mobile 会把所有必要的样式应用在列表上,使其成为易于触摸的控件。当触摸点击列表项时,jQuery Mobile 会触发该列表项里的第一个链接,并通过 Ajax 请求链接的 URL 地址,在 DOM 中创建一个新的页面并产生页面转场效果。

当为列表元素添加了 data-role="list"属性之后,jQuery Mobile 能够将本地所有的 HTML 列表(<ul>或<ol>)自动增强为一个优化的移动视图。在默认情况下,在显示增强后的列表时会占据整个屏幕。如果列表条目包含链接,则会以易于触摸的按钮方式来显示,而且会带有一个右对齐的箭头图标。在默认情况下,列表会使用调色板颜色"c"(灰色)来样式化。要应用其他主题,则需要为列表元素或列表条目(<li>)添加 data-theme 属性。

在接下来的内容中,将通过一个具体实例的实现过程,详细讲解在 iOS 中使用列表的方法。

具体实现

实例文件 basic.html 的具体实现代码如下所示。

```html
<!DOCTYPE html>
<html>
  <head>
  <meta charset="utf-8">
  <title>Lists</title>
  <meta name="viewport" content="width=device-width, minimum-scale=1.0, maximum-scale=1.0">
  <link rel="stylesheet" href="http://code.jquery.com/mobile/1.0/jquery.mobile-1.0.min.css" />
  <script src="http://code.jquery.com/jquery-1.16.4.min.js"></script>
  <script src="http://code.jquery.com/mobile/1.0/jquery.mobile-1.0.min.js"></script>
  </head>
<body>

<div data-role="page">
<div data-role="header">
    <h1>使用列表</h1>
</div>

<div data-role="content">
    <ul data-role="listview" data-theme="c">
        <li><a href="#">AAA</a></li>
        <li><a href="#">BBB</a></li>
        <li><a href="#">CCC</a></li>
        <li><a href="#">DDD</a></li>
        <li><a href="#">EEE</a></li>
        <li><a href="#">FFF</a></li>
        <li><a href="#">GGG</a></li>
        <li><a href="#">HHH</a></li>
        <li><a href="#">IIIII</a></li>
    </ul>
```

```
        </div>
    </div>

    </body>
</html>
```

在上述实例代码中，使用"ul"和"ui"标记简单实现了一个列表效果。执行后的效果如图 13-17 所示。

图 13-17　执行效果

实例 266　在 iOS 中使用两列表格

实例 266	在 iOS 中使用两列表格
源码路径	\daima\266\

实例说明

在 jQuery Mobile 应用中，要想构建两栏的布局（50%，50%），需要先构建一个父容器，并在里面添加一个名字为"ui-grid-a"的 class，以在内部设置两个子容器，给第一个子容器添加 class:ui-block-a，第二个子容器添加 class:ui-block-b。例如：

```
<div class="ui-grid-a">
 <div class="ui-block-a"><strong>I'm Block A</strong> and text inside will wrap</div>
 <div class="ui-block-b"><strong>I'm Block B</strong> and text inside will wrap</div>
</div><!-- /grid-a -->
```

上述代码的执行效果如图 13-1 所示。

图 13-18　执行效果

如上图 13-18 的执行效果所示，默认的两栏没有样式，并行排列。分栏的 class 可以应用到任何类型的容器上。而在图 13-19 所示的效果中，给表单的 fieldset 添加 class="ui-grid-a"，然后给两个 button 所在的子容器添加属性 class="ui-block-a"和 class="ui-block-b"，设置使两个容器各自占 50%的宽。

在图 13-20 所示的区块中增加了两个 class，增加 ui-bar 的 class 给默认的 bar padding，增加的 ui-bar-e 的 class 应用背景渐变以及工具栏的主题 e 的字体样式。然后在每个网格的标签内增加 style="height:120px"的属性来设置高度。

图 13-19　执行效果

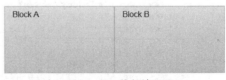

图 13-20　执行效果图

具体实现

实例文件 2col.html 的具体实现代码如下所示。

```html
<!DOCTYPE html>
<html>
    <head>
        <meta charset="utf-8">
        <title>Grid Example</title>
        <!--<meta name="viewport" content="width=device-width, initial-scale=1">-->
        <meta name="viewport" content="width=device-width, maximum-scale=1">
        <link rel="stylesheet" href="http://code.jquery.com/mobile/1.0/jquery.mobile-1.0.min.css" />
        <script src="http://code.jquery.com/jquery-1.6.4.min.js"></script>
        <script src="http://code.jquery.com/mobile/1.0/jquery.mobile-1.0.min.js"></script>
    </head>
    <body>

<div data-role="page" id="home">
    <div data-role="header">
        <h1>两列的表格</h1>
    </div>

    <div data-role="content" >
        <div class="ui-grid-a">
            <div class="ui-block-a"><strong>块A</strong><br>The text will wrap within the grid.</div>
            <div class="ui-block-b"><strong>块B</strong><br>More text.</div>
        </div>
    </div>
</div>
</body>
</html>
```

在上述实例代码中，外层表格（Outer Grid）使用 CSS 表格属性 ui-grid-a 进行配置。然后添加了两个内层块，第一个块被分配了一个 CSS 属性 ui-block-a，第二个块被分配了一个 CSS 属性 ui-block-b。执行结果如图 13-21 所示。列是等间距、无边界的，而且每个块内的文本在必要时会换行显示。作为一个额外的优点，iQuery Mobile 内的表格相当灵活，而且会根据不同的屏幕显示尺寸以自适应的方式进行呈现。

图 13-21　执行效果

实例 267　在 iOS 中实现可折叠内容块效果

实例 267	在 iOS 中实现可折叠内容块效果
源码路径	\daima\267\

实例说明

在 jQuery Mobile 应用中，在创建可折叠的内容块时需要如下所示的两个元素。

（1）创建一个容器并添加 data-role="collapsible"属性。

可以通过添加 data-collapsed 属性的方式，将容器配置为折叠的或展开的。在默认情况下，可折叠的区域块将会以展开方式显示(data-collpased="false")。为了在最初以折叠方式显示区域块，需要为容器添加属性 data-collpased="true"属性。

（2）在容器内，添加任意的页眉元素（H1～H6）。jQuery Mobile 框架会对页眉进行样式化，使其看起来就像是一个带有左对齐的加号图标或减号图标的可单击按钮，其中加号图标或减号图标用来指示该容器是否是展开的。

在页眉之后，可以为可折叠的区域块添加任何 HTML 标记。jQuery Mobile 框架会将该标记包含在容器内，当用户轻敲页眉时，该容器或者是展开，或者是折叠。通过为可折叠的容器添加 data-theme 和 data-content-theme 属性，可以分别主题化可折叠的块和与其相关联的按钮。

在接下来的内容中，将通过一个具体实例的实现过程，详细讲解在 iOS 中实现可折叠内容块效果的方法。

具体实现

实例文件 block.html 的具体实现代码如下所示。

```html
<div data-role="page" id="home" data-theme="b">
  <div data-role="header" data-theme="a">
      <h1>设置</h1>
  </div>

  <div data-role="content">

      <div data-role="collapsible" data-collapsed="true" data-theme="a" data-content-theme="b">
          <h3>无线</h3>
          <ul data-role="listview" data-inset="true">
              <li><a href="#"><img src="images/cloud-default.png" height="22" width="22">MM</a></li>
              <li><a href="#"><img src="images/cloud-default.png" height="22" width="22">NN</a></li>
          </ul>
      </div>

      <div data-role="collapsible" data-theme="a" data-content-theme="b">
          <h3>程序应用</h3>
          <ul data-role="listview" data-inset="true">
              <li><a href="#"><img src="images/cloud-default.png" height="22" width="22">AA</a></li>
              <li><a href="#"><img src="images/cloud-default.png" height="22" width="22">BB</a></li>
              <li><a href="#"><img src="images/cloud-default.png" height="22" width="22">CC</a></li>
          </ul>
      </div>

      <div data-role="collapsible" data-collapsed="true" data-theme="a" data-content-theme="b">
          <h3>显示</h3>
          <ul data-role="listview" data-inset="true">
              <li><a href="#">DD</a></li>
              <li><a href="#">EE</a></li>
          </ul>
      </div>

      <div data-role="collapsible" data-collapsed="true" data-theme="a" data-content-theme="b">
          <h3>声音</h3>
          <ul data-role="listview" data-inset="true">
              <li><a href="#">FF</a></li>
              <li><a href="#">GG</a></li>
          </ul>
      </div>

      <div data-role="collapsible" data-collapsed="true" data-theme="a" data-content-theme="b">
          <h3>安全</h3>
          <ul data-role="listview" data-inset="true">
              <li><a href="#">HH</a></li>
              <li><a href="#">XX</a></li>
          </ul>
      </div>
  </div>
</div>
```

```
</div>
```
在上述实例代码中，除了默认情况下为展开状态的"程序应用"区域块之外，其他所有的内容块都已经显式设置为折叠状态。执行后的效果如图 13-22 所示。

图 13-22　执行效果

实例 268　搭建 PhoneGap 开发环境

实例 268	搭建 PhoneGap 开发环境
源码路径	无

实例说明

在使用 PhoneGap 进行移动 Web 开发之前，需要先搭建 PhoneGap 开发环境。在安装 PhoneGap 开发环境之前，需要先安装如下所示的框架。

- Java SDK。
- Eclipse。
- iOS SDK。
- ADT Plugin。

具体实现

（1）登录 PhoneGap 的官方网站：http://phonegap.com/download/，如图 13-23 所示。

（2）单击最新版本下方的 ![download] 按钮，下载 PhoneGap 开发包，下载成功后的压缩包名为 "phonegap-2.9.0.zip"。

图 13-23　PhoneGap 的官方网站

（3）解压缩文件 phonegap-2.9.0.zip，假设解压到本地硬盘的"D"目录下，解压后的根目录名是"phonegap-2.9.0"，双击打开后的效果如图 13-24 所示。

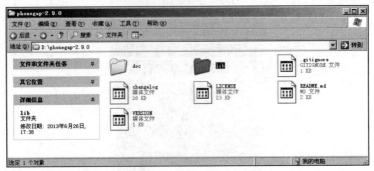

图 13-24 "phonegap-2.9.0"的根目录

对图 13-24 中各个子目录的具体说明如下所示。
- "doc"：在里面包含了 PhoneGap 的源代码文档，如图 13-25 所示。
- "lib"：在里面包含了 PhoneGap 支持的各种平台，如图 13-26 所示。

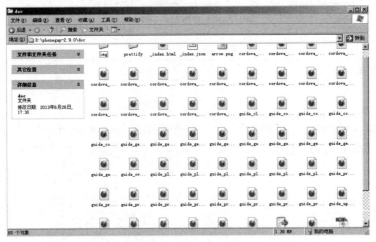

图 13-25 "doc"目录

图 13-26 "lib"目录

- "changelog"：一个日志文件，保存了更改历史记录信息和作者信息等。
- "LICENSE"：Apache 软件许可证（v2 版本）。
- "VERSION"：版本信息。
- "README.md"：帮助文档。

实例 269　在 iOS 平台创建基于 PhoneGap 的程序

实例 269	在 iOS 平台创建基于 PhoneGap 的程序
源码路径	\daima\269\

实例说明

在接下来的内容中，将创建第一个基于 iOSC 系统的 PhoneGap 实例。首先，利用 HTML、CSS 和 JavaScript 来搭建一个标准的 Web 应用程序，然后用 PhoneGap 封装来访问移动设备的基本信息，在 iOS 模拟器上调试成功后，最后部署到实体机。为了在不同的设备上得到一样的渲染效果，将采用 jQuery Mobile 来设计应用程序界面。

具体实现

（1）在开始之前需要先准备集成开发环境 Xcode，必须先安装 iOS SDK 以及 PhoneGap。如果应用程序仅在模拟器中运行，则不需要准备开发者证书。

（2）利用 Xcode 中的模板创建一个空项目，将整个目录结构主要分为 3 个部分：项目文件夹（以项目名称为文件夹名称，这里是 HelloWorld）、Frameworks 和 Products。Frameworks 中包含该应用可能用到的所有库文件，一般不需要修改。Products 文件夹包含了编译成功后的 .app 文件。HelloWorld 文件夹包含项目的主体文件，其中 Cordova.framework 引入了 Cordova 静态库，Resources 目录包含图片和国际化有关的资源。Classes 目录包含了应用程序委派的头文件和可执行文件、主界面控制器的头文件和可执行文件。Plugins 中包含了可能添加的插件头文件和可执行文件。Supporting Files 中的文件 .plist 类似于项目的 properties，包含项目基本信息（如名称和图标），InfoPlist.strings 包含国际化 info.plist 键值对。

图 13-27　目录结构

（3）把系统生成的 www 文件夹添加到 HelloWorld 中，具体做法是右键单击 HelloWorld 项目，在弹出的快捷菜单中选择"添加文件到 HelloWorld"菜单，然后选择 www 目录，最后点击 Finish 按钮。此时可以看到 www 文件夹出现在项目的文件列表下，并且文件夹的图标是蓝色的，表示该文件夹已经成为文件引用类型，而不是虚拟的目录。

创建后的目录结构如图 13-27 所示。

在 www 目录下编写测试的网页文件 index.html，具体代码如下所示。

```
<!DOCTYPE html>
<html>
<head>
  <meta charset="utf-8">
  <meta name="viewport" content="width=device-width, initial-scale=1">
  <title>index.html</title>
  <link rel="stylesheet" href="jquery.mobile-1.0.1.min.css" />
  <script type="text/javascript" charset="utf-8" src="jquery.js"></script>
  <script type="text/javascript" charset="utf-8" src="jquery.mobile-1.0.1.min.js"></script>
  <script type="text/javascript" charset="utf-8" src="cordova.js" ></script>
  <script type="text/javascript" charset="utf-8">

  $( function() {

  });
  $(document).ready(function(){
```

```
        console.log("jquery ready");
        document.addEventListener("deviceready", onDeviceReady, false);
        console.log("register the listener");
    });

    function onDeviceReady()
    {
        console.log("onDeviceReady");
        $(".content").html("<ul
data-role='listview'><li>"+device.name+"</li><li>"+device.cordova+"</li><li>"+device.
platform+"</li><li>"+device.version+"</li><li>"+device.uuid+"</li></ul>");
    }

    </script>
</head>
<body>
<!-- begin first page -->
<div id="page1" data-role="page" >
<header data-role="header"><h1>Hello World</h1></header>
<div data-role="content" class="content">
<h3>设备信息</h3>

</ul>
</div>
<footer data-role="footer"><h1>Footer</h1></footer>
</div>
<!-- end first page -->
</body>
</html>
```

在 iOS 模拟器中的执行效果如图 13-28 所示。

图 13-28　执行效果

实例 270　使用通知 API

实例 270	使用通知 API
源码路径	\daima\270

实例说明

　　作为一个良好的 PhoneGap 应用程序，应该具有良好的交互性，能够在恰当的时刻给予用户必要的通知或反馈。不论这样的信息是关于操作出错，还是寻求确认，抑或是提示操作正在进行，在 PhoneGap 应用中，均提供了统一的通知 API 来解决此类问题。

　　在下面的内容中，将详细讲解通知 API 的基本用法。

　　提示对话框通知 notification.alert()。大多数 PhoneGap 使用本地对话框实现该功能。然而，一些平台只是简单地使用浏览器的 alert 函数,而这种方法通常是不能定制的。notification.alert()的函数签名如下所示:

```
navigator.notification.alert(message, alertCallback, [title], [buttonName]);
```

各个参数的具体说明如下所示。
- message:对话框信息(字符串类型)。
- alertCallback:当警告对话框被忽略时调用的回调函数(函数类型)。
- title:对话框标题(字符串类型)(可选项,默认值为"Alert")。
- buttonName:按钮名称(字符串类型)(可选项,默认值为"OK")。

具体实现

下面的实例文件 270.html 演示了 notification.alert()的基本用法。

```html
<!DOCTYPE html>
<html>
<head>
  <meta charset="utf-8">
  <meta name="viewport" content="width=device-width, initial-scale=1">
  <title>index.html</title>
  <script type="text/javascript" charset="utf-8" src="cordova.js" ></script>
    <script type="text/javascript" charset="utf-8">

// 等待加载 PhoneGap
document.addEventListener("deviceready", onDeviceReady, false);

// PhoneGap 加载完毕
function onDeviceReady() {
    //空
}

// 显示定制警告框
function showAlert() {
    navigator.notification.alert(
        'You are the winner!',     // 显示信息
        'Game Over',               // 标题
        'Done'                     // 按钮名称
    );
}
</script>
</head>
<body>
  <p><a href="#" onclick="showAlert(); return false;">Show Alert</a></p>
</body>
</html>
```

执行后将在页面中显示一个"Show Alert"链接,点击后将弹出一个警告框。执行效果如图 13-29 所示。

图 13-29 执行效果

实例 271　使用确认 API

实例 271	使用确认 API
源码路径	\daima\271

实例说明

对于那些可能出现用户误操作或者涉及严重影响的用户交互场景，需要使用 notification.confirm() 来得到用户的确认，然后再进行之后的处理。notification.confirm() 的函数签名如下所示。

```
navigator.notification.confirm(message, confirmCallback, [title],
[buttonLabels]);
```

各个参数的具体说明如下所示。
- message：对话框信息，是字符串类型。
- confirmCallback：按下按钮后触发的回调函数，返回按下按钮的索引（1、2 或 3），函数类型。
- title：对话框标题，是字符串类型，这是一个可选项，默认值为"Confirm"。
- buttonLabels：逗号分隔的按钮标签字符串，是字符串类型，这是一个可选项，默认值为"OK、Cancel"。

notification.confirm 函数显示一个定制性比浏览器的 confirm 函数更好的本地对话框。下面是其用法演示。

```
// 处理确认对话框返回的结果
function onConfirm(button) {
    alert('You selected button ' + button);
}

// 显示一个定制的确认对话框
function showConfirm() {
    navigator.notification.confirm(
        'You are the winner!',  // 显示信息
        onConfirm,              // 按下按钮后触发的回调函数，返回按下按钮的索引
        'Game Over',            // 标题
        'Restart,Exit'          // 按钮标签
    );
}
```

具体实现

下面的实例文件 271.html 演示了 notification.confirm() 的基本用法。

```
<!DOCTYPE html>
<html>
<head>
  <meta charset="utf-8">
  <meta name="viewport" content="width=device-width, initial-scale=1">
  <title>index.html</title>
  <script type="text/javascript" charset="utf-8" src="cordova.js" ></script>
  <script type="text/javascript" charset="utf-8">

  // 等待加载 PhoneGap
  document.addEventListener("deviceready", onDeviceReady, false);

  // PhoneGap 加载完毕
  function onDeviceReady() {
      // 空
  }

  // 处理确认对话框返回的结果
```

```
        function onConfirm(button) {
            alert('You selected button ' + button);
        }

        // 显示一个定制的确认对话框
        function showConfirm() {
            navigator.notification.confirm(
                'You are the winner!',    // 显示信息
                onConfirm,                // 按下按钮后触发的回调函数,返回按下按钮的索引
                'Game Over',              // 标题
                'Restart,Exit'            // 按钮标签
            );
        }
    </script>
</head>
<body>
    <p><a href="#" onclick="showConfirm(); return false;">Show
    Confirm</a></p>
</body>
</html>
```

执行后将在页面中显示一个"Show Confirm"链接,点击后将弹出一个警告框。执行效果如图13-30所示。

图13-30 执行效果

第 14 章　Swift 实战

Swift 是 Apple 公司在 WWDC 2014 所发布的一门编程语言，用来编写 OS X 和 iOS 应用程序。苹果公司在设计 Swift 语言时，就有意将其和 Objective-C 共存。Objective-C 是 Apple 操作系统在导入 Swift 前使用的编程语言。本章将通过几个典型实例的实现过程，详细介绍使用 Swift 语言开发 iOS 项目的基本知识。

实例 272　使用 Xcode 创建 Swift 程序

实例 272	使用 Xcode 创建 Swift 程序
源码路径	\daima\272

实例说明

当苹果公司推出 Swift 编程语言时，建议使用 Xcode 6 来开发 Swift 程序。在实例的内容中，将详细讲解使用 Xcode 6 创建 Swift 程序的方法。

具体实现

（1）打开 Xcode 6，单击"Create a new Xcode Project"新创建一个工程文件，如图 14-1 所示。

图 14-1　新创建一个工程文件

（2）在弹出界面的左侧栏目中选择"Application"，在右侧选择"Command Line Tool"，然后单击"Next"按钮，如图 4-2 所示。

（3）在弹出的界面中设置各个选项值，在"Language"选项中设置编程语言为"Swift"，然后单击"Next"按钮，如图 14-3 所示。

实例 272　使用 Xcode 创建 Swift 程序

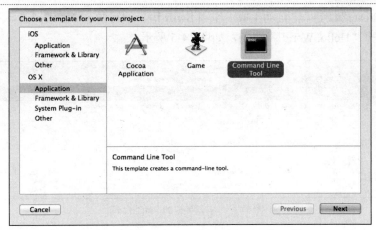

图 14-2　创建一个 "Command Line Tool" 工程

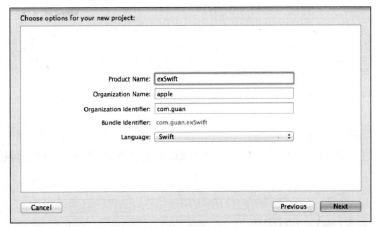

图 14-3　设置编程语言为 "Swift"

（4）在弹出的界面中设置当前工程的保存路径，如图 14-4 所示。

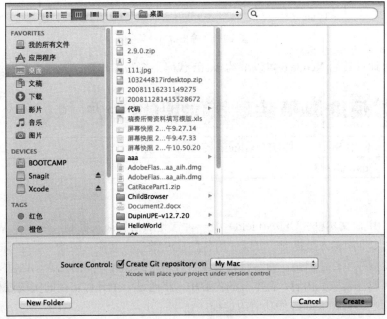

图 14-4　设置保存路径

527

第 14 章 Swift 实战

（5）单击"Create"按钮，将自动生成一个用 Swift 语言编写的 iOS 工程。在工程文件 main.swift 中会自动生成一个"Hello, World!"语句，如图 14-5 所示。

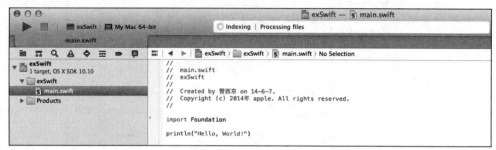

图 14-5　自动生成的 Swift 代码

文件 main.swift 的代码是自动生成的，具体代码如下所示。

```
//
//  main.swift
//  exSwift
//
//  Created by admin on 14-14-7.
//  Copyright (c) 2014年 apple. All rights reserved.
//

import Foundation

println("Hello, World!")
```

单击图 14-5 左上角的 ▶ 按钮运行工程，会在 Xcode 6 下方的控制台中输出运行结果，如图 14-6 所示。

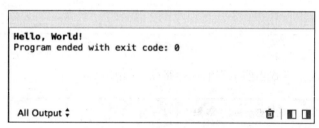

图 14-6　输出执行结果

由此可见，通过使用 Xcode 6 可以节约 Swift 代码的编写工作量，提高了开发效率。

实例 273　使用 Swift 实现 UITextField 控件

实例 273	使用 Swift 实现 UITextField 控件
源码路径	\daima\273

实例说明

在 iOS 应用中，文本框（UITextField）是一种常见的信息输入机制，类似于 Web 表单中的表单字段。当在文本框中输入数据时，可以使用各种 iOS 键盘将其输入限制为数字或文本。和按钮一样，文本框也能响应事件，但是，通常将其实现为被动（passive）界面元素，这意味着视图控制器可随时通过 text 属性读取其内容。在本节的内容中，将通过一个具体实例的实现过程，详细讲解基于 Swift 语言实现 UITextField 控件功能的过程。

具体实现

(1) 打开 Xcode 6.1 或 Xcode 6.2,然后创建一个名为 "LTBouncyPlaceholderDemo" 的工程,工程的最终目录结构如图 14-7 所示。

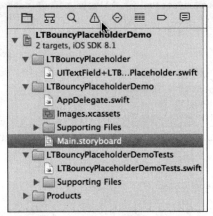

图 14-7　Xcode 工程的最终目录结构

(2) 打开 Main.storyboard,为本工程设计两个视图界面,在第二个视图中实现 UITextField 控件效果,如图 14-8 所示。

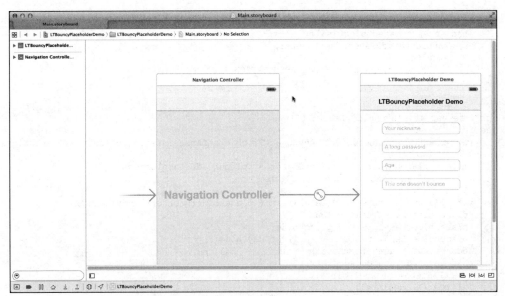

图 14-8　Main.storyboard 设计视图

(3) 本实例是用 Swift 语言编写的 UITextField 扩展,Dribbble 目录中有一个 QuartzComposer 文件。在文件 UITextField+LTBouncyPlaceholder.swift 中编写 UITextField extension(扩展),并且实现了 PlaceHolder 的弹出动画效果。具体实现代码如下所示。

```
import Foundation
import UIKit
import QuartzCore

var kAlwaysBouncePlaceholderPointer: Void?
var kAbbreviatedPlaceholderPointer: Void?
var kPlaceholderLabelPointer: Void?
var kRightPlaceholderLabelPointer: Void?

let kAnimationDuration: CFTimeInterval = 0.6
```

```swift
extension UITextField {
    
    public var alwaysBouncePlaceholder: Bool {
    get {
        var _alwaysBouncePlaceholderObject : AnyObject?
            = objc_getAssociatedObject(self, &kAlwaysBouncePlaceholderPointer)
        if let _alwaysBouncePlaceholder = _alwaysBouncePlaceholderObject?.boolValue {
            return _alwaysBouncePlaceholder
        }
        return false
    }
    set {
        lt_placeholderLabel.hidden = !newValue
        objc_setAssociatedObject(self,
            &kAlwaysBouncePlaceholderPointer,
            newValue,
            objc_AssociationPolicy(OBJC_ASSOCIATION_RETAIN_NONATOMIC))
    }
    }
    
    public var abbreviatedPlaceholder: String? {
    get {
        var _abbreviatedPlaceholderObject: AnyObject? = objc_getAssociatedObject(self, &kAbbreviatedPlaceholderPointer)
        if let _abbreviatedPlaceholder: AnyObject = _abbreviatedPlaceholderObject {
            return _abbreviatedPlaceholder as? String
        }
        return nil
    }
    set {
        lt_rightPlaceholderLabel.text = newValue
        objc_setAssociatedObject(self,
            &kAbbreviatedPlaceholderPointer,
            newValue,
            objc_AssociationPolicy(OBJC_ASSOCIATION_RETAIN_NONATOMIC))
    }
    }
    
    private var lt_placeholderLabel: UILabel {
    get {
        var _placeholderLabelObject: AnyObject? = objc_getAssociatedObject(self, &kPlaceholderLabelPointer)
        if let _placeholderLabel : AnyObject = _placeholderLabelObject {
            return _placeholderLabel as UILabel
        }
        var _placeholderLabel = UILabel(frame: placeholderRectForBounds(bounds))
        _placeholderLabel.font = font
        _placeholderLabel.text = placeholder
        _placeholderLabel.textColor = .lightGrayColor()
        addSubview(_placeholderLabel)
        objc_setAssociatedObject(self,
            &kPlaceholderLabelPointer,
            _placeholderLabel,
            objc_AssociationPolicy(OBJC_ASSOCIATION_RETAIN_NONATOMIC))
        return _placeholderLabel
    }
    }
    
    private var lt_rightPlaceholderLabel: UILabel {
    get {
        var _rightPlaceholderLabelObject: AnyObject? = objc_getAssociatedObject(self, &kRightPlaceholderLabelPointer)
        if let _rightPlaceholderLabel: AnyObject = _rightPlaceholderLabelObject {
            return _rightPlaceholderLabel as UILabel
        }
        var _rightPlaceholderLabel = UILabel(frame: placeholderRectForBounds(bounds))
        _rightPlaceholderLabel.font = font
        _rightPlaceholderLabel.textColor = .lightGrayColor()
        _rightPlaceholderLabel.layer.opacity = 0.0
        addSubview(_rightPlaceholderLabel)
        objc_setAssociatedObject(self,
```

```swift
            &kRightPlaceholderLabelPointer,
            _rightPlaceholderLabel,
            objc_AssociationPolicy(OBJC_ASSOCIATION_RETAIN_NONATOMIC))
        return _rightPlaceholderLabel
    }
}

func _drawPlaceholderInRect(rect: CGRect) {
    println("swizzled default method")
}

override public func willMoveToSuperview(newSuperview: UIView!) {
    if (nil != newSuperview) {
        lt_placeholderLabel.setNeedsDisplay()
        
        struct TokenHolder {
            static var token: dispatch_once_t = 0;
        }
        
        dispatch_once(&TokenHolder.token) {
            var originMethod: Method = class_getInstanceMethod(object_getClass(self), Selector("drawPlaceholderInRect:"))
            var swizzledMethod: Method = class_getInstanceMethod(object_getClass(self), Selector("_drawPlaceholderInRect:"))
            method_exchangeImplementations(originMethod, swizzledMethod)
            
        }
        
        NSNotificationCenter.defaultCenter().addObserver(self,
            selector: Selector("_didChange:"),
            name: UITextFieldTextDidChangeNotification,
            object: nil)
    } else {
        NSNotificationCenter.defaultCenter().removeObserver(self,
            name: UITextFieldTextDidChangeNotification,
            object: nil)
    }
}

func _didChange (notification: NSNotification) {
    if notification.object === self {
        if text.lengthOfBytesUsingEncoding(NSUTF8StringEncoding) > 0 {
            if alwaysBouncePlaceholder {
                
                _animatePlaceholder(toRight: true)
            } else {
                lt_placeholderLabel.hidden = true
            }
        } else {
            if alwaysBouncePlaceholder {
                _animatePlaceholder(toRight: false)
            } else {
                lt_placeholderLabel.hidden = false
            }
        }
    }
}

private var _widthOfAbbr: Float {
get {
    let rightPlaceholder: String? = !abbreviatedPlaceholder!.isEmpty ? abbreviatedPlaceholder : placeholder
    
    if let _rightPlaceholder = rightPlaceholder {
        let attributes = [NSFontAttributeName: lt_rightPlaceholderLabel.font]
        var abbrSize = _rightPlaceholder.sizeWithAttributes(attributes)
        return Float(abbrSize.width)
    }
    return 0
}
}
```

第 14 章 Swift 实战

```swift
    private func _bounceKeyframes(#toRight: Bool) -> NSArray {
        let steps = 100
        var values = [Double]()
        var value: Double
        let e = 2.5
        let distance = Float(placeholderRectForBounds(bounds).size.width) - _widthOfAbbr
        for t in 0..<steps {
            value = Double(distance)
                * (toRight ? -1 : 1)
                * Double(pow(e, -0.055 * Double(t)))
                * Double(cos(0.1 * Double(t)))
                + (toRight ? Double(distance) : 0)
            values.append(value)
        }
        return values
    }

    private func _animatePlaceholder (#toRight: Bool) {
        if let abbrPlaceholder = abbreviatedPlaceholder {
            if (toRight) {
                if lt_rightPlaceholderLabel.layer.presentationLayer().opacity > 0 {
                    return
                }

                lt_placeholderLabel.layer.removeAllAnimations()
                lt_rightPlaceholderLabel.layer.removeAllAnimations()

                let bounceToRight = CAKeyframeAnimation(keyPath: "position.x")
                bounceToRight.timingFunction = CAMediaTimingFunction(name: kCAMediaTimingFunctionLinear)
                bounceToRight.duration = kAnimationDuration
                bounceToRight.values = _bounceKeyframes(toRight: true)
                bounceToRight.fillMode = kCAFillModeForwards
                bounceToRight.additive = true
                bounceToRight.removedOnCompletion = false

                let fadeOut = CABasicAnimation(keyPath: "opacity")
                fadeOut.timingFunction = CAMediaTimingFunction(name: kCAMediaTimingFunctionLinear)
                fadeOut.fromValue = 1
                fadeOut.toValue = 0
                fadeOut.duration = kAnimationDuration / 3
                fadeOut.fillMode = kCAFillModeBoth
                fadeOut.removedOnCompletion = false
                lt_placeholderLabel.layer.addAnimation(bounceToRight, forKey: "bounceToRight")
                lt_placeholderLabel.layer.addAnimation(fadeOut, forKey: "fadeOut")

                let fadeIn = CABasicAnimation(keyPath: "opacity")
                fadeIn.timingFunction = CAMediaTimingFunction(name: kCAMediaTimingFunctionLinear)
                fadeIn.fromValue = 0
                fadeIn.toValue = 1
                fadeIn.duration = kAnimationDuration / 3
                fadeIn.fillMode = kCAFillModeForwards
                fadeIn.removedOnCompletion = false

                lt_rightPlaceholderLabel.layer.addAnimation(bounceToRight, forKey: "bounceToRight")
                lt_rightPlaceholderLabel.layer.addAnimation(fadeIn, forKey: "fadeIn")
            } else {
                lt_placeholderLabel.layer.removeAllAnimations()
                lt_rightPlaceholderLabel.layer.removeAllAnimations()

                let bounceToLeft = CAKeyframeAnimation(keyPath: "position.x")
                bounceToLeft.timingFunction = CAMediaTimingFunction(name: kCAMediaTimingFunctionLinear)
                bounceToLeft.duration = kAnimationDuration
                bounceToLeft.values = _bounceKeyframes(toRight: false)
                bounceToLeft.fillMode = kCAFillModeForwards
                bounceToLeft.additive = true
                bounceToLeft.removedOnCompletion = false

                let fadeIn = CABasicAnimation(keyPath: "opacity")
                fadeIn.timingFunction = CAMediaTimingFunction(name: kCAMediaTimingFunctionEaseIn)
                fadeIn.duration = kAnimationDuration / 3
                fadeIn.fillMode = kCAFillModeForwards
```

实例 273　使用 Swift 实现 UITextField 控件

```swift
            fadeIn.fromValue = 0
            fadeIn.toValue = 1
            fadeIn.removedOnCompletion = false
            lt_placeholderLabel.layer.addAnimation(fadeIn, forKey: "fadeIn")
            lt_placeholderLabel.layer.addAnimation(bounceToLeft, forKey: "bounceToLeft")
            
            let fadeOut = CABasicAnimation(keyPath: "opacity")
            fadeOut.timingFunction = CAMediaTimingFunction(name: kCAMediaTiming FunctionEaseIn)
            fadeOut.duration = kAnimationDuration / 3
            fadeOut.fillMode = kCAFillModeForwards
            fadeOut.fromValue = 1
            fadeOut.toValue = 0
            fadeOut.removedOnCompletion = false
            lt_rightPlaceholderLabel.layer.addAnimation(fadeOut, forKey: "fadeOut")
            lt_rightPlaceholderLabel.layer.addAnimation(bounceToLeft, forKey: "bounceToLeft")
        }
    } else {
        lt_placeholderLabel.layer.removeAllAnimations()
        if toRight {
            let bounceToRight = CAKeyframeAnimation(keyPath: "position.x")
            bounceToRight.timingFunction = CAMediaTimingFunction(name: kCAMediaTimingFunction Linear)
            bounceToRight.duration = kAnimationDuration
            bounceToRight.values = _bounceKeyframes(toRight: true)
            bounceToRight.fillMode = kCAFillModeForwards
            bounceToRight.additive = true
            bounceToRight.removedOnCompletion = false
            lt_placeholderLabel.layer.addAnimation(bounceToRight, forKey: "bounceToRight")
        } else {
            let bounceToLeft = CAKeyframeAnimation(keyPath: "position.x")
            bounceToLeft.timingFunction = CAMediaTimingFunction(name: kCAMediaTimingFunctionLinear)
            bounceToLeft.duration = kAnimationDuration
            bounceToLeft.values = _bounceKeyframes(toRight: false)
            bounceToLeft.fillMode = kCAFillModeForwards
            bounceToLeft.additive = true
            bounceToLeft.removedOnCompletion = false
            lt_placeholderLabel.layer.addAnimation(bounceToLeft, forKey: "bounceToLeft")
        }
    }
    
    }
}
```

执行后将实现一个具有弹出动画功能的 UITextField 控件效果，如图 14-9 所示。

图 14-9　执行效果

第 14 章　Swift 实战

实例 274　基于 Swift 使用 UITextView 控件

实例 274	基于 Swift 使用 UITextView 控件
源码路径	\daima\274

实例说明

在 iOS 应用中，UITextView 是一个类。在 Xcode 中当使用 IB 给视图放上去一个文本框后，选中文本框后可以在 Attribute Inspector 中设置其各种属性。在本节的内容中，将通过一个具体实例的实现过程，详细讲解基于 Swift 语言实现显示 UITextView 文本的过程。

具体实现

（1）打开 Xcode 6.1 或 Xcode 6.2，然后新创建一个名为 "Placeholder Test" 的工程，工程的最终目录结构如图 14-10 所示。

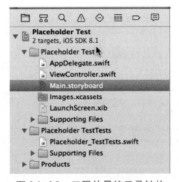

图 14-10　工程的最终目录结构

（2）打开 Main.storyboard，为本工程设计一个 View Controller 视图界面，如图 14-11 所示。

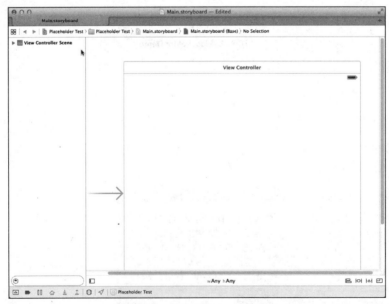

图 14-11　Main.storyboard 记事板

（3）本实例的程序文件是 ViewController.swift，功能是通过 UITextView 在屏幕中显示指定的文本内容，具体实现代码如下所示。

```
class ViewController: UIViewController, UITextViewDelegate {

    @IBOutlet weak var textView: UITextView!

    override func viewDidLoad() {
        super.viewDidLoad()
        textView.delegate = self
        if (textView.text == "") {
            textViewDidEndEditing(textView)
        }
        var tapDismiss = UITapGestureRecognizer(target: self, action: "dismissKey board")
        self.view.addGestureRecognizer(tapDismiss)
    }

    func dismissKeyboard(){
        textView.resignFirstResponder()
    }

    override func didReceiveMemoryWarning() {
        super.didReceiveMemoryWarning()
        // Dispose of any resources that can be recreated.
    }

    func textViewDidEndEditing(textView: UITextView) {
        if (textView.text == "") {
            textView.text = "Placeholder"
            textView.textColor = UIColor.lightGrayColor()
        }
        textView.resignFirstResponder()
    }

    func textViewDidBeginEditing(textView: UITextView){
        if (textView.text == "Placeholder"){
            textView.text = ""
            textView.textColor = UIColor.blackColor()
        }
        textView.becomeFirstResponder()
    }

}
```

执行后将在屏幕中显示指定的文本内容，如图 14-12 所示。

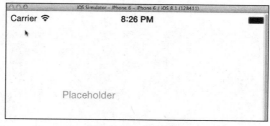

图 14-12　执行效果

实例 275　使用 Swift 实现 UISlider 控件效果

实例 275	使用 Swift 实现 UISlider 控件效果
源码路径	\daima\275

实例说明

在 iOS 应用中，滑块为用户提供了一种可见的针对一定范围的调整方法，可以通过拖动一个滑动条改变它的值,并且可以对其配置上合适的不同值域。可以设置滑块值的范围,也可以在两端加上图片,以及进行各种调整让它更美观。滑块非常适合用于表示在很大范围(但不精确)的数值中进行选择,如音量设置、灵敏度控制等用途。在本节的内容中,将通过一个具体实例的实现过程,详细讲解用 Swift 语言实现的 UISlider 控件效果的过程。

具体实现

(1)打开 Xcode,然后创建一个名为 "Fibonacci" 的工程,工程的最终目录结构如图 14-13 所示。

图 14-13 工程的目录结构

(2)打开 Main.storyboard,为本工程设计一个视图界面,在里面分别插入 Horizontal Slider 控件、Label 控件和 Text 控件,如图 14-14 所示。

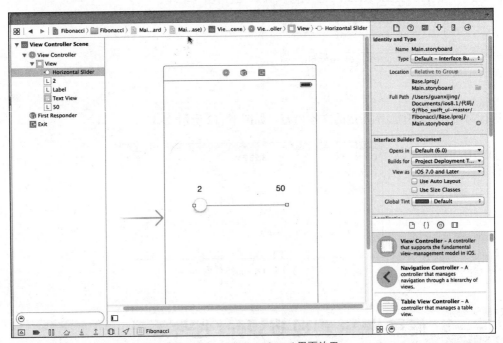

图 14-14 Main.storyboard 界面效果

(3)编写类文件 FibonacciModel.swift,通过 calculateFibonacciNumbers 计算斐波那契数值,具体实现代码如下所示。

```
import Foundation

public class FibonacciModel {
```

```swift
    public init () {}

    public func calculateFibonacciNumbers (minimum2 endOfSequence:Int) -> Array<Int> {

        //初始值属性
        var sequence : [Int] = [1,1]

        for number in 2..<endOfSequence {

            var newFibonacciNumber = sequence[number-1] + sequence[number-2]
            sequence.append(newFibonacciNumber)
        }

        return sequence
    }

}
```

(4) 编写文件 ViewController.swift，监听滑动条数值的变动，并及时显示滑块中的更新值。文件 ViewController.swift 的具体实现代码如下所示。

```swift
import UIKit

class ViewController: UIViewController {
    @IBOutlet weak var theSlider: UISlider!

    @IBOutlet weak var outputTextView: UITextView!
    @IBOutlet weak var selectedValueLabel: UILabel!
    var fibo: FibonacciModel = FibonacciModel()

    override func viewDidLoad() {
        super.viewDidLoad()
    }

    override func didReceiveMemoryWarning() {
        super.didReceiveMemoryWarning()
        // Dispose of any resources that can be recreated.
    }


    func addASlider() {
    }


    @IBAction func sliderValueDidChange(sender: UISlider) {

    //func sliderValueDidChange () {

        var returnedArray: [Int] = []
        var formattedOutput:String = ""

        //显示更新的滑块值
        self.selectedValueLabel!.text = String(Int(theSlider!.value))

        //Calculate the Fibonacci elements based on the new slider value
        returnedArray = self.fibo.calculateFibonacciNumbers(minimum2: Int(theSlider!.value))

        //Put the elements in a nicely formatted array
        for number in returnedArray {

            formattedOutput = formattedOutput + String(number) + ", "
        }

        //Update the textfield with the formatted array
        self.outputTextView!.text = formattedOutput
    }
}
```

本实例执行后将在屏幕中实现一个滑动条效果，如图14-15所示。

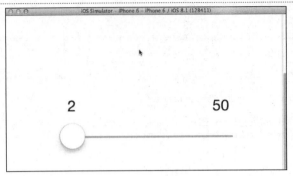

图 14-15 执行效果

实例 276　使用 Swift 实现 Imageview 控件效果

实例 276	使用 Swift 实现 UIImageview 控件效果
源码路径	\daima\276

实例说明

在 iOS 应用中，图像视图(UIImageView)用于显示图像。可以将图像视图加入到应用程序中，并用于向用户呈现信息。UIImageView 实例还可以创建简单的基于帧的动画，其中包括开始、停止和设置动画播放速度的控件。在使用 Retina 屏幕的设备中，图像视图可利用其高分辨率屏幕。令开发人员兴奋的是，我们无需编写任何特殊代码，无需检查设备类型，而只需将多幅图像加入到项目中，而图像视图将在正确的时间加载正确的图像。在本节的内容中，将通过一个具体实例的实现过程，详细讲解基于 Swift 语言使用 UIImageView 控件的过程。

具体实现

（1）打开 Xcode，然后创建一个名为"ButtonWithImageAndTitleDemo"的工程，工程的最终目录结构如图 14-16 所示。

图 14-16　工程的目录结构

（2）编写类文件 ButtonWithImageAndTitleExtension.swift，功能是设置为 UIButton 和按钮图像设置标题，并为每个图像按钮设置对应的标题。在本实现文件中通过 case 语句处理了 Top、Bottom、Left 和 Right 等 4 种位置的图标按钮。文件 ButtonWithImageAndTitleExtension.swift 的具体实现代码如下所示。

实例 276 使用 Swift 实现 Imageview 控件效果

```swift
import UIKit

extension UIButton {
    @objc func set(image anImage: UIImage?, title: NSString!, titlePosition: UIViewContentMode, additionalSpacing: CGFloat, state: UIControlState){
        self.imageView?.contentMode = .Center
        self.setImage(anImage?, forState: state)

        positionLabelRespectToImage(title!, position: titlePosition, spacing: additionalSpacing)

        self.titleLabel?.contentMode = .Center
        self.setTitle(title?, forState: state)
    }

    private func positionLabelRespectToImage(title: NSString, position: UIViewContentMode, spacing: CGFloat) {
        let imageSize = self.imageRectForContentRect(self.frame)
        let titleFont = self.titleLabel?.font!
        let titleSize = title.sizeWithAttributes([NSFontAttributeName: titleFont!])

        var titleInsets: UIEdgeInsets
        var imageInsets: UIEdgeInsets

        switch (position){
        case .Top:
            titleInsets = UIEdgeInsets(top: -(imageSize.height + titleSize.height + spacing), left: -(imageSize.width), bottom: 0, right: 0)
            imageInsets = UIEdgeInsets(top: 0, left: 0, bottom: 0, right: -titleSize.width)
        case .Bottom:
            titleInsets = UIEdgeInsets(top: (imageSize.height + titleSize.height + spacing), left: -(imageSize.width), bottom: 0, right: 0)
            imageInsets = UIEdgeInsets(top: 0, left: 0, bottom: 0, right: -titleSize.width)
        case .Left:
            titleInsets = UIEdgeInsets(top: 0, left: -(imageSize.width * 2), bottom: 0, right: 0)
            imageInsets = UIEdgeInsets(top: 0, left: 0, bottom: 0, right: -(titleSize.width * 2 + spacing))
        case .Right:
            titleInsets = UIEdgeInsets(top: 0, left: 0, bottom: 0, right: -spacing)
            imageInsets = UIEdgeInsets(top: 0, left: 0, bottom: 0, right: 0)
        default:
            titleInsets = UIEdgeInsets(top: 0, left: 0, bottom: 0, right: 0)
            imageInsets = UIEdgeInsets(top: 0, left: 0, bottom: 0, right: 0)
        }

        self.titleEdgeInsets = titleInsets
        self.imageEdgeInsets = imageInsets
    }
}
```

（3）文件 ViewController.swift 的功能是调用类文件 ButtonWithImageAndTitleExtension.swift，通过 viewDidLoad() 根据屏幕位置载入对应的按钮图像。文件 ViewController.swift 的具体实现代码如下所示。

```swift
import UIKit

class ViewController: UIViewController {
    @IBOutlet weak var button: UIButton!
    @IBOutlet weak var thirdButton: UIButton!

    override func viewDidLoad() {
        super.viewDidLoad()
        // Do any additional setup after loading the view, typically from a nib.
        button.set(image: UIImage(named: "shout"), title: "Shout", titlePosition: .Top, additionalSpacing: 30.0, state: .Normal)
        thirdButton.set(image: UIImage(named: "shout"), title: "This is an XIB button", titlePosition: .Bottom, additionalSpacing: 6.0, state: .Normal)

        var secondButton = UIButton.buttonWithType(.System) as UIButton
        secondButton.frame = CGRectMake(0, 50, 100, 400)
```

```
        secondButton.center = CGPointMake(view.frame.size.width/2, 50)
        secondButton.set(image:    UIImage(named:    "settings"),    title:    "Settings",
titlePosition: .Left, additionalSpacing: 0.0, state: .Normal)
        view.addSubview(secondButton)
    }

    override func didReceiveMemoryWarning() {
        super.didReceiveMemoryWarning()
        // Dispose of any resources that can be recreated.
    }
}
```

本实例执行后将分别在屏幕顶部、中间和底部显示不同的图标，如图 14-17 所示。

顶部按钮　　　　　　中间按钮　　　　　　底部按钮

图 14-17　执行效果

实例 277　基于 Swift 控制是否显示密码明文

实例 277	基于 Swift 控制是否显示密码明文
源码路径	\daima\277

实例说明

在大多数传统桌面应用程序中，通过复选框和单选按钮来实现开关功能。在 iOS 中，Apple 放弃了这些界面元素，取而代之的是开关和分段控件。在 iOS 应用中，使用开关控件（UISwitch）来实现"开/关"UI 元素，它类似于传统的物理开关。在本节的内容中，将通过一个具体实例的实现过程，详细讲解基于 Swift 语言控制是否显示密码明文的过程。

具体实现

（1）打开 Xcode，然后创建一个名为"DKTextField.Swift"的工程，工程的最终目录结构如图 14-18 所示。

图 14-18　工程的目录结构

（2）打开 Main.storyboard，为本工程设计一个视图界面，在里面添加一个 Switch 控件，此控件作为控制是否显示密码明文的开关，如图 14-19 所示。

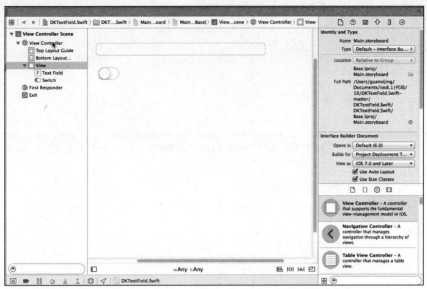

图 14-19　Main.storyboard 界面

（3）由于系统的 UITextField 控件在切换到密码状态时会清除之前的输入文本，于是特意编写类文件 DKTextField.swift，DKTextField 继承于 UITextField，并且不影 UITextFiel 的 Delegate。文件 DKTextField.swift 的具体实现代码如下所示。

```swift
import UIKit

class DKTextField: UITextField {
    
    required init(coder aDecoder: NSCoder) {
        super.init(coder: aDecoder)
        
    }
    override init(frame: CGRect) {
        super.init(frame: frame)
        self.awakeFromNib()
        
    }
    private var password:String = ""
    
    private var beginEditingObserver:AnyObject!
    
    private var endEditingObserver:AnyObject!
    
    override func awakeFromNib() {
        super.awakeFromNib()
        
      //  unowned var that=self
        
        self.beginEditingObserver                                                      = NSNotificationCenter.defaultCenter().addObserverForName(UITextFieldTextDidBeginEditingNotification, object: nil, queue: nil, usingBlock: {
            [unowned self](note:NSNotification!) in
            
            if self == note.object as DKTextField && self.secureTextEntry {
                self.text = ""
                self.insertText(self.password)
            }
        })
        
        self.endEditingObserver = NSNotificationCenter.defaultCenter().addObserver ForName(UITextFieldTextDidEndEditingNotification, object: nil, queue: nil, usingBlock: {
            [unowned self](note:NSNotification!) in
```

```
            if self == note.object as DKTextField {
                self.password = self.text
            }
            
        })
    }
    deinit{
    
        NSNotificationCenter.defaultCenter().removeObserver(self.beginEditingObserver)
        NSNotificationCenter.defaultCenter().removeObserver(self.endEditingObserver)
    }
    override var secureTextEntry: Bool{
        get {
            return super.secureTextEntry
        }
        set{
            self.resignFirstResponder()
            super.secureTextEntry = newValue
            self.becomeFirstResponder()
        }
    }
}
```

（4）编写文件 ViewController.swift，功能是通过 switchChanged 监听 UISwitch 控件的开关状态，并根据监听到的状态设置密码的显示样式。文件 ViewController.swift 的具体实现代码如下所示。

```
import UIKit

class ViewController: UIViewController {

    @IBOutlet weak var textField: DKTextField!
    
    override func viewDidLoad() {
        super.viewDidLoad()
        // Do any additional setup after loading the view, typically from a nib.
    }

    override func didReceiveMemoryWarning() {
        super.didReceiveMemoryWarning()
        // Dispose of any resources that can be recreated.
    }
    
    @IBAction func switchChanged(sender: AnyObject) {
    
        self.textField.secureTextEntry = (sender as UISwitch).on
    
    }
}
```

下面看执行后的效果，如果打开 UISwitch 控件则显示密码，如图 14-20 所示。

图 14-20　显示密码

如果关闭 UISwitch，则显示密码明文，如图 14-21 所示。

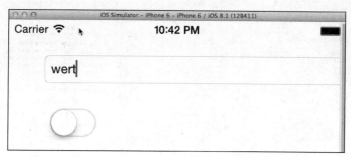

图 14-21　显示明文

实例 278　基于 Swift 使用 UISegmentedControl 控件

实例 278	基于 Swift 使用 UISegmentedControl 控件
源码路径	\daima\278

实例说明

在 iOS 应用中，当用户输入的不仅仅是布尔值时，可使用分段控件 UISegmentedControl 实现我们需要的功能。分段控件提供一栏按钮（有时称为按钮栏），但只能激活其中一个按钮。在本节的内容中，将通过一个具体实例的实现过程，详细讲解基于 Swift 语言使用 UISegmentedControl 控件的过程。

具体实现

（1）打开 Xcode 6，然后创建一个名为"GolangStudy"的工程，工程的最终目录结构如图 14-22 所示。

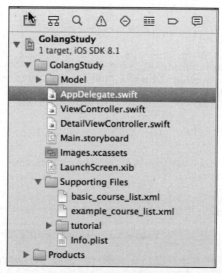

图 14-22　工程的目录结构

（2）打开 Main.storyboard，为本工程设计一个视图界面，在视图顶部通过 UISegmentedControl 控制显示类别，在底部通过 TableView 显示某类别的详细内容，如图 14-23 所示。

第 14 章 Swift 实战

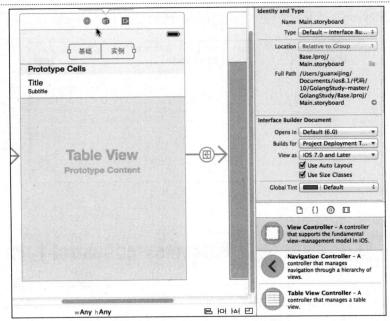

图 14-23 Main.storyboard 设计界面

（3）编写文件 ViewController.swift 实现主视图功能，在顶部显示 UISegmentedControl 控制面板，在底部列表显示某类别下的所有信息。文件 ViewController.swift 的具体实现代码如下所示。

```
import UIKit
class ViewController: UIViewController,UITableViewDataSource,UITableViewDelegate {

    @IBOutlet weak var tableView: UITableView!

    var courseSet:[Course] = []

    override func viewDidLoad() {
        super.viewDidLoad()
        courseSet = XmlParseUtil(name: "basic_course_list").courseSet
        self.navigationItem.backBarButtonItem = UIBarButtonItem(title: "返回", style: .Plain,
target:
nil, action: nil)
    }

    @IBAction func segmentChanged(sender: UISegmentedControl) {
        if sender.selectedSegmentIndex == 0 {
            courseSet = XmlParseUtil(name: "basic_course_list").courseSet
        }else if sender.selectedSegmentIndex == 1 {
            courseSet = XmlParseUtil(name: "example_course_list").courseSet
        }
        tableView.reloadData()
    }

    func tableView(tableView: UITableView, numberOfRowsInSection section: Int) -> Int {
        return courseSet.count;
    }

    func tableView(tableView: UITableView, cellForRowAtIndexPath indexPath: NSIndexPath)
-> UITableViewCell {
        var cell:UITableViewCell = tableView.dequeueReusableCellWithIdentifier("Cell")!
as UITableViewCell
        var course:Course = courseSet[indexPath.row] as Course
        cell.textLabel.text = course.getName()
        cell.detailTextLabel?.text = course.getDesc()
        return cell
    }

    override func prepareForSegue(segue: UIStoryboardSegue, sender: AnyObject?) {
```

实例 279　基于 Swift 使用 UIScrollView 控件

```
        if segue.identifier == "show" {
            let viewController = segue.destinationViewController as DetailViewController
            var course:Course = courseSet[tableView.indexPathForSelectedRow()!.row] asCourse
            viewController.pathname = course.getChapter()
            viewController.title = course.getName()
            viewController.courseSet = courseSet
            viewController.index = tableView.indexPathForSelectedRow()!.row
        }
    }

    override func didReceiveMemoryWarning() {
        super.didReceiveMemoryWarning()
        // Dispose of any resources that can be recreated.
    }

}
```

到此为止，整个实例介绍完毕。单击 UI 主界面中"UISegmentedControl"列表项后，执行效果如图 14-24 所示。

图 14-24　执行效果

实例 279　基于 Swift 使用 UIScrollView 控件

实例 279	基于 Swift 使用 UIScrollView 控件
源码路径	\daima\279

实例说明

大家肯定使用过这样的应用程序，它显示的信息在一屏中容纳不下。在这种情况下，使用可滚动视图控件（UIScrollView）来解决。顾名思义，可滚动的视图提供了滚动功能，可显示超过一屏的信息。但是，在让我们能够通过 Interface Builder 将可滚动视图加入项目中方面，iOS 做的并不完美。我们可以添加可滚动视图，但要想让它实现滚动效果，必须在应用程序中编写一行代码。在本节的内容中，将通过一个具体实例的实现过程，详细讲解基于 Swift 语言使用 UIScrollView 控件的过程。

具体实现

（1）打开 Xcode，然后创建一个名为"UIScrollView-Sample"的工程，工程的最终目录结构如图 14-25 所示。

545

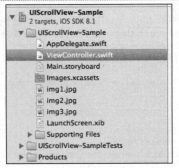

图 14-25　工程的目录结构

（2）编写文件 ViewController.swift，功能是在视图中追加显示指定位置的 3 幅图像，使用 UIScrollView 控件来滚动显示展示的图片。文件 ViewController.swift 的主要实现代码如下所示。

```swift
import UIKit

class ViewController: UIViewController {

    override func viewDidLoad() {
        super.viewDidLoad()

        //设置 UIImage 的素材位置
        let img1 = UIImage(named:"img1.jpg");
        let img2 = UIImage(named:"img2.jpg");
        let img3 = UIImage(named:"img3.jpg");

        //UIImageView 中添加图像
        let imageView1 = UIImageView(image:img1)
        let imageView2 = UIImageView(image:img2)
        let imageView3 = UIImageView(image:img3)

        //UIScrollView 滚动
        let scrView = UIScrollView()

        //表示位置
        scrView.frame = CGRectMake(50, 50, 240, 240)

        //所有视图大小
        scrView.contentSize = CGSizeMake(240*3, 240)

        //UIImageView 坐标位置
        imageView1.frame = CGRectMake(0, 0, 240, 240)
        imageView2.frame = CGRectMake(240, 0, 240, 240)
        imageView3.frame = CGRectMake(480, 0, 240, 240)


        //在 view 追加图像
        self.view.addSubview(scrView)
        scrView.addSubview(imageView1)
        scrView.addSubview(imageView2)
        scrView.addSubview(imageView3)

        // 设置图像边界
        scrView.pagingEnabled = true

        //设置 scroll 画面的初期位置
        scrView.contentOffset = CGPointMake(0, 0);

    }
    override func didReceiveMemoryWarning() {
        super.didReceiveMemoryWarning()
    }
}
```

执行后将在屏幕中显示指定位置的图像，效果如图 14-26 所示。

图 14-26　执行效果

左右触摸屏幕中的图像时，会展示另外的素材图片，如图 14-27 所示。

图 14-27　显示另外的图片

实例 280　基于 Swift 使用 UIPageControl 控件

实例 280	基于 Swift 使用 UIPageControl 控件
源码路径	\daima\280

实例说明

在开发 iOS 应用程序的过程中，经常需要翻页功能来显示内容过多的界面，其作用和滚动控件类似，iOS 应用程序中的翻页控件是 PageControll。在本节的内容中，将通过一个具体实例的实现过程，详细讲解基于 Swift 语言联合使用 UIPageControl 和 UIScrollView 控件的过程。

具体实现

（1）打开 Xcode，然后创建一个名为"MyFirstSwiftTest"的工程，工程的最终目录结构如图 14-28 所示。

第14章 Swift 实战

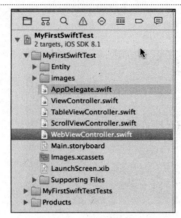

图14-28　工程的目录结构

（2）编写文件 ViewController.swift，本实例是一个综合实例，分别演示了 UIPageControl、UIWebView、UIScrollView、UITableView、UIButton 和 UILabel 等控件的基本用法，在主视图中列表显示了上述控件的名称，并监听用户单击触摸列表的选项，根据触摸选项来到第二个界面，并显示上述单个控件的用法。文件 ViewController.swift 的具体实现代码如下所示。

```swift
import UIKit

class ViewController: UIViewController,UITextFieldDelegate
{
    var a:Int = 0
    var isuse:Bool = false
    var authButton :UIButton!
    override func viewDidLoad()
    {
        super.viewDidLoad()
        // Do any additional setup after loading the view, typically from a nib.
        self.view.backgroundColor = UIColor.lightGrayColor()
        self.title = "首页"
        
        // UIButton
        authButton = UIButton.buttonWithType(UIButtonType.Custom) as? UIButton
        authButton.frame = CGRect(x: 10, y: 70, width: 150, height: 30)
        authButton.setTitle("这是一个全局按钮", forState: UIControlState. Normal)
        authButton.setTitleColor(UIColor.redColor(), forState: UIControlState.Normal)
        authButton.setTitleColor(UIColor.blueColor(), forState: UIControlState. Highlighted)
        authButton.addTarget(self, action: Selector("btnClick:"), forControlEvents:UIControlEvents.TouchUpInside)
        authButton.tag = 1000;
        self.view.addSubview(authButton)
        
        var btn2:UIButton! = UIButton.buttonWithType(UIButtonType.Custom) as? UIButton
        btn2.frame = CGRectMake(10, 100, 105, 30)
        btn2.setTitle("局部按钮", forState: UIControlState.Normal)
        btn2.setTitleColor(UIColor.yellowColor(), forState: UIControlState.Normal)
        btn2.tag = 1001;
        btn2.addTarget(self, action: Selector("btnClick:"), forControlEvents: UIControlEvents.TouchUpInside)
        self.view.addSubview(btn2)
        
        //UILabel
        var firstLabel:UILabel! = UILabel(frame: CGRect(x: 10, y: 130, width: 105, height: 20))
        firstLabel.backgroundColor = UIColor.clearColor()
        firstLabel.textColor = UIColor(red: 0, green: 174, blue: 232, alpha: 1)
        firstLabel.textAlignment = NSTextAlignment.Left
        firstLabel.font = UIFont.boldSystemFontOfSize(16)
        firstLabel.text = "这是一个 Label"
        self.view.addSubview(firstLabel)
```

```swift
        var secondLabel = UILabel()
        secondLabel.frame = CGRectMake(10, 150, 105, 20)
        secondLabel.backgroundColor = UIColor.clearColor()
        secondLabel.textColor = UIColor(red: 100, green: 174, blue: 232, alpha: 1)
        secondLabel.textAlignment = NSTextAlignment.Center
        secondLabel.font = UIFont.systemFontOfSize(12)
        secondLabel.text = "这是第二个 Label"
        secondLabel.lineBreakMode = NSLineBreakMode.ByWordWrapping
        secondLabel.sizeToFit()
        self.view.addSubview(secondLabel)


        //UITextField
        var firstTextField = UITextField()
        firstTextField.backgroundColor = UIColor.whiteColor()
        firstTextField.frame = CGRectMake(10, 170, 150, 20)
        firstTextField.textColor = UIColor.blackColor()
        firstTextField.autocapitalizationType = UITextAutocapitalizationType.None
        //首字母自动大写
        firstTextField.autocorrectionType = UITextAutocorrectionType.No
        //自动纠错
        firstTextField.borderStyle = UITextBorderStyle.RoundedRect
        //边框样式
        firstTextField.placeholder = "请输入内容"
        firstTextField.font = UIFont(name: "Arial", size: 7.0)
        //字体
        firstTextField.clearButtonMode = UITextFieldViewMode.Always;
        //输入框中是否有个叉号,在什么时候显示,用于一次性删除输入框中的内容
        firstTextField.secureTextEntry = false
        //每输入一个字符就变成点,用语密码输入
        firstTextField.clearsOnBeginEditing = true
        //再次编辑就清空
        firstTextField.contentVerticalAlignment =
        UIControlContentVerticalAlignment.Center  //内容的垂直对齐方式
        firstTextField.adjustsFontSizeToFitWidth = false
        //设置为 true 时文本会自动缩小以适应文本窗口大小,默认是保持原来大小,而让长文本滚动
        firstTextField.keyboardType = UIKeyboardType.Default
        //设置键盘样式
        firstTextField.returnKeyType = UIReturnKeyType.Done
        //return 键变成什么键
        firstTextField.keyboardAppearance=UIKeyboardAppearance.Default
        //键盘外观
        firstTextField.delegate = self
        self.view.addSubview(firstTextField)


        var titleArray :NSArray = ["UITableView","UIScrollView","UIWebView"]

        for var index=0; index<titleArray.count; index++
        {
            var btn3:UIButton! = UIButton.buttonWithType(UIButtonType.System) as? UIButton
            btn3.frame = CGRect(x: 10, y:190+index*30, width:150, height:30)
            var btnStr = "点击进入\(titleArray.objectAtIndex(index))"
            btn3.setTitle(btnStr, forState: UIControlState.Normal)
            btn3.titleLabel?.font = UIFont.systemFontOfSize(14)
            btn3.setTitleColor(UIColor.blueColor(), forState: UIControlState.Normal)
            btn3.tag = 1002+index;
            btn3.addTarget(self, action: Selector("btnClick:"), forControlEvents: UIControlEvents.TouchUpInside)
            self.view.addSubview(btn3)
        }

    }

    //按钮点击方法
    func btnClick(sender:UIButton!)
    {
        var btn:UIButton = sender
        switch(btn.tag){
```

```
        case 1000:
            a++
            if(a>100)
            {
                a=1
            }
            println("按钮被点击了\(a)次")
            authButton.setTitle("全局按钮被点击了\(a)次", forState: UIControlState.Normal)
        case 1001:
            println("局部按钮被点击了")
        case 1002:
            println("点击进入UITableView")
            var tableVC:TableViewController = TableViewController()
            self.navigationController?.pushViewController(tableVC, animated: true)
        case 1003:
            println("点击进入UIScrollView")
            var scrollVC:ScrollViewController = ScrollViewController()
            self.navigationController?.pushViewController(scrollVC, animated: true)
        case 1004:
            println("点击进入UIWebView")
            var webVC:WebViewController = WebViewController()
            self.navigationController?.pushViewController(webVC, animated: true)
        default:
            println("无操作")
        }
    }

    //UITextFieldDelegate
    func textFieldShouldReturn(textField: UITextField) -> Bool
    {
        textField.resignFirstResponder()
        return true
    }

    override func didReceiveMemoryWarning()
    {
        super.didReceiveMemoryWarning()
        // Dispose of any resources that can be recreated.
    }
}
```

主界面执行后将列表显示几个常用的控件名，如图14-29所示。

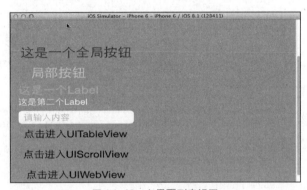

图14-29　主界面列表视图

（3）文件ScrollViewController.swift功能是，当点击条目中的UIScrollView控件名称时，会在新界面中演示这个控件的功能，并且同时演示了UIPageControl控件的基本用法。文件ScrollViewController.swift的主要实现代码如下所示。

```
import UIKit
class ScrollViewController: UIViewController,UIScrollViewDelegate {
    var slideArray:NSArray!
    var pageControl :UIPageControl!
    var labelScrollView:UIScrollView!
```

```swift
var mainScreenWidthUse:CGFloat = 0
var mainScreenHeightUse:CGFloat = 0

override func viewDidLoad() {
    super.viewDidLoad()
    self.title = "滚动页面 UIScrollView"
    self.view.backgroundColor = UIColor.whiteColor()
    // Do any additional setup after loading the view.
    
    slideArray = ["1","2","3","4","5"]
    
    let mainScreenWidth = self.view.frame.size.width
    mainScreenWidthUse = mainScreenWidth
    let mainScreenHeight = self.view.frame.size.height
    mainScreenHeightUse = mainScreenHeight
    
    //UIScrollView
    labelScrollView = UIScrollView()
    labelScrollView.frame = self.view.frame
    labelScrollView.bounces = false
    labelScrollView.backgroundColor = UIColor.clearColor()
    labelScrollView.pagingEnabled = true
    //决定 ScrollView 第一页显示的内容
    labelScrollView.contentOffset = CGPoint(x: mainScreenWidthUse, y: 0)
    labelScrollView.contentSize = CGSize(width :(CGFloat)(slideArray.count+2) * mainScreenWidth, height :mainScreenHeight-100)
    labelScrollView.showsHorizontalScrollIndicator = false
    labelScrollView.showsVerticalScrollIndicator = false
    labelScrollView.delegate = self
    self.view.addSubview(labelScrollView)
    
    //为了实现循环滚动
    //最后一页
    var startLabel:UILabel! = UILabel(frame: CGRect(x: 0, y: 0, width: mainScreenWidth, height: mainScreenHeight))
    startLabel.backgroundColor = UIColor.clearColor()
    startLabel.textColor = UIColor.blackColor()
    startLabel.textAlignment = NSTextAlignment.Center
    startLabel.font = UIFont.boldSystemFontOfSize(32)
    startLabel.text = "第\(slideArray.objectAtIndex(slideArray.count-1))页"
    labelScrollView.addSubview(startLabel)
    
    for var i=0;i<slideArray.count;i++
    {
        var useLabel:UILabel! = UILabel(frame: CGRect(x: (CGFloat)(i + 1)*mainScreenWidth, y: 0, width: mainScreenWidth, height: mainScreenHeight))
        useLabel.backgroundColor = UIColor.clearColor()
        useLabel.textColor = UIColor.blackColor()
        useLabel.textAlignment = NSTextAlignment.Center
        useLabel.font = UIFont.boldSystemFontOfSize(32)
        useLabel.text = "第\(slideArray.objectAtIndex(i))页"
        labelScrollView.addSubview(useLabel)
    }
    
    //第一页
    var endLabel:UILabel! = UILabel(frame: CGRect(x: (CGFloat)(slideArray.count + 1)* mainScreenWidth, y: 0, width: mainScreenWidth, height: mainScreenHeight))
    endLabel.backgroundColor = UIColor.clearColor()
    endLabel.textColor = UIColor.blackColor()
    endLabel.textAlignment = NSTextAlignment.Center
    endLabel.font = UIFont.boldSystemFontOfSize(32)
    endLabel.text = "第\(slideArray.objectAtIndex(0))页"
    labelScrollView.addSubview(endLabel)
    
    //UIPageControl
    pageControl = UIPageControl()
    pageControl.frame = CGRect(x: 0, y: mainScreenHeight-100, width: mainScreenWidth, height: 20)
    pageControl.pageIndicatorTintColor = UIColor.redColor()
    pageControl.currentPageIndicatorTintColor = UIColor.blackColor()
```

```swift
        pageControl.numberOfPages = slideArray.count
        pageControl.addTarget(self, action: Selector("pageControlAction"), forControlEvents: UIControlEvents.TouchUpInside)
        self.view.addSubview(pageControl)
        
    }
    
    //点击 pagecontrol 响应事件
    func pageControlAction(){
        var page:Int = pageControl.currentPage
        labelScrollView.setContentOffset(CGPoint(x: mainScreenWidthUse*(CGFloat)(page+1), y: 0), animated: true)
    }
    
    //UIScrollViewDelegate
    func scrollViewDidEndDecelerating(scrollView: UIScrollView)
    {
        //当前页
        var currentPage = Int(labelScrollView.contentOffset.x/mainScreenWidthUse)
//        var currentPage = Int((labelScrollView.contentOffset.x - mainScreenWidthUse / ( CGFloat(slideArray.count) + 2)) / mainScreenWidthUse + 1)
        
        if (currentPage == 0)
        {
            labelScrollView.scrollRectToVisible(CGRect(x: mainScreenWidthUse * CGFloat(slideArray.count), y: 0, width: mainScreenWidthUse, height: mainScreenHeightUse), animated: false)
        }
        else if (currentPage == slideArray.count + 1)
        {
            labelScrollView.scrollRectToVisible(CGRect(x: mainScreenWidthUse, y: 0, width: mainScreenWidthUse, height: mainScreenHeightUse), animated: false)
        }
        
    }
    
    func scrollViewDidScroll(scrollView: UIScrollView) {
        var page:Int = Int(labelScrollView.contentOffset.x/mainScreenWidthUse)-1
        pageControl.currentPage = page;
    }
    
    override func didReceiveMemoryWarning() {
        super.didReceiveMemoryWarning()
        // Dispose of any resources that can be recreated.
    }
}
```

到此为止，整个实例介绍完毕。单击 UI 主界面中"UIScrollView"列表项后的执行效果如图 14-30 所示。

图 14-30 执行效果

实例 281　基于 Swift 使用 UIAlertView 控件

实例 281	基于 Swift 使用 UIAlertView 控件
源码路径	\daima\281

实例说明

　　iOS 应用程序是以用户为中心的，这意味着它们通常不在后台执行功能或在没有界面的情况下运行。它们让用户能够处理数据、玩游戏、通信或执行众多其他的操作。当应用程序需要发出提醒、提供反馈或让用户做出决策时，它总是以相同的方式进行。Cocoa Touch 通过各种对象和方法来引起用户注意，这包括 UIAlertView 和 UIActionSheet。这些控件不同于本书前面介绍的其他对象，需要我们使用代码来创建他们。在本节的内容中，将通过一个具体实例的实现过程，详细讲解基于 Swift 语言使用 UIAlertView 控件的过程。

具体实现

　　（1）打开 Xcode，然后创建一个名为"AlertController---Swift"的工程，工程的最终目录结构如图 14-31 所示。

图 14-31　工程的目录结构

　　（2）在 Xcode 6 的 Main.storyboard 面板中设置 UI 界面，在里面添加了 11 个 Label 文本框，如图 14-32 所示。

图 14-32　设置 Main.storyboard 面板

（3）编写文件 ViewController.swift，功能是在 UI 中添加 11 个文本信息，并监听用户对文本的触摸操作，根据监听事件来显示不同风格、不同样式的 UIAlertView 控件效果，在屏幕中展示了不同的对话框样式。文件 ViewController.swift 的具体实现代码如下所示。

```swift
import UIKit

class ViewController: UIViewController {
    @IBAction func Btn_UIAlertView_DefaultStyle(sender: UIButton) {
        //常规对话框，最简单的 UIAlertView 使用方法
        var alertView = UIAlertView()
        
        alertView.delegate = self
        alertView.title = "常规对话框"
        alertView.message = "常规对话框风格"
        alertView.addButtonWithTitle("取消")
        alertView.addButtonWithTitle("确定")
        
        alertView.show()
        
        //只有一个按钮的 swift 初始化
//var alertView = UIAlertView(title: "常规对话框", message: "常规对话框风格", delegate: self,
//cancelButtonTitle: "取消")
//alertView.show()
    }
    
    @IBAction func Btn_UIAlertView_PlainTextStyle(sender: UIButton) {
        //文本对话框，带有一个文本框
        var alertView = UIAlertView()
        
        alertView.delegate = self
        alertView.title = "文本对话框"
        alertView.message = "请输入文字："
        alertView.addButtonWithTitle("取消")
        alertView.addButtonWithTitle("确定")
        alertView.alertViewStyle = UIAlertViewStyle.PlainTextInput
        
        alertView.show()
    }
    
    @IBAction func Btn_UIAlertView_SecureTextStyle(sender: UIButton) {
        //密码对话框，带有一个拥有密码安全保护机制的密码文本框
        var alertView = UIAlertView()
        
        alertView.delegate = self
        alertView.title = "密码对话框"
        alertView.message = "请输入密码："
        alertView.addButtonWithTitle("取消")
        alertView.addButtonWithTitle("确定")
        alertView.alertViewStyle = UIAlertViewStyle.SecureTextInput
        
        alertView.show()
    }
    
    @IBAction func Btn_UIAlertView_LoginAndPasswordStyle(sender: UIButton) {
        //登录对话框，仿照登录框的效果制作，拥有两个文本框，其中一个是密码文本框
        var alertView = UIAlertView()
        
        alertView.delegate = self
        alertView.title = "登录对话框"
        alertView.message = "请输入用户名和密码："
        alertView.addButtonWithTitle("取消")
        alertView.addButtonWithTitle("登录")
        alertView.alertViewStyle = UIAlertViewStyle.LoginAndPasswordInput
        
        alertView.show()
    }
```

```swift
@IBAction func Btn_UIAlertController_BasicAlertStyle(sender: UIButton) {
    //基本对话框，使用 iOS 8 创建的 UIAlertController 类，同 UIAlertView 的常规对话框相同
    var alertController = UIAlertController(title: "基本对话框", message: "带有基本按钮的对话框",
    preferredStyle: UIAlertControllerStyle.Alert)
    
    var cancelAction = UIAlertAction(title: "取消", style: UIAlertActionStyle.Cancel, handler: nil)
    var okAction = UIAlertAction(title: "确定", style: UIAlertActionStyle.Default, handler: nil)
    
    alertController.addAction(cancelAction)
    alertController.addAction(okAction)
    
    self.presentViewController(alertController, animated: true, completion: nil)
}

@IBAction func Btn_UIAlertController_DestructiveActions(sender: UIButton) {
    //重置对话框，带有一个醒目的"毁坏"样式的按钮
    var alertController = UIAlertController(title: "重置对话框", message: "带有"毁坏"样式按钮的对话框", preferredStyle: UIAlertControllerStyle.Alert)
    
    var resetAction = UIAlertAction(title: "重置", style: UIAlertActionStyle.Destructive,handler:nil)
    var cancelAction = UIAlertAction(title: "取消", style: UIAlertActionStyle.Cancel, handler: nil)
    
    alertController.addAction(resetAction)
    alertController.addAction(cancelAction)
    
    self.presentViewController(alertController, animated: true, completion: nil)
}

@IBAction func Btn_UIAlertController_LoginAndPasswordStyle(sender: UIButton) {
//登录对话框，必须要输入 3 个字符以上才能激活"登录"按钮，会调用 alertTextFieldDidChange:函数
    var alertController = UIAlertController(title: "登录对话框", message: "请输入用户名或密码：",
    preferredStyle: UIAlertControllerStyle.Alert)
    
    alertController.addTextFieldWithConfigurationHandler { (textField: UITextField!) -> Void in
        textField.placeholder = "用户名"
        NSNotificationCenter.defaultCenter().addObserver(self, selector: Selector("alertTextFieldDidChange:"), name: UITextFieldTextDidChangeNotification, object: textField)
    }
    
    alertController.addTextFieldWithConfigurationHandler { (textField: UITextField!) -> Void in
        textField.placeholder = "密码"
        textField.secureTextEntry = true
    }
    
    var cancelAction = UIAlertAction(title: "取消", style: UIAlertActionStyle.Cancel) { (action: UIAlertAction!) -> Void in
        NSNotificationCenter.defaultCenter().removeObserver(self, name: UITextFieldTextDidChangeNotification, object: nil)
    }
    
    var loginAction = UIAlertAction(title: "登录", style: UIAlertActionStyle.Default) { (action: UIAlertAction!) -> Void in
        NSNotificationCenter.defaultCenter().removeObserver(self, name: UITextFieldTextDidChangeNotification, object: nil)
    }
    
    loginAction.enabled = false
    
    alertController.addAction(cancelAction)
    alertController.addAction(loginAction)
    
    self.presentViewController(alertController, animated: true, completion: nil)
}

func alertTextFieldDidChange(notification: NSNotification){
    var alertController = self.presentedViewController as UIAlertController?
    
    if alertController != nil {
        var login = alertController!.textFields?.first as UITextField
```

```
            var loginAction = alertController!.actions.last as UIAlertAction
            loginAction.enabled = countElements(login.text) > 2
        }
    }

    @IBAction func Btn_UIAlertController_ActionSheet(sender: UIButton) {
        //上拉菜单，使用 UIPopoverPresentationController 来防止 iPad 上运行时异常
        var alertController = UIAlertController(title: "保存或删除数据", message: "注意：删除操作无法恢复！", preferredStyle: UIAlertControllerStyle.ActionSheet)

        var cancelAction = UIAlertAction(title: "取消", style: UIAlertActionStyle.Cancel, handler: nil)
        var deleteAction = UIAlertAction(title: "删除", style: UIAlertActionStyle.Destructive, handler: nil)
        var archiveAction = UIAlertAction(title: "保存", style: UIAlertActionStyle.Default, handler: nil)

        alertController.addAction(cancelAction)
        alertController.addAction(deleteAction)
        alertController.addAction(archiveAction)

        var popover = alertController.popoverPresentationController
        if popover != nil {
            popover?.sourceView = sender
            popover?.sourceRect = sender.bounds
            popover?.permittedArrowDirections = UIPopoverArrowDirection.Any
        }

        self.presentViewController(alertController, animated: true, completion: nil)
    }
    override func viewDidLoad() {
        super.viewDidLoad()
        // Do any additional setup after loading the view, typically from a nib.
    }

    override func didReceiveMemoryWarning() {
        super.didReceiveMemoryWarning()
        // Dispose of any resources that can be recreated.
    }
}
```

执行后的初始效果如图 14-33 所示。

图 14-33　初始执行效果

单击"常规对话框"后的效果如图 14-34 所示。

图 14-34　单击"常规对话框"后的效果

单击"文本对话框"后的效果如图 14-35 所示。

图 14-35　单击"文本对话框"后的效果

单击"密码对话框"后的效果如图 14-36 所示。

图 14-36　单击"密码对话框"后的效果

实例 282　基于 Swift 在表视图中使用其他控件

实例 282	基于 Swift 在表视图中使用其他控件
源码路径	\daima\282

实例说明

使用表视图可以在屏幕上显示一个单元格列表,每个单元格都可以包含多项信息,但仍然是一个整体。并且可以将表视图划分成多个区(section),以便从视觉上将信息分组。表视图控制器是一种只能显示表视图的标准视图控制器,可以在表视图占据整个视图时使用这种控制器。通过使用标准视图控制器可以根据需要在视图中创建任意尺寸的表,我们只需将表的委托和数据源输出口连接到视图控制器类即可。在本节的内容中,将通过一个具体实例的实现过程,详细讲解在表视图中使用其他控件的过程。

具体实现

(1)打开 Xcode,创建一个名为"ILoveSwift"的工程,工程的最终目录结构如图 14-37 所示。

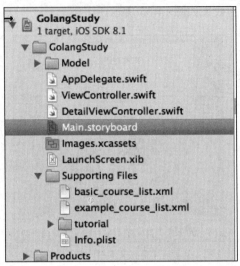

图 14-37　工程的目录结构

(2)在 Xcode 6 的 Main.storyboard 面板中设置 UI 界面,其中一个视图界面是通过 Table View 实现的,在第二个界面中插入了 Label 控件和 Button 控件,如图 14-38 所示。

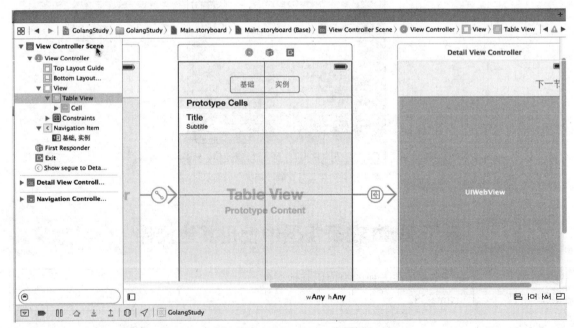

图 14-38　Main.storyboard 面板

(3)第一个界面的实现文件是 ViewController.swift,其实现过程已经在本章前面的范例中进行了讲解,在此不再介绍。第一个界面的执行效果如图 14-39 所示。

实例 282　基于 Swift 在表视图中使用其他控件

图 14-39　第一个界面的执行效果

（4）编写文件 DetailViewController.swift，功能是当单击底部列表中的某一个选项后会显示这个选项的详细信息。文件 DetailViewController.swift 的具体实现代码如下所示。

```swift
import UIKit
class DetailViewController: UIViewController,UIWebViewDelegate{
    @IBOutlet weak var webView: UIWebView!
    @IBOutlet weak var nextBtn: UIBarButtonItem!
    var courseSet:[Course] = []
    var pathname:String!
    var index:Int = 0
    override func viewDidLoad() {
        super.viewDidLoad()
        webView.delegate = self
        loadData(pathname)
    }
    func loadData(pathname:String){
        let path = NSBundle.mainBundle().pathForResource(pathname, ofType: "html")!
        let requestURL = NSURL(fileURLWithPath: path)!
        let request = NSURLRequest(URL: requestURL)
        webView.loadRequest(request)
    }
    func webViewDidStartLoad(webView: UIWebView) {
        UIApplication.sharedApplication().networkActivityIndicatorVisible = true
    }
    func webViewDidFinishLoad(webView: UIWebView) {
        UIApplication.sharedApplication().networkActivityIndicatorVisible = false
    }
    @IBAction func nextBtn(sender: AnyObject) {
        index = index+1
        if index < courseSet.count {
            var course = courseSet[index] as Course
            self.title = course.getName()
            pathname = course.getChapter()
            loadData(pathname)
        }else{
            nextBtn.enabled = false
        }
    }
    override func didReceiveMemoryWarning() {
        super.didReceiveMemoryWarning()
    }
}
```

到此为止，整个实例介绍完毕。单击 UI 主界面中"UISegmentedControl"列表项后会显示列表中某个标题选项的详细信息，例如"内置基础数据类型"选项的详情界面效果如图 14-40 所示。

第 14 章 Swift 实战

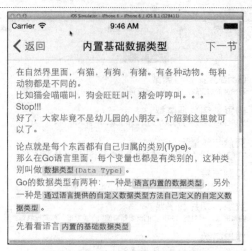

图 14-40 第二个界面的详情效果

实例 283 使用 Swift 实现自定义进度条效果

实例 283	使用 Swift 实现自定义进度条效果
源码路径	\daima\283

实例说明

在 iOS 应用中，通过 UIProgressView 来显示进度效果，如音乐、视频的播放进度，和文件的上传下载进度等。在本节的内容中，将通过一个具体实例的实现过程，详细讲解基于 Swift 语言实现一个自定义进度条效果的过程。

具体实现

（1）打开 Xcode，然后创建一个名为 "KYCircularProgress" 的工程，工程的最终目录结构如图 14-41 所示。

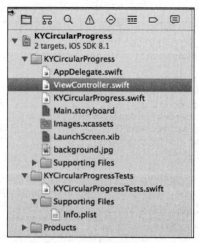

图 14-41 工程的目录结构

（2）再看 LaunchScreen.xib 设计界面，创建了一个 UIViewController 试图界面，如图 14-42 所示。

实例 283　使用 Swift 实现自定义进度条效果

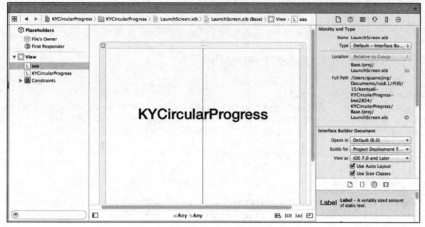

图 14-42　LaunchScreen.xib 设计界面

（3）编写文件 ViewController.swift，功能是在视图界面中创建了 3 种进度条样式 circularProgress1、circularProgress2 和 circularProgress3，然后分别通过函数 setupKYCircularProgress1()、setupKYCircularProgress2()和 setupKYCircularProgress3()分别设置了上述 3 种进度条的具体样式，第一种是环形显示进度数字样式，第二种是环形不显示进度数字样式，第三种是绘制五角星样式。文件 ViewController.swift 的具体实现代码如下所示。

```swift
import UIKit

class ViewController: UIViewController {
    
    var circularProgress1: KYCircularProgress!
    var circularProgress2: KYCircularProgress!
    var circularProgress3: KYCircularProgress!
    var progress: UInt8 = 0
    
    override func viewDidLoad() {
        super.viewDidLoad()
        
        setupKYCircularProgress1()
        setupKYCircularProgress2()
        setupKYCircularProgress3()
        
        NSTimer.scheduledTimerWithTimeInterval(0.03, target: self, selector: Selector("updateProgress"),
            userInfo: nil, repeats: true)
    }
    
    override func didReceiveMemoryWarning() {
        super.didReceiveMemoryWarning()
    }
    
    func setupKYCircularProgress1() {
        circularProgress1 = KYCircularProgress(frame: CGRectMake(0, 0, self.view.frame.size.width,
            self.view.frame.size.height/2))
        let center = (CGFloat(160.0), CGFloat(200.0))
        circularProgress1.path = UIBezierPath(arcCenter: CGPointMake(center.0, center.1), radius:
CGFloat(circularProgress1.frame.size.width/3.0), startAngle: CGFloat(M_PI), endAngle: CGFloat(0.0),
clockwise: true)
        circularProgress1.lineWidth = 8.0
        
        let textLabel = UILabel(frame: CGRectMake(circularProgress1.frame.origin.x + 120.0, 170.0,
            80.0, 32.0))
        textLabel.font = UIFont(name: "HelveticaNeue-UltraLight", size: 32)
        textLabel.textAlignment = .Center
        textLabel.textColor = UIColor.greenColor()
        textLabel.alpha = 0.3
        self.view.addSubview(textLabel)
        
        circularProgress1.progressChangedClosure({ (progress: Double, circularView: KYCircularProgress) in
```

```swift
            println("progress: \(progress)")
            textLabel.text = "\(Int(progress * 100.0))%"
        })
        
        self.view.addSubview(circularProgress1)
    }
    
    func setupKYCircularProgress2() {
        circularProgress2 = KYCircularProgress(frame: CGRectMake(0, circularProgress1.frame.size.height, self.view.frame.size.width/2, self.view.frame.size.height/3))
        circularProgress2.colors = [0xA6E39D, 0xAEC1E3, 0xAEC1E3, 0xF3C0AB]
        
        self.view.addSubview(circularProgress2)
    }
    
    func setupKYCircularProgress3() {
        circularProgress3 = KYCircularProgress(frame: CGRectMake(circularProgress2.frame.size.width*1.25, circularProgress1.frame.size.height*1.15, self.view.frame.size.width/2, self.view.frame.size.height/2))
        circularProgress3.colors = [0xFFF77A, 0xF3C0AB]
        circularProgress3.lineWidth = 3.0
        
        let path = UIBezierPath()
        path.moveToPoint(CGPointMake(50.0, 2.0))
        path.addLineToPoint(CGPointMake(84.0, 86.0))
        path.addLineToPoint(CGPointMake(6.0, 33.0))
        path.addLineToPoint(CGPointMake(96.0, 33.0))
        path.addLineToPoint(CGPointMake(17.0, 86.0))
        path.closePath()
        circularProgress3.path = path
        
        self.view.addSubview(circularProgress3)
    }
    
    func updateProgress() {
        progress = progress &+ 1
        let normalizedProgress = Double(progress) / 255.0
        
        circularProgress1.progress = normalizedProgress
        circularProgress2.progress = normalizedProgress
        circularProgress3.progress = normalizedProgress
    }
}
```

（4）文件的 **KYCircularProgress.swift** 功能是实现进度条的进度绘制功能，分别通过变量 startAngle 和变量 endAngle 设置进度条的起始点。文件 **KYCircularProgress.swift** 的主要实现代码如下所示。

```swift
import Foundation
import UIKit

// MARK: - KYCircularProgress
class KYCircularProgress: UIView {
    typealias progressChangedHandler = (progress: Double, circularView: KYCircularProgress) -> ()
    private var progressChangedClosure: progressChangedHandler?
    private var progressView: KYCircularShapeView!
    private var gradientLayer: CAGradientLayer!
    var progress: Double = 0.0 {
        didSet {
            let clipProgress = max( min(oldValue, 1.0), 0.0)
            self.progressView.updateProgress(clipProgress)
            
            if let progressChanged = progressChangedClosure {
                progressChanged(progress: clipProgress, circularView: self)
            }
        }
    }
    var startAngle: Double = 0.0 {
        didSet {
            self.progressView.startAngle = oldValue
```

```swift
        }
    }
    var endAngle: Double = 0.0 {
        didSet {
            self.progressView.endAngle = oldValue
        }
    }
    var lineWidth: Double = 8.0 {
        willSet {
            self.progressView.shapeLayer().lineWidth = CGFloat(newValue)
        }
    }
    var path: UIBezierPath? {
        willSet {
            self.progressView.shapeLayer().path = newValue?.CGPath
        }
    }
    var colors: [Int]? {
        didSet {
            updateColors(oldValue)
        }
    }
    var progressAlpha: CGFloat = 0.55 {
        didSet {
            updateColors(self.colors)
        }
    }
    
    required init(coder aDecoder: NSCoder) {
        super.init(coder: aDecoder)
        setup()
    }
    
    override init(frame: CGRect) {
        super.init(frame: frame)
        setup()
    }
    
    private func setup() {
        self.progressView = KYCircularShapeView(frame: self.bounds)
        self.progressView.shapeLayer().fillColor = UIColor.clearColor().CGColor
        self.progressView.shapeLayer().path = self.path?.CGPath
        
        gradientLayer = CAGradientLayer(layer: layer)
        gradientLayer.frame = self.progressView.frame
        gradientLayer.startPoint = CGPointMake(0, 0.5);
        gradientLayer.endPoint = CGPointMake(1, 0.5);
        gradientLayer.mask = self.progressView.shapeLayer();
        gradientLayer.colors = self.colors ?? [colorHex(0x9ACDE7).CGColor!, colorHex(0xE7A5C9).CGColor!]
        
        self.layer.addSublayer(gradientLayer)
        self.progressView.shapeLayer().strokeColor = self.tintColor.CGColor
    }
    
    func progressChangedClosure(completion: progressChangedHandler) {
        progressChangedClosure = completion
    }
    
    private func colorHex(rgb: Int) -> UIColor {
        return UIColor(red: CGFloat((rgb & 0xFF0000) >> 16) / 255.0,
                     green: CGFloat((rgb & 0xFF00) >> 8) / 255.0,
                      blue: CGFloat(rgb & 0xFF) / 255.0,
                     alpha: progressAlpha)
    }
    
    private func updateColors(colors: [Int]?) -> () {
        var convertedColors: [AnyObject] = []
        if let inputColors = self.colors {
            for hexColor in inputColors {
                convertedColors.append(self.colorHex(hexColor).CGColor!)
            }
```

```swift
        } else {
            convertedColors = [self.colorHex(0x9ACDE7).CGColor!, self.colorHex(0xE7A5C9).CGColor!]
        }
        self.gradientLayer.colors = convertedColors
    }
}

// MARK: - KYCircularShapeView
class KYCircularShapeView: UIView {
    var startAngle = 0.0
    var endAngle = 0.0
    
    override class func layerClass() -> AnyClass {
        return CAShapeLayer.self
    }
    
    private func shapeLayer() -> CAShapeLayer {
        return self.layer as CAShapeLayer
    }
    
    required init(coder aDecoder: NSCoder) {
        super.init(coder: aDecoder)
    }
    
    override init(frame: CGRect) {
        super.init(frame: frame)
        self.updateProgress(0)
    }
    
    override func layoutSubviews() {
        super.layoutSubviews()
        
        if self.startAngle == self.endAngle {
            self.endAngle = self.startAngle + (M_PI * 2)
        }
        self.shapeLayer().path = self.shapeLayer().path ?? self.layoutPath().CGPath
    }
    
    private func layoutPath() -> UIBezierPath {
        var halfWidth = CGFloat(self.frame.size.width / 2.0)
        return UIBezierPath(arcCenter: CGPointMake(halfWidth, halfWidth), radius: halfWidth - self.shapeLayer().lineWidth, startAngle: CGFloat(self.startAngle), endAngle: CGFloat(self.endAngle), clockwise: true)
    }
    
    private func updateProgress(progress: Double) {
        CATransaction.begin()
        CATransaction.setValue(kCFBooleanTrue, forKey: kCATransactionDisableActions)
        self.shapeLayer().strokeEnd = CGFloat(progress)
        CATransaction.commit()
    }
}
```

到此为止，整个实例全部介绍完毕。执行后将在屏幕中显示 3 种不同样式的进度条效果，如图 14-43 所示。

图 14-43　执行效果

实例 284　基于 Swift 使用 UIViewController 控件

实例 284	基于 Swift 使用 UIViewController 控件
源码路径	\daima\284

实例说明

UIViewController 的主要功能是控制画面的切换，其中的 view 属性（UIView 类型）管理整个画面的外观。在开发 iOS 应用程序时，其实不使用 UIViewController 也能编写出 iOS 应用程序，但是这样整个代码会看起来将非常凌乱。如果可以将不同外观的画面进行整体地切换显然更合理，UIViewController 正是用于实现这种画面切换方式的。在本实例中，演示了在 Swift 程序中使用 UIViewController 控件的基本过程。

具体实现

（1）打开 Xcode，然后创建一个名为"TSwift"的工程，工程的最终目录结构如图 14-44 所示。

图 14-44　工程的目录结构

（2）打开 Main.storyboard，为本工程设计一个 ViewController 视图界面，如图 14-45 所示。

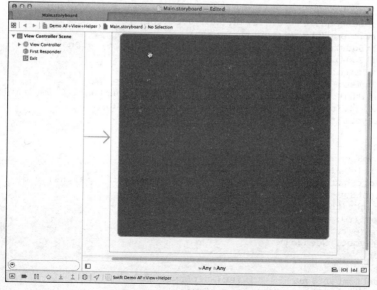

图 14-45　Main.storyboard 视图界面

第 14 章 Swift 实战

(3) 编写文件 ViewController.swift,功能是根据 X 点和 Y 点坐标在 UIView 视图中分别绘制虚线框、红圈、左上方、右上方、左下角、右下角、中心圆和中心圆阴影图像。文件 ViewController.swift 的具体实现代码如下所示。

```swift
import Foundation
import UIKit
import QuartzCore

class ViewController: UIViewController {

    let redCircle = UIView(autoLayout:true)

    override func viewDidLoad() {

        super.viewDidLoad()

        view.backgroundColor = UIColor(white: 0.5, alpha: 1)

        //虚线框
        let width = view.smallestSideLength() * 0.8
        var dashedBox = UIView(autoLayout:true)
        view.addSubview(dashedBox)
        dashedBox.backgroundColor = UIColor(white: 1, alpha: 0.5)
        dashedBox
            .width(width)
            .height(to: dashedBox, attribute: .Width)
            .center(to:view)
            .layoutIfNeeded()
        dashedBox
            .borderWithDashPattern([2, 6], borderWidth: 4, borderColor: UIColor.whiteColor(),
              cornerRadius: 6)
            .shadow(color: UIColor.blackColor(), offset: CGSize(width: 0, height: 0),
              radius: 6, opacity: 2, isMasked: false)

        //红圈
        view.addSubview(redCircle)
        redCircle.backgroundColor = UIColor.redColor()
        redCircle
            .size(to: dashedBox, constant: CGSize(width: -50, height: -50))
            .center(to: view)
            .layoutIfNeeded()
        redCircle.roundCornersToCircle(borderColor: UIColor.whiteColor(), borderWidth: 12)
        redCircle.clipsToBounds = true

        //左上方
        var topLeftSquare = UIView(autoLayout:true)
        redCircle.addSubview(topLeftSquare)
        topLeftSquare.backgroundColor = UIColor(white: 0, alpha: 0.3)
        topLeftSquare
            .left(to: redCircle)
            .top(to: redCircle)
            .width(to: redCircle, attribute: .Width, constant: 0, multiplier: 0.48)
            .height(to: topLeftSquare, attribute: .Width)
            .layoutIfNeeded()

        //右上方
        var topRightSquare = UIView(autoLayout:true)
        redCircle.addSubview(topRightSquare)
        topRightSquare.backgroundColor = UIColor(white: 0, alpha: 0.3)
        topRightSquare
            .right(to: redCircle)
            .top(to: redCircle)
            .size(to: topLeftSquare)
            .layoutIfNeeded()

        //左下角
        var bottomLeftSquare = UIView(autoLayout:true)
        redCircle.addSubview(bottomLeftSquare)
        bottomLeftSquare.backgroundColor = UIColor(white: 0, alpha: 0.3)
```

```swift
        bottomLeftSquare
            .left(to: redCircle)
            .bottom(to: redCircle)
            .size(to: topLeftSquare)
            .layoutIfNeeded()


        //右下角
        var bottomRightSquare = UIView(autoLayout:true)
        redCircle.addSubview(bottomRightSquare)
        bottomRightSquare.backgroundColor = UIColor(white: 0, alpha: 0.3)
        bottomRightSquare
            .right(to: redCircle)
            .bottom(to: redCircle)
            .size(to: topLeftSquare)
            .layoutIfNeeded()
        //中心圆
        var centerCircle = UIView(autoLayout:true)
        redCircle.addSubview(centerCircle)
        centerCircle.backgroundColor = UIColor(white: 1, alpha: 0.7)
        centerCircle
            .size(to: redCircle, constant: CGSize(width: 0, height: 0), multiplier: 0.6)
            .center(to: view)
            .layoutIfNeeded()
        centerCircle.roundCornersToCircle()

        //中心圆阴影
        var redCircleShadow = UIView(autoLayout:true)
        view.insertSubview(redCircleShadow, belowSubview: redCircle)
        redCircleShadow.backgroundColor = UIColor.blackColor()
        redCircleShadow
            .size(to: redCircle)
            .center(to: redCircle)
            .layoutIfNeeded()
        redCircleShadow.cornerRadius(redCircleShadow.width()/2)
        redCircleShadow.shadow(color: UIColor.blackColor(), offset: CGSize(width: 0, height: 0), radius: 6, opacity: 1, isMasked: false)


        // X点
        let dotsViewX = UIView(autoLayout:true)
        view.addSubview(dotsViewX)
        dotsViewX
            .left(0)
            .centerY(to: view)
            .width(to: view)
            .height(10)
            .layoutIfNeeded()
        var dots = [UIView]()
        for i in 1..<11 {
            var dot = UIView(autoLayout:true)
            dot.backgroundColor = UIColor(white: 0.8, alpha: 0.8)
            dotsViewX.addSubview(dot)
            dot.cornerRadius(5)
            dot.clipsToBounds = true
            dot.layoutIfNeeded()
            dot.tag = i
            dots.append(dot)
        }
        dotsViewX.spaceSubviewsEvenly(dots, size: CGSize(width: 10, height: 10))

        // Y点
        let dotsViewY = UIView(autoLayout:true)
        view.addSubview(dotsViewY)
        dotsViewY
            .top(0)
            .centerX(to: view)
            .height(to: view)
            .width(10)
            .layoutIfNeeded()
        dots = [UIView]()
```

```
        for i in 1..<11 {
            var dot = UIView(autoLayout:true)
            dot.backgroundColor = UIColor.darkGrayColor()
            dotsViewY.addSubview(dot)
            dot.cornerRadius(5)
            dot.clipsToBounds = true
            dot.layoutIfNeeded()
            dot.tag = i
            dots.append(dot)
        }
        dotsViewY. spaceSubviewsEvenly(dots, size: CGSize(width: 10, height: 10),axis:
        .Vertical)

    }
    override  func  viewWillTransitionToSize(size:  CGSize,  withTransitionCoordinator
    coordinator: UIViewControllerTransitionCoordinator) {

        let transitionToWide = size.width > size.height

        coordinator.animateAlongsideTransition({
            context in
            //创建一个过渡和内容相匹配的持续时间
            let transition = CATransition()
            transition.duration = context.transitionDuration()

            transition.timingFunction = CAMediaTimingFunction(name: kCAMediaTimingFunction
            EaseInEaseOut)
            }, completion: nil)
    }
}
```

本实例执行后的效果如图 14-46 所示。

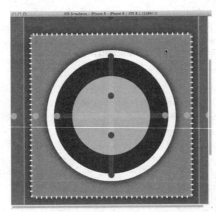

图 14-46 执行效果

实例 285　基于 Swift 综合使用界面视图

实例 285	基于 Swift 综合使用界面视图控件
源码路径	\daima\285

实例说明

在本节的实例中,演示了在 Swift 程序中综合使用界面视图控件创建 iOS 应用程序的基本过程,本实例演示了如下 4 种类型视图的处理过程。

- UIAlerts。
- UIViewController.presentViewController。

- UINavigationController.navigationController.pushViewController。
- UIPopoverController。

具体实现

（1）打开 Xcode，然后创建一个名为"Popping"的工程，工程的最终目录结构如图 14-47 所示。

图 14-47　工程的目录结构

（2）打开 Main.storyboard，为本工程设计 4 个不同的视图界面，如图 14-48 所示。

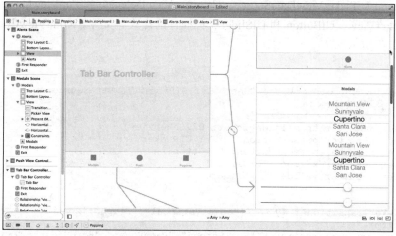

图 14-48　Main.storyboard 视图界面

（3）文件 FirstViewController.swift 的功能是设置 UIAlert 提示框的显示样式和内容，具体实现代码如下所示。

```swift
import UIKit

class FirstViewController: UIViewController {

    @IBOutlet weak var alertButton: UIButton!
    @IBOutlet weak var actionsheetButton: UIButton!
    @IBAction func clickAlertButton(sender: AnyObject) {
        var style:UIAlertControllerStyle
        var styleName:String
        if (sender as NSObject == alertButton) {
            style = .Alert
            styleName = "Alert"
        } else {
            style = .ActionSheet
            styleName = "ActionSheet"
```

```
        }
        let alertController = UIAlertController(title: "Example \(styleName)", message:
            "This is an \(styleName).", preferredStyle:style )
        alertController.popoverPresentationController?.sourceView = self.actionsheetButton
        alertController.popoverPresentationController?.sourceRect = self.actionsheet
        Button. bounds

        let callAction = UIAlertAction(title: "OK", style: .Default, handler: {
            action in
            println("hit alert")
        }
        )
        alertController.addAction(callAction)

        presentViewController(alertController, animated: true, completion: nil)
    }
    override func viewDidLoad() {
        super.viewDidLoad()
        // Do any additional setup after loading the view, typically from a nib.
    }

    override func didReceiveMemoryWarning() {
        super.didReceiveMemoryWarning()
        // Dispose of any resources that can be recreated.
    }

}
```

（4）在文件 SecondViewController.swift 中定义类 SecondViewController，此类继承与 UIViewController、UIPickerViewDataSource、UIPickerViewDelegate 和 PopViewControllerDelegate，并设置了 UIPickerView、UIButton 和 UISlider 3 种控件的样式。文件 SecondViewController.swift 的具体实现代码如下所示。

```
import UIKit

class SecondViewController: UIViewController,UIPickerViewDataSource, UIPickerViewDelegate,
PopViewControllerDelegate {

    @IBOutlet weak var transitionPicker: UIPickerView!
    @IBOutlet weak var presentationPicker: UIPickerView!
    @IBOutlet weak var presentButton: UIButton!
    @IBOutlet weak var xSlider: UISlider!
    @IBOutlet weak var ySlider: UISlider!

    let pickerTransitionArray = ["CoverVertical","FlipHorizontal", "CrossDissolve",
"PartialCurl"]
    let pickerPresentationArray = ["FullScreen","PageSheet","FormSheet","CurrentContext",
"Custom","OverFullScreen","OverCurrentContext","Popover"]

    var popupViewController:PopViewController?

    override func viewDidLoad() {
        super.viewDidLoad()
        transitionPicker.dataSource = self
        transitionPicker.delegate = self
        presentationPicker.dataSource = self
        presentationPicker.delegate = self
    }
    //MARK: pickerViewDelegate
    func numberOfComponentsInPickerView(pickerView: UIPickerView) -> Int {
        return 1
    }
    func pickerView(pickerView: UIPickerView, numberOfRowsInComponent component: Int) ->
Int {
        if (pickerView == transitionPicker) {
            return pickerTransitionArray.count //UIModalTransitionStyle
        } else {
            return pickerPresentationArray.count //UIModalPresentationStyle
```

```swift
    }
}
func pickerView(pickerView: UIPickerView, titleForRow row: Int, forComponent component:
Int) -> String! {
    if (pickerView == transitionPicker) {
        return(pickerTransitionArray[row])
    } else {
        return(pickerPresentationArray[row])
    }
}
//MARK: present

@IBAction func clickPresent(sender: AnyObject) {
    let sb = UIStoryboard(name: "Main", bundle: nil)
    popupViewController = (sb.instantiateViewControllerWithIdentifier("popper")! as
PopViewController)
    popupViewController!.delegate = self

    //you'd think there'd be an easier way to hook up an enum to a picker...

    var trans:UIModalTransitionStyle
    switch transitionPicker.selectedRowInComponent(0) {
    case 0:
        trans = .CoverVertical
    case 1:
        trans = .FlipHorizontal
    case 2:
        trans = .CrossDissolve
    case 3:
        trans = .PartialCurl
    default:
        trans = .CoverVertical
    }
    popupViewController!.modalTransitionStyle = trans

    var pres:UIModalPresentationStyle
    switch presentationPicker.selectedRowInComponent(0) {
    case 0:
        pres = .FullScreen
    case 1:
        pres = .PageSheet
    case 2:
        pres = .FormSheet
    case 3:
        pres = .CurrentContext
    case 4:
        pres = .Custom
    case 5:
        pres = .OverFullScreen
    case 6:
        pres = .OverCurrentContext
    case 7:
        pres = .Popover
    default:
        pres = .FullScreen
    }
    popupViewController!.modalPresentationStyle = pres

    if (pres == .FormSheet && trans == .PartialCurl) {
        let alertController = UIAlertController(title: "Ooops!", message: "Sorry, that
        combination is not available!", preferredStyle:.Alert )
        let callAction = UIAlertAction(title: "OK", style: .Default, handler: {
            action in
            println("hit alert")
        })
        alertController.addAction(callAction)
        presentViewController(alertController, animated: true, completion: nil)
        return
    }

    popupViewController!.preferredContentSize = CGSize(width:self.view.frame.width *
```

第 14 章 Swift 实战

```swift
            CGFloat(xSlider.value / 100.0), height:self.view.frame.height * CGFloat(ySlider.value / 100.0))
        popupViewController!.popoverPresentationController?.sourceView = self.presentButton.imageView
        popupViewController!.popoverPresentationController?.sourceRect = self.presentButton.bounds
        
        self.presentViewController(popupViewController!, animated: true, completion: {})
    }
    
    func closePop(sender:AnyObject) {
        self.dismissViewControllerAnimated(true, completion: {})
    }
    override func didReceiveMemoryWarning() {
        super.didReceiveMemoryWarning()
        // Dispose of any resources that can be recreated.
    }

}
```

（5）文件 PopViewController.swift 实现了 PopViewController 视图界面效果，具体实现代码如下所示。

```swift
import UIKit

protocol PopViewControllerDelegate {
    func closePop(sender:AnyObject)
}

class PopViewController: UIViewController {
    var delegate:PopViewControllerDelegate?
    override func viewDidLoad() {
        super.viewDidLoad()

        // Do any additional setup after loading the view.
    }

    override func didReceiveMemoryWarning() {
        super.didReceiveMemoryWarning()
        // Dispose of any resources that can be recreated.
    }
    
    @IBAction func clickCloseButton(sender: AnyObject) {
        self.delegate?.closePop(self)
    }

}
```

（6）文件 PushViewController.swift 实现了 PushViewController 视图界面效果，具体实现代码如下所示。

```swift
import UIKit

class PushViewController: UIViewController,UIPickerViewDataSource, UIPickerViewDelegate {

    @IBOutlet weak var transitionPicker: UIPickerView!
    
    var pusherViewController:UIViewController!
    let pickerTransitionArray = ["CurlDown","CurlUp","FlipFromLeft", "FlipFromRight","None"] //UIViewAnimationTransition
    override func viewDidLoad() {
        super.viewDidLoad()
        
        transitionPicker.dataSource = self
        transitionPicker.delegate = self
    }
    func numberOfComponentsInPickerView(pickerView: UIPickerView) -> Int {
        return 1
    }
```

```swift
        func pickerView(pickerView: UIPickerView, numberOfRowsInComponent component: Int) ->
        Int {
            return pickerTransitionArray.count
        }
        func pickerView(pickerView: UIPickerView, titleForRow row: Int, forComponent component:
        Int) -> String! {
            return(pickerTransitionArray[row])
        }
        @IBAction func clickPushButton(sender: AnyObject) {
            let sb = UIStoryboard(name: "Main", bundle: nil)
            pusherViewController = (sb.instantiateViewControllerWithIdentifier("pusher")! as
UIViewController)
            self.navigationController?.pushViewController(pusherViewController!, animated: true)
        }
        @IBAction func clickPushChangeAnimation(sender: AnyObject) {

            let sb = UIStoryboard(name: "Main", bundle: nil)
            pusherViewController = (sb.instantiateViewControllerWithIdentifier("pusher")! as
UIViewController)

            var trans:UIViewAnimationTransition
            switch transitionPicker.selectedRowInComponent(0) {
            case 0:
                trans = .CurlDown
            case 1:
                trans = .CurlUp
            case 2:
                trans = .FlipFromLeft
            case 3:
                trans = .FlipFromRight
            default:
                trans = .None
            }

            var navigationController = UINavigationController()
            UIView.animateWithDuration(0.75, animations: {
                UIView.setAnimationCurve(.EaseInOut)
                self.navigationController?.pushViewController(self.pusherViewController!,
                animated: false)
                UIView.setAnimationTransition(trans, forView: self.navigation Controller!.
                view, cache: false)

            })
        }
        override func didReceiveMemoryWarning() {
            super.didReceiveMemoryWarning()
        }
    }
```

(7) 文件 PopperViewController.swift 实现了 PopperViewController 视图界面效果,具体实现代码如下所示。

```swift
import UIKit

class PopperViewController: UIViewController, PopViewControllerDelegate {
    var popover:UIPopoverController!
    override func viewDidLoad() {
        super.viewDidLoad()
    }
    override func didReceiveMemoryWarning() {
        super.didReceiveMemoryWarning()
    }
    @IBAction func goPopover(sender: AnyObject) {

        let sb = UIStoryboard(name: "Main", bundle: nil)
        let popoverViewController = (sb.instantiateViewControllerWithIden tifier("popper")!
        as PopViewController)
        popoverViewController.delegate = self

        popover=UIPopoverController(contentViewController: popoverViewController)
```

```
        popover!.presentPopoverFromRect(sender.frame, inView: self.view, permittedArrow
        Directions: .Any, animated: true)
    }

    func closePop(sender:AnyObject) {
        popover!.dismissPopoverAnimated(true)
    }
}
```

执行后的主界面效果如图 14-49 所示。

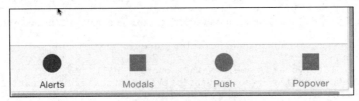

图 14-49　执行效果

单击 Alerts 选项后的效果如图 14-50 所示。

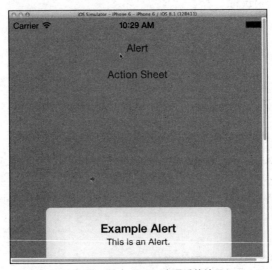

图 14-50　单击 Alerts 选项后的效果

单击 Modals 选项后的效果如图 14-51 所示。

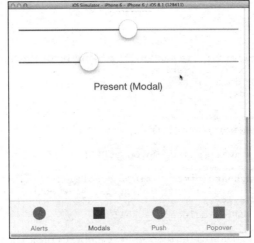

图 14-51　单击 Modals 选项后的效果

单击 Push 选项后的效果如图 14-52 所示。

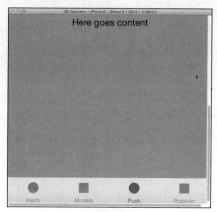

图 14-52　单击 Push 选项后的效果

实例 286　基于 Swift 语言实现 ImagePicker 功能

实例 286	基于 Swift 语言实现 ImagePicker 功能
源码路径	\daima\286

实例说明

图像选择器(UIImagePickerController)的工作原理与 MPMediaPickerController 类似，但不是显示一个可用于选择歌曲的视图，而是显示用户的照片库。用户选择照片后，图像选择器会返回一个相应的 UIImage 对象。与 MPMediaPickerController 一样，图像选择器也以模态方式出现在应用程序中。因为这两个对象都实现了自己的视图和视图控制器，所以几乎只需调用 presentModalViewController 就能显示它们。在本节的内容中，将通过一个具体实例的实现过程，详细讲解基于 Swift 语言实现 ImagePicker 控件功能的过程。

具体实现

（1）打开 Xcode，然后创建一个名为"CustomImagePicker"的工程，工程的最终目录结构如图 14-53 所示。

图 14-53　工程的目录结构

（2）打开 Main.storyboard，为本工程设计一个视图界面，在里面插入了 UIScrollView 控件，如

图 14-54 所示。

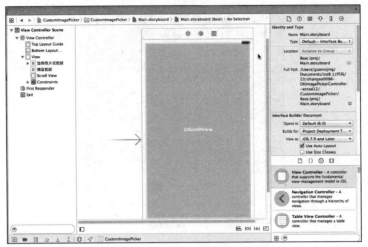

图 14-54　Main.storyboard 设计界面

（3）本实例借助了 DKImagePickerController 类库，这是用 Swift 编写的类文件，功能是实现了一个简单 ImagePickerController 功能，实现此类文件需要用到 AssetsLibrary.framework 库。文件 DKImagePickerController.swift 是一个开源文件，核心代码如下所示。

```
import UIKit
import AssetsLibrary


//声明
protocol DKImagePickerControllerDelegate : NSObjectProtocol {
    /// Called when right button is clicked.
    ///
    /// :param: images Images of selected
    func imagePickerControllerDidSelectedAssets(images: [DKAsset]!)
    /// Called when cancel button is clicked.
    func imagePickerControllerCancelled()
}

////////////////////////////////////////////////////////////////////////////////

// 单元标识符
let GroupCellIdentifier = "GroupCellIdentifier"
let ImageCellIdentifier = "ImageCellIdentifier"

// 提醒
let DKImageSelectedNotification = "DKImageSelectedNotification"
let DKImageUnselectedNotification = "DKImageUnselectedNotification"

// 模型组
class DKAssetGroup : NSObject {
    var groupName: NSString!
    var thumbnail: UIImage!
    var group: ALAssetsGroup!
}

// 配置模型
class DKAsset: NSObject {
    var thumbnailImage: UIImage?
    lazy var fullScreenImage: UIImage? = {
        return UIImage(CGImage: self.originalAsset.defaultRepresentation().fullScreenImage().takeUnretainedValue())
    }()
    lazy var fullResolutionImage: UIImage? = {
```

```swift
        return UIImage(CGImage: self.originalAsset.defaultRepresentation().fullResolution
Image().takeUnretainedValue())
        }()
    var url: NSURL?
    
    private var originalAsset: ALAsset!
    
    // Compare two assets
    override func isEqual(object: AnyObject?) -> Bool {
        let other = object as DKAsset!
        return self.url!.isEqual(other.url!)
    }
}

//内部
extension UIViewController {
    var imagePickerController: DKImagePickerController? {
        get {
            let nav = self.navigationController
            if nav is DKImagePickerController {
                return nav as? DKImagePickerController
            } else {
                return nil
            }
        }
    }
}

////////////////////////////////////////////////////////////////////////////////
/////
// 显示 Group 组中的所有图片
////////////////////////////////////////////////////////////////////////////////
/////

class DKImageGroupViewController: UICollectionViewController {
    
    class DKImageCollectionCell: UICollectionViewCell {
        var thumbnail: UIImage! {
            didSet {
                self.imageView.image = thumbnail
            }
        }
        
        override var selected: Bool {
            didSet {
                checkView.hidden = !super.selected
            }
        }
        
        private var imageView = UIImageView()
        private var checkView = UIImageView(image: UIImage(named: "photo_checked"))
        
        override init(frame: CGRect) {
            super.init(frame: frame)
            imageView.frame = self.bounds
            self.contentView.addSubview(imageView)
            self.contentView.addSubview(checkView)
        }
        
        required init(coder aDecoder: NSCoder) {
            fatalError("init(coder:) has not been implemented")
        }
        
        override func layoutSubviews() {
            super.layoutSubviews()
            
            imageView.frame = self.bounds
            checkView.frame.origin = CGPoint(x: self.contentView.bounds.width - checkView.
            bounds.width, y: 0)
        }
    }
```

```swift
    var assetGroup: DKAssetGroup!
    private lazy var imageAssets: NSMutableArray = {
        return NSMutableArray()
    }()
    
    override init() {
        let layout = UICollectionViewFlowLayout()
        
        let interval: CGFloat = 3
        layout.minimumInteritemSpacing = interval
        layout.minimumLineSpacing = interval
        
        let screenWidth = UIScreen.mainScreen().bounds.width
        let itemWidth = (screenWidth - interval * 3) / 4
        
        layout.itemSize = CGSize(width: itemWidth, height: itemWidth)
        super.init(collectionViewLayout: layout)
    }
    
    required init(coder aDecoder: NSCoder) {
        fatalError("init(coder:) has not been implemented")
    }
    
    override func viewDidLoad() {
        super.viewDidLoad()
        assert(assetGroup != nil, "assetGroup is nil")
        
        self.title = assetGroup.groupName
        
        self.collectionView.backgroundColor = UIColor.whiteColor()
        self.collectionView.allowsMultipleSelection = true
        self.collectionView.registerClass(DKImageCollectionCell.self,
forCellWithReuseIdentifier: ImageCellIdentifier)
        
        assetGroup.group.enumerateAssetsUsingBlock {[unowned self](result: ALAsset!,
index: Int, stop: UnsafeMutablePointer<ObjCBool>) in
            if result != nil {
                let asset = DKAsset()
                asset.thumbnailImage = UIImage(CGImage:result.thumbnail().takeUnretainedValue())
                asset.url = result.valueForProperty(ALAssetPropertyAssetURL) as? NSURL
                asset.originalAsset = result
                self.imageAssets.addObject(asset)
            } else {
                self.collectionView.reloadData()
                dispatch_async(dispatch_get_main_queue()) {
                    self.collectionView.scrollToItemAtIndexPath(NSIndexPath(forRow:
                    self.imageAssets.count-1, inSection: 0),
                        atScrollPosition: UICollectionViewScrollPosition.Bottom,
                        animated: false)
                }
            }
        }
    }
    
    //Mark: - UICollectionViewDelegate, UICollectionViewDataSource methods
    override func numberOfSectionsInCollectionView(collectionView: UICollectionView) -> Int {
        return 1
    }
    
    override func collectionView(collectionView: UICollectionView, numberOfItemsInSection section: Int) -> Int {
        return imageAssets.count
    }
    
    override func collectionView(collectionView: UICollectionView, cellForItemAtIndexPath indexPath: NSIndexPath) -> UICollectionViewCell {
        let cell = collectionView.dequeueReusableCellWithReuseIdentifier (ImageCellIdentifier,
        forIndexPath: indexPath) as DKImageCollectionCell
```

```swift
        let asset = imageAssets[indexPath.row] as DKAsset
        cell.thumbnail = asset.thumbnailImage
        
        if find(self.imagePickerController!.selectedAssets, asset) != nil {
            cell.selected = true
            collectionView.selectItemAtIndexPath(indexPath, animated: false, scrollPosition:
            UICollectionViewScrollPosition.None)
        } else {
            cell.selected = false
            collectionView.deselectItemAtIndexPath(indexPath, animated: false)
        }
        
        return cell
    }
    
    override func collectionView(collectionView: UICollectionView, didSelectItemAtIndexPath indexPath: NSIndexPath) {
        NSNotificationCenter.defaultCenter().postNotificationName(DKImageSelectedNotification, object: imageAssets[indexPath.row])
    }
    
    override func collectionView(collectionView: UICollectionView, didDeselectItemAtIndexPath indexPath: NSIndexPath) {
        NSNotificationCenter.defaultCenter().postNotificationName(DKImageUnselectedNotification, object: imageAssets[indexPath.row])
    }
}

////////////////////////////////////////////////////////////////////////////////////
// MARK: 显示所有组
////////////////////////////////////////////////////////////////////////////////////

class DKAssetsLibraryController: UITableViewController {
    
    lazy private var groups: NSMutableArray = {
        return NSMutableArray()
    }()
    
    lazy private var library: ALAssetsLibrary = {
        return ALAssetsLibrary()
    }()
    
    private var noAccessView: UIView!
    
    override func viewDidLoad() {
        super.viewDidLoad()
        
        self.tableView.registerClass(UITableViewCell.self,      forCellReuseIdentifier:
        GroupCellIdentifier)
        self.view.backgroundColor = UIColor.whiteColor()
        
        library.enumerateGroupsWithTypes(0xFFFFFFFF, usingBlock: {(group: ALAssetsGroup!,
        stop: UnsafeMutablePointer<ObjCBool>) in
            if group != nil {
                if group.numberOfAssets() != 0 {
                    let groupName = group.valueForProperty(ALAssetsGroupPropertyName) as
                    NSString
                    
                    let assetGroup = DKAssetGroup()
                    assetGroup.groupName = groupName
                    assetGroup.thumbnail   =   UIImage(CGImage:  group.posterImage().take
                    UnretainedValue())
                    assetGroup.group = group
                    self.groups.insertObject(assetGroup, atIndex: 0)
                }
            } else {
                self.tableView.reloadData()
            }
```

```swift
        }, failureBlock: {(error: NSError!) in
            self.noAccessView.frame = self.view.bounds
            self.tableView.scrollEnabled = false
            self.tableView.separatorStyle = UITableViewCellSeparatorStyle.None
            self.view.addSubview(self.noAccessView)
        })
    }

    // MARK: - UITableViewDelegate, UITableViewDataSource methods
    override func numberOfSectionsInTableView(tableView: UITableView) -> Int {
        return 1
    }

    override func tableView(tableView: UITableView, numberOfRowsInSection section: Int) -> Int {
        return groups.count
    }

    override func tableView(tableView: UITableView, cellForRowAtIndexPath indexPath: NSIndexPath) -> UITableViewCell {
        let cell = tableView.dequeueReusableCellWithIdentifier(GroupCellIdentifier, forIndexPath: indexPath) as UITableViewCell

        let assetGroup = groups[indexPath.row] as DKAssetGroup
        cell.textLabel.text = assetGroup.groupName
        cell.imageView.image = assetGroup.thumbnail

        return cell
    }

    override func tableView(tableView: UITableView, didSelectRowAtIndexPath indexPath: NSIndexPath) {
        tableView.deselectRowAtIndexPath(indexPath, animated: true)

        let assetGroup = groups[indexPath.row] as DKAssetGroup
        let imageGroupController = DKImageGroupViewController()
        imageGroupController.assetGroup = assetGroup
        self.navigationController?.pushViewController(imageGroupController, animated: true)
    }
}

/////////////////////////////////////////////////////////////////////////////////
// MARK: - 主控制器视图
/////////////////////////////////////////////////////////////////////////////////

class DKImagePickerController: UINavigationController {

    /// The height of the bottom of the preview
    var previewHeight: CGFloat = 80
    var rightButtonTitle: String = "确定"
    /// Displayed when denied access
    var noAccessView: UIView = {
        let label = UILabel()
        label.text = "用户拒绝访问"
        label.textAlignment = NSTextAlignment.Center
        label.textColor = UIColor.lightGrayColor()
        return label
    }()

    class DKPreviewView: UIScrollView {
        let interval: CGFloat = 5
        private var imageLengthOfSide: CGFloat!
        private var assets = [DKAsset]()
        private var imagesDict: [DKAsset : UIImageView] = [:]

        override func layoutSubviews() {
            super.layoutSubviews()

            imageLengthOfSide = self.bounds.height - interval * 2
```

实例 286 基于 Swift 语言实现 ImagePicker 功能

```swift
    }

    func imageFrameForIndex(index: Int) -> CGRect {
        return CGRect(x: CGFloat(index) * imageLengthOfSide + CGFloat(index + 1) *
            interval, y: (self.bounds.height - imageLengthOfSide)/2,
            width: imageLengthOfSide, height: imageLengthOfSide)
    }

    func insertAsset(asset: DKAsset) {
        let imageView = UIImageView(image: asset.thumbnailImage)
        imageView.frame = imageFrameForIndex(assets.count)

        self.addSubview(imageView)
        assets.append(asset)
        imagesDict.updateValue(imageView, forKey: asset)
        setupContent(true)
    }

    func removeAsset(asset: DKAsset) {
        imagesDict.removeValueForKey(asset)
        let index = find(assets, asset)
        if let toRemovedIndex = index {
            assets.removeAtIndex(toRemovedIndex)
            setupContent(false)
        }
    }

    private func setupContent(isInsert: Bool) {
        if isInsert == false {
            for (index,asset) in enumerate(assets) {
                let imageView = imagesDict[asset]!
                imageView.frame = imageFrameForIndex(index)
            }
        }
        self.contentSize = CGSize(width: CGRectGetMaxX((self.subviews.last as UIView).
frame) + interval, height: self.bounds.height)
    }
}

class DKContentWrapperViewController: UIViewController {
    var contentViewController: UIViewController
    var bottomBarHeight: CGFloat = 0
    var showBottomBar: Bool = false {
        didSet {
            if self.showBottomBar {
                self.contentViewController.view.frame.size.height = self.view.bounds.
                size.
                height - self.bottomBarHeight
            } else {
                self.contentViewController.view.frame.size.height = self.view.bounds.
                size.height
            }
        }
    }

    init(_ viewController: UIViewController) {
        contentViewController = viewController

        super.init(nibName: nil, bundle: nil)
        self.addChildViewController(viewController)

        contentViewController.addObserver(self, forKeyPath: "title", options: NSKey
        ValueObservingOptions.New, context: nil)
    }

    deinit {
        contentViewController.removeObserver(self, forKeyPath: "title")
    }

    required init(coder aDecoder: NSCoder) {
        fatalError("init(coder:) has not been implemented")
    }
```

```swift
        override func observeValueForKeyPath(keyPath: String, ofObject object: AnyObject,
change: [NSObject : AnyObject], context: UnsafeMutablePointer<Void>) {
            if keyPath == "title" {
                self.title = contentViewController.title
            }
        }
        
        override func viewDidLoad() {
            super.viewDidLoad()
            
            self.view.backgroundColor = UIColor.whiteColor()
            self.view.addSubview(contentViewController.view)
            contentViewController.view.frame = view.bounds
        }
    }
    
    internal var selectedAssets: [DKAsset]!
    internal weak var pickerDelegate: DKImagePickerControllerDelegate?
    lazy internal var imagesPreviewView: DKPreviewView = {
        let preview = DKPreviewView()
        preview.hidden = true
        preview.backgroundColor = UIColor.lightGrayColor()
        return preview
    }()
    lazy internal var doneButton: UIButton = {
        let button = UIButton.buttonWithType(UIButtonType.Custom) as UIButton
        button.setTitle("", forState: UIControlState.Normal)
        button.setTitleColor(self.navigationBar.tintColor, forState: UIControlState.Normal)
        button.reversesTitleShadowWhenHighlighted = true
        button.addTarget(self, action: "onDoneClicked", forControlEvents: UIControlEvents.TouchUpInside)
        return button
    }()
    
    convenience override init() {
        var libraryController = DKAssetsLibraryController()
        var wrapperVC = DKContentWrapperViewController(libraryController)
        self.init(rootViewController: wrapperVC)
        libraryController.noAccessView = noAccessView
        wrapperVC.bottomBarHeight = previewHeight
        
        selectedAssets = [DKAsset]()
    }
    
    deinit {
        NSNotificationCenter.defaultCenter().removeObserver(self)
    }
    
    override func viewDidLoad() {
        super.viewDidLoad()
        
        imagesPreviewView.frame = CGRect(x: 0, y: view.bounds.height - previewHeight,
                            width: view.bounds.width, height: previewHeight)
        imagesPreviewView.autoresizingMask = UIViewAutoresizing.FlexibleWidth | UIViewAutoresizing.FlexibleTopMargin
        
        view.addSubview(imagesPreviewView)
        
        NSNotificationCenter.defaultCenter().addObserver(self, selector: "selectedImage:",
            name: DKImageSelectedNotification, object: nil)
        NSNotificationCenter.defaultCenter().addObserver(self, selector: "unselectedImage:",
            name: DKImageUnselectedNotification, object: nil)
    }
    
    override func pushViewController(viewController: UIViewController, animated: Bool) {
        var wrapperVC = DKContentWrapperViewController(viewController)
        wrapperVC.bottomBarHeight = previewHeight
        wrapperVC.showBottomBar = !imagesPreviewView.hidden
        
        super.pushViewController(wrapperVC, animated: animated)
```

```
        self.topViewController.navigationItem.rightBarButtonItem = UIBarButtonItem (custom
        View: self.doneButton)

        if self.viewControllers.count == 1 && self.topViewController?.navigationItem.
        leftBar ButtonItem == nil {
            self.topViewController.navigationItem.leftBarButtonItem = UIBarButtonItem
            (barButton
            SystemItem: UIBarButtonSystemItem.Cancel,
                target: self,
                action: "onCancelClicked")
        }
    }

    // MARK: - Delegate methods
    func onCancelClicked() {
        if let delegate = self.pickerDelegate {
            delegate.imagePickerControllerCancelled()
        }
    }

    func onDoneClicked() {
        if let delegate = self.pickerDelegate {
            delegate.imagePickerControllerDidSelectedAssets(self.selectedAssets)
        }
    }

    // MARK: - Notifications
    func selectedImage(noti: NSNotification) {
        if let asset = noti.object as? DKAsset {
            selectedAssets.append(asset)
            imagesPreviewView.insertAsset(asset)
            imagesPreviewView.hidden = false

            (self.viewControllers as [DKContentWrapperViewController]).map {$0. showBottomBar
            = !self.imagesPreviewView.hidden}
            self.doneButton.setTitle(rightButtonTitle + "(\(selectedAssets.count))", forState:
            UIControl State.Normal)
            self.doneButton.sizeToFit()
        }
    }

    func unselectedImage(noti: NSNotification) {
        if let asset = noti.object as? DKAsset {
            selectedAssets.removeAtIndex(find(selectedAssets, asset)!)
            imagesPreviewView.removeAsset(asset)

            self.doneButton.setTitle(rightButtonTitle + "(\(selectedAssets.count))", forState:
            UIControlState.Normal)
            self.doneButton.sizeToFit()
            if selectedAssets.count <= 0 {
                imagesPreviewView.hidden = true

                (self.viewControllers     as      [DKContentWrapperViewController]).map
                {$0.showBottomBar = !self.imagesPreviewView.hidden}
                self.doneButton.setTitle("", forState: UIControlState.Normal)
            }
        }
    }
}
```

（4）在文件 ViewController.swift 中初始化了整个工程，增加了通过系统的 Controller 选择图片或播放视频，调用 DKImagePickerController 类实现了对指定图片的展示功能，也就是实现了 ImagePicker 控件浏览图片的功能效果。文件 ViewController.swift 的具体实现代码如下所示。

```
import UIKit
import MobileCoreServices
import MediaPlayer

class ViewController: UIViewController,UINavigationControllerDelegate, UIImagePicker
Controller Delegate, DKImagePickerControllerDelegate {
    @IBOutlet var imageScrollView: UIScrollView!
```

```swift
    var player: MPMoviePlayerController?
    var videoURL: NSURL?
    
    override func viewDidLoad() {
        super.viewDidLoad()
        
    }
    
    override func didReceiveMemoryWarning() {
        super.didReceiveMemoryWarning()
        // Dispose of any resources that can be recreated.
    }
    
    // 使用系统的图片选取器
    func showSystemController() {
        let pickerController = UIImagePickerController()
        pickerController.delegate = self
        pickerController.sourceType = UIImagePickerControllerSourceType.PhotoLibrary
        pickerController.mediaTypes = [kUTTypeImage!,kUTTypeMovie!]
        
        self.presentViewController(pickerController, animated: true) {}
    }
    
    // 使用自定义的图片选取器
    func showCustomController() {
        let pickerController = DKImagePickerController()
        pickerController.pickerDelegate = self
        self.presentViewController(pickerController, animated: true) {}
    }
    
    @IBAction func showImagePicker() {
//        showSystemController()
        showCustomController()
    }
    
    // 使用系统的播放器播放视频
    @IBAction func playVideo() {
        if let videoURL = self.videoURL {
            NSNotificationCenter.defaultCenter().addObserver(self, selector: "exitPlayer:",
                name: MPMoviePlayerPlaybackDidFinishNotification, object: nil)
            
            let player = MPMoviePlayerController(contentURL: videoURL)
            player.movieSourceType = MPMovieSourceType.File
            player.controlStyle = MPMovieControlStyle.Fullscreen
            player.fullscreen = true
            player.scalingMode = MPMovieScalingMode.Fill
            
            player.view.frame = view.bounds
            view.addSubview(player.view)
            
            player.prepareToPlay()
            player.play()
            
            self.player = player
        }
    }
    
    // 退出播放器
    func exitPlayer(notification: NSNotification) {
        let reason = (notification.userInfo!)[MPMoviePlayerPlaybackDidFinishReasonUserInfoKey] as NSNumber!
        if reason.integerValue == MPMovieFinishReason.UserExited.rawValue {
            NSNotificationCenter.defaultCenter().removeObserver(self)
            self.player?.view.removeFromSuperview()
            self.player = nil
        }
    }
    
    // MARK: - UIImagePickerControllerDelegate methods
    func imagePickerController(picker: UIImagePickerController, didFinishPickingMediaWithInfo info: [NSObject : AnyObject]) {
```

```swift
        let mediaType = info[UIImagePickerControllerMediaType] as NSString!
        println(mediaType)
        if mediaType.isEqualToString(kUTTypeImage) {
            let selectedImage = info[UIImagePickerControllerOriginalImage] as UIImage!
            imageScrollView.subviews.map(){$0.removeFromSuperview()}
            let imageView = UIImageView(image: selectedImage)
            imageView.contentMode = UIViewContentMode.ScaleAspectFit
            imageView.frame = imageScrollView.bounds
            imageScrollView.addSubview(imageView)
        } else {
            self.videoURL = info[UIImagePickerControllerMediaURL] as NSURL!
            let alert = UIAlertView(title: "选择的视频 URL", message: videoURL!.absoluteString,
                delegate: nil, cancelButtonTitle: "确定")
            alert.show()
        }
        
        picker.dismissViewControllerAnimated(true, completion: nil)
    }
    
    // MARK: - DKImagePickerControllerDelegate methods
    // 取消时的回调
    func imagePickerControllerCancelled() {
        self.dismissViewControllerAnimated(true, completion: nil)
    }
    
    // 选择图片并确定后的回调
    func imagePickerControllerDidSelectedAssets(assets: [DKAsset]!) {
        imageScrollView.subviews.map(){$0.removeFromSuperview}
        
        for (index, asset) in enumerate(assets) {
            let imageHeight: CGFloat = imageScrollView.bounds.height / 2
            
            let imageView = UIImageView(image: asset.thumbnailImage)
            imageView.contentMode = UIViewContentMode.ScaleAspectFit
            imageView.frame = CGRect(x: 0, y: CGFloat(index) * imageHeight, width: image
            ScrollView.bounds.width, height: imageHeight)
            imageScrollView.addSubview(imageView)
            
        }
        imageScrollView.contentSize.height = CGRectGetMaxY((imageScrollView.subviews.last
        as UIView).frame)
        
        self.dismissViewControllerAnimated(true, completion: nil)
    }
}
```

本实例执行后将在主视图界面中列表显示不同的类型组,如图 14-55 所示。

单击列表中的某一个组后会来到照片选择界面,在此可以勾选不同的需要的照片,如图 14-56 所示。

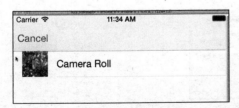

图 14-55 列表显示

图 14-56 选择照片

选择照片完毕后,单击"确定"按钮后会展示这组中的所有照片,如图 14-57 所示。

第 14 章　Swift 实战

图 14-57　展示组中的所有照片

实例 287　基于 Swift 实现一个音乐播放器

实例 287	基于 Swift 实现一个音乐播放器
源码路径	\daima\287

实例说明

在本节的内容中，将通过一个具体实例的实现过程，详细讲解基于 Swift 语言实现一个音乐播放器的过程。读者在调试运行本实例之前，需要先确保在程序指定的路径下存在音频文件。

具体实现

（1）打开 Xcode，然后创建一个名为 "MusicPlayer" 的工程，工程的最终目录结构如图 14-58 所示。

图 14-58　工程的目录结构

（2）在 Xcode 6 的 Main.stoyboard 面板中设计 UI 界面，在主界面中列表 iTunes 市场中的专辑名称，如图 14-59 所示。

实例 287　基于 Swift 实现一个音乐播放器

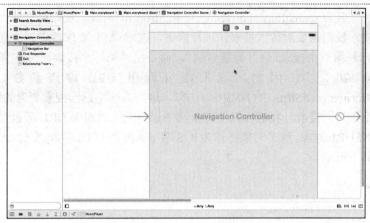

图 14-59　第一个界面

在第二个界面中，通过 Table View 视图在底部显示搜索结果，在顶部显示搜索表单，如图 14-60 所示。

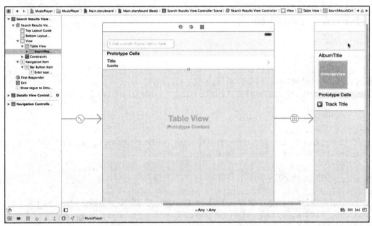

图 14-60　第二个界面

在第三个界面显示搜索结果列表中某个专辑的详细信息，顶部使用 UIImageView 控件显示专辑图片，下方通过 Track 显示专辑中的各首歌曲名称，如图 14-61 所示。

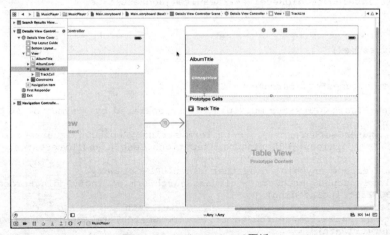

图 14-61　Main.stoyboard 面板

（3）文件 APIController.swift 的功能是实现对 iTunes 搜索 API 的调用，并让一个定制的代理来

第 14 章　Swift 实战

接收响应。创建一个名为 searchItunesFor(searchTerm:String)的函数。我们用它来实现对任意搜索词的网络请求。首先，我们需要对传入搜索词进行修正，搜索 API 要求搜索词的格式为"第一个搜索词+第二个搜索词+第三个搜索词+其他搜索词"，而不是"第一个搜索词%20 第二个搜索词%20 第三个搜索词%20……"。因此，我们没有调用 URL 编码函数，而是调用了 sringByReplacingOccurenesOfString 的 NSString 方法。这个方法将返回搜索变量的修正版本，使用+号替代了其中的空格符。接着，我们转义了搜索词中那些无法识别为 URL 所包含的字符。再接下来的两行定义了 NSURL 对象，这个对象将作为 iOS 网络 API 的 URL 参数。文件 APIController.swift 的具体实现代码如下所示。

```swift
import Foundation

protocol APIControllerProtocol {
    func didReceiveAPIResults(results: NSDictionary)
}

class APIController {
    
    var delegate: APIControllerProtocol
    
    init(delegate: APIControllerProtocol) {
        self.delegate = delegate
    }
    
    func get(path: String){
        let url = NSURL(string: path)
        let session = NSURLSession.sharedSession()
        let task = session.dataTaskWithURL(url!, completionHandler: {data, response, error -> Void in
            println("Task completed")
            if(error != nil) {
                // If there is an error in the web request, print it to the console
                println(error.localizedDescription)
            }
            var err: NSError?
            
            var jsonResult = NSJSONSerialization.JSONObjectWithData(data, options: NSJSONReadingOptions.MutableContainers, error: &err) as NSDictionary
            if(err != nil) {
                // If there is an error parsing JSON, print it to the console
                println("JSON Error \(err!.localizedDescription)")
            }
            let results: NSArray = jsonResult["results"] as NSArray
            
            self.delegate.didReceiveAPIResults(jsonResult)
            
        })
        
        task.resume()
    }
    
    func searchItunesFor(searchTerm: String) {
        
        // The iTunes API wants multiple terms separated by + symbols, so replace spaces with + signs
        let itunesSearchTerm = searchTerm.stringByReplacingOccurrencesOfString(" ", withString: "+", options: NSStringCompareOptions.CaseInsensitiveSearch, range: nil)
        
        // Now escape anything else that isn't URL-friendly
        if let escapedSearchTerm = itunesSearchTerm.stringByAddingPercentEscapesUsingEncoding(NSUTF8StringEncoding) {
            let urlPath = "https://itunes.apple.com/search?term=\(escapedSearchTerm)&media= music&entity=album"
            get(urlPath)
        }
    }
```

```swift
    func lookupAlbum(collectionId: Int) {
        get("https://itunes.apple.com/lookup?id=\(collectionId)&entity=song")
    }
}
```

(4)文件 SearchResultsViewController.swift 的功能是根据用户输入的关键词显示搜索结果列表，具体实现代码如下所示。

```swift
import UIKit
import QuartzCore

class SearchResultsViewController: UIViewController, UITableViewDataSource, UITableViewDelegate, UITextFieldDelegate, APIControllerProtocol {
    
    @IBOutlet weak var searchField: UITextField!
    @IBOutlet var appsTableView : UITableView?
    
    var albums = [Album]()
    var api : APIController?
    var imageCache = [String : UIImage]()
    let kCellIdentifier: String = "SearchResultCell"
    
    override func viewDidLoad() {
        super.viewDidLoad()
        // Do any additional setup after loading the view, typically from a nib.
        api = APIController(delegate: self)
        UIApplication.sharedApplication().networkActivityIndicatorVisible = true
        api!.searchItunesFor("Beatles")
    }
    
    override func didReceiveMemoryWarning() {
        super.didReceiveMemoryWarning()
        // Dispose of any resources that can be recreated.
    }
    
    // UITextFieldDelegate
    func textFieldShouldReturn(textField: UITextField!) -> Bool {
        var searchText = self.searchField.text.stringByTrimmingCharactersInSet (NSCharacterSet.whitespaceAndNewlineCharacterSet())
        if searchText.utf16Count > 0{
            println(searchText)
            api!.searchItunesFor(searchText)
            
//            self.toDoItem = ToDoItem(name: self.searchField.text)
        }
        
        println("done!")
        textField.resignFirstResponder()
        return false
    }
    
    // MARK: UITableViewDataSource
    func tableView(tableView: UITableView, numberOfRowsInSection section: Int) -> Int {
        return albums.count
    }
    
    
    func tableView(tableView: UITableView, cellForRowAtIndexPath indexPath: NSIndexPath) -> UITableViewCell {
        
        let cell: UITableViewCell = tableView.dequeueReusableCellWithIdentifier (kCellIdentifier) as UITableViewCell
        
        // Add a check to make sure this exists
        let album = self.albums[indexPath.row]
        cell.textLabel.text = album.title
        cell.imageView.image = UIImage(named: "Blank52")
        
        // Get the formatted price string for display in the subtitle
        let formattedPrice = album.price
```

```swift
        // Jump in to a background thread to get the image for this item

        // Grab the artworkUrl60 key to get an image URL for the app's thumbnail
        let urlString = album.thumbnailImageURL

        // Check our image cache for the existing key. This is just a dictionary of UIImages
        //var image: UIImage? = self.imageCache.valueForKey(urlString) as? UIImage
        var image = self.imageCache[urlString]

        if( image == nil ) {
            // If the image does not exist, we need to download it
            var imgURL: NSURL = NSURL(string: urlString)!

            // Download an NSData representation of the image at the URL
            let request: NSURLRequest = NSURLRequest(URL: imgURL)
            NSURLConnection.sendAsynchronousRequest(request,                         queue:
NSOperationQueue.mainQueue(),   completionHandler:   {(response:   NSURLResponse!,data:
NSData!,error: NSError!) -> Void in
                if error == nil {
                    image = UIImage(data: data)

                    // Store the image in to our cache
                    self.imageCache[urlString] = image
                    dispatch_async(dispatch_get_main_queue(), {
                        if let cellToUpdate = tableView.cellForRowAtIndexPath(indexPath) {
                            cellToUpdate.imageView.image = image
                        }
                    })
                }
                else {
                    println("Error: \(error.localizedDescription)")
                }
            })

        }
        else {
            dispatch_async(dispatch_get_main_queue(), {
                if let cellToUpdate = tableView.cellForRowAtIndexPath(indexPath) {
                    cellToUpdate.imageView.image = image
                }
            })
        }

        cell.detailTextLabel?.text = formattedPrice

        return cell
    }

    func   tableView(tableView:   UITableView,   willDisplayCell   cell:   UITableViewCell,
forRowAtIndexPath indexPath: NSIndexPath) {
        cell.layer.transform = CATransform3DMakeScale(0.1,0.1,1)
        UIView.animateWithDuration(0.25, animations: {
            cell.layer.transform = CATransform3DMakeScale(1,1,1)
        })
    }

    override func prepareForSegue(segue: UIStoryboardSegue, sender: AnyObject?) {
        var detailsViewController: DetailsViewController = segue.destinationViewController
        as DetailsViewController
        var albumIndex = appsTableView!.indexPathForSelectedRow()!.row
        var selectedAlbum = self.albums[albumIndex]
        detailsViewController.album = selectedAlbum
    }

    func didReceiveAPIResults(results: NSDictionary) {
        var resultsArr: NSArray = results["results"] as NSArray
        dispatch_async(dispatch_get_main_queue(), {
            self.albums = Album.albumsWithJSON(resultsArr)
            self.appsTableView!.reloadData()
            UIApplication.sharedApplication().networkActivityIndicatorVisible = false
```

        })
    }
}

（5）为了方便地传递专辑信息，需要创建一个表示专辑的模型。创建一个新的 Swift 文件 Album.swift，这是一个非常简单的类，它只为我们提供了专辑的几个属性。我们创建了类型为可选字符串的 6 个不同的属性，增加了一个在使用这个对象之前对其展开的初始化方法。初始化方法非常简单，它仅仅依据所提供参数设置所有属性。文件 Album.swift 的具体实现代码如下所示。

```swift
import Foundation

class Album {
    var title: String
    var price: String
    var thumbnailImageURL: String
    var largeImageURL: String
    var itemURL: String
    var artistURL: String
    var collectionId: Int

    init(name: String, price: String, thumbnailImageURL: String, largeImageURL: String,
    itemURL: String, artistURL: String, collectionId: Int) {
        self.title = name
        self.price = price
        self.thumbnailImageURL = thumbnailImageURL
        self.largeImageURL = largeImageURL
        self.itemURL = itemURL
        self.artistURL = artistURL
        self.collectionId = collectionId
    }

    class func albumsWithJSON(allResults: NSArray) -> [Album] {

        // Create an empty array of Albums to append to from this list
        var albums = [Album]()

        // Store the results in our table data array
        if allResults.count>0 {

            // Sometimes iTunes returns a collection, not a track, so we check both for the 'name'
            for result in allResults {

                var name = result["trackName"] as? String
                if name == nil {
                    name = result["collectionName"] as? String
                }

                // Sometimes price comes in as formattedPrice, sometimes as collectionPrice..
                //and sometimes it's a float instead of a string. Hooray!
                var price = result["formattedPrice"] as? String
                if price == nil {
                    price = result["collectionPrice"] as? String
                    if price == nil {
                        var priceFloat: Float? = result["collectionPrice"] as? Float
                        var nf: NSNumberFormatter = NSNumberFormatter()
                        nf.maximumFractionDigits = 2
                        if priceFloat != nil {
                            price = "$"+nf.stringFromNumber(priceFloat!)!
                        }
                    }
                }

                let thumbnailURL = result["artworkUrl60"] as? String ?? ""
                let imageURL = result["artworkUrl100"] as? String ?? ""
                let artistURL = result["artistViewUrl"] as? String ?? ""

                var itemURL = result["collectionViewUrl"] as? String
                if itemURL == nil {
                    itemURL = result["trackViewUrl"] as? String
                }
```

```
                var collectionId = result["collectionId"] as? Int
                var newAlbum = Album(name: name!, price: price!, thumbnailImageURL:
thumbnailURL, largeImageURL: imageURL, itemURL: itemURL!, artistURL: artistURL,
collectionId: collectionId!)
                albums.append(newAlbum)
            }
        }
        return albums
    }
}
```

（6）文件 DetailsViewController.swift 实现了显示唱片集详细信息的功能，这是一个新的视图。首先我们创建 DetailsViewControlle 类，添加一个名为 DetailsViewController.swift 并继承自 UIViewController 的文件。此视图控制器将非常简单，在此只需添加一个 album，然后实现 UIViewController 的 init 和 viewDidLoad 方法即可。文件 DetailsViewController.swift 的具体实现代码如下所示。

```
import UIKit
import MediaPlayer
import QuartzCore

class DetailsViewController: UIViewController, APIControllerProtocol, UITableView
Delegate, UITableViewDataSource {

    var album: Album?
    var tracks = [Track]()

    @IBOutlet weak var titleLabel: UILabel!
    @IBOutlet weak var albumCover: UIImageView!
    @IBOutlet weak var tracksTableView: UITableView!
    lazy var api : APIController = APIController(delegate: self)
    var mediaPlayer: MPMoviePlayerController = MPMoviePlayerController()

    required init(coder aDecoder: NSCoder) {
        super.init(coder: aDecoder)
    }

    override func viewDidLoad() {
        super.viewDidLoad()
        titleLabel.text = self.album?.title
        albumCover.image = UIImage(data: NSData(contentsOfURL: NSURL(string: self.album!.
        largeImageURL)!)!)

        // Load in tracks
        if self.album != nil {
            api.lookupAlbum(self.album!.collectionId)
        }
    }

    func tableView(tableView: UITableView, numberOfRowsInSection section: Int) -> Int {
        return tracks.count
    }

    func tableView(tableView: UITableView, cellForRowAtIndexPath indexPath: NSIndexPath)
    -> UITableViewCell {
        let cell = tableView.dequeueReusableCellWithIdentifier("TrackCell") as TrackCell
        let track = tracks[indexPath.row]
        cell.titleLabel.text = track.title
        cell.playIcon.text = "▶ "

        return cell
    }

    func tableView(tableView: UITableView, didSelectRowAtIndexPath indexPath: NSIndexPath)
{
        var track = tracks[indexPath.row]
        mediaPlayer.stop()
        mediaPlayer.contentURL = NSURL(string: track.previewUrl)
```

```swift
            mediaPlayer.play()
            if let cell = tableView.cellForRowAtIndexPath(indexPath) as? TrackCell {
                cell.playIcon.text = "  "
            }
        }
    }
    
    func tableView(tableView: UITableView, willDisplayCell cell: UITableViewCell, forRow
    AtIndexPath indexPath: NSIndexPath) {
        cell.layer.transform = CATransform3DMakeScale(0.1,0.1,1)
        UIView.animateWithDuration(0.25, animations: {
            cell.layer.transform = CATransform3DMakeScale(1,1,1)
        })
    }
    
    // MARK: APIControllerProtocol
    func didReceiveAPIResults(results: NSDictionary) {
        var resultsArr: NSArray = results["results"] as NSArray
        dispatch_async(dispatch_get_main_queue(), {
            self.tracks = Track.tracksWithJSON(resultsArr)
            self.tracksTableView.reloadData()
            UIApplication.sharedApplication().networkActivityIndicatorVisible = false
        })
    }
}
```

（7）文件 Track.swift 的功能是列表显示同一专辑中的歌曲列表，具体实现代码如下所示。

```swift
import Foundation
class Track {
    
    var title: String
    var price: String
    var previewUrl: String
    
    init(title: String, price: String, previewUrl: String) {
        self.title = title
        self.price = price
        self.previewUrl = previewUrl
    }
    
    class func tracksWithJSON(allResults: NSArray) -> [Track] {
        
        var tracks = [Track]()
        
        if allResults.count>0 {
            for trackInfo in allResults {
                // Create the track
                if let kind = trackInfo["kind"] as? String {
                    if kind=="song" {
                        
                        var trackPrice = trackInfo["trackPrice"] as? String
                        var trackTitle = trackInfo["trackName"] as? String
                        var trackPreviewUrl = trackInfo["previewUrl"] as? String
                        
                        if(trackTitle == nil) {
                            trackTitle = "Unknown"
                        }
                        else if(trackPrice == nil) {
                            println("No trackPrice in \(trackInfo)")
                            trackPrice = "?"
                        }
                        else if(trackPreviewUrl == nil) {
                            trackPreviewUrl = ""
                        }
                        
                        var track = Track(title: trackTitle!, price: trackPrice!, previewUrl: trackPreviewUrl!)
                        tracks.append(track)
                        
                    }
                }
            }
        }
```

```
        }
        return tracks
    }
}
```

执行后的初始效果会列表显示默认专辑信息,如图 14-62 所示。

图 14-62 初始执行效果

专辑详情界面效果如图 14-63 所示。

图 14-63 专辑详情界面效

播放专辑中某首音乐的效果如图 14-64 所示。

图 14-64 播放专辑中某首音乐